NATURAL HOSTS OF SIV

IMPLICATION IN AIDS

NATURAL HOSTS OF SIV

IMPLICATION IN AIDS

Edited by

AFTAB A. ANSARI, PHD
Professor
Department of Pathology & Laboratory Medicine
Emory University School of Medicine
Atlanta, GA, USA

GUIDO SILVESTRI, M.D.
Professor
Department of Pathology and Laboratory Medicine
Emory University School of Medicine & Yerkes National Primate Research Center Atlanta, GA, USA

ELSEVIER

AMSTERDAM • BOSTON • HEIDELBERG • LONDON
NEW YORK • OXFORD • PARIS • SAN DIEGO
SAN FRANCISCO • SINGAPORE • SYDNEY • TOKYO
Academic Press is an imprint of Elsevier

Academic Press is an imprint of Elsevier
32 Jamestown Road, London NW1 7BY, UK
225 Wyman Street, Waltham, MA 02451, USA
525 B Street, Suite 1800, San Diego, CA 92101-4495, USA

British Library Cataloguing-in-Publication Data
A catalogue record for this book is available from the British Library.

Library of Congress Cataloging-in-Publication Data
A catalog record for this book is available from the Library of Congress.

ISBN: 978-0-12-404734-1

For information on all Academic Press publications
visit our website at elsevierdirect.com

Typeset by TNQ Books and Journals
www.tnq.co.in

14 15 16 17 10 9 8 7 6 5 4 3 2 1

Contents

7. Gastrointestinal Immunity in Natural Hosts of Simian Immunodeficiency Virus

MOLLY R. PERKINS, JASON M. BRENCHLEY

8. Innate Immunity in Simian Immunodeficiency Virus Infection

ROGER KEITH REEVES, STEVEN E. BOSINGER

9. Humoral Immune Responses in SIV Infection of Sooty Mangabeys: Implications for HIV-1 Pathogenesis and Vaccine Design

CYNTHIA A. DERDEYN

10. Cellular Immune Responses in Natural and Non-natural Hosts of Simian Immunodeficiency Virus Infection

NAMITA ROUT, AMITINDER KAUR

11. Lack of Chronic Immune Activation During Natural Host SIV Infection

VASUDHA SUNDARAVARADAN, MELANIE A. GASPER, KIRAN D. MIR, DONALD L. SODORA

12. Distinct Cellular Targets of SIV Infection in Natural and Non-natural Hosts of SIV

THOMAS H. VANDERFORD, MIRKO PAIARDINI

13. SIV Transmission in Natural Hosts

ANN CHAHROUDI, SALLIE PERMAR, IVONA PANDREA

14. Immunogenetics of NK Cell Receptors and MHC Class I Ligands in Non-human Primates

LUTZ WALTER

15. The Different Modes of Resistance to AIDS: Lessons from HIV/SIV Controllers and SIV Natural Hosts

LISA A. CHAKRABARTI

16. Humans Resembling Natural Hosts of SIV

AMALIO TELENTI

Epigraphy

"An accomplished scientist does not always come up with the correct answers to complex biological problems BUT does put together knowledge in a logical order that serves the scientific community a foundation to formulate the right questions."

Ansari, 2014

Contributors

Daniel C. Anderson Yerkes National Primate Research Center, Emory University, Atlanta, GA, USA

Ann Chahroudi Department of Pediatrics, Emory University School of Medicine, Atlanta, GA, USA; The Yerkes National Primate Research Center, Emory Vaccine Center, Atlanta, GA, USA

Aftab A. Ansari Department of Pathology and Laboratory Medicine, Emory University School of Medicine, and Yerkes National Primate Research Center, Atlanta, GA, USA

Cristian Apetrei Center for Vaccine Research, University of Pittsburgh, Pittsburgh, PA, USA

Steven E. Bosinger Division of Microbiology and Immunology, Emory Vaccine Center, Yerkes National Primate Research Center, Atlanta, GA, USA

Jason M. Brenchley Program in Tissue Immunity and Repair, Laboratory of Molecular Microbiology, NIAID, NIH, Bethesda, MD, USA

Lisa A. Chakrabarti Unité de Pathogénie Virale, Institut Pasteur, Paris, France

Joyce K. Cohen Yerkes National Primate Research Center, Emory University, Atlanta, GA, USA; Department of Psychiatry and Behavioral Sciences, Emory University School of Medicine, Atlanta, GA, USA

Cynthia L. Courtney Yerkes National Primate Research Center, Emory University, Atlanta, GA, USA; Department of Pathology and Laboratory Medicine, Emory University School of Medicine, Atlanta, GA, USA

Cynthia A. Derdeyn Department of Pathology and Laboratory Medicine, Emory University, Atlanta, GA, USA

Anapatricia Garcia Yerkes National Primate Research Center, Emory University, Atlanta, GA, USA; Department of Pathology and Laboratory Medicine, Emory University School of Medicine, Atlanta, GA, USA

Melanie A. Gasper Seattle Biomedical Research Institute, Seattle, WA, USA

Edward J.D. Greenwood Department of Veterinary Medicine, University of Cambridge, Cambridge, UK

Sanjeev Gumber Yerkes National Primate Research Center, Emory University, Atlanta, GA, USA; Department of Pathology and Laboratory Medicine, Emory University School of Medicine, Atlanta, GA, USA

Jonathan L. Heeney Department of Veterinary Medicine, University of Cambridge, Cambridge, UK

Amitinder Kaur Division of Immunology, New England Primate Research Center, Harvard Medical School, Southborough, MA, USA

Frank Kirchhoff Institute of Molecular Virology, Ulm University Medical Center, Ulm, Germany

Florian Liegeois UMI233, TransVIHMI, Institut de Recherche pour le Développement (IRD), Montpellier, France and Université Montpellier 1, Montpellier, France

Dongzhu Ma Center for Vaccine Research, University of Pittsburgh, Pittsburgh, PA, USA

Preston A. Marx Chair, Division of Microbiology, Tulane National Primate Research Center, Tulane University, Covington, LA, USA

Kiran D. Mir Emory Vaccine Center, Yerkes Primate Center, Atlanta, GA, USA

Mirko Paiardini Department of Pathology and Laboratory Medicine, Emory University, School of Medicine, Atlanta, GA, USA; Division of Microbiology and Immunology, Yerkes National Primate Research Center, Emory University, Atlanta, GA, USA

Ivona Pandrea Department of Pathology, School of Medicine, University of Pittsburgh, Pittsburgh, PA, USA; Center for Vaccine Research, University of Pittsburgh, Pittsburgh, PA, USA

Martine Peeters UMI233, TransVIHMI, Institut de Recherche pour le Développement (IRD), Montpellier, France and Université Montpellier 1, Montpellier, France

Molly R. Perkins Program in Tissue Immunity and Repair, Laboratory of Molecular Microbiology, NIAID, NIH, Bethesda, MD, USA

Sallie Permar Human Vaccine Institute, Duke University Medical Center, Durham, NC, USA

Roger Keith Reeves Division of Immunology, New England Primate Research Center, Harvard Medical School, Southborough, MA, USA

Namita Rout Division of Immunology, New England Primate Research Center, Harvard Medical School, Southborough, MA, USA

Daniel Sauter Institute of Molecular Virology, Ulm University Medical Center, Ulm, Germany

Fabian Schmidt Department of Veterinary Medicine, University of Cambridge, Cambridge, UK

Prachi Sharma Yerkes National Primate Research Center, Emory University, Atlanta, GA, USA; Department of Pathology and Laboratory Medicine, Emory University School of Medicine, Atlanta, GA, USA

Guido Silvestri Department of Pathology and Laboratory Medicine, Emory University School of Medicine, and Yerkes National Primate Research Center, Atlanta, GA, USA

Donald L. Sodora Seattle Biomedical Research Institute, Seattle, WA, USA

Elizabeth Strobert Yerkes National Primate Research Center, Emory University, Atlanta, GA, USA

Vasudha Sundaravaradan Seattle Biomedical Research Institute, Seattle, WA, USA

Amalio Telenti Institute of Microbiology, University Hospital, University of Lausanne, Lausanne, Switzerland

Thomas H. Vanderford Department of Pathology and Laboratory Medicine, Emory University, School of Medicine, Atlanta, GA, USA; Division of Microbiology and Immunology, Yerkes National Primate Research Center, Emory University, Atlanta, GA, USA

Francois Villinger Yerkes National Primate Research Center, Emory University, Atlanta, GA, USA; Department of Pathology and Laboratory Medicine, Emory University School of Medicine, Atlanta, GA, USA

Lutz Walter Primate Genetics Laboratory, German Primate Center, Leibniz Institute for Primate Research, Kellnerweg, Göttingen, Germany

Introduction

Natural non-human primate hosts of the simian immunodeficiency virus (SIV) do not develop AIDS despite carrying viral loads that normally lead to pathology and death in non-natural hosts. A large body of data that documents a variety of immunological and virological differences between SIV-infected natural versus non-natural hosts indicate that the clues to disease resistance are mostly host related and have evolved over 100 if not thousands of years. We now face the daunting task of identifying which of these differences (and by what mechanisms) contribute to disease resistance in the natural hosts and ultimately exploit these findings for the design of novel interventions to treat or prevent HIV infection of humans.

CHAPTER

1

Comparative Studies of Natural and Non-natural Hosts of SIV—An Overview

Aftab A. Ansari, Guido Silvestri

Department of Pathology and Laboratory Medicine, Emory University School of Medicine, and Yerkes National Primate Research Center, Atlanta, GA, USA

OVERVIEW

The overall objective of this book is to provide readers with a single reference resource that contains a series of chapters comprising results, views, and concepts formulated by various investigators involved in the comparative studies of the natural and non-natural hosts of primate lentiviruses such as human immunodeficiency virus (HIV) and simian immunodeficiency virus (SIV). It is reasoned that such a summary will provide a valuable foundation upon which future studies can be planned, discussions initiated, and critical experiments performed that will move this field forward.

As the titles indicate, the initial chapters include a fascinating story of how SIV was accidentally transmitted in the 1970s from the disease-resistant natural hosts to the disease-susceptible non-natural hosts as an unintended consequence of scientific experimentation by the Nobel laureate Dr Carlton Gajdusek and his colleagues working on "kuru," a rare disease transmitted by brain tissue in a population in New Guinea who practiced cannibalism. Another example of unintended consequences were studies being performed by scientists that were attempting to find a model other than the armadillo for human leprosy and thought that they had developed such a non-human primate model in the early 1980s, i.e., before HIV/SIV were isolated and identified. Concealed within these

Natural Hosts of SIV.
http://dx.doi.org/10.1016/B978-0-12-404734-1.00001-2

historical descriptions is the finding of the unusual natural transmission of *Mycobacterium leprae* and *Coccidiodes immitis* infections in one of the natural hosts of SIV (sooty mangabey, a species from West Africa) [1–4]. Thus, a sooty mangabey was found to spontaneously acquire *M. leprae* infection and when tissues from this animal were injected into other mangabeys, the recipients developed a lepromatous form of *M. leprae*, which was at the time touted as the first description of a non-human primate (NHP) model to study human leprosy and a major advance in the field of leprosy research. The finding of *C. immitis* infection coupled with the *M. leprae* infections unique to this species prompts the obvious question as to whether the presence of the SIV infection of this natural host makes this species uniquely susceptible or whether the immune system of this species has evolved to confer this unique susceptibility to such microorganisms. This issue still remains to be addressed.

This interesting historical perspective is followed by a chapter on the prevalence and molecular epidemiology of SIVs in the wild, with the use of innovative sequencing techniques that facilitate the understanding of the evolution of these viruses. In addition, it is clear from this chapter that it is likely that >90% of the approximately 70 Old World NHP species in Africa are likely to be infected with species-specific SIVs, but we have knowledge of only 45 such species. It is also clear from this and other chapters that both SIV and the NHP species that harbor such viruses have co-evolved, and adaptation is one of the key elements that need to be recognized. A cautionary note was also expressed by the authors with regard to the pathogenicity of SIVs in natural hosts, citing the more recent studies of the evidence for pathogenicity of SIVcpz in chimpanzees in the wild. Thus, the blanket statement of disease resistance of natural hosts of SIV that are being raised in captivity at the various primate centers needs to be punctuated with the realization that detailed studies of its potential to cause disease in the wild in the same natural hosts are lacking at present. The idea being conveyed here is that there is likely to be continued evolution of both SIVs and the natural hosts that results in survival of only those within the species that co-evolve mechanisms that protect them from developing disease.

Along similar lines, the book includes a chapter devoted to SIVcpz in chimpanzees, with a more detailed description of the various chimpanzee species and the characterization of viruses from such species. The authors describe the differences and similarities between other natural hosts of SIV and the SIVcpz that infects chimpanzees. As cautioned above, while the previous paradigm predicated that the chimpanzees have been infected for thousands of years and are to a large extent disease resistant, emerging data have questioned this view. Thus, based on the fact that there is evidence for pathogenicity of SIVcpz infection of chimpanzees, the argument that transmission of SIV in this species is a more recent occurrence has strengthened, also explaining the presence of pathology as a result of the lack of sufficient time to reach a peaceful co-adaptation between host and virus. Further studies are clearly in order to address this issue.

These findings are logically followed by a detailed review of our current knowledge on the role of the various viral proteins, including the "accessory" proteins, that have been identified and sequenced and their functional consequences described. Thus, the repertoire of these virally encoded accessory proteins, including Nef, Vpu, and Vpx, that have evolved and the pathogenic consequences of such evolution are elegantly outlined. What is most striking is the list of unanswered questions that still remain with regard to these accessory proteins. In addition, a discussion of the role of the corresponding host antiviral restriction factors, which include TRIM-5α, tetherin, APOBEC3G, and SAMHD1, are presented with an emphasis that the role of such host proteins is still in its infancy and there are likely to be additional host factors that have yet to be identified. Understanding how such viral and host proteins have co-evolved will provide not only some important insights on the molecular mechanisms of host/parasite relationships but also crucial information that could be potentially harnessed for the design of candidate vaccines against HIV-1.

One of the issues that has garnered serious attention during the past several years is the phenomenon termed "microbial translocation," a term that has entered our lexicon in HIV/AIDS research. Indeed, a key biological event identified to date that differentiates pathogenic from non-pathogenic HIV/SIV infection, natural versus non-natural hosts of SIV, and the kinetics of disease progression is the occurrence of chronic immune activation (CIMA). Thus CIMA has been heralded as one of the most important correlates of pathogenic infection and the best marker for disease progression. It has been shown that CIMA is at least in part the result of microbial translocation that results from the loss of integrity of the gastrointestinal tissue barrier that separates the microbial flora from the lumen of the intestine to the systemic circulation. The chronic flow of microbiota and its products from the gut to the circulation induce both innate and adaptive immune cell activation with resulting toxic levels of cytokines, which dysregulate multiple biological systems, including, eventually, the coagulation cascade, leading to multiple organ system failure and death. Thus, there are two chapters within this book that concern CIMA. One of the chapters is focused on outlining the physiological and immunological mechanisms that maintain mucosal integrity and how its breach leads to microbial translocation resulting in CIMA in the disease-susceptible hosts—while, interestingly, such a breach occurs only mildly and transiently in the natural hosts. An important side note of this chapter is an emphasis on the fact that the mechanisms that lead to reversal and subsequent maintenance of mucosal integrity in one natural host may not in fact be similar to those in another host. The second chapter expands on the overall concepts laid out in the previous chapter by outlining all the cellular characters that are involved in the dialogue between the pro- and anti-inflammatory effects of the host immune system, with a focus on those that are functional at the mucosal interface. Thus, the basic tenets of this chapter are that natural hosts have developed a remarkable array of regulatory mechanisms that can dampen/mute proinflammatory responses and maintain homeostasis. It is clear from the gist of this chapter that these unique regulatory mechanisms that have been acquired by the natural hosts of SIV are the ones we need to exploit and target as part of novel therapeutic strategies aimed to minimize the damage that SIV (and, by implication, HIV) mediates against the gut mucosa. A related chapter with regard to the mechanisms by which the natural hosts protect themselves from the development of disease is the important finding of differences in the phenotypically defined subset of cells that SIV targets in the natural hosts. Thus, the findings outlined in this chapter suggest that the natural hosts best studied so far, which include the sooty mangabeys and the African green monkeys (AGMs), each have naturally evolved a set of mechanisms that modulate the major receptors for SIV and promote the infection of a subset of cells that are more dispensable for the host. This deviation allows for the survival and function of the subset that is critical for the host to maintain antigen-specific recall responses, immune homeostasis, mucosal barrier integrity, and, importantly, lymphoid tissue architecture [5]. Obviously, more detailed studies are required in efforts to exploit this finding for therapeutic purposes because it is not clear how this deviation evolved in the natural hosts over the millennia and is in fact mediated at the molecular level. Once again, it does appear that the mechanisms are species specific; and thus, while there is a common endpoint (disease resistance), the pathways are quite distinct, which can be viewed as hurdles or can be viewed as multiple targets available for identifying therapeutic strategies.

The next few chapters sequentially deal with the characteristics of the innate and acquired humoral/cellular responses; each chapter covers not only the subtle differences in the phenotype of cell lineages that execute these functions, but also how these functions differ between the natural hosts and the non-natural hosts of SIV. A comprehensive description of the various cell lineages that comprise the innate immune system and their role in either influencing the quality and quantity of virus-specific immune responses or regulating viral loads is described. Among the highlights of this chapter is the description of the role of the plasmacytoid dendritic cells (pDCs), whose trafficking from the periphery to the gut tissues and their subsequent accumulation within the gut was shown to lead to high levels of IFN-α in the non-natural hosts, associated with disease progression. On the contrary, such trafficking of pDCs in the natural hosts was found to be transient following acute infection. These findings suggest a potentially important difference in the role of pDCs in natural versus non-natural hosts. The coincident depletion of Th17/IL-22 synthesizing cells, also uniquely in the SIV-infected non-natural but not the natural hosts, suggests perhaps a linkage between these two events. Attempts to convert the infection and disease profile of natural hosts to that of non-natural hosts with the administration of IFN-α to promote innate immune signaling, however, while inducing a transient increase in the activation of CD8$^+$ T cells, failed to show CD4$^+$ T-cell depletion and showed no major effects on viral loads. An opposite strategy with the use of TLR7/9 antagonist to block IFN-α signaling in SIV-infected non-natural hosts not only had no effect on the depletion of CD4$^+$ or CD8$^+$ T cells, but—of interest—did not affect the expression of interferon-stimulating genes, suggesting that such manipulations in vivo perhaps are not effective because they do not function individually and/or they represent a downstream event. Similarly, attempts to administer lipopolysaccharides to the SIV-infected natural hosts to mimic the effect of microbial translocation in the non-natural hosts induced a transient increase in plasma viral loads but did not lead to disease, suggesting either that such manipulations do not faithfully replicate the conditions noted in the non-natural hosts or that the disease-inducing events are complex and require multiple system manipulations. Another cell lineage that appears to be important is the innate lymphoid cells (unique lineage only present in gut tissues). Their depletion/absence in the SIV-infected non-natural hosts but continued presence in the natural SIV-infected host, coupled with a role of these cells to maintain Th17/IL-22 in the gut mucosa, prompts the need for further detailed study of this difference. Of interest are the studies on the role of NK cells. Depletion studies during acute infection using either in vivo depleting antibodies or the use of a JAK3 inhibitor while showing varying levels of depletion of this cell lineage failed to show any major difference in viral loads. However, depletion of NK cell lineage using the JAK3 inhibitor during acute infection did appear to influence plasma and cellular viral loads during chronic infection by an as-yet-unidentified mechanism [6]. A role of this cell lineage in mediating ADCC activity and to serve as enhancers of immune responses by adjuvants has also been highlighted in this chapter. In all, it appears that much has yet to be learned on the role of the innate immune network on influencing the course of disease in the natural and non-natural hosts of SIV.

A summary of the humoral anti-SIV response in the natural host with a focus on the sooty mangabeys and a comparative analysis of the antibody response between sooty mangabeys and rhesus macaques is outlined in the next chapter. To a large extent, it appears that natural SIV hosts do not mount a robust virus-specific neutralizing antibody response and the elimination of B cells in the natural hosts does not lead to increases in plasma viral loads. While the author is clear that these are not likely contributing to the disease resistance of the natural hosts, it is indeed possible that the blunting of vigorous virus-neutralizing antibody responses contributes to the disease-protective mechanisms in these natural hosts by preventing the development of hyperimmune activation, dysregulation, and perturbation of lymphoid tissue architecture that is characteristic of the non-natural hosts. One is thus left with the idea that the natural hosts have developed a potent regulatory mechanism that prevents the development of vigorous humoral virus-specific immune responses and its corollary: that vigorous virus-specific immune responses are contributing to the development of disease. It is requested that the reader keep this thought in mind as we try to put this in context with all the other findings displayed in the chapters.

The comparative analysis of SIV-specific humoral responses is followed by a chapter on the virus-specific cellular responses, including the responses of cells that form a bridge between humoral and cell-mediated immune responses. The findings of this chapter suggest that while the virus-specific cellular immune responses are noted in both the natural and non-natural hosts of SIV, the response in the natural hosts may in fact be more effective in limiting tissue damage and resulting immunopathology, particularly in secondary lymphoid tissues and in specific $CD4^+$ T-cell subsets. The emphasis is also being placed on the importance of regulatory mechanisms and specific molecules (PD-1, CTLA-4, LAG-2, 2B4, CD160 and GP149) that mediate such regulatory function of adaptive immune responses that have evolved selectively with enhanced function evolved in the natural hosts and, of importance, a constellation of mechanisms that maintain the presence of normal frequencies and function of these so-called cell lineages that fall within the cusp of humoral and cellular immune responses (NK regs, NK-T cells and KIRs) are illustrated in Figure 1.1.

This is followed by two chapters that address additional important roles in the field of natural and non-natural SIV infections, including the mechanisms of viral transmission and the role of genetic factors. With regard to the studies of viral transmission, it does seem ironic that even after three decades of research in the field of HIV/SIV, we still do not fully comprehend the impact of the route of transmission on disease outcome. Thus, while it is clear that the mucosal route is the major route of natural transmission, we do not fully understand how the different routes affect the generation of virus-specific humoral and cellular responses and the impact of the innate immune system in the localized environment that influences the outcome in natural versus non-natural hosts of SIV. This is because this subject encompasses multiple routes and involves not only the transmission between adults but also mother-to-child transmission and how this differs between natural and non-natural hosts. There does seem to be a hierarchy, however, in the dose of virus required for transmission via the different routes in adults. Thus, the dose of virus required to transmit infection increases, in order, via

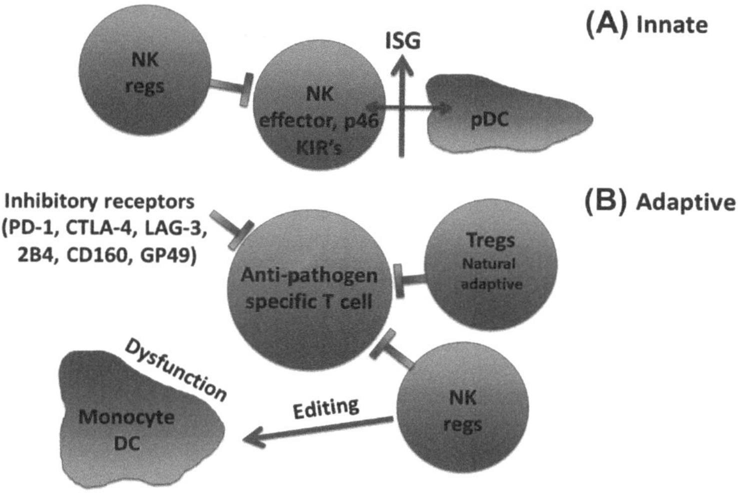

FIGURE 1.1 Mechanisms that contribute to distinct outcomes of SIV infection in natural versus non-natural non-human primate hosts. Nature has built in a series of redundant effector regulatory mechanisms to insure that the immune responses of the host stay in check and do not result in pathology. This regulatory process occurs during both (A) innate and (B) adaptive immune responses. For innate immune responses, regulation of the synthesis of interferon-stimulated genes is one good example. In the case of adaptive immune responses, the regulatory molecules include the gradual expression and upregulation of the inhibitory receptors (PD-1, CTLA-4, LAG-3, 2B4, CD160, and gp49), which generate negative signals upon ligation, and the activation of regulatory T cells (Tregs) and regulatory NK cells (NKregs), which also function to kill immature dendritic cells to promote muting of the immune responses. Some of these regulatory mechanisms function systemically and others function in select tissues and organs, providing a multi-pronged safeguard to fine-tune the host anti-pathogen relationship. The natural hosts of SIV have evolved to optimally utilize such regulatory mechanisms, whereas dysregulation of such regulatory mechanisms characterizes the non-natural hosts of SIV.

blood, rectal tissues, vaginal tissues, and oral tissues, respectively. It is also clear that vaginal transmission is a rare event, which has prompted the use of low-dose repeated intravaginal transmission as a model to better mimic HIV-1 transmission in humans. Germane to the major thrust of this book are studies of the differences between the natural versus the non-natural hosts of SIV. A number of points appear to be clear from the studies performed to date. First of all, the majority of SIV transmitted in the natural hosts is via the sexual route, and there is high concordance between age of the natural hosts and seroconversion and virus positivity. Secondly, one of the most important differences between the natural and non-natural hosts of SIV is the relative resistance from mother-to-child transmission in the natural hosts as compared with the non-natural hosts of SIV. This is true not only for natural epidemiologic studies of animals in the wild but also for experimental studies in which the mother's milk was experimentally manipulated to contain high viral levels. However, what is lacking in these studies are the precise mechanisms involved in the resistance of the infants of the natural hosts. One of the suggested mechanisms implicates the differences in the cell lineages that are targets for infection in the natural hosts, similar to the argument advanced above as a mechanism of disease protection. However, the molecular mechanisms responsible for these differences in the type of target cells between natural and non-natural hosts remain incompletely understood.

The subject of the role of genetic factors in differences in susceptibility to transmission, infection, and disease progression has been, unfortunately, the most difficult to study. This is because most studies being performed at primate centers involve a limited number of animals of each species and genetic association studies clearly require studies of large cohorts. This problem is further compounded by the lack of detailed knowledge of the sequences of the genes involved and the nature of the polymorphisms of the select genes, particularly in the natural hosts. Thus, basic studies of the detailed characterization of the degree of MHC class I and II polymorphisms in the non-natural hosts are still lacking and there is very limited knowledge of the MHC genes of the natural hosts. Nonetheless, since there is now clear evidence for the role of MHC class I/II and those that involve NK cell function such as NK cell receptors, KIRs, and FcR on the pathogenesis of human HIV-1, it was important to include a chapter on this subject, which is very ably summarized herein.

Finally, a very fascinating topic is covered by the last two chapters of this book. These chapters summarize what we know at present with regard to HIV-1-infected humans that show characteristics similar to natural SIV hosts. While there are clear differences between "elite controllers" of human HIV-1 infection and the natural hosts of SIV, which include low viremia and potent antiviral T-cell immune responses in the former but not the latter, the number of similarities is of great interest. This includes the absence of chronic immune activation, a downregulated IFN-α response, lower viral loads in lymph nodes, and relative sparing of the central memory and potentially the follicular helper $CD4^+$ T cells. A similar theme is projected in the second chapter on this subject but also includes the remarkable description of a rare but interesting subset of HIV-1-infected humans that maintain high viral loads for a substantial time period, but with no loss of $CD4^+$ T cells and no signs of disease progression (i.e., viremic nonprogressors or VNP). The authors also highlight our limited understanding of the biology and pathogenesis of HIV-2 infection in humans and that more detailed studies of these individuals, in concert with studies of elite controllers, VNPs, highly HIV-1-exposed but noninfected humans, and normal and/or fast progressors, provide the entire spectrum of HIV-1-infected humans. A study of the differences in the pathogenic mechanisms of these cohorts is clearly in order and likely to be highly informative.

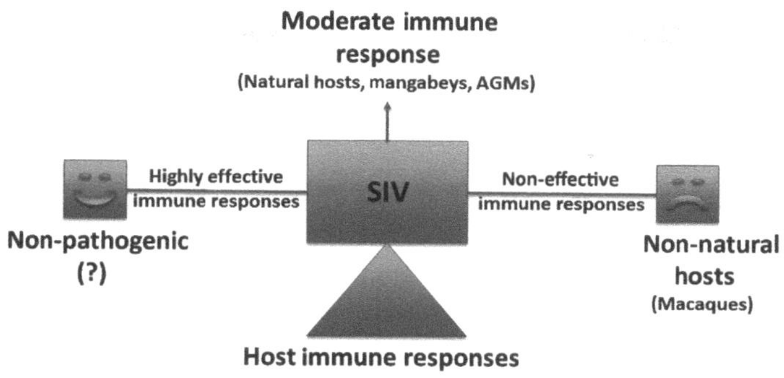

FIGURE 1.2 The balancing act between SIV (the pathogen) and the immune response of the hosts. There are basically three distinct outcomes. Thus, either the immune response mediated by the host is not effective in eliminating the pathogen (as in the case of non-natural hosts, macaques), or the immune response is highly effective in eliminating the pathogen (not yet observed in the case of SIV), or the quality and quantity of the immune response has evolved such that the pathogen does not harm the host and a balance is achieved between the immune response and the degree of immunopathology (as in the case of the natural hosts, sooty mangabeys and AGMs), which is herein defined as host accommodation.

We have taken the prerogative as editors to highlight some of the issues that need emphasis with regard to the potential mechanisms that may be involved in the lack of disease progression, with the hope that the issues highlighted provide food for thought for the next generation of scientists involved in attempts to define the mechanisms by which the natural hosts of SIV actually live quite a long and healthy life despite chronic SIV infection and with no detectable pathologic sequelae. A summary of what we understand with regard to net results of virus/host interaction in natural versus non-natural hosts (distinguishing effective from non-effective and moderate or muted immune responses) is depicted in Figure 1.2. Thus, highly effective immune responses result in lack of pathogenesis and presence of strong but ineffective immune responses leads to a pathogenic outcome, while more moderate and focused immune responses have evolved in the natural hosts and successfully achieved a non-pathogenic balance between virus and host.

As stated above, the issues that we would like to emphasize include:

1. *In vivo viral passage*: A virological issue that needs to be *emphasized* concerns the fact that viral isolates from the natural hosts of SIV generally need to be passaged in vivo in the non-natural hosts to identify isolates that can replicate efficiently in the new non-natural hosts and cause disease. Thus, not all SIV isolates from sooty mangabeys are equally pathogenic in macaques and not all SIV isolates from macaques replicate efficiently in activated $CD4^+$ T cells from mangabeys. These findings suggest that SIV undergoes "*species-specific adaptation*" and that these are characteristics unique to both virus and host.

2. *Role of "founder viruses"*: The fact that sexual transmission of HIV-1 involves the transmission of only a single or a select few viral species, leading to the term "founder viruses," clearly prompts us to determine whether this is also true in the natural hosts of SIV. Within this context, it is important to perform these studies not only in the natural hosts raised in captivity but also in those in the wild, a topic not easy to study. In addition, the role of "env glycosylation" in viral transmission in the natural versus the non-natural hosts and in MCT studies between the natural and non-natural hosts needs to be addressed.
3. *Differences among various species of natural hosts*: Another issue that is important to keep in mind is that there are differences in the immunological and virological findings even among the various natural hosts of SIV, which requires one to analyze the data obtained within this context. Thus, the mechanisms of transmission may differ among each natural host of SIV and the findings from one species may not be amenable to extrapolation. This subject is also discussed in detail in several chapters.
4. *Gastrointestinal dysfunction*: Reports of severe gastrointestinal dysfunction with villous atrophy and crypt cell proliferation in SIV-infected macaques and HIV-1-infected humans in the early 1990s established the occurrence of gastrointestinal pathology in SIV- and HIV-infected individuals [7,8]. While initial SIV infection of both natural and non-natural hosts leads to the same degree of gastrointestinal pathology (with enteropathy and intestinal permeability) during acute infection, this pathology is largely reversible in disease-resistant natural hosts but irreversible in non-natural hosts [9]. Notably, administration of antiviral therapy to macaques as early as 1 week post infection did not influence the loss of $CD4^+$ T cells in the gut tissues but enhanced rapid restoration with memory $CD4^+$ T cells [10]. The data show that during the early stages of infection there is not only a massive loss of $CD4^+$ T cells but also the apoptosis of epithelial cells that line the small and large intestines of untreated SIV-infected macaques, which is likely to be a driving force for gastrointestinal pathology [10]. Select studies have documented the loss of Tregs [11] and Th17cells in the mucosal tissue [12] of SIV infected macaques but not sooty mangabeys [13]. These studies have been followed by the introduction of the concept of microbial translocation [14] as a mechanism to explain the differences in the outcome of SIV infection between natural and non-natural hosts (discussed in several chapters). Attempts to use gene profiling [15] and investigation of the dysfunction of the intracellular molecular pathways in SIV-infected macaques have been performed [16,17]. However, the factors that lead to regeneration and reconstitution of gastrointestinal tissues in the natural but not non-natural hosts

require further investigation. Future studies should examine intestinal dendritic cells that are involved in polarizing anti-inflammatory responses inducing tolerance, gut-homing NK cells, and progenitor stem cells that give rise to gut epithelial cells [18]. In addition, it is possible that more knowledge of the proteins involved in regulating intestinal permeability and intestinal epithelial regeneration, particularly in the natural hosts of SIV, may pave the way for identifying novel therapeutic strategies for HIV-infected humans.

5. *Models that attempt to explain the divergent clinical outcome of SIV infection in the natural versus the non-natural hosts:*

   a. *Role of immune activation.*

      Although the contribution of CIMA to disease progression in HIV-infected humans and SIV-infected non-natural hosts is a reasonable and logical hypothesis, the lack of significant CIMA is probably not the only mechanism underlying the distinct clinical outcome in natural hosts. This view is supported by our finding that hyperimmunization of a cohort of six SIV-infected sooty mangabeys at weekly intervals for >12 months sequentially with adjuvant-incorporated KLH, tetanus toxoid, a mixture of allogeneic cells, and intranasal exposure to immunogenic doses of influenza virus led to marked increases in the frequencies and absolute numbers of in vivo activated $CD4^+$ and $CD8^+$ T cells. Such levels of experimental immune activation of the natural host did not induce detectable disease or a major increase in plasma viral loads. These findings suggest that it may be the quality and/or differences in the consequences of CIMA that play a role in the different outcomes of SIV infection in the natural versus the non-natural hosts of SIV. The experimental manipulation of the natural host that leads to CIMA, which results in progression to disease, is required to underscore the role of CIMA in disease progression.

   b. *Role of host accommodation.*

      The symbiotic relationships that can exist between a vertebrate host and certain microorganisms were first acknowledged many years ago [19]. There are numerous examples in nature, highlighted by the relationship between host and microbe within the gastrointestinal tract under normal physiological conditions. There is no detectable immune response against foreign organisms at this site. Microorganisms establish a symbiotic relationship in a natural host, in which it is not pathogenic, but when they infect another species, they cause pathologic effects. For example, the deer mouse is the natural host for the Hantaviruses, which often cause a fatal disease in humans termed Hantavirus pulmonary syndrome. The infected mice carry high levels of the virus in their respiratory tract and have antibodies against the virus, but they have a normal lifespan [20]. Herpes B virus, which causes a fatal encephalitis in humans, is commonly found in >70% of rhesus macaques, and >80% of those that are infected also have detectable antibodies and remain healthy. However, the mechanisms that underlie the accommodation of such pathogenic organisms by their natural hosts, and the disease resistance of the hosts, remain largely undefined. *It is of importance to note that the "MAJOR" decision making tissue site involved in recognition of self versus nonself is at the gastrointestinal/liver intersection and the liver in essence could be playing the critical role in dictating host/virus outcome in the natural versus the non-natural hosts of SIV.* The potential pathways that result in polarized outcomes as a consequence of viral infection resulting in viral clearance versus virus induced pathology (death) are by reason distinct from those that result in chronic infection and in turn the consequence of chronic infection range from slow progressive disease (non-natural hosts) and a lack of detectable disease (natural hosts) that is associate with host accommodation is depicted in Figure 1.3. The question arises as to whether we can harness knowledge of how such "natural" host adaptation occurs and then attempt to "experimentally" induce such host adaptation between HIV and humans to limit the development of immunodeficiency disease.

It is of interest to note that in fact there have been numerous examples of such "experimental" adaptation of host and foreign microorganisms or antigens. Thus, the field of organ transplantation has provided us with a plethora of such examples. The term "host accommodation" [21–23] refers to successful engraftment of an organ or tissue in an allogeneic host in the presence of readily detectable antibodies against the donor tissue. In a general sense, host accommodation (HA) involves the induction of a coordinated series of biological changes in the host that enable donor cells and tissues to persist in the presence of antibodies against donor antigens recognized as foreign by the host. The successful transplantation of ABO-incompatible donor organs in hosts that have high level of anti-AB antibodies is a good example of such HA. Thus, a similar relationship likely exists between SIV and its natural hosts in that the host lives a normal healthy life with the foreign antigen in the presence of readily detectable levels of antiviral antibodies.

We propose that host accommodation could be a viable explanation for the absence of pathogenic disease in natural NHP hosts of SIV. We propose that the trafficking of NK cells and select subset of dendritic cells to the gastrointestinal tract during acute SIV infection in the presence of a low frequency of CD4+CCR5+ cells regulates the level of inflammation and creates a Th2-like environment that could contribute to the host accommodation in natural hosts. This symbiotic host–virus relationship leads to readily detectable antibodies and persistent viremia, but no overt pathology.

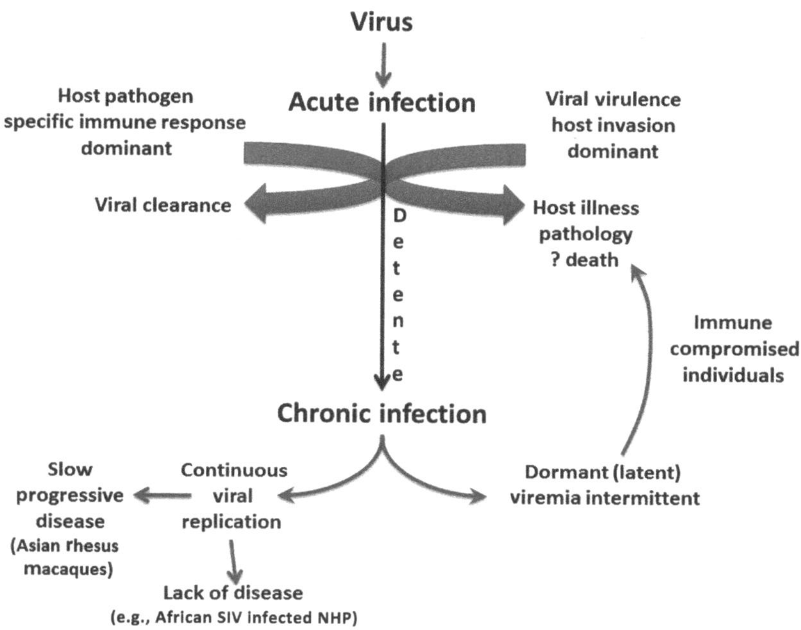

FIGURE 1.3 The evolution of host accommodation for SIV infection: During acute infection, virus replicates and if the host virus-directed innate followed by adaptive immune response is dominant, viral clearance is achieved. If the virus has developed a dominant means to evade host innate immune mechanisms, leading to dysregulated virus-specific adaptive immunity, such infection leads to death of the host. For select viruses including SIV, neither the host response nor the virus-mediated pathology is dominant, which leads to chronic infection. Thereafter, either (1) the virus becomes "dormant" and can revert back to induce pathology and death if the host becomes immune compromised, or (2) there is continuous virus replication, which can cause a slow, debilitating disease (as noted for Asian macaques), or (3) there is continuous virus replication to varying degrees, with no discernible host pathology. SIV infection of natural African hosts is an example of the last alternative. What we are attempting to unravel is what has taken perhaps a million years of co-evolution to achieve, as is the case with herpes viruses.

OUTLOOK

While considerable advances have been made in identifying both phenotypic and functional differences in the immune responses of SIV-infected AIDS-resistant natural hosts as compared with SIV-infected AIDS-susceptible non-natural non-human primate hosts, unfortunately, at present the mechanisms responsible for such distinct clinical outcome remain incompletely understood. Select parameters such as high viral loads and/or degree of $CD4^+$ T-cell depletion are not by themselves sufficient to lead to disease. Clearly, additional studies are required to define the mechanisms by which the natural hosts of SIV avoid disease progression and the non-natural hosts develop AIDS following SIV infection. One of the prevailing hypotheses for such distinct clinical outcomes concerns the role of chronic immune activation, which occurs in the SIV-infected non-natural hosts but not the natural hosts. It is reasoned that such CIMA leads eventually to immune exhaustion and loss of immune competence, which results in increased susceptibility to opportunistic infections and, finally, death. While this is a reasonable model, we submit that lack of CIMA alone does not explain the disease resistance of the natural hosts. Instead, host accommodation, broadly defined as the ability of the natural host to mount a controlled innate and noninflammatory SIV-specific humoral and cellular immune response, which has evolved during a long evolutionary time period, is an additional potential explanation for such distinct clinical outcome. A list of immunological features of SIV infection in natural versus non-natural hosts is illustrated (Table 1.1). This list is not by any means complete but serves as a foundation for future additions and modifications as additional data become available.

There continue to be significant gaps in our understanding of the biological features that distinguish pathogenic from non-pathogenic SIV infection in the non-human primate hosts. These issues are summarized in Box 1.1. These include the need to define the relative efficiency of virus replication within $CD4^+$ T cells from the natural versus the non-natural hosts. Thus, do the $CD4^+$ T cells from the natural hosts produce more SIV per cell than the non-natural hosts, which leads to the protection of a higher frequency of $CD4^+$ T cells from lysis and dysfunction? One of the characteristics of the SIV-infected natural hosts that has been reasoned to contribute to disease resistance is the lower frequency of $CD4^+$, $CCR5^+$ cells. However, the molecular mechanisms that regulate the expression of CCR5 remain unclear. While both natural and non-natural hosts demonstrate essentially similar gastrointestinal pathology during acute infection, there is significant gradual recovery of the damage in the natural but not the non-natural hosts during chronic infection. The mechanism(s) for such differential response to gastrointestinal tissue injury remains to be defined. Also, while it is reasonable to assume that qualitative and quantitative aspects of the immunological events that occur during the acute infection period may play a major role in setting the stage for disease resistance and/or susceptibility, further studies are required to reach a satisfactory understanding of the pathogenic role of these early events.

What seems to be critical to solve many of these remaining issues is to determine whether and how we can experimentally convert the non-natural host of SIV into a disease-resistant host, a process that nature has accomplished over a long period of time—or, vice versa, whether and how we can induce AIDS in a natural AIDS-resistant host. Attempts to accomplish this by the in vivo depletion of cell lineages, while informative, has not recapitulated the outcome of SIV infection in natural hosts

TABLE 1.1 Comparative Features of SIV Infection in Natural Disease-Resistant versus Non-natural Disease-Susceptible Non-human Primates

Features	Natural Hosts	Non-natural Hosts
Plasma viral loads	Similar	Similar
Cellular viral loads	Similar	Similar
Cell target for virus	$CD4^+$, macrophages	$CD4^+$, macrophages
Antibodies against SIV	Lower ⇑⇑⇑	High ⇑⇑⇑⇑
Antibodies against SIVgag	±	+
Depletion of $CD4^+$ T cells (acute infection)	⇓	⇓⇓⇓
Depletion of $CD4^+$ T cells (chronic infection)	⇓	⇓⇓⇓⇓
Depletion of $CD4^+$ T cells (gut, acute infection)	⇓⇓⇓	⇓⇓⇓
Restoration of $CD4^+$ T cells (gut)	⇑⇑⇑	–
Levels of $CD4^+CCR5^+$ cells in PBMC	Low 3–7%	10–15%
Levels of $CD4^+CCR5^+$ cells in gut	Low 2–5%	40–60%
CD4 TH17 cells post SIV	Preserved	Depleted
SIV-specific cellular response	+ or ±	+++
SIV-specific TH1/TH2 response	TH2	TH1
Gastrointestinal tissue pathology post SIV (acute infection)	Marked	Marked
Gastrointestinal tissue pathology post SIV (chronic infection)	75% Reversed	Continued
IFN-α (acute to chronic infection)	High to low	High to high
Chronic immune activation	–	⇑⇑⇑⇑
Lymph node tissue architecture	Normal	Dysregulated
Susceptibility to apoptosis of $CD4^+$ T cells	Low	High
Susceptibility to apoptosis of bystander cells	Low	Medium
Potential to undergo anergy	Resistant	Susceptible

as compared with non-natural hosts, as summarized in Table 1.2. Clearly, examples exist for such experimental manipulation, in particular, in studies of tissue and organ transplantation. If and when such objectives are achieved, the question then arises as to how we can harness such knowledge in the formulation of novel approaches to treat and prevent HIV-1 infection in humans. While experimental induction of selective adaptive

BOX 1.1

1. Can we determine the viral: burst size (amount of virus produced by unit number of $CD4^+$, $CCR5^+$ T cells) from natural hosts such as sooty mangabeys and the non-natural hosts such as macaques, and can the mechanisms that regulate the expression of CCR5 by $CD4^+$ T cells be defined, which may shed light on the differences in the efficiency of virus replication in the target cells of the natural versus the non-natural hosts of SIV and its role in disease resistance?
2. Can one identify the cellular and molecular mechanisms that lead to reversible versus nonreversible pathology of the gastrointestinal tissues of the SIV-infected natural versus the non-natural hosts?
3. Can SIV-specific cellular tolerance be experimentally achieved in the non-natural hosts; and, if so, does this lead to the conversion of the disease-susceptible SIV-infected macaques into a disease-resistant state, much like mangabeys or AGMs?
4. Can events that occur during acute infection, such as the induction and subsequent regulation of innate immune responses that distinguish non-pathogenic from pathogenic infection, be identified; and, if so, can the mechanisms unique to non-pathogenic hosts be experimentally induced in the non-natural hosts and lead to disease resistance?
5. Can strategies aimed at preventing and/or attenuating gastrointestinal tissue pathology in the non-natural hosts during acute SIV infection alone and/or in combination with the institution of effective antiretroviral therapy lead to the generation of long-term nonprogressor status; and, if so, can this paradigm be utilized for HIV-1-infected humans?

TABLE 1.2 In Vivo Cell Depletion Studies in Natural versus Non-natural Hosts of SIV

Cell Lineage Depleted	Effect on Natural Host	Effect on Non-natural Host
CD8	½ log ⇑ in VL or no effect. No evidence of disease	1–3 log ⇑⇑⇑ in VL, accelerated disease progression
CD4	⇓⇓⇓⇓ in VL but transient, no evidence of disease	No reports
CD20	No difference in VL, no evidence of disease	Play a role during acute infection
CD16	No reports	Transient ⇑ in VL, no effect on disease progression

VL, viral load.

immune response tolerance maybe difficult to achieve, we submit that therein lies one of the challenges for the formulation of an effective HIV-1 vaccine.

Acknowledgments

Supported by NIH NIAID RO1 AI078773 and AI098628 (AAA).

References

[1] Gormus BJ, Martin LN, Baskin GB. A brief history of the discovery of natural simian immunodeficiency virus (SIV) infections in captive sooty mangabey monkeys. Front Biosci 2004;9:216–24.

[2] Gormus BJ, Wolf RH, Baskin GB, Ohkawa S, Gerone PJ, Walsh GP, et al. A second sooty mangabey monkey with naturally acquired leprosy: first reported possible monkey-to-monkey transmission. Int J Lepr Other Mycobact Dis 1988;56:61–5.

[3] Meyers WM, Walsh WM, Brown HL, Binford CH, Imes Jr GD, Hadfield TL, et al. Leprosy in a mangabey monkey–naturally acquired infection. Int J Lepr Other Mycobact Dis 1985;53:1–14.

[4] Pappagianis D, Vanderlip J, May B. Coccidioidomycosis naturally acquired by a monkey, *Cercocebus atys*, in Davis, California. Sabouraudia 1973;11:52–5.

[5] Chahroudi A, Bosinger SE, Vanderford TH, Paiardini M, Silvestri G. Natural SIV hosts: showing AIDS the door. Science 2012;335:1188–93.

[6] Takahashi Y, Byrareddy SN, Albrecht C, Bremeier M, Walter L, Mayne AE, et al. In vivo administration of a JAK3 inhibitor during acute SIV infection leads to significant increases in viral load during chronic infection. PLoS Pathog 2014;10:e1003929.

[7] Heise C, Miller CJ, Lackner A, Dandekar S. Primary acute simian immunodeficiency virus infection of intestinal lymphoid tissue is associated with gastrointestinal dysfunction. J Infect Dis 1994;169:1116–20.

[8] Heise C, Vogel P, Miller CJ, Halsted CH, Dandekar S. Simian immunodeficiency virus infection of the gastrointestinal tract of rhesus macaques. Functional, pathological, and morphological changes. Am J Pathol 1993;142:1759–71.

[9] Brenchley JM, Douek DC. The mucosal barrier and immune activation in HIV pathogenesis. Curr Opin HIV AIDS 2008;3:356–61.

[10] Verhoeven D, Sankaran S, Silvey M, Dandekar S. Antiviral therapy during primary simian immunodeficiency virus infection fails to prevent acute loss of $CD4^+$ T cells in gut mucosa but enhances their rapid restoration through central memory T cells. J Virol 2008;82:4016–27.

[11] Chase AJ, Sedaghat AR, German JR, Gama L, Zink MC, Clements JE, et al. Severe depletion of $CD4^+$ $CD25^+$ regulatory T cells from the intestinal lamina propria but not peripheral blood or lymph nodes during acute simian immunodeficiency virus infection. J Virol 2007;81:12748–57.

[12] Cecchinato V, Trindade CJ, Laurence A, Heraud JM, Brenchley JM, Ferrari MG, et al. Altered balance between Th17 and Th1 cells at mucosal sites predicts AIDS progression in simian immunodeficiency virus-infected macaques. Mucosal Immunol 2008;1:279–88.

[13] Brenchley JM, Paiardini M, Knox KS, Asher AI, Cervasi B, Asher TE, et al. Differential Th17 CD4 T-cell depletion in pathogenic and nonpathogenic lentiviral infections. Blood 2008;112:2826–35.

[14] Brenchley JM, Price DA, Schacker TW, Asher TE, Silvestri G, Rao S, et al. Microbial translocation is a cause of systemic immune activation in chronic HIV infection. Nat Med 2006;12:1365–71.

[15] George MD, Verhoeven D, McBride Z, Dandekar S. Gene expression profiling of gut mucosa and mesenteric lymph nodes in simian immunodeficiency virus-infected macaques with divergent disease course. J Med Primatol 2006;35:261–9.
[16] Lackner AA, Mohan M, Veazey RS. The gastrointestinal tract and AIDS pathogenesis. Gastroenterology 2009;136:1965–78.
[17] Mohan M, Aye PP, Borda JT, Alvarez X, Lackner AA. CCAAT/enhancer binding protein beta is a major mediator of inflammation and viral replication in the gastrointestinal tract of simian immunodeficiency virus-infected rhesus macaques. Am J Pathol 2008;173:106–18.
[18] van der Flier LG, Clevers H. Stem cells, self-renewal, and differentiation in the intestinal epithelium. Annu Rev Physiol 2009;71:241–60.
[19] Lyte M. The microbial organ in the gut as a driver of homeostasis and disease. Med Hypotheses 2009;18:634–8.
[20] Easterbrook JD, Klein SL. Immunological mechanisms mediating hantavirus persistence in rodent reservoirs. PLoS Pathog 2008;4:e1000172.
[21] Bach FH, Hancock WW, Ferran C. Protective genes expressed in endothelial cells: a regulatory response to injury. Immunol Today 1997;18:483–6.
[22] Soares MP, Lin Y, Sato K, Stuhlmeier KM, Bach FH. Accommodation. Immunol Today 1999;20:434–7.
[23] Tang AH, Platt JL. Accommodation of grafts: implications for health and disease. Hum Immunol 2007;68:645–51.

CHAPTER

2

History of Simian Immunodeficiency Virus Discovery

Preston A. Marx

Chair, Division of Microbiology, Tulane National Primate Research Center, Tulane University, Covington, LA, USA

SERENDIPITOUS PATHWAYS TO AN AIDS MODEL AND SIV'S AFRICAN ORIGIN

Important discoveries are often serendipitous and obey the Law of Unintended Consequences, which states, "The purposeful actions of people and especially of governments always have effects that are unintended" [1]. The pathway to discovering the African natural history of the simian immunodeficiency virus (SIV) and development of the macaque (mac) AIDS models began with a lymphoma outbreak in 1969 of unknown cause at the California National Primate Research Center (CNPRC) macaque colony located in Davis, California. This epizootic unintentionally led to the literal creation of SIVmac, which was eventually identified at the New England Primate Research Center (NEPRC) in Southborough, Massachusetts. This discovery led in turn to the discovery at the Tulane National Primate Research Center (TNPRC) in Covington, Louisiana that AIDS most likely had a simian origin on the African continent.

All of the above began with unrelated research projects at these three primate centers. The SIVs and derivatives commonly used in AIDS animal model research resulted from the unintended consequences of non-human primate (NHP) model research unrelated to AIDS carried out by scientists at three primate centers who were unaware of SIV's existence; in fact the research took place before AIDS was reported in humans in 1981 [2]. After SIV's discovery, interest in the AIDS model grew exponentially. The

Natural Hosts of SIV.
http://dx.doi.org/10.1016/B978-0-12-404734-1.00002-4

number of publications on AIDS NHP models exceeds 6500 and increases every day. The use of the SIV macaque model, in particular SIV in rhesus (Rh) macaques of Indian origin, has led to important discoveries even before they were known in HIV-1-infected human beings. For example, early immunopathogenesis of SIV was characterized in SIV-infected Rh [3,4] before the same studies were done in humans [5].

This chapter will present a historical perspective of how the major SIVs were serendipitously discovered by alert scientists and how these discoveries led to the development of the most-used AIDS animal model and ultimately to determining the origins of HIV-1 and HIV-2 in the Congo river basin and West Africa, respectively.

SIVMAC EMERGED FROM THE CALIFORNIA NATIONAL PRIMATE RESEARCH CENTER MACAQUE COLONY

The origin of SIVmac can be traced to the largest outbreak of lymphomas ever recorded in macaques. The lymphomas occurred in 23 rhesus macaques and one stump-tailed macaque between February 1969 and March 1971 at the California National Primate Center in Davis, California [6]. Macaque lymphomas had never been previously seen in over 11,500 macaques that had been housed at the California center from 1961 until the outbreak in 1969. This macaque disease was not reported as AIDS, or even as an immunodeficiency, since the outbreak occurred a decade before AIDS in humans was reported. However in a University of California Davis PhD thesis research project on this lymphoma outbreak, the syndrome, which included lymphoma and opportunistic infections, was characterized as an "immune defect in cell-mediated immunity… immune incompetence was heightened by some environmental factor (virus or chemical)" [7].

SURVIVORS OF THE MACAQUE LYMPHOMA OUTBREAK SENT TO THE NEW ENGLAND AND WASHINGTON PRIMATE RESEARCH CENTERS

Primate centers frequently sell and exchange monkeys among themselves for the purpose of expanding their colonies, increasing genetic diversity and obtaining monkeys needed for research projects. In this regard, apparently healthy rhesus macaques that had been involved in the California lymphoma outbreak and harbored silent SIV infections were transferred to the New England and Washington National Primate Research Centers. The silent SIV infections manifested themselves as lymphomas in wild-caught macaques at the NEPRC that had not been used in any experiments. These macaques had been obtained from "another institution," according to a report from New England researchers [8]. This other institution was likely the CNPRC. The lymphoma was serially passaged into rhesus macaques of Indian-origin and then subsequently to more Indian-origin rhesus macaques, resulting in the highly pathogenic viruses, SIVmac251 from passage 4 and SIVmac239 from passage 7. This serial passage is well documented by Mansfield and colleagues [9]. The origin of SIVmac251 and SIVmac239 at the NEPRC was made abundantly clear as originating at the CNPRC by phylogenetic studies of Apetrei and Marx [10] (Figure 2.1). Much later it was shown that SIVmac239 was somewhat less pathogenic in rhesus macaques of Chinese origin, most likely due to its serial adaption in rhesus macaques of Indian origin [13]. It is important to note that primary unpassaged SIVsm isolates are not as pathogenic in macaques as SIVmac239.

Surprisingly, in some cases rhesus macaques receiving tissues directly from SIV-infected sooty mangabeys (SMs), a species from West Africa, were documented to have cleared these SIVsm infections, with no detectable viral loads for over 1 year [14,15]. Figure 2.2 outlines the serial passage at the New England Primate Center adapted from Mansfield et al. [9]. When HIV AIDS was described in 1981 [21], the disease in these rhesus macaques was finally recognized as AIDS [8,22].

SIVmne, a less pathogenic member of the SIVmac group, surfaced in pig-tailed macaques (*Macaca nemestrina*) at the WNPRC (Figure 2.1). SIVmne is less pathogenic compared to SIVmac239 because SIVmne was not serially passaged to the same extent as SIVmac239 [9]. An additional virus derived from SIVmac, termed SIV32H [23], was derived from SIVmac251 by a research group in the UK. The phylogenetic relationships between sooty mangabeys at the California NPRC, SIVmac at the NEPRC, and SIVmne at the WNPRC are illustrated in Figure 2.1.

Sequence data of SIVmac142, also isolated from Indian-origin rhesus macaques housed at the New England Primate Center, were shown to be derived from the same series of animal passages at the NEPRC that generated the other SIVmac viruses [9]. SIVmac142 is not widely used, probably because the molecular clone was less pathogenic and did not protect rhesus macaques from superinfection with pathogenic SIVmac251 [24].

WHAT'S IN A NAME? SIMIAN IMMUNODEFICIENCY VIRUS (AKA STLV-III, STLV-IIIAGM, AND HTLV-IV)

The nomenclature for SIV has always followed HIV, albeit a winding quasi-political path that plagued HIV nomenclature in the early days. SIVmac was first named STLV-III following the naming of HIV as HTLV-III, given in the first publication by the RC Gallo laboratory that was at the time housed at the

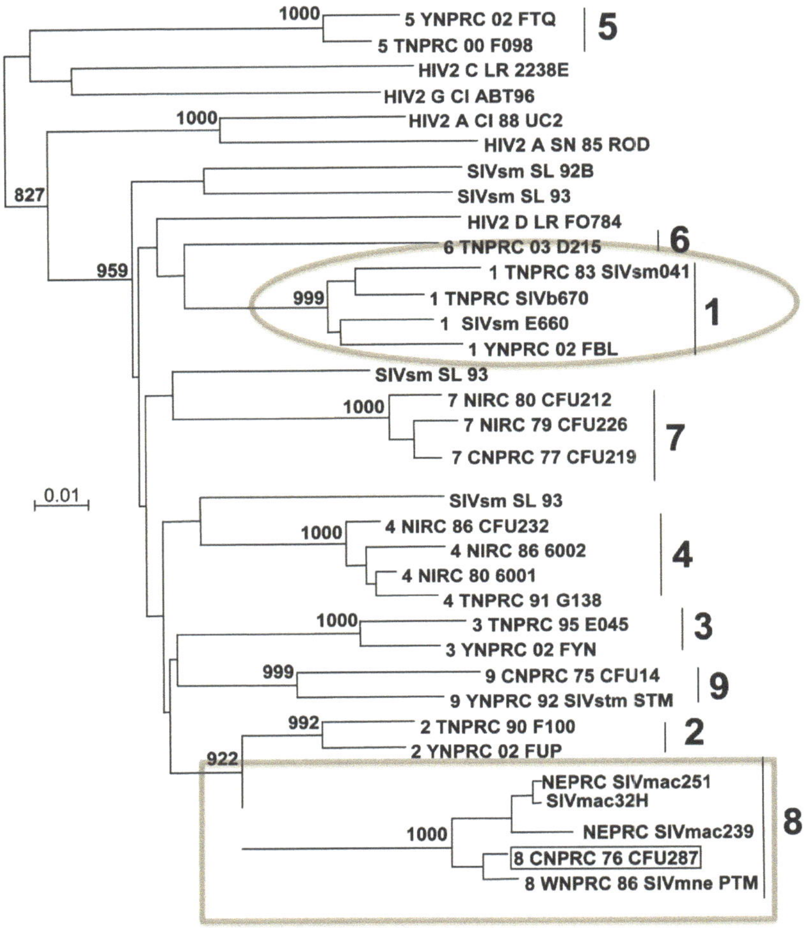

FIGURE 2.1 The phylogenetic tree shown is the origins of SIVmac (boxed) and SIVb670 groups. Each divergent group of SIVsm and derived viruses were given genetic lineage numbers 1–9. The source of SIVma251 and SIVmac239 is a sooty mangabey monkey number CFU287 1976 from the California National Primate Research (boxed). This mangabey was housed at the California center in the 1970s. SIVmne, first found in a pig-tailed macaque [12] with lymphoma at the Washington National Primate Research Center, also originated from this same CNPRC mangabey CFU287. *Modified from Ref. [11].*

NIH [25]. At that time, the human AIDS virus had two names, HTLV-III and LAV, before the name HIV was agreed to by researchers in the field [26]. After this publication, the HTLV-III name was replaced by HIV and the STLV-III name quickly followed suit in favor of SIV. However, there was a long-lasting effect in that the history of the nomenclature of SIV did result in significant confusion among scientists and the press.

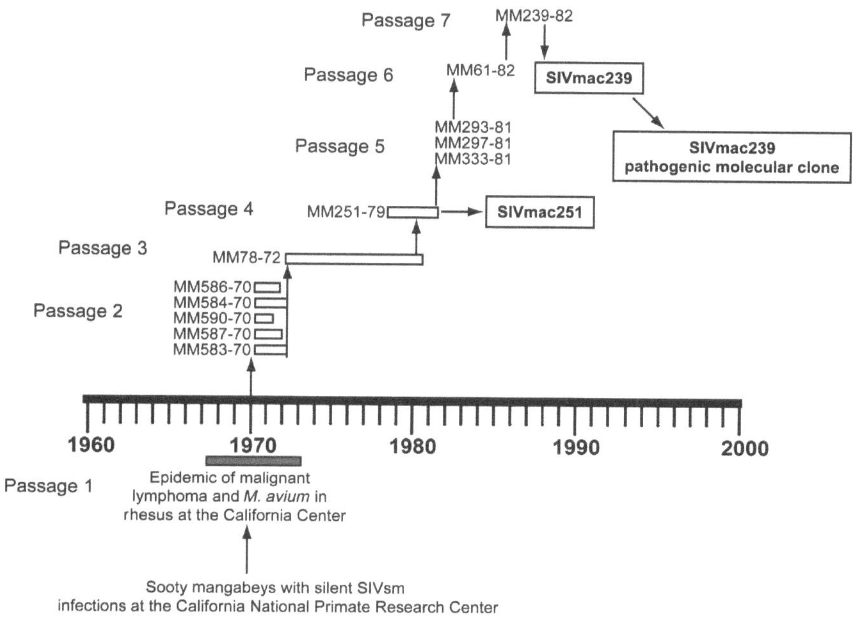

FIGURE 2.2 Serial passage history of simian immunodeficiency virus macaque isolates (SIVmac) commonly used in rhesus models of AIDS. The SIVmac group originated from one sooty mangabey (SM) at the California National Primate Research Center [9] (see Figure 2.1). The SMs had received clinical material from kuru patients [16–19]. Rhesus (Rh) macaques were used to serially passage tissue from the kuru-tissue-inoculated SMs. These rhesus macaques developed AIDS-related B-cell lymphomas and *Micobacterium avium* opportunistic infections at the California National Primate Research Center in the late 1960s and early 1970s [6]. Rhesus macaques that were still healthy but harboring silent SIVsm infections were transferred to the New England Primate Research Center, where lymphomas soon developed and the unknown agent was passaged six additional times to generate SIVmac251 (4th passage) and SIVmac239 (7th passage). The causative agent of AIDS (SIVsm) was passaged in Rh macaques, resulting in highly pathogenic viruses that were identified later in 1985 as SIV [20]. *Figure adapted from Ref. [9], with permission.*

HIV-1 was discovered by a group of scientists and physicians at the Pasteur Institute, Hopitals Pitié-Salpêtrière and Bichat Claude Bernard, all in Paris [27]. Francoise Barré-Sinoussi and Jean Claude Chermann of the Department of Virology at Pasteur Institute did the majority of the laboratory work. Luc Montagnier chaired the department. The manuscript was received by *Science* magazine and published in less than a month. Barré-Sinoussi and Montagnier were awarded the Nobel Prize in Physiology or Medicine in 2008 for this discovery. Professor Barré-Sinoussi, in her Nobel lecture, gave additional credit to Jean-Claude Chermann, since it was Chermann [28] who suggested that the AIDS virus was cytotoxic, an uncommon thought about retroviruses, and that the addition of fresh lymphocytes would facilitate this first isolation of the AIDS virus. The French group did not initially name the new virus in their *Science* publication but did name it lymphadenopathy-associated virus (LAV) in subsequent publications because the isolate was from a patient with lymphadenopathy who had not yet progressed to AIDS [27]. At about the same time LAV was named, the first publication on HIV by the Gallo NIH research group named the AIDS virus human T-cell lymphotropic virus-III (HTLV-III) in the erroneous conclusion that HTLV-III was a member of the human T-cell lymphoma/leukemia virus I (HTLV-I), genus *Deltaretrovirus* [25].

Following suit, the first publication of the simian relative of HIV was named simian T-cell lymphotropic virus III (STLV-III) because, by analogy with HTLV-III, STLV-III would also be related to simian T-cell leukemia virus group [20]. STLV-III was so named even though there was as yet no STLV-II, thus skipping the nomenclature ahead to claim a jargon-relationship with HTLV-III with the obvious intention that a genetic relationship would soon follow. Ironically, the original STLV-III publication in *Science* [20] made multiple references to how unlike STLV-III was to the HTLV group, yet the virus was still named as a member of the HTLV family. Undoubtedly this naming was the result of the influence of coauthors who had reported one year earlier that antibody from rhesus macaques with lymphomas at the New England Primate Center reacted with HTLV-membrane antigens [29]. Both the nomenclature and the science were further confused by the isolation of a virus from African green monkey cells that was named STLV-IIIagm [30]. In the following year this same group reported finding the STLV-IIIagm-related virus in persons in West Africa [31]. This "human virus" was renamed HTLV-IV. In 1989, the Desrosiers group at the NEPRC cleared up this confusion by showing with nucleotide sequence data that both HTLV-IV and STLV-IIIagm were laboratory contaminations by SIVmac251 [32].

The STLV-IIIagm publication was featured on the cover of *Science* magazine and was responsible for the widely held and persistent view by the press, some scientific writers (and their lay readers), and even some scientists and physicians that African green monkeys were the source of HIV. In the well-documented book *The River,* it was proposed that oral poliovirus vaccines (OPV) were contaminated with SIV and that OPV immunization campaigns in Africa launched the AIDS epidemic [33]. Edward Hooper, the author of *The River*, was initially misled by the STLVagmIII publication [30] and/or by Tom Curtis of the *Rolling Stone* who wrote a series on the AGM-OPV link to AIDS [34]. The thinking was that OPV must have involved STLVagmIII in some way that led to the AIDS epidemic. Two problems were quickly pointed out. First the real SIVagm, if it contaminated oral polio vaccines, was not ancestral to either HIV-1 or HIV-2. Hooper thus went on to focus his OPV theory to include chimpanzee tissues being used for preparation of oral polio vaccine. It was eventually reported that there were no evidence of SIV or chimpanzee tissue in stocks of OPV [35]. Another flaw that is not often discussed is that SIVsm is ancestral to HIV-2, and for the Hooper oral polio vaccine to be viable, it needed to explain both HIV epidemics, HIV types 1 and 2. It follows that OPV had to also contain sooty mangabey tissues and perhaps even gorilla tissue to be involved in the origins of the HIVs. There is no evidence that anything other than macaque monkey kidneys were used, and macaques are not naturally infected with SIV.

The first bona fide SIVagm from African green monkeys was found 2 years later at the Institute for Primate Research in Karen, Kenya by the Hayami group of Japan [36]. When HIV and SIV were sequenced, neither were related to the genus *Deltaretrovirus* (HTLV/STLV), but belonged to the entirely different *Lentivirus* genus [37–40]. LAV/HTLV-III was renamed HIV and STLV-III was renamed SIV [41]. The terms STLV-IIIagm and HTLV-IV were discarded and have even faded from the memory of researchers who experienced these bizarre events.

ORIGINS OF SIVMAC FROM KURU EXPERIMENTS USING ASIAN RHESUS MONKEYS

The precise experiments that unintentionally serially passaged SIV from SMs to macaques and induced the lymphoma outbreak in California can be traced to experiments by D. Carlton Gajdusek's laboratory (NIH), which used sooty mangabeys (SMs) in attempts to create an animal model for kuru [16–19]. In these experiments involving multiple species of NHP, SMs were inoculated with brain tissue homogenates from kuru patients. Tissues from the SMs were then passaged into rhesus and stump-tailed macaques (*Macaca arctoides*) at the California primate center [17,18]. It was after these macaque experiments that the lymphoma outbreak occurred [16]. Figure 2.1 shows the phylogenetic relationships that conclusively link SIVmac251 to SIVsm found in SM number CFU287 1976 at the California Primate Center at the time when the Gajdusek group was using rhesus, stump-tailed macaques [17] and sooty mangabeys [18] for kuru research. The passage of SM tissues to both stump-tailed [6,42] and Indian-origin rhesus macaques [6] at the California center resulted in the largest known lymphoma outbreak in macaques. It is worth noting that SIVmne, originally found in a pig-tailed macaque with lymphoma at the Washington National Primate Research Center, was also traced to the SIVsm/SIVmac lineage group 8 at the California Primate Center [12] (Figure 2.1).

The SIVmac group of viruses was isolated by the Desrosiers group in 1984 and published in *Science* magazine in 1985 [20]. The different isolates of the SIVmac virus were named for the New England Primate Center colony monkey numbers of the macaques in which SIV were found.

SIVmac251 and SIVmac239, the most commonly used SIVmac viruses, were named for rhesus macaque number 251-79 and SIVmac239 from rhesus macaque number 239-82. These viruses are the same except that SIVmac239 is an isolate from a later in vivo passage [9]. Rh251-79 was infected by inoculation with a homogenate of retro-orbital lymphoma tissue from Rh78-72. The tumor in Rh78-72 was reported to be a naturally occurring tumor [8]. The lymphomas appeared to be "spontaneous" because Rh78-72 had not been inoculated with lymphoma tissue. The SIVmac lineage was derived from a cohort of monkeys shipped to the NEPRC from the California NPRC. This cohort of five rhesus monkeys (Figure 2.2) harbored silent infections with SIVmac from sooty mangabeys at the California NPRC. How 78-72 became infected with the California-derived SIVsm is not clear. Some type of parenteral exposure, vertical transmission, or close physical contact between 78 and 72 and the California NPRC cohort must have occurred at the New England Primate Center. Macaque B-cell lymphomas were later found not to be naturally occurring in macaques when the SIVmac isolates [20] were characterized and the connection to human AIDS was made. The sequence of lymphoma tissue passages that involved seven passages of SIV at the NEPRC is unraveled in Mansfield et al. (Figure 2.2) [9], before being isolated from Rh 239-82 (Figure 2.1) as SIVmac239. Unknown to the Desrosiers group, all of the SIVmac strains were derived from a single source, a sooty mangabey number CFU287 at the California National Primate Research center (Figures 2.1 and 2.3) [10], thus explaining why the Desrosiers group initially believed that SIV was a macaque virus. There were no sooty mangabeys ever housed at the NEPRC [9]. This fact is important, because the African origin of SIV was actually discovered at the Tulane National Primate Research Center [2,37].

In summary, the research pathway that led to SIVmac and SIVsm becoming a pathogenic virus for macaques was circuitous beginning in California, a continent away from New England and a decade before SIVmac was discovered at the NEPRC. The major finding that SIVs derived from sooty mangabeys could cause AIDS in macaques was the central point of the new model. However, to a large extent SIVs are not highly pathogenic for macaques on first passage [14,15].

BOB GORMUS AT THE TULANE NATIONAL PRIMATE RESEARCH CENTER UNCOVERS THE AFRICAN ORIGIN OF SIV

Although SIVs have their natural origins in healthy African monkeys and apes, the first known SIV, SIVmac, was isolated from rhesus macaques (Rh, *Macaca mulatta)* of Indian origin at the New England Primate Research Center where no sooty mangabeys were housed [20]. Although there are a

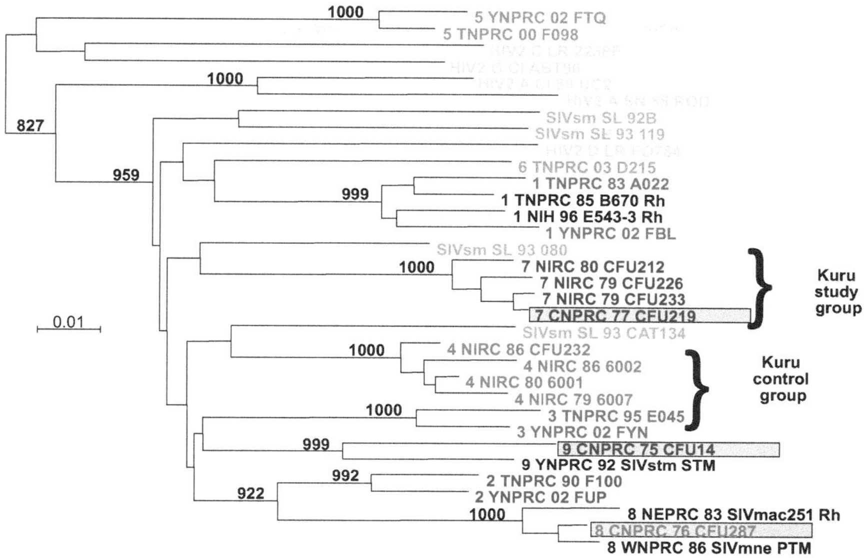

FIGURE 2.3 **Molecular epidemiologic evidence of SIVsm serial transmission during kuru experiments at the California National Primate Research Center.** SIVsm lineages segregate according to experimental and control group in kuru research. Lineages 1–9 (designated by the number in front of the virus) are clusters of SIVsm derived from a single mangabey and are highly related and branch together, in the same way that HIV-1 groups M–P are from different founder chimpanzees. Different SIVsm lineages display the divergent genetic patterns like HIV-2 subtypes or groups. Lineages 8 and 9 are strains that were cross-species transmitted to macaques at the California National Primate Research Center. SIVsm lineages 4 and 7 at the New Iberia Primate Research (NIPRC) were used as kuru and control experimental groups. These SMs were transferred from California to the NIPRC. Lineage 8, the origin of SIVmac and SIVmne, were contemporary to kuru experiments at the California center. The assigned lineage, the Primate Center of origin, the year of sample collection, SM identification, and the macaque species for macaque strains are given. *From Apetrei et al., Ref. [10], published with permission.*

few notable exceptions [43,44], SIV infections are generally not pathogenic in African species [11,37,45,46]. Therefore SIV existed as silent infections in sooty mangabeys and went undiscovered for many years in SM colonies at the Yerkes National Primate Research Center, the TNPRC, and the New Iberia Primate Research center in New Iberia, Louisiana. SIV infections induce frank disease in Asian macaque species including the lymphomas of B-cell origin and *Micobacterium avium* infections [6,8,9,20,41]. B-cell lymphomas are the hallmark of SIV-AIDS, because the occurrence of this tumor in macaques is exceedingly rare, and probably nonexistent, in macaques that are SIV negative [6].

The African monkey species origin of SIVmac had to be uncovered by a researcher passaging sooty mangabey tissue to rhesus macaques and observing the telltale lymphomas. This researcher was Bob Gormus at the Delta Regional Primate Research Center in Covington, Louisiana. The Delta Primate Center is now the Tulane National Primate Research Center. The original Delta Primate Center name reflected the intention that this primate center would be a consortium between multiple universities in the Mississippi River Delta (Peter Gerone, former TNPRC director, personal communication). The SIV found at this center in rhesus macaque number B670 was named STLV-III/Delta for the Delta primate center and STLV-III, the name of SIV at that time. The STLV-III/Delta name fell into disuse and the virus was renamed SIVb670 for the infected rhesus house number B670, from which this SIV was first isolated and characterized [11]. SIVsm in sooty mangabey tissue from the Tulane sooty mangabey colony caused AIDS in two of four rhesus macaques in its first passage through Indian-origin rhesus macaques (Figure 2.4) and in eight of nine after the second passage [2]. SIVb670 was from the second passage and became the key to unlocking the African origin of SIV.

SIVmac, having been found in Asian macaques at the New England Primate Center, was not recognized as a virus that originated from African monkeys. The primate center had failed to make a connection between SIVmac and the sooty mangabeys and macaques at the California Primate center. Bob Gormus, on the other hand, had used sooty mangabeys in a leprosy model [2] and he realized that the source of SIV was not macaques but sooty mangabeys. Gormus and colleagues wrote retrospectively, "Experimental leprosy studies using *Mycobacterium leprae* inoculum isolated from a sooty mangabey monkey (SMM) resulted in the accidental discovery that SMMs asymptomatically carry simian immunodeficiency virus (SIV) that is pathogenic in macaques" [2].

Therefore the discovery that SIV was an African virus resulted from the unintended consequences of a leprosy project. A research group at the Tulane Primate Center had developed a leprosy model using the sooty mangabey colony in which a naturally occurring case of leprosy appeared in SM number A022 in the colony (Figure 2.4) [49,50]. SIV was recognized by Gormus as having an African origin when leprosy material was passaged from SM A015 (Figure 2.4) to another Tulane SM 022 that was SIV positive. Tissues from SM 022 tissue were subsequently passaged into four rhesus macaques. One of the four, Rh 8664, developed lymphoma and six more rhesus macaques were inoculated with tissue from Rh 8664 (Figure 2.4). Five of the six developed AIDS [2]. In the second passage, one of the rhesus macaque with SIV, number B670 developed AIDS (Figure 2.4), and this rhesus was the subject of the *Nature* publication reporting that SIV has an African origin. Gormus recognized that the virus, SIVb670, came from SM AO22 and was not from rhesus macaques. Thus the African origin of SIV was established.

The SIVb670/H4/H9 group of SIVs were derived from three sooty mangabeys in the colony at the TNPRC (Figure 2.1, lineage group 1). This lineage

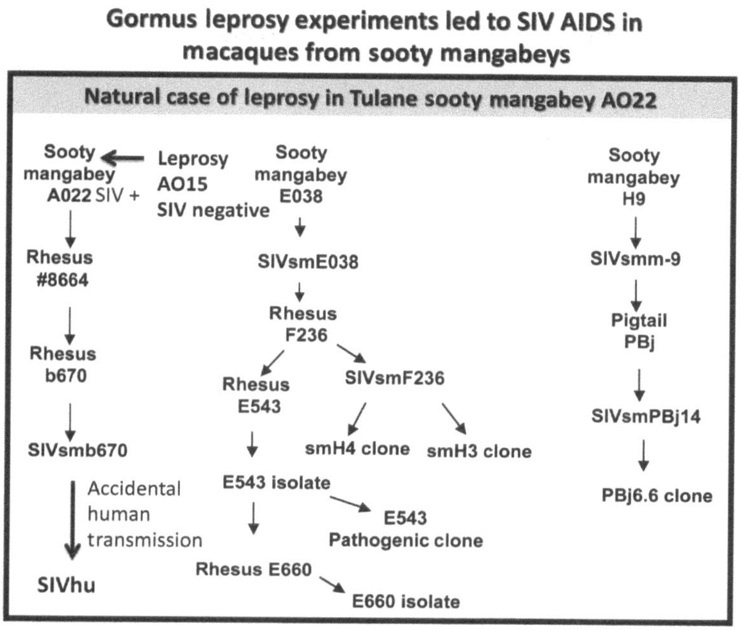

FIGURE 2.4 **Experiments at the Tulane Primate Center that led to the discovery of the African origins of SIV.** SIVb670 was created during a leprosy project at the Tulane National Primate Research Center (TNPRC) in which naturally occurring leprosy was found in wild-caught SM number A015. The experiments were designed to adapt the leprosy *Bacillus* to rhesus monkeys. SM A015 was SIV negative. Lepromatous tissue from A015 was serially passaged to another SM A022 that was SIV positive. Cutaneous lepromatous leprosy tissue was inoculated in four rhesus monkeys, two of which remained clinically normal and two developed leprosy and lymphoma. Lymphoma tissue from one of the four, Rh 8664, was passaged into six additional Rh monkeys, including B670. SIV was isolated from rhesus B670 and named SIVdelta, and this virus is the subject of Ref. [37] in which the African origin of SIV was first reported. The SIVe660 group was derived from SMe038, also at the TNPRC. The SIVpbj group was derived from the Yerkes National Primate Research Center SM colony. All of these SIVs are phylogenetic lineage 1 viruses (see Figure 2.1) SIVb670 was accidentally transmitted to a laboratory worker who had dermatitis and worked with SIVb670 without glove protection. SIVb670 was isolated from blood of the exposed laboratory worker and named SIVhu [47]. This human infection with SIVb670 was transient and not pathogenic. *Adapted from Ref. [48], with permission.*

group 1 is a separate lineage group from SIVmac, because it originates from an SM naturally infected with a divergent SIVsm [10]. As in the kuru experiments by the Gajdusek group [16], the Tulane group passaged SIV-infected tissues from sooty mangabeys to rhesus macaques with the goal of furthering their leprosy model without knowledge of SIV or AIDS. Commonly used members of the lineage 1 group are SIVb670 [11]; SIV-E660 [51], an uncloned isolate that is pathogenic and moderately susceptible to neutralizing antibody [51–53]; and SIVpbj [54]. A pathogenic molecular clone of lineage 1, E543-3, is also available [51]. Finally, it was considered that SIV-infected mangabeys at American primate centers were not natural infections of sooty mangabeys and had been inadvertently experimentally induced at American primate centers. The true African origin of SIV in sooty mangabeys was proven by studies of SIV in SMs in West Africa [45]. The names b670, E660, pbj, and E543-3 are all derived from the monkey numbers.

SIV H4 virus was derived from SM 038, an SM in the TNPRC colony. SIV H4 was passaged in rhesus macaques (Figure 2.4) and was the first SIV sequenced from a West African sooty mangabey [37]. This seminal paper showed that SIVsm was the most likely source of HIV-2 in West Africa.

HIV-1 ANCESTRY DISCOVERED IN AN EX-PET CHIMPANZEE IN GABON

This historical presentation may seem out of order, presenting HIV-2 before HIV-1 because the first AIDS virus discovered was HIV-1 group M from the main group of AIDS viruses [27]. This HIV-1 group nomenclature is worth further mentioning because the M was later added for main group to distinguish it from group O for outlier, which the second HIV-1 group found. HIV-1 group N was found in one person in Cameroon and was named N to fill in the alphabetical gap between M and O. The letter N designation was suggested by the manuscript reviewer as a bit of humor (P. Marx, personal communication). The N designation became official when HIV-1 N went to publication in *Nature*. There is now a HIV-1 group P, the naming of which followed the alphabetical order uncharacteristically beginning from the middle of the alphabet.

However, presenting HIV-2 ancestry first is correct since this is a historical presentation following the chronology of SIV/HIV discoveries. This order reflects the history of SIVsm that opened the path for the origin of HIV-1 somewhere in some simian species in Africa. That somewhere was identified by Martine Peeters and her group when a chimpanzee was brought to the Center International de Recherches Medicales de Franceville in Gabon (English International Center for Medical Research of Franceville, http://www.ird.fr/les-partenariats/nos-principaux-partenaires/cirmf). This primate center is known by its French abbreviation, CIRMF.

The CIRMF is a state-of-the-art primate research center in southeastern Gabon in the city of Franceville. As is the case in all of Africa, whenever there is a non-human primate facility nearby, the local population will bring non-human primates pets to the center when they no longer want them. The Peeters group isolated SIVcpzGab1 from one such wild-caught chimpanzee and showed it to be in the HIV-1 group [55]. There are many more contributions of historic significance from the Peeters group. The finding of a chimpanzee troop in southeastern Cameroon that is naturally infected with a SIVcpz that is closely related to HIV-1 group M is one such example [46]. The SIVcpz studies and others discoveries of natural SIV infections by the Peeters group in Africa are presented in Chapter 3 and are not further described here.

WAS SIVMAC THE IDEAL MODEL FOR AIDS?

There has been much discussion in the literature regarding the value of AIDS animal models [56–58]. The viewpoint as to what constitutes a "perfect" AIDS model is obviously a matter of opinion. But most will agree that an ideal model for any infectious disease is the induction of the human disease in a common laboratory animal with a virus isolated from human tissue. Although ideal model conditions exist for some human diseases, the perfect model criteria are not rigorously met by any of the AIDS animal models. The most common AIDS virus is HIV-1 group M, but it does not reproducibly infect or cause AIDS in any readily available animal species, although many species were tested [59–61]. Although HIV-1 readily infects chimpanzees [62–65], AIDS is a relatively rare outcome [66], and chimpanzees are hardly a common laboratory animal. There are reports of HIV-1 infection in pig-tailed macaques and HIV-2 infection in baboons and cynomolgus macaques. These animals are readily available. HIV-1 infection in pig-tailed macaques has been of limited value due to the transient nature of the HIV infection [67]. Infection for baboons and cynomolgus macaques with HIV-2 has not been widely used because of transient infections [68–70].

SIVmac in macaques represents an animal model compromise because SIVmac is related to HIV-2, the less prevalent human AIDS virus and SIVmac is not a human virus. Nevertheless, SIVs derived from sooty mangabeys readily induce AIDS in macaques, and macaques are a common laboratory species. Therefore, the SIV-macaque model is widely acknowledged as the "best available" model for human AIDS. The combination of the close genetic relationship between SIV and HIV-2 [37] and the induction of AIDS in commonly available laboratory monkey species, led to the development of SIV infection in macaques (SIVmac) as the animal model most widely used for AIDS research.

ANCIENT ORIGINS OF THE SIV FAMILY IN AFRICA RAISE MORE QUESTIONS

The SIV family is ancient in Africa, existing since before the Pleistocene epoch [71,72]. The hunting practices that have exposed hunters to bush meat and thus to SIV are well established as existing from at least 10,000

BCE until the present [73], and may extend to 50,000 BCE since human habitation likely dates from that period in West Africa. The question raised by this fact is, if SIV is ancient and human SIV exposure is ancient, why did AIDS emerge only in the twentieth century and not before? The conditions for potential AIDS emergence had existed for millennia. Therefore AIDS cannot be a zoonosis as reported in a high profile journal [74]. Zoonosis means that the disease is transmitted from an animal species to man, and this is not the case because AIDS would be ancient if it were transmitted as a zoonosis. Rabies is the best known of the zoonoses, and when rabies virus is transmitted to humans, the outcome is disease. The continued use of the term "zoonosis" to describe AIDS is unfortunate, since it is unscientific and medically incorrect.

The OPV theory of the origin of HIV/AIDS as described above is not consistent with the facts known about HIV emergence. However Edward Hooper, author of *The River*, the book on the OPV theory, deserves full credit for publicizing the importance of HIV emergence theories and for promoting interest in this unsolved question [33]. The more interesting of the remaining viable theories on how SIVs transitioned to HIV-1 and HIV-2 have been described [75–77]. Future work is needed to unravel this mystery because animal viruses are emerging and reemerging in new places every year. Unless we understand virus emergence, how can we prevent new epidemics?

References

[1] Norton R. The concise encyclopedia of economics. 2nd ed. Indianapolis. In: Library Fund Incorporated.

[2] Gormus BJ, Martin LN, Baskin GB. A brief history of the discovery of natural simian immunodeficiency virus (SIV) infections in captive sooty mangabey monkeys. Front Biosci 2004;9:216–24.

[3] Chalifoux LV, Ringler DJ, King NW, Sehgal PK, Desrosiers RC, Daniel MD, et al. Lymphadenopathy in macaques experimentally infected with the simian immunodeficiency virus (SIV). Am J Pathol 1987;128:104–10.

[4] Wyand MS, Ringler DJ, Naidu YM, Mattmuller M, Chalifoux LV, Sehgal PK, et al. Cellular localization of simian immunodeficiency virus in lymphoid tissues. II. In situ hybridization. Am J Pathol 1989;134:385–93.

[5] Pantaleo G, Graziosi C, Fauci AS. New concepts in the immunopathogenesis of human immunodeficiency cell depletion and viral replication in SIV infection. Science 1993;280:427–31.

[6] Stowell RE, Smith EK, España C, Nelson VG. Outbreak of malignant lymphoma in rhesus monkeys. Lab Invest 1971;25:470–9.

[7] Lowenstine LJ. Interactions between macaque macrophages and *Mycobacterium avium* [Ph.D. dissertation]. Davis: University of California; 1983.

[8] Hunt RD, Blake BJ, Chalifoux LV, Sehgal PK, King NW, Letvin NL. Transmission of naturally occurring lymphoma in macaque monkeys. Proc Natl Acad Sci USA 1983;80:5085–9.

[9] Mansfield KG, Lerche NW, Gardner MB, Lackner AA. Origins of simian immune-deficiency virus infection in macaques at the New England Regional Primate Research Center. J Med Primatol 1995;24:116–22.

[10] Apetrei C, Kaur A, Lerche NW, Metzger M, Pandrea I, Hardcastle J, et al. Molecular epidemiology of SIVsm in US primate centers unravels the origin of SIVmac and SIVstm. J Virol 2005;79:8991–9005.

[11] Murphey-Corb M, Martin LN, Rangan SR, Baskin GB, Gormus BJ, Wolf RH, et al. Isolation of an HTLV-III-related retrovirus from macaques with simian AIDS and its possible origin in asymptomatic mangabeys. Nature 1986;321:435–7.

[12] Benveniste RE, Arthur LO, Tsai CC, Sowder R, Copeland TD, Henderson LE, et al. Isolation of a lentivirus from a macaque with lymphoma: comparison with HTLV-III/LAV and other lentiviruses. J Virol 1986;60:483–90.

[13] Ling B, Veazey RS, Luckay A, Penedo C, Xu K, Lifson JD, et al. SIV(mac) pathogenesis in rhesus macaques of Chinese and Indian origin compared with primary HIV infections in humans. AIDS 2002;16:1489–96.

[14] Chen Z, Telfer P, Reed P, Zhang L, Getti A, Ho DD, et al. Isolation and characterization of the first simian immunodeficiency virus from a feral sooty mangabey (*Cercocebus atys*) in West Africa. J Med Primatol 1995;24:108–15.

[15] Gaufin T, Gautam R, Kasheta M, Ribeiro R, Ribka E, Barnes M, et al. Limited ability of humoral immune responses in control of viremia during infection with SIVsmmD215 strain. Blood 2009;113:4250–61.

[16] Apetrei C, Lerche NW, Pandrea I, Gormus B, Silvestri G, Kaur A, et al. Kuru experiments triggered the emergence of pathogenic SIVmac. AIDS 2006;20:317–21.

[17] Espana C, Gajdusek DC, Gibbs CJ Jr, Osburn BI, Gribble DH, Cardinet GH, et al. Transmission of Creutzfeldt-Jakob disease to the stumptail macaque (*Macaca arctoides*). Proc Soc Exp Biol Med 1975;149:723–4.

[18] Masters CL, Alpers MP, Gajdusek DC, Kakulas BA. Experimental kuru in the gibbon and sooty mangabey and Creutzfeldt-Jakob disease in the pigtailed macaque. With a summary of the host range of the subacute spongiform virus encephalopathies. J Med Primatol 1976;5:205–9.

[19] Gajdusek DC, Gibbs Jr CJ. Transmission of kuru from man to rhesus monkey (*Macaca mulatta*) 8 and one-half years after inoculation. Nature 1972;240:351.

[20] Daniel MD, Letvin NL, King NW, Kannagi M, Sehgal PK, Hunt RD, et al. Isolation of T-cell tropic HTLV-III-like retrovirus from macaques. Science 1985;228:1201–4.

[21] Gottlieb MS, Schroff R, Schanker HM, Weisman JD, Fan PT, Wolf RA, et al. *Pneumocystis carinii* pneumonia and mucosal candidiasis in previously healthy homosexual men: evidence of a new acquired cellular immunodeficiency. N Engl J Med 1981;305:1425–31.

[22] Letvin NL, Aldrich WR, King NW, Blake BJ, Daniel MD, Hunt RD. Experimental transmission of macaque AIDS by means of inoculation of macaque lymphoma tissue. Lancet 1983;2:599–602.

[23] Rud EW, Cranage M, Yon J, Quirk J, Ogilvie L, Cook N, et al. Molecular and biological characterization of simian immunodeficiency virus macaque strain 32H proviral clones containing nef size variants. J Gen Virol 1994;75:529–43.

[24] Denesvre C, Le Grand R, Boissin-Cans F, Chakrabarti L, Hurtrel B, Vaslin B, et al. Highly attenuated SIVmac142 is immunogenic but does not protect against SIV- mac251 challenge. AIDS Res Hum Retrov 1995;11:1397–406.

[25] Popovic M, Sarngadharan MG, Read E, Gallo RC. Detection, isolation, and continuous production of cytopathic retroviruses (HTLV-III) from patients with AIDS and pre-AIDS. Science 1984;224:497–500.

[26] Coffin J, Haase A, Levy JA, Montagnier L, Oroszlan S, Teich N, et al. Human immunodeficiency viruses. Science 1986;232:697.

[27] Barré-Sinoussi F, Chermann JC, Rey F, Nugeyre MT, Chamaret S, Gruest J, et al. Isolation of a T lymphotropic retrovirus from a patient at risk for acquired immune deficiency syndrome (AIDS). Science 1983;220:868–71.

[28] Barré-Sinoussi F. Nobel Lecture. http://www.nobelprize.org/nobel_prizes/medicine/laureates/2008/barre-sinoussi_lecture.pdf and personal communication.

[29] Homma T, Kanki PJ, King Jr NW, Hunt RD, O'Connell MJ, Letvin NL, et al. Lymphoma in macaques: association with virus of human T lymphotrophic family. Science 1984;225:716–8.
[30] Kanki PJ, Alroy J, Essex M. Isolation of T-lymphotropic retrovirus related to HTLV-III/LAV from wild-caught African green monkeys. Science 1985;230:951–4.
[31] Kanki PJ, Barin F, M'Boup S, Allan JS, Romet-Lemonne JL, Marlink R, et al. New human T-lymphotropic retrovirus related to simian T-lymphotropic virus type III (STLV-IIIAGM). Science 1986;232:238–43.
[32] Kestler III HW, Yen LI, Naidu YM, Butler CV, Ochs MF, Jaenel G, et al. Comparison of simian immunodeficiency virus isolates. Nature 1988;331:619–22.
[33] Hooper E. The river. Little, Brown and Company Publishers; 1999.
[34] Tom Curtis. The origin of AIDS: a startling new theory attempts to answer the question "was it an act of God or an act of man?". Roll Stone 1992;626:54–59, 61, 106, 108. http://www.uow.edu.au/~bmartin/dissent/documents/AIDS/Curtis92.html.
[35] Dickson D. Tests fail to support claims for origin of AIDS in polio vaccine. Nature 2000;407:117.
[36] Ohta Y, Masuda T, Tsujimoto H, Ishikawa K, Kodama T, Morikawa S, et al. Isolation of simian immunodeficiency virus from African green monkeys and seroepidemiologic survey of the virus in various non-human primates. Int J Cancer 1988;41:115–22.
[37] Hirsch VM, Olmsted RA, Murphey-Corb M, Purcell RH, Johnson PR. An African primate lentivirus (SIVsm) closely related to HIV-2. Nature 1989;339:389–92.
[38] Chakrabarti L, Guyader M, Alizon M, Daniel MD, Desrosiers RC, Tiollais P, et al. Sequence of simian immunodeficiency virus from macaque and its relationship to other human and simian retroviruses. Nature 1987;328:543–7.
[39] Luciw PA, Potter SJ, Steimer K, Dina D, Levy JA. Molecular cloning of AIDS- associated retrovirus. Nature 1984;312:760–3.
[40] Ratner L, Fisher A, Jagodzinski LL, Mitsuya H, Liou R-S, Gallo RC, et al. Complete nucleotide sequences of functional clones of the AIDS virus. AIDS Res Hum Retrov 1987;3:57–69.
[41] Feichtinger H, Putkonen P, Parravicini C, Li SL, Kaaya EE, Böttiger D, et al. Malignant lymphomas in cynomolgus monkeys infected with simian immunodeficiency virus. Am J Pathol 1990;137:1311–5.
[42] Khan AS, Galvin TA, Lowenstine LJ, Jennings MB, Gardner MB, Buckler CE. A highly divergent simian immunodeficiency virus (SIVstm) recovered from stored stump-tailed macaque tissues. J Virol 1991;65:7061–5.
[43] Ling B, Apetrei C, Pandrea I, Veazey RS, Lackner AA, Gormus B. Classic AIDS in a sooty mangabey after an 18-year natural infection. J Virol 2004;78:8902–8.
[44] Keele BF, Jones JH, Terio KA, Estes JD, Rudicell RS, Wilson ML, et al. Increased mortality and AIDS-like immunopathology in wild chimpanzees infected with SIVcpz. Nature 2009;460:515–9.
[45] Marx PA, Li Y, Lerche NW, Sutjipto S, Gettie A, Yee JA, et al. Isolation of a simian immunodeficiency virus related to human immunodeficiency virus type 2 from a West African pet sooty mangabey. J Virol 1991;65:4480–5.
[46] Peeters M, Fransen K, Delaporte E, Van den Haesevelde M, Gershy-Damet G-M, Kestens L, et al. Isolation and characterization of a new chimpanzee lentivirus (simian immunodeficiency virus isolate cpz-ant) from a wild-captured chimpanzee. AIDS 1992;6:447–51.
[47] Khabbaz RF, Heneine W, George JR, Parekh B, Rowe T, Woods T, et al. Brief report: infection of a laboratory worker with simian immunodeficiency virus. N Engl J Med January 20, 1994;330(3):172–7.
[48] Voevodin A, Marx P. Simian virology. Wiley-Blackwell publishers; 2009.
[49] Wolf RH, Gormus BJ, Martin LN, Baskin GB, Walsh GP, Meyers WM, et al. Experimental leprosy in three species of monkeys. Science 1985;227:529–31.

[50] Fukunishi Y, Meyers WM, Binford CH, Walsh GP, Johnson FB, Gerone PJ, et al. Electron microscopic study of leprosy in a mangabey monkey (natural infection). Int J Lepr Other Mycobact Dis 1984;52:203–7.
[51] Hirsch V, Adger-Johnson D, Campbell B, Goldstein S, Brown C, Elkins WR, et al. A molecularly cloned, pathogenic, neutralization-resistant simian immunodeficiency virus, SIVsmE543-3. Virology 1997;71:1608–20.
[52] Hirsch VM, Fuerst TR, Sutter G, Carroll MW, Yang LC, Goldstein S, et al. Patterns of viral replication correlate with outcome in simian immunodeficiency virus (SIV)-infected macaques: effect of prior immunization with a trivalent SIV vaccine in modified vaccinia virus Ankara. J Virol 1996;70:3741–52.
[53] Hirsch VM, Johnson PR. Pathogenic diversity of simian immunodeficiency viruses. Virus Res 1994;32:183–203.
[54] Fultz PN, Zack PM. Unique lentivirus–host interactions: SIVsmmPBj14 infection of macaques. Virus Res 1994;32:205–25.
[55] Peeters M, Honoré C, Huet T, Bedjabaga L, Ossari S, Bussi P, et al. Isolation and partial characterization of an HIV-related virus occurring naturally in chimpanzees in Gabon. AIDS 1989;3:625–30.
[56] Fauci AS, Johnston MI, Dieffenbach CW, Burton DR, Hammer SM, Hoxie JA, et al. HIV vaccine research: the way forward. Science 2008;321:530–2.
[57] Watkins DI, Burton DR, Kallas EG, Moore JP, Koff WC. Nonhuman primate models and the failure of the Merck HIV-1 vaccine in humans. Nat Med 2008;14:617–21.
[58] Koff WC, Burton DR, Johnson PR, Walker BD, King CR, Nabel GJ, et al. Accelerating next-generation vaccine development for global disease prevention. Science 2013;340:1232910.
[59] Adachi A, Gendelman HE, Koenig S, Folks T, Willey R, Rabson A, et al. Production of acquired immunodeficiency syndromeassociated retrovirus in human and non-human cells transfected with an infectious molecular clone. J Virol 1986;59:284–91.
[60] Levy JA, Cheng-Mayer C, Dina D, Luciw PA. AIDS retrovirus (ARV-2) clone replicates in transfected human and animal fibroblasts. Science 1986;232:998–1001.
[61] Morrow WJ, Wharton M, Lau D, Levy JA. Small animals are not susceptible to human immunodeficiency virus infection. J Gen Virol 1987;68:2253–7.
[62] Alter HJ, Eichberg JW, Masur H, Saxinger WC, Gallo R, Macher AB, et al. Transmission of HTLV-III infection from human plasma to chimpanzees: an animal model for AIDS. Science 1984;226:549–52.
[63] Fultz PN, McClure HM, Swenson RB, McGrath CR, Brodie A, Getchell JP, et al. Persistent infection of chimpanzees with human T-lymphotropic virus type III/lymphadenopathy-associated virus: a potential model for acquired immunodeficiency syndrome. J Virol 1986;58:116–24.
[64] Prince AM, Allan J, Andrus L, Brotman B, Eichber J, Fouts R, et al. Virulent HIV strains, chimpanzees, and trial vaccines. Science 1999;283:1117–8.
[65] Saksela K, Muchmore E, Girard M, Fultz P, Baltimore D. High viral load in lymph nodes and latent human immunodeficiency virus (HIV) in peripheral blood cells of HIV-1-infected chimpanzees. J Virol 1993;67:7423–7.
[66] O'Neil SP, Novembre FJ, Hill AB, Suwyn C, Hart CE, Evans-Strickfaden T, et al. Progressive infection in a subset of HIV-1-positive chimpanzees. J Infect Dis 2000;182:1051–62.
[67] Kent SJ, Corey L, Agy MB, Morton WR, McElrath MJ, Greenberg PD. Cytotoxic and proliferative T cell responses in HIV-1-infected *Macaca nemestrina*. J Clin Invest 1995;95:248–56.
[68] Barnett SW, Murthy KK, Herndier BG, Levy JA. An AIDS-like condition induced in baboons by HIV-2. Science 1994;266:642–6.
[69] Looney DJ, McClure J, Kent SJ, Radaelli A, Kraus G, Schmidt A, et al. A minimally replicative HIV-2 live-virus vaccine protects *M. nemestrina* from disease after HIV-2(287) challenge. Virology 1998;242:150–60.

[70] Walther-Jallow L, Nilsson C, Söderlund J, ten Haaft P, Mäkitalo B, Biberfeld P, et al. Cross protection against mucosal simian immunodeficiency virus (SIVsm) challenge in human immunodeficiency virus type 2-vaccinated cynomolgus monkeys. J Gen Virol 2001;82:1601–12.
[71] Worobey M, Telfer P, Souquière S, Hunter M, Coleman CA, Metzger MJ, et al. Island biogeography reveals the deep history of SIV. Science 2010;329:1487.
[72] Ma D, Jasinska A, Kristoff J, Grobler JP, Turner T, Jung Y, et al. SIVagm infection in wild African green monkeys from South Africa: epidemiology, natural history, and evolutionary considerations. For the International Vervet Research Consortium. PLoS Pathog 2013;9:e1003011.
[73] McIntosh SK. From stone to metal: new perspectives on the later prehistory of West Africa. J World Prehist 1988;2:89–133.
[74] Sharp PM, Rayner JC, Hahn Evolution BH. Great apes and zoonoses. Science 2013;340:284–6.
[75] Marx PA, Alcabes PG, Drucker E. Serial human passage of SIV by unsterile injecting and the emergence of epidemic HIV in Africa. Philos Trans R Soc Ser B 2001;356:911–20.
[76] Drucker E, Alcabes PG, Marx PA. The injection century: consequences of massive unsterile injecting for the emergence of human pathogens. Lancet 2001;358:1989.
[77] de Sousa JD, Alvarez C, Vandamme AM, Müller V. Enhanced heterosexual transmission hypothesis for the origin of pandemic HIV-1. Viruses 2012;4:1950–83.

CHAPTER

3

Simian Immunodeficiency Virus Infections in the Wild

Martine Peeters[1], Dongzhu Ma[2], Florian Liegeois[1], Cristian Apetrei[2]

[1]UMI233, TransVIHMI, Institut de Recherche pour le Développement (IRD), Montpellier, France and Université Montpellier 1, Montpellier, France; [2]Center for Vaccine Research, University of Pittsburgh, Pittsburgh, PA, USA

INTRODUCTION

Simian immunodeficiency viruses (SIV) naturally infect African nonhuman primate (NHP) hosts and are of special interest and importance because they are the root causes of the AIDS pandemic, one of the most threatening infectious diseases to have emerged in the twentieth century. Human immunodeficiency virus type 1 and 2 (HIV-1 and HIV-2), the infectious agents causing AIDS in humans, emerged respectively after cross-species transmissions of SIVs from chimpanzees and gorillas in west central Africa and from SIVs infecting sooty mangabeys in West Africa [1–3]. Since the description of the first AIDS cases in the 1980s, the estimated cumulative number of HIV infections worldwide is almost 60 million [4]. The aim of this chapter is to describe the actual knowledge on SIV infections in wild African NHPs.

HISTORY

Shortly after the identification of HIV-1 as the cause of AIDS in 1983, the first simian lentivirus, SIVmac, was isolated at the New England Primate Research Center (NEPRC) in 1984 from captive rhesus macaques (*Macaca mulatta*) with a clinical presentation similar to AIDS [5–7]. Retrospective studies showed that this SIVmac virus was introduced

Natural Hosts of SIV.
http://dx.doi.org/10.1016/B978-0-12-404734-1.00003-6

at NEPRC by other rhesus monkeys that were previously housed at the California National Primate Research Center (CNPRC), where they survived an earlier (late-1960s) disease outbreak that was also characterized by immune suppression and opportunistic infections [8]. The infected rhesus macaques had been in contact with wild-caught and healthy sooty mangabeys at the CNPRC that have been retrospectively shown to be infected with a closely related virus, SIVsmm. The close phylogenetic relationship between SIVmac and SIVsmm identified mangabeys as the source of SIV in macaques [8–10]. SIVs have since been isolated from many wild African NHP species, but not from wild Asian or New World NHPs [11–13]. Moreover, SIVs appear to be nonpathogenic for their natural host in the wild, in contrast to captive Asian macaque species, suggesting that Asian NHPs are not natural hosts for SIVs. However, it is important to note that no large surveys have been conducted yet on wild NHP species in these continents.

SIV CLASSIFICATION

Currently, SIV infection has been documented in at least 45 different African NHP species (Table 3.1). SIVs are named according to the host species, and a three-letter code refers to the common name of the corresponding NHP species; e.g., SIVrcm for SIVs from red-capped mangabeys, SIVgsn for greater spot-nosed monkeys, etc. When different subspecies of the same species are infected, a designation referring to the name of the subspecies is added to the letter code; for example, SIVcpz*Ptt* and SIVcpz*Pts* differentiates between the SIVcpz lineages obtained from two different chimpanzee subspecies, *Pan troglodytes troglodytes* and *P. troglodytes schweinfurthii*, respectively.

All primate lentiviruses have a common genomic structure, which consists of long terminal repeats (LTR) that flank both ends of the genome, three structural genes, *gag*, *pol*, and *env*, and five accessory genes, *vif*, *vpr*, *tat*, *rev*, and *nef*. Some SIVs carry an additional regulatory gene, *vpx* or *vpu*, and based on these different genomic organizations, we can distinguish between three patterns. SIVagm, SIVsyk, SIVmnd-1, SIVlho, SIVsun, SIVcol, SIVtal, SIVdeb, SIVwrc, SIVolc, SIVkrc, and SIVkrtg display the basic structure with three major and five accessory genes [14–23]. SIVcpz, SIVgor, SIVgsn, SIVmus, SIVmon, and SIVden harbor an additional accessory gene, *vpu*, like HIV-1, but SIVcpz and SIVgor differ from the other members of this group by the fact that *env* and *nef* genes are not overlapping [3,24–27]. SIVsmm, SIVrcm, SIVmnd-2, SIVdrl, and SIVagi harbor a supplemental accessory gene, *vpx*, like HIV-2 [2,28–30]. For the remaining SIVs, full-length

TABLE 3.1 African Non-human Primates Infected with SIV

Genus	Species/Subspecies	Common Name	SIV Lineage	SIV Prevalence[b]	Geographic Distribution
Pan	*troglodytes troglodytes*	Central African chimpanzee	SIVcpz*Ptt*	0–40%	West Central: Cameroon, Congo, West Central African Republic, Gabon, Equatorial Guinea
	troglodytes schweinfurthii	Eastern chimpanzee	SIVcpz*Pts*	0–40%	East: DRC, Uganda, Tanzania
Gorilla	*gorilla gorilla*	Western lowland gorilla	SIVgor	0–20%	West Central: Cameroon, Gabon, Congo, Central African Republic
Colobus	*guereza*	Mantled guereza	SIVcol, SIVkmgr	18–60%	West central: Nigeria to Ethiopia/Tanzania
	angolensis	Angolan colobus	?	na	West central: Congo Basin
	satanus	Black colobus	SIVblc	30%	West central: South-west Cameroon to Congo River, Bioko
Piliocolobus	*badius badius*	Western red colobus	SIVwrc*Pbb*	50–80%	West central: Guinea to Ghana
	badius temminckii	Temminck's red colobus	SIVwrc*Pbt*	10%	West central: Senegal, Gambia
	tholloni	Tshuapa red colobus	SIVtrc[a]	24%	Central: Below Congo River
	rufomitratus tephrosceles	Ugandan red colobus	SIVkrc	23–60%	East: Uganda
Procolobus	*verus*	Olive colobus	SIVolc	na	West central: Sierra-Leone to Ghana
Lophocebus	*albigena*	Gray-cheeked mangabey	?	na	Central: Nigeria to Uganda/Burundi
	aterrimus	Black crested mangabey	SIVbkm[a]	na	Central: DRC

Continued

TABLE 3.1 African Non-human Primates Infected with SIV — cont'd

Genus	Species/Subspecies	Common Name	SIV Lineage	SIV Prevalence[b]	Geographic Distribution
Papio	*anubis*	Olive baboon	?	na	West to east: Mali to Ethiopia
	cynocephalus	Yellow baboon	[SIVagm-ver][a]	na	Central: Angola to Tanzania
	ursinus	Chacma baboon	[SIVagm-ver][a]	na	South: Southern Angola to Zambia
Cercocebus	*atys*	Sooty mangabey	SIVsmm	50%	West central: Senegal to Ghana
	torquatus	Red-capped mangabey	SIVrcm	50%	West central: Nigeria, Cameroon, Gabon
	agilis	Agile mangabey	SIVagi	na	Central: northeast Gabon to northeast Congo
Mandrillus	*sphinx*	Mandrill	SIVmnd-1, mnd-2	33%	West central: Southern Cameroon, Cameroon (south of the Sanaga) to Gabon, Congo
	leucophaeus	Drill	SIVdrl	22%	West central: Southeast Nigeria to Cameroon (north of Sanaga), Bioko
Allenopithecus	*nigroviridis*	Allen's swamp monkey	SIVasm[a]	6%	Central: Congo
Miopithecus	*talapoin*	Angolan talapoin	SIVtal[a]	na	West central: east coast of Angola into DRC
	ogouensis	Gabon talapoin	SIVtal	17%	West central: Cameroon (south of the Sanaga), Gabon
Erythrocebus	*patas*	Patas monkey	[SIVagm-sab][a]	7%	West to east: Senegal to Ethiopia, Tanzania

Chlorocebus	*sabaeus*	Sabaeus monkey	SIVagm-sab	47%	West central: Senegal to Volta River
	aethiops	Grivet	SIVagm-gri	na	East: Sudan, Eritrea, Ethiopia
	tantalus	Tantalus monkey	SIVagm-tan	50%	Central: Ghana to Uganda
	pygerythrus	Vervet monkey	SIVagm-ver	na	South: South Africa to Somalia and Angola
Cercopithecus	*diana*	Diana monkey	?	na	West central: Sierra Leone to Ivory Coast
	nictitans	Greater spot-nosed monkey	SIVgsn	0–7%	Central: Forest blocks from West Africa to DRC
	mitis	Blue monkey	SIVblu[a]	na	East central: East Congo to Rift Valley
	albogularis	Sykes's monkey	SIVsyk	46%	East: Somalia to eastern Cape
	mona	Mona monkey	SIVmon	na	West Central: Niger delta to Cameroon (north of Sanaga)
	lowei	Lowe's mona monkey	?	na	West Central: Liberia to Ivory Coast
	campbelli	Campbell's monkey	?	na	West Central: Gambia to Liberia
	pogonias	Crowned guenon	?	na	West central: Cross-river in Nigeria to Congo (east)
	denti	Dent's mona monkey	SIVden	na	Central: South of Congo river
	cephus	Mustached guenon	SIVmus	1%	West central: Cameroon (south of Sanaga) to east of Congo river

Continued

TABLE 3.1 African Non-human Primates Infected with SIV — cont'd

Genus	Species/Subspecies	Common Name	SIV Lineage	SIV Prevalence[b]	Geographic Distribution
	erythrotis	Red-eared monkey	SIVery[a]	33%	West central: Cross river in Nigeria to Sanaga in Cameroon, Bioko
	ascanius whitesidi	Red-tailed monkey	SIVasc[a]	25%	Central: Congo Basin
	ascanius schmidti	Kibale red tailed guenon	SIVkrtg	38%	Central: North Congo to Uganda
	lhoesti	l'Hoest's monkey	SIVlho	na	Central: Eastern DRC to western Uganda
	solatus	Sun-tailed monkey	SIVsun	na	West central: Tropical forest of Gabon
	preussi	Preuss's monkey	SIVpre[a]	22%	West central: Cross river in Nigeria to Sanaga in Cameroon, Bioko
	hamlyni	Owl-faced monkey	?	na	Central: Eastern DRC to Ruanda
	neglectus	De Brazza's monkey	SIVdeb	20–40%	Central: Angola, Cameroon, Gabon to Uganda, western Kenya
	wolfi	Wolf's monkey	SIVwol	12%	Central: Congo basin, below Congo river

?, only serological evidence for SIV infection.
[ ], SIV infections resulting from cross-species transmissions of local African green monkey species.
na, not available.
[a] *Only partial sequences are available.*
[b] *Prevalence observed in wild NHP primate populations are shown.*

sequences are not yet available. Therefore, information on accessory genes is lacking [31].

EVOLUTION OF SIVs

As shown in Table 3.1, serological evidence of SIV infection is observed for 45 species, and SIV infection has been confirmed by sequence analysis in 39 species. A high genetic diversity is observed and, in general, each primate species is infected with a species-specific virus, which forms monophyletic lineages in phylogenetic trees (Figure 3.1). The genetic diversity among NHP lentiviruses is extremely complex. SIV

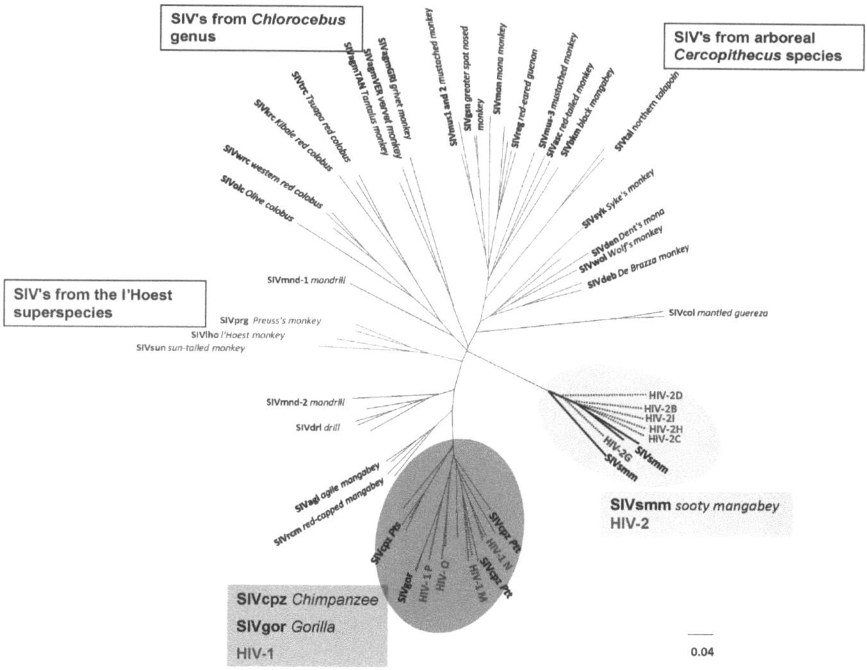

FIGURE 3.1 Genetic diversity and evolutionary history of the different lineages of HIV/SIV. Phylogenetic tree analysis using the neighbor joining method on a 294 bp fragment from the *pol* gene of different SIVs infecting non-human primates and HIVs infecting humans. Branch lengths are drawn to scale (the scale bar indicates 0.04 substitutions per site). Each NHP is infected with a species-specific SIV lineage. The different colors are used for clarity to discriminate the major SIV lineages. The different HIV-1 and HIV-2 lineages, which are interspersed with the SIVcpz/SIVgor and SIVsmm lineages, respectively, are indicated in red and blue circles, respectively. (See color plate at the back of the book.)

phylogenetic clusters are only partially superimposable on trees representing phylogenetic relations between NHP hosts. Therefore, there are examples of co-evolution between the virus and the host, cross-species transmission, and recombination between distant SIV lineages, and certain species can even harbor different SIV lineages. However, cross-species transmissions could give erroneous impressions of co-evolution, especially when chances for efficient host switch are higher among genetically closely related species [32].

Co-evolution over long periods of time is the case for the four different African green monkey species from the *Chlorocebus* genus. The four species live in geographically separate and nonoverlapping areas across Africa and each is infected with a species-specific SIV, i.e., SIVagmVer in vervets, SIVagmGri in grivets, and SIVagmTan and SIVagmSab in tantalus and sabaeus monkeys, respectively [33–37]. The SIVs from the l'hoesti superspecies, (i.e., SIVlho from *Cercopithecus lhoesti*, SIVsun from *Cercopithecus solatus*, and SIVprg from *Cercopithecus preussi*) and SIVs from arboreal *Cercopithecus* species each also form separate clusters in the phylogenetic tree of SIVs [14,15,18,38] (Figure 3.1).

On the other hand, there are also numerous examples of cross-species transmission of SIVs between NHP species with overlapping habitats or among species that live in polyspecific associations. For example, in Senegal, West Africa, SIVagmSAB from sabaeus monkeys (*Chlorocebus sabaeus*) has been transmitted to patas monkeys (*Erythrocebus patas*) and in South Africa, SIVagmVER from vervets has been transmitted to yellow and chacma baboons [39–41]. A recent study showed that a wild-caught, but captive, agile mangabey (*Cercocebus agilis*) in Cameroon was infected with SIVagi, a virus that is closely related to SIVrcm from red-capped mangabeys [28]. Both species inhabit west central Africa and their habitats overlap only in a small geographic area in southwest Cameroon [42]. However, among more than 180 wild agile mangabeys, no SIV infection was confirmed by polymerase chain reaction [43]. Therefore we cannot totally exclude that this agile mangabey became infected with an SIVrcm strain in captivity, and more studies on wild red-capped and agile mangabeys are thus necessary to clarify whether agile mangabeys are naturally infected with an SIV and whether their SIV is closely related to SIVrcm or not. Similarly, a black mangabey (*Lophocebus aterrimus*) has been described to be infected with a virus, SIVbkm, that falls within the cluster of the SIVs from arboreal *Cercopithecus* species, and is most closely related to SIVs from red-tailed guenons (*Cercocebus ascanius)* [44,45]. Both species are endemic in the Democratic Republic of Congo (DRC) and have overlapping geographic ranges, but the SIVbkm virus was obtained from a captive monkey in the zoo from Kinshasa, the capital city of DRC, and a cross-species transmission in captivity cannot be excluded. Finally, the phylogenetic relationship between SIVgor from western lowland gorillas (*Gorilla gorilla gorilla*) and SIVcpz from chimpanzees (*P. t. troglodytes*) suggest that gorillas acquired the virus also via cross-species transmission [3,46].

There are also examples of cross-species transmission of SIVs that are followed by recombination between SIV species-specific lineages. Greater spot-nosed and mustached monkeys live in polyspecific associations and are infected with SIVgsn and SIVmus, respectively, which are distinct but closely related species-specific lineages [24]. In Cameroon, mustached monkeys are infected with two distinct SIVs, referred to as SIVmus-1 and SIVmus-2, the latter being a recombinant lineage between SIVmus and SIVgsn and another unknown SIV [47]. The recent full-length genome characterization of new SIVmus strains in mustached monkeys from Gabon confirmed a complex evolutionary history among SIVs from the SIVgsn/mus/mon lineage with additional cross-species events followed by different recombination patterns [48]. One of the most striking examples of cross-species transmission, followed by recombination, is SIVcpz in chimpanzees. The 5′ region of SIVcpz is most similar to SIVrcm from red-capped mangabeys, and the 3′ region is closely related to SIVgsn from greater spot-nosed monkeys [49]. Chimpanzees are known to hunt monkeys for food and the recombination of these monkey viruses occurred most probably within chimpanzees. This recombinant virus is the common ancestor of today's SIVcpz lineages, which in turn are transmitted to gorillas [3,46].

Finally, some NHPs are infected with more than one SIV lineage. Mandrills are infected with SIVmnd-1 in southern Gabon, south of the Ogooue River, and with SIVmnd-2 in northern Gabon and Cameroon [30]. Mustached monkeys are infected with two different variants, indicated as SIVmus-1, 2 documented in Cameroon and an additional complex recombinant SIVmus strain that has been documented in Gabon [47,48]. In contrast to mandrills, the different viruses that co-circulate in the mustached monkeys are not separated by geographical barriers. Similarly, mantled guerezas in the Kibale forest in east Africa are also infected with two different SIVs [50].

HOW LONG WERE SIVs PRESENT IN NON-HUMAN PRIMATES

The widespread presence of SIVs in numerous African NHPs suggests that SIV is very old; however, molecular clock methods indicate a time scale of centuries to 2000 years only [51]. A recent report studied SIVs in NHPs from Bioko Island, Equatorial Guinea, which was isolated from the African mainland 10,000–12,000 years ago with the elevation of the sea levels. Phylogenetic analysis showed that SIVs from NHPs from Bioko and SIVs from related species on the continent are infected with related viruses, suggesting that they evolved independently since Bioko became isolated. By using the date of the separation of Bioko Island to calibrate molecular clock analysis, it was shown that SIVs have been present in African primates for more than 32,000 years [38]. A more recent study of the diversity of SIVagmVer in South Africa documented different clustering patterns of the SIVs infecting vervets on the two sides of the Drakensberg Mountain. By recalibrating the trees to consider both the Bioko sequences and the sequences from the vervets in South Africa, the origin of SIVs could be retraced to 800,000–2,500,000 years ago [37]. Finally, the discovery of an endogenous lentivirus in the genome of the gray mouse lemur (*Microcebus murinus*), from Madagascar, suggests that SIV could even be older [52]. Phylogenetic analysis shows that the gray mouse lemur prosimian immunodeficiency virus (pSIVgml) is basal to all known primate lentiviruses. NHPs from Madagascar and Africa have been separated for at least 14 million years [53]. Low or absence of pathogenicity of SIVs is likely a consequence of long-term host–virus co-evolution.

Alternatively, the interaction between host and virus on a molecular level can also be used to estimate the extent to which primates and lentiviruses have coexisted. A recent study showed that the region of APOBEC3G targeted by Vif is adaptively diversifying in independent primate lineages. The observations from this study suggest that the minimum age for the association between Old World monkeys and SIV is around 5–6 million years and could possibly extend to 12 million years [54].

CHALLENGES TO STUDY SIV INFECTION IN WILD NON-HUMAN PRIMATE POPULATIONS

At present, SIVs have been identified in 45 different NHP species from Africa, but little, if any, information on the prevalence and molecular epidemiology is available for most of them. Almost all SIV lineages were discovered based on cross-reactivity with HIV-1 or HIV-2 antigens from commercially available assays; therefore SIV infection is most likely underestimated [31,55–58]. To increase sensitivity, in-house SIV-specific enzyme-linked immunosorbent assay tests have been developed, using SIV antigens from different SIV lineages [43,44,59–61]. However, each time a new species-specific SIV lineage is described or when a wide genetic diversity within SIV lineages is observed in current or future studies in the field, these tests need to be updated. The importance of including a wide variety of SIV antigens was clearly illustrated by the identification of SIVtrc in Tshuapa red colobus samples from DRC, which yielded high rates of false negative results in assays based on HIV-1/2 cross-reactivity [44,62].

Another factor that makes studies on SIV infection in wild NHPs not easy is the fact that wild NHP populations live in isolated forest regions and are difficult to localize, especially when they are arboreal. The initial studies on SIV have been done on blood samples from captive NHPs in primate centers or zoos, or pet animals, but these populations do not reflect the situation in wild populations [13,63]. Given the difficult access to primates in their natural habitats, and the endangered status of several NHP species, it was thus necessary to develop methods that allow antibody or viral detection utilizing body fluids that can be collected noninvasively. Significant efforts have been made over the last decade to optimize the detection of antibodies and viral RNA in fecal samples, although at lower sensitivities than in blood [64–66]. Large-scale molecular epidemiological studies have been successfully conducted with this approach to study SIV infection in chimpanzees and gorillas across Africa. However, the technique seems to be not equally efficient on all NHP species: for example, no antibodies could be detected in feces from western red and olive colobus monkeys in West Africa due to an absence of IgG in the fecal samples from these species, and screening by viral RNA amplification is necessary [67]. Collecting and analyzing fecal samples is labor-intensive in the field, especially for arboreal NHP species, but also in the laboratory because the host species has to be confirmed by sequence analysis of the mitochondrial DNA and microsatellite analysis needs to be done to enumerate individuals for the estimation of prevalence [68]. Yet noninvasive sampling is the strategy of choice for characterizing SIVs in the wild, as numerous African NHP species that are natural hosts of SIV are highly endangered.

SIV PREVALENCE AND MOLECULAR EPIDEMIOLOGY IN AFRICAN NON-HUMAN PRIMATES

Molecular Epidemiology of SIVs in African Apes and Origin of HIV-1

The NHP species that have been the primary focus of studies designed to understand SIV prevalence and genetic diversity are the chimpanzees because they harbor the ancestors of HIV-1 strains in humans. Large-scale molecular epidemiological studies were initiated across Africa, covering almost the entire geographic range of the four different chimpanzee subspecies. A noninvasive approach was used and several thousand fecal samples were collected from chimpanzees. These studies revealed a heterogeneous prevalence, depending on the localities, with rates ranging from 0% to more than 40% of SIVcpz*Ptt* in *P. t. troglodytes* from west central Africa and also for SIVcpz*Pts* in *P. t. schweinfurthii* from East Africa [56,58,69,70]. No SIV infection has been detected yet in the other two chimpanzee subspecies, *P. t. ellioti* (previously *P. t. vellerosus*) and *P. t. verus* [71–73]. These studies also showed that the two SIVcpz-infected chimpanzee subspecies harbor subspecies-specific SIVcpz lineages, i.e., SIVcpz*Ptt* for *P. t. troglodytes* and SIVcpz*Pts* for *P. t. schweinfurthii*. Importantly, SIVcpz*Ptt* strains are significantly more closely related to HIV-1 strains from humans [56,58,69,70,74]. More in-depth analysis of SIVcpz strains from numerous different locations showed phylogeographic clustering within both SIVcpz*Ptt* and SIVcpz*Pts* lineages [58,70]. Based on this phylogeographic clustering, the ancestors of HIV-1 M and N could be traced to distinct chimpanzee communities in southeast and south-central Cameroon, respectively [58,69].

Chimpanzees and gorillas are sympatric species, i.e., their habitats overlap. Fecal samples have also been collected from gorillas and revealed that western lowland gorillas (*G. g. gorilla*) are infected with an SIVgor that forms a monophyletic group within the HIV-1/SIVcpz*Ptt* radiation, supporting a chimpanzee origin for SIVgor [3]. Interestingly, all SIVgor strains are most closely related to HIV-1 group O and P [46,55,68,75]. The overall SIVgor prevalence was three times lower than that observed in chimpanzees in the same areas and range from 0% to 20% in certain gorilla groups [55]. Similar to SIVcpz, phylogeographic clustering was also seen for SIVgor in gorillas and HIV-1 P is most likely of gorilla origin, but the reservoirs of the direct ancestors of HIV-1 O have not yet been identified [75,76].

Bonobos, the third ape species in Africa, are endemic in DRC and their range is below the Congo River and does not overlap with that of gorillas and chimpanzees. To date, almost 1000 fecal samples have been analyzed, but no SIV infection has been documented [70] (Ahuke-Mundeke, personal communication).

These large-scale noninvasive studies provided important data on SIV prevalence and genetic diversity in wild ape populations, but more importantly, they allowed important advances in the understanding of the origin of HIV-1. Because the four HIV-1 groups fall within the HIV-1/SIVcpz*Ptt*/SIVgor radiation, the four cross-species transmissions giving rise to HIV-1 occurred most likely in western equatorial Africa, corresponding to the geographic area of *P. t. troglodytes* chimpanzees and western lowland gorillas (*G. g. gorilla*). The spread of the four HIV-1 groups is very unbalanced, with a pandemic group (HIV-1 M) and an epidemic one (HIV-1 O), and with the other two (HIV-1 N and P) only occurring in a handful of patients [77].

Molecular Epidemiology of SIVs in Sooty Mangabeys and Origin of HIV-2

The genetic diversity among SIVsmm isolates has been documented on isolates obtained from captive sooty mangabey populations in primate centers in the United States, pets, or mangabey bushmeat in rural areas from Sierra Leone and Liberia, and by a noninvasive survey among wild NHP populations in Côte d'Ivoire [57,78,79]. A high genetic diversity is seen among SIVsmm isolates, and the ancestors of the HIV-2 group A and B viruses, responsible for the HIV-2 epidemic in West Africa, were identified in wild sooty mangabey populations in the Taï forest in Côte d'Ivoire, close to the border with Liberia [57]. The occurrence of at least nine cross-species transmissions from sooty mangabeys to humans in West Africa have been documented thus far, leading to the nine HIV-2 groups [77,80]. Similar to HIV-1, these HIV-2 variants also have different epidemiological histories; only groups A and B have epidemic potential since they were able to spread to some extent across West Africa. The other HIV-2 groups have remained restricted to a few cases of human infections, most of which are located in the area of emergence [77].

SIV Prevalence and Genetic Diversity in Other African Monkeys

The other NHP species have been less studied, because currently no human strains close to SIVs other than those from chimpanzees, gorillas, and mangabeys have been reported. However, given the ongoing and increasing contact between a wide diversity of NHP species and humans in Africa, through hunting and butchering, it is likely that SIV and other simian viruses are still being transmitted to humans [81–84]. The description in 2009 of HIV-1 group P in two Cameroonian patients and the recent report on a ninth cross-species transmission of SIVsmm in rural Ivory Coast illustrates that our knowledge on HIV diversity and possible cross-species transmission is still incomplete [75,80]. Some SIVs indeed have the ability to replicate in vitro in cultures of activated human peripheral blood mononuclear cells (PBMCs) [11].

Therefore, it is important to have data on the diversity of SIVs and their prevalence in NHPs that are frequently hunted. An alternative approach to determine the SIV prevalence and to measure simultaneously the extent of SIV exposure is to analyze tissue and/or blood samples collected from NHP bushmeat, although without encouraging further hunting. Studies on bushmeat samples from different forest regions in Cameroon, the DRC, Equatorial Guinea, and Gabon revealed an overall SIV seroprevalence in NHP bushmeat, all species confounded, that ranged from 3% in Cameroon to approximately 20% in the other countries [38,43,44,57,59]. These studies also showed significant differences in prevalence rates per species (from 0% to >40%), and variations within species according to the sampling site. Low prevalence was observed in *Cercopithecus nictitans* and *Cercopithecus cephus* in Cameroon, but varied according to sampling sites from 0% to 7% [43]. On the other hand, SIVdeb is widely present in De Brazza monkeys across Central Africa, 20–40% prevalence, and genetic diversity seems lower, although strains from animals in Cameroon, DRC, and Uganda form phylogeographic clusters [85]. Similarly, SIV infection is also widespread among the different red colobus species and subspecies [22,28,86,87].

To date, data from different studies has shown that a high frequency of animals of certain species are infected and almost any survey of such species leads to the finding of SIV-positive animals, which is the case for red colobus species and subspecies, African green monkeys, sooty mangabeys, red-capped mangabeys, De Brazza monkeys, or mandrills [31]. On the other hand, there are species in which the prevalence of SIV is heterogeneous and can be high in certain populations, like chimpanzees or gorillas [77]. Finally there are species in which the prevalence of SIV is low and heterogeneous, as for example in greater spot-nosed or mustached monkeys [43]. The different SIV prevalence across different NHP species can be related to several factors, which include the behavior, population structure, viral, or host characteristics.

TRANSMISSION OF SIMIAN RETROVIRUSES IN THEIR NATURAL HOSTS

For SIV, horizontal transmission by sexual contact or biting as well as vertical transmission have been demonstrated, but which mode of transmission predominates will most likely depend on the behavior and social structure of the different NHP species [30,37,56,57,68,88–92].

Seroepidemiological surveys in AGMs, sooty mangabeys, and mandrills revealed higher prevalence levels in adult monkeys than in juveniles, indicating a horizontal route of transmission. The prevalence of SIVagm in adult AGMs from multiple locations in South Africa is very high (80–90% in adult females), up to 10-fold higher than in infants and juveniles [37]. This is consistent with an earlier study on vervet monkeys residing within the Awash National Park in Ethiopia, East Africa, in which all animals identified as juveniles by dentition were SIV seronegative [92].

In addition to the sexual route of transmission, a non-negligible proportion of SIV transmission in the wild likely occurs through biting or fighting, as injuries are frequent; for example, 10% of vervets in South Africa showed signs of recent injuries that could result in exposure, such as deep lacerations [37]. Injuries preponderantly occur during the mating season, due to both male transfer and contests for dominance between males [93], and are usually cleaned by licking. In support of this route of transmission are the results of the survey in South Africa that identified near-identical virus strains in vervets, which were not direct mating partners [37]. Monkeys are highly susceptible to SIV infection through oral exposure [94,95], and the high levels of systemic viral replication documented in the wild [37,96] may facilitate oral transmission during wound cleaning. SIV transmission through biting has been described in captive AGMs [92], sooty mangabeys [97], and chimpanzees belonging to two different subspecies [98]. SIVsmm has also been reported to be transmitted among macaques by biting [63].

The rates of maternal-to-infant transmission in African NHP species that are natural hosts of SIVs are significantly lower than the 35–40% mother-to-infant transmission rates reported in HIV infection (http://www.unaids.org) or 40–70% transmission rates in rhesus macaques [99] and are likely the result of virus-host co-adaptation [100]. The exact route of transmission (in utero, perinatally, or via breast milk) has not been identified. In a prospective study, experimental mother-to-offspring transmission by breastfeeding was not observed in mandrills [101], while another study did not demonstrate vertical transmission in AGMs [91]. Some recent studies also suggest vertical transmission as a potential mechanism of SIVsm transmission [57,78].

Epidemiologic patterns of seroconversion in wild primate populations endemically infected with lentiviruses suggest these agents are most efficiently transmitted during adult contacts. Considering that it is well established that HIV is spread by sexual contact primarily via mucosal exposure, it is likely that naturally occurring viruses are spread via this route as well. While maternal-to-offspring transmissions have been reported in primates, this is relatively rare compared to horizontal transmission.

HISTORY OF SIV IN THEIR NATURAL HOSTS

In spite of the active viral replication and high prevalence levels, it is generally assumed that SIV infection is nonpathogenic in natural hosts [102–106]. Yet, over the last decade, sporadic cases of progression to AIDS were reported in African NHP natural hosts of SIVs upon infection with either their species-specific viruses [107–109] or cross-species-transmitted SIVs [110]. These cases of clinical disease only occurred in a handful of monkeys that were all outliving the lifespan of the species and had an infection that was documented for decades [107–109]. This observation supports a paradigm in which the postulated lack of SIV disease progression in African NHPs is due to an active control of the deleterious consequences of SIV infection by the hosts, as the viruses retained their pathogenic potential [108]. As such, SIV infection in natural hosts is a persistent infection with an incubation period that exceeds the normal lifespan of the naturally infected animal [111].

In general, information on the natural history of SIV infection in the wild is very scarce, mostly due to the lack of research infrastructure for pathogenesis studies in the wild. The use of alternative samples, such as bushmeat and noninvasive samples (i.e., feces), while permitting the characterization of viral diversity and prevalence, does not permit the study of the virological and immunophenotypic markers necessary to assess the natural history of SIV infection in the wild. Therefore, the major focus of studies carried out thus far in wild African NHP species involved identification and characterization of the species-specific SIVs [102,112]. Most of the data supporting the lack of pathogenicity of SIVs in their natural hosts were derived from the study of captive monkeys and of only three species (AGMs, sooty mangabeys, and mandrills) [102–106].

To draw definitive conclusions about the pathogenicity of SIVs in their natural hosts in the wild, large-scale studies and a long follow-up are clearly needed. A case in point to support this statement is the recent documentation of the pathogenic potential of SIVcpz infection in wild chimpanzees. For decades, SIVcpz infection was considered to be nonpathogenic in chimps, based on studies performed in a very limited number of captive SIVcpz-infected chimpanzees [113,114]. It was only after establishment of the habituated but wild-living populations of naturally SIVcpz-infected chimpanzees (*P. t. schweinfurthii*) at the Gombe National Park in Tanzania and a long-term follow-up that this paradigm shifted. SIVcpz infection was reported to be pathogenic in naturally infected chimpanzees [90,115], being associated with a 10- to 16-fold increase in age-corrected risk of death and reduced fertility in SIV-positive females, in terms of both their birth rate and the survival of the offspring [115]. Immunohistochemistry and in situ hybridization of post mortem spleen and lymph node samples showed lower $CD4^+$ T-cell counts in SIV-positive versus SIV-negative individuals [90]. Similarly, a report of a naturally SIV-infected *P. t. troglodytes* chimpanzee confiscated in Cameroon in 2003 suggests clinical progression to an AIDS-like disease in this animal [116]. This observation is important as SIVcpz*Ptt* is the direct ancestor of HIV-1.

The fact that such a major effect of SIVcpz infection went undetected for decades supports the need for systematic assessment of the natural history of SIV infection in African NHPs in their native habitat, and not in captive environments (where health status is controlled, nutrition is monitored, and exposure to adventitious agents that may impact the clinical status is limited) to provide significant information regarding the clinical outcome of SIV infection in natural hosts.

Yet, only scattered assessments of the viral loads (VLs) in either a limited number of wild mandrills or recently captured vervets are available [117,118]. Only recently, a large-scale study in wild AGMs (vervets) reported on the features of the natural history of SIV infection in the wild [37].

Clinical and Biological Data

None of the SIV-infected monkey samples thus far in the wild presented with any of the clinical signs associated with AIDS, including fever, weight loss, lymphadenopathy, or opportunistic infections. Although the cross-sectional nature of the studies in the wild generally precludes assessment of changes in monkey weight and thus a direct assessment of weight loss, comparative analyses of the body mass index between SIV-positive and SIV-negative monkeys were performed on vervets in South Africa and reported that SIV status does not impact on the normal weight of wild vervets [37,96].

SIV Receptor Use and Tropism

Most of the SIVs naturally infecting NHPs in Africa use the same set of receptors used by HIV-1, i.e., CD4 as a primary binding receptor and chemokine receptors serving as viral co-receptors [119,120]. Differently from HIV-1, which can use both CCR5 and CXCR4 [121], most of the SIVs naturally infecting African NHPs exclusively use CCR5 as the main co-receptor [119,120,122] and no correlation can be established between the co-receptor usage and the pathogenesis in vivo in NHPs [120]. Several SIV strains isolated from African NHPs were reported to use CXCR4 (SIVmnd-1, SIVagmSab, and several SIVsmm strains), but no pathologic correlation has been described in these monkey species [123–125]. SIVrcm uses CCR2b as the co-receptor for viral entry [29,126], because the CCR5 gene of red-capped mangabeys contains a 24-base pair deletion in the *env* binding region [126]. A fraction (8%) of sooty mangabeys have also been shown to carry a CCR5 allele containing a 2-base pair deletion (Δ2) encoding a truncated molecule that is not expressed on the cell surface and does not support SIV entry in vitro [127].

Natural hosts of SIVs express lower levels of CCR5 on memory $CD4^+$ T cells in PBMC and mucosal tissues, compared to progressive hosts [128]. This restriction of CCR5 expression may help preserve the homeostasis of central memory $CD4^+$ T cells despite high levels of viremia [129], reduce the homing of activated $CD4^+$ T cells to inflamed tissues [128,130], and influence SIV transmission [100,101]. Indeed, in wild sabeus monkeys from Gambia the levels of CCR5 expression by the $CD4^+$ T cells are strongly associated with the prevalence of SIVsab [96].

SIV Replication in Wild Animals

The levels of plasma viremia during chronic infection have been thought to be highly predictive of the outcome of human HIV-1 infection [131] and, as we previously reported, may also predict the outcome of

SIV infection in natural hosts, as the handful of African NHPs reported to progress to AIDS exhibited higher chronic VLs than nonprogressors [132]. To date, VLs in natural hosts of SIVs such as AGMs, sooty mangabeys, and mandrills have only been assessed in either experimentally SIV-infected monkeys [101,117,118,124,133–141] or captive naturally infected monkeys [118,132,139,142]. Only one study reported VLs in wild African monkeys on a very limited number of samples [118].

The levels of viral replication in a large cohort of naturally SIV-infected vervets in the wild ranged from 10^4 to 10^7 SIVagmVer RNA copies/ml (Figure 3.2), higher in the juvenile than in adult AGMs, but this difference is not statistically significant. In this study, while VLs of the majority of SIVagm-infected animals were relatively within a close range (10^4–10^5 SIVagmVer RNA copies/ml), VLs were higher (10^6–10^8 SIVagmVer RNA copies/ml) in approximately 10% of vervets, which were further documented to be recently infected (Fiebig II) [143], indicating a very active transmission of SIVagmVer in the wild vervet population.

Interestingly, about half of the SIVagm-infected vervet dams were documented as lactating at the time of sampling. The VLs of these lactating dams are detailed in Figure 3.2 and demonstrate that the offspring are exposed to high levels of virus during the lactation period. Meanwhile, the prevalence of SIV infection in AGM females may reach levels of 80–90% and, as such, there is a large in utero exposure of AGM offspring to SIVagm. However, despite massive exposure to SIVagm both in utero and through breastfeeding, the prevalence of SIVagm is very low in infant vervets in the wild.

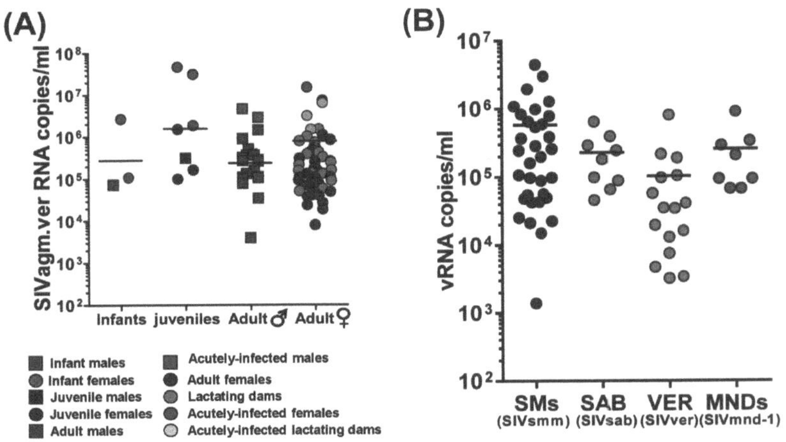

FIGURE 3.2 Comparison between SIV plasma viral loads in naturally infected vervets in the wild (A) and experimentally infected African NHPs of different species (B). In (A) VLs of lactating dams are depicted in lighter colors. Detection limit: 100 copies/ml. (See color plate at the back of the book.)

The levels of viral replication in wild African NHP hosts of SIVs are in the range of those reported during experimental studies [137] (Figure 3.2). As such, by corroborating the assessments of viral replication in captive and wild African NHP hosts of SIV, we may conclude that the levels of SIV replication in wild NHPs are one order of magnitude higher than those observed in chronically HIV-1-infected patients (3×10^4 HIV-1 RNA copies/ml) [144]. Assuming that the shedding of the HIV and SIV in the seminal and vaginal fluids of humans and AGMs is similar, the higher chronic SIV VL levels in natural hosts are probably significantly higher than the threshold above which the virus can be effectively transmitted through sexual contact (i.e., in humans, large-scale epidemiological studies established the transmission threshold of plasma VLs at 1500 copies/ml, one order of magnitude lower than the average chronic VLs [144]). Note, however, that no study investigated the VLs in seminal fluid in any of the natural hosts of SIVs, including AGMs. As such, the assessment of viral replication in wild NHPs, together with the relevant data showing a dramatic increase in SIV prevalence in monkeys that reach adult status and become sexually active, strongly suggests that the main route of SIV transmission in the wild is through sexual contact.

The high levels of chronic viral replication and high prevalence in the wild may also suggest a more efficient mucosal transmission of SIV in natural hosts. Yet, the low CCR5 expression on the mucosal $CD4^+$ T cells of the natural hosts may compensate for the high levels of virus replication and thus account for similarity in the mechanisms of SIV transmission observed in pathogenic hosts. Indeed, characterization of the viral diversity in acutely infected AGMs in the wild demonstrated a genetic bottleneck of virus transmission, which was highly similar to those reported in humans and macaques [145,146], with one to three SIVagmSab variants being identified in acutely infected sabeus monkeys from Gambia [96].

Immunopathology of SIV Infection in Wild Natural Hosts

A key factor in the pathogenesis of the HIV-induced immunodeficiency is the failure of the lymphoid regenerative capacity [147–152]. In particular, bone marrow suppression, reduced thymic output, and loss of naïve T cells have all been observed in pathogenic SIV- and HIV-infected individuals [153–156]. In addition to the loss of $CD4^+$ T cells, which is the hallmark of HIV infection, an overall immune dysfunction occurs in HIV infection, with alterations of the effectors of adaptive immune response as well as that of innate immune effectors (B cells, dendritic cells, natural killer cells, or monocytes). Altogether, this pan-dysfunction of the immune effectors is critical for driving disease progression to AIDS.

Unfortunately, to date there is no published study on the immunophenotypic changes that occur in naturally SIV-infected African non-human primates in the wild. We recently had, however, the opportunity to perform immunophenotypic characterization in SIV-infected and uninfected wild sabeus monkeys from Gambia and we did not observe any significant difference between the two groups with regard to the frequency of $CD4^+$ T cells, $CD8^+$ T cells, B cells, monocytes, natural killer cells, or myeloid or plasmacytoid dendritic cells [96].

It is currently thought that the main reason for the lack of disease progression observed in the vast majority of SIV-infected African NHPs that are natural hosts of SIV is their ability to resolve chronic immune activation during the transition from acute to chronic infection [136,157–160]. The main factor behind this control of immune activation is that African NHP hosts have the ability to maintain the integrity of the mucosal barrier and thus to control microbial translocation [136,161], which has been reported to be the main factor behind increased immune activation that drives disease progression in pathogenic HIV/SIV infections [147,162]. It is critical to assess the integrity of the mucosal barrier in a natural host in the wild, where monkeys are exposed to many pathogens in a significantly more hostile environment than captive animals, which may impact the ability of SIV-infected monkeys to maintain the barrier.

While a direct assessment of the mucosal barrier in the wild is virtually impossible, the levels of sCD14 were tested in wild AGMs as a surrogate biomarker of microbial translocation and showed to be unchanged in SIV-infected vervets compared to uninfected ones (Figure 3.3). Even the samples collected from acutely infected monkeys did not show a significant increase in the levels of sCD14, in agreement with previous results from experimentally infected AGMs [135,136] and suggesting that wild SIVagm-infected AGMs maintain the integrity of the mucosal barrier.

Assessment of the levels of chronic immune activation in wild African NHPs is equally difficult and data are only available from the

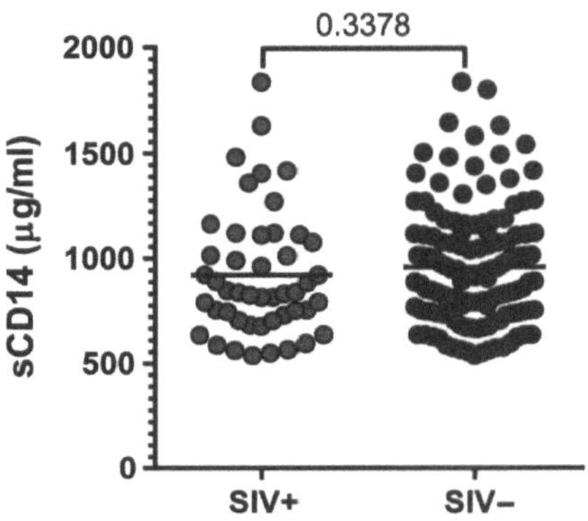

FIGURE 3.3 No significant difference is noted in the plasma levels of sCD14 (used as a surrogate marker for microbial translocation) in uninfected and SIVagmVer-infected vervets from South Africa.

same cohort of sabeus monkeys in Gambia. In that study, no difference was observed between SIVagmSab-infected and uninfected AGMs [96]. Furthermore, cytokine and chemokine testing as markers of chronic immune activation in vervet and sabeus AGMs from South Africa and Gambia showed similar levels of proinflammatory cytokines and chemokines in SIV-infected and uninfected monkeys, again supporting the nonpathogenic nature of SIVagm infection in the wild [37,96].

CONCLUSIONS

Our knowledge on retroviral evolution remains limited because to date only a small number of viral strains have been characterized from the different NHP species and only from limited geographic areas. Moreover, several simian retroviruses have only been described from captive animals and can thus bias the overall picture. The number of species infected with SIV is most likely underestimated, since at least one-third of the more than 70 recognized Old World monkey and ape species in sub-Saharan Africa have yet to be tested or only a very few individuals of a given species have been tested. Knowing that almost 90% of the primate species tested has been shown to be infected with an SIV, many of the remaining African NHP species can be expected to harbor SIV infection. Identification of additional SIV from more species and more geographic areas will allow a more detailed characterization of the evolution of SIVs in general. It is also worthwhile to increase efforts to study more Asian and New World primates. Importantly, studies should be done as much as possible on wild NHP populations. This is clearly not easy, because of the endangered status of many NHP species, but the development of noninvasive techniques to detect viral infection in fecal samples makes access to wild NHP populations possible.

To date, information on the natural history of SIV infection in wild NHPs in their natural environment is very scarce, mostly due to the lack of research infrastructure for pathogenesis studies in the wild and the limitations of noninvasive fecal samples to study clinical and immunological evolution. Although the characterization of the pathogenesis of SIV in the wild is at its inception, noninvasive monitoring of wild habituated chimpanzees showed that SIVcpz is pathogenic for its natural host and studies on African green monkeys support a nonprogressive outcome of SIVagm infection in the wild. To draw definitive conclusions about pathogenicity of SIVs in their natural hosts in the wild, large-scale and long-term follow-up studies are needed. There is also a need to develop tools or identify markers that can monitor immune status and clinical evolution of SIV infection in fecal samples.

References

[1] Gao F, Bailes E, Robertson DL, Chen Y, Rodenburg CM, Michael SF, et al. Origin of HIV-1 in the chimpanzee *Pan troglodytes troglodytes*. Nature 1999;397:436–41.

[2] Hirsch VM, Olmsted RA, Murphey-Corb M, Purcell RH, Johnson PR. An African primate lentivirus (SIVsm) closely related to HIV-2. Nature 1989;339:389–92.

[3] Van Heuverswyn F, Li Y, Neel C, Bailes E, Keele BF, Liu W, et al. Human immunodeficiency viruses: SIV infection in wild gorillas. Nature 2006;444:164.

[4] www.unaids.org.

[5] Daniel MD, Letvin NL, King NW, Kannagi M, Sehgal PK, Hunt RD, et al. Isolation of T-cell tropic HTLV-III-like retrovirus from macaques. Science 1985;228:1201–4.

[6] Letvin NL, Daniel MD, Sehgal PK, Desrosiers RC, Hunt RD, Waldron LM, et al. Induction of AIDS-like disease in macaque monkeys with T-cell tropic retrovirus STLV-III. Science 1985;230:71–3.

[7] Kanki PJ, McLane MF, King Jr NW, Letvin NL, Hunt RD, Sehgal P, et al. Serologic identification and characterization of a macaque T-lymphotropic retrovirus closely related to HTLV-III. Science 1985;228:1199–201.

[8] Apetrei C, Kaur A, Lerche NW, Metzger M, Pandrea I, Hardcastle J, et al. Molecular epidemiology of SIVsm in US Primate Centers unravels the origin of SIVmac and SIVstm. J Virol 2005;79:8991–9005.

[9] Murphey-Corb M, Martin LN, Rangan SR, Baskin GB, Gormus BJ, Wolf RH, et al. Isolation of an HTLV-III-related retrovirus from macaques with simian AIDS and its possible origin in asymptomatic mangabeys. Nature 1986;321:435–7.

[10] Fultz PN, McClure HM, Anderson DC, Swenson RB, Anand R, Srinivasan A. Isolation of a T-lymphotropic retrovirus from naturally infected sooty mangabey monkeys (*Cercocebus atys*). Proc Natl Acad Sci USA 1986;83:5286–90.

[11] Apetrei C, Robertson DL, Marx PA. The history of SIVs and AIDS: epidemiology, phylogeny and biology of isolates from naturally SIV infected non-human primates (NHP) in Africa. Front Biosci 2004;9:225–54.

[12] Lowenstine LJ, Pedersen NC, Higgins J, Pallis KC, Uyeda A, Marx P, et al. Seroepidemiologic survey of captive old-world primates for antibodies to human and simian retroviruses, and isolation of a lentivirus from sooty mangabeys (*Cercocebus atys*). Int J Cancer 1986;38:563–74.

[13] Ohta Y, Masuda T, Tsujimoto H, Ishikawa K, Kodama T, Morikawa S, et al. Isolation of simian immunodeficiency virus from African green monkeys and seroepidemiologic survey of the virus in various non-human primates. Int J Cancer 1988;41:115–22.

[14] Beer BE, Bailes E, Goeken R, Dapolito G, Coulibaly C, Norley SG, et al. Simian immunodeficiency virus (SIV) from sun-tailed monkeys (*Cercopithecus solatus*): evidence for host-dependent evolution of SIV within the *C. lhoesti* superspecies. J Virol 1999;73:7734–44.

[15] Bibollet-Ruche F, Bailes E, Gao F, Pourrut X, Barlow KL, Clewley J, et al. A new simian immunodeficiency virus lineage (SIVdeb) infecting de Brazza's monkeys (*Cercopithecus neglectus*): evidence for a Cercopithecus monkey virus clade. J Virol 2004;78:7748–62.

[16] Courgnaud V, Pourrut X, Bibollet-Ruche F, Mpoudi-Ngole E, Bourgeois A, Delaporte E, et al. Characterization of a novel simian immunodeficiency virus from guereza colobus monkeys (*Colobus guereza*) in Cameroon: a new lineage in the nonhuman primate lentivirus family. J Virol 2001;75:857–66.

[17] Fukasawa M, Miura T, Hasegawa A, Morikawa S, Tsujimoto H, Miki K, et al. Sequence of simian immunodeficiency virus from African green monkey, a new member of the HIV/SIV group. Nature 1988;333:457–61.

[18] Hirsch VM, Campbell BJ, Bailes E, Goeken R, Brown C, Elkins WR, et al. Characterization of a novel simian immunodeficiency virus (SIV) from L'Hoest monkeys (*Cercopithecus l'hoesti*): implications for the origins of SIVmnd and other primate lentiviruses. J Virol 1999;73:1036–45.

[19] Hirsch VM, Dapolito GA, Goldstein S, McClure H, Emau P, Fultz PN, et al. A distinct African lentivirus from Sykes' monkeys. J Virol 1993;67:1517–28.

[20] Liegeois F, Courgnaud V, Switzer WM, Murphy HW, Loul S, Aghokeng A, et al. Molecular characterization of a novel simian immunodeficiency virus lineage (SIVtal) from northern talapoins (*Miopithecus ogouensis*). Virology 2006;349:55–65.

[21] Liegeois F, Lafay B, Formenty P, Locatelli S, Courgnaud V, Delaporte E, et al. Full-length genome characterization of a novel simian immunodeficiency virus lineage (SIVolc) from olive Colobus (*Procolobus verus*) and new SIVwrcPbb strains from Western Red Colobus (*Piliocolobus badius badius*) from the Tai Forest in Ivory Coast. J Virol 2009;83:428–39.

[22] Locatelli S, Lafay B, Liegeois F, Ting N, Delaporte E, Peeters M. Full molecular characterization of a simian immunodeficiency virus, SIVwrcpbt from Temminck's red colobus (*Piliocolobus badius temminckii*) from Abuko Nature Reserve, the Gambia. Virology 2008;376:90–100.

[23] Tsujimoto H, Hasegawa A, Maki N, Fukasawa M, Miura T, Speidel S, et al. Sequence of a novel simian immunodeficiency virus from a wild-caught African mandrill. Nature 1989;341:539–41.

[24] Courgnaud V, Abela B, Pourrut X, Mpoudi-Ngole E, Loul S, Delaporte E, et al. Identification of a new simian immunodeficiency virus lineage with a *vpu* gene present among different cercopithecus monkeys (*C. mona, C. cephus,* and *C. nictitans*) from Cameroon. J Virol 2003;77:12523–34.

[25] Courgnaud V, Salemi M, Pourrut X, Mpoudi-Ngole E, Abela B, Auzel P, et al. Characterization of a novel simian immunodeficiency virus with a *vpu* gene from greater spot-nosed monkeys (*Cercopithecus nictitans*) provides new insights into simian/human immunodeficiency virus phylogeny. J Virol 2002;76:8298–309.

[26] Dazza MC, Ekwalanga M, Nende M, Bin Shamamba K, Bitshi P, Saragosti S. Characterization of novel simian immunodeficiency virus from *Cercopithecus mona denti* (SIVden) from the Democratic Republic of Congo. J Virol 2005;79:8560–71.

[27] Vanden Haesevelde MM, Peeters M, Jannes G, Janssens W, van der Groen G, Sharp PM, et al. Sequence analysis of a highly divergent HIV-1-related lentivirus isolated from a wild captured chimpanzee. Virology 1996;221:346–50.

[28] Ahuka-Mundeke S, Liegeois F, Ayouba A, Foupouapouognini Y, Nerrienet E, Delaporte E, et al. Full-length genome sequence of a simian immunodeficiency virus (SIV) infecting a captive agile mangabey (*Cercocebus agilis*) is closely related to SIVrcm infecting wild red-capped mangabeys (*Cercocebus torquatus*) in Cameroon. J Gen Virol 2010;91:2959–64.

[29] Beer BE, Foley BT, Kuiken CL, Tooze Z, Goeken RM, Brown CR, et al. Characterization of novel simian immunodeficiency viruses from red-capped mangabeys from Nigeria (SIVrcmNG409 and -NG411). J Virol 2001;75:12014–27.

[30] Souquiere S, Bibollet-Ruche F, Robertson DL, Makuwa M, Apetrei C, Onanga R, et al. Wild *Mandrillus sphinx* are carriers of two types of lentivirus. J Virol 2001;75: 7086–96.

[31] Locatelli S, Peeters M. Cross-species transmission of simian retroviruses: how and why they could lead to the emergence of new diseases in the human population. AIDS 2012;26:659–73.

[32] Charleston MA, Robertson DL. Preferential host switching by primate lentiviruses can account for phylogenetic similarity with the primate phylogeny. Syst Biol 2002;51:528–35.

[33] Apetrei C, Gaufin T, Gautam R, Vinton C, Hirsch VM, Lewis M, et al. Pattern of SIVagm infection in patas monkeys suggests that host adaptation to SIV infection may result in resistance to infection and virus extinction. J Infect Dis 2010;202 (Suppl. 3):S371–6.
[34] Hirsch VM, McGann C, Dapolito G, Goldstein S, Ogen-Odoi A, Biryawaho B, et al. Identification of a new subgroup of SIVagm in tantalus monkeys. Virology 1993;197:426–30.
[35] Miura T, Tsujimoto H, Fukasawa M, Ohta Y, Honjo S, Hayami M. Genetic analysis and infection of SIVAGM and SIVMND. J Med Primatol 1989;18:255–9.
[36] Wertheim JO, Worobey M. A challenge to the ancient origin of SIVagm based on African green monkey mitochondrial genomes. PLoS Pathog 2007;3:e95.
[37] Ma D, Jasinska A, Kristoff J, Grobler JP, Turner T, Jung Y, et al. SIVagm infection in wild African green monkeys from South Africa: epidemiology, natural history, and evolutionary considerations. PLoS Pathog 2013;9:e1003011.
[38] Worobey M, Telfer P, Souquière S, Hunter M, Coleman CA, Metzger MJ, et al. Island biogeography reveals the deep history of SIV. Science 2010;330:1487.
[39] Jin MJ, Rogers J, Phillips-Conroy JE, Allan JS, Desrosiers RC, Shaw GM, et al. Infection of a yellow baboon with simian immunodeficiency virus from African green monkeys: evidence for cross-species transmission in the wild. J Virol 1994;68: 8454–60.
[40] van Rensburg EJ, Engelbrecht S, Mwenda J, Laten JD, Robson BA, Stander T, et al. Simian immunodeficiency viruses (SIVs) from eastern and southern Africa: detection of a SIVagm variant from a chacma baboon. J Gen Virol 1998;79:1809–14.
[41] Bibollet-Ruche F, Galat-Luong A, Cuny G, Sarni-Manchado P, Galat G, Durand JP, et al. Simian immunodeficiency virus infection in a patas monkey (*Erythrocebus patas*): evidence for cross-species transmission from African green monkeys (*Cercopithecus aethiops sabaeus*) in the wild. J Gen Virol 1996;77:773–81.
[42] Gautier-Hion A, Colyn M, Gautier JP. Histoire naturelle des primates d'Afrique Centrale; 1999. Libreville, Gabon.
[43] Aghokeng AF, Ayouba A, Mpoudi-Ngole E, Loul S, Liegeois F, Delaporte E, et al. Extensive survey on the prevalence and genetic diversity of SIVs in primate bushmeat provides insights into risks for potential new cross-species transmissions. Infect Genet Evol 2010;10:386–96.
[44] Ahuka-Mundeke S, Ayouba A, Mbala-Kingebeni P, Liegeois F, Esteban A, Lunguya-Metila O, et al. Novel multiplexed HIV/simian immunodeficiency virus antibody detection assay. Emerg Infect Dis 2011;17:2277–86.
[45] Takemura T, Ekwalanga M, Bikandou B, Ido E, Yamaguchi-Kabata Y, Ohkura S, et al. A novel simian immunodeficiency virus from black mangabey (*Lophocebus aterrimus*) in the Democratic Republic of Congo. J Gen Virol 2005;86:1967–71.
[46] Takehisa J, Kraus MH, Ayouba A, Bailes E, Van Heuverswyn F, Decker JM, et al. Origin and biology of simian immunodeficiency virus in wild-living western gorillas. J Virol 2009;83:1635–48.
[47] Aghokeng AF, Bailes E, Loul S, Courgnaud V, Mpoudi-Ngolle E, Sharp PM, et al. Full-length sequence analysis of SIVmus in wild populations of mustached monkeys (*Cercopithecus cephus*) from Cameroon provides evidence for two co-circulating SIVmus lineages. Virology 2007;360:407–18.
[48] Liegeois F, Boue V, Butel C, Ngari P, Mouacha F, Mve Ondo B, et al. Full-length genome analyses of two highly divergent SIVmus strains isolated in mustached monkeys (*C. cephus*) hunted for bushmeat in Gabon. In: 20th conference on retroviruses and opportunistic infections; 2013. Atlanta, GA.
[49] Bailes E, Gao F, Bibollet-Ruche F, Courgnaud V, Peeters M, Marx PA, et al. Hybrid origin of SIV in chimpanzees. Science 2003;300:1713.

[50] Switzer W, Lauck M, Shankar A, Hyeroba D, Tumukunde A, Chapman C, et al. Discovery of novel and highly divergent SIV in three wild monkey species in Kibale National Park, Uganda. In: 20th conference on retroviruses and opportunistic infections; 2013. Atlanta, GA.
[51] Sharp PM, Bailes E, Gao F, Beer BE, Hirsch VM, Hahn BH. Origins and evolution of AIDS viruses: estimating the time-scale. Biochem Soc Trans 2000;28: 275–82.
[52] Gifford RJ, Katzourakis A, Tristem M, Pybus OG, Winters M, Shafer RW. A transitional endogenous lentivirus from the genome of a basal primate and implications for lentivirus evolution. Proc Natl Acad Sci USA 2008;105:20362–7.
[53] Perelman P, Johnson WE, Roos C, Seuanez HN, Horvath JE, Moreira MA, et al. A molecular phylogeny of living primates. PLoS Genetics 2011;7(3):e1001342.
[54] Compton AA, Emerman M. Convergence and divergence in the evolution of the APOBEC3G-Vif interaction reveal ancient origins of simian immunodeficiency viruses. PLoS Pathog 2013;9:e1003135.
[55] Neel C, Etienne L, Li Y, Takehisa J, Rudicell RS, Bass IN, et al. Molecular epidemiology of simian immunodeficiency virus infection in wild-living gorillas. J Virol 2010;84:1464–76.
[56] Rudicell RS, Piel AK, Stewart F, Moore DL, Learn GH, Li Y, et al. High prevalence of simian immunodeficiency virus infection in a community of savanna chimpanzees. J Virol 2011;85:9918–28.
[57] Santiago ML, Range F, Keele BF, Li Y, Bailes E, Bibollet-Ruche F, et al. Simian immunodeficiency virus infection in free-ranging sooty mangabeys (*Cercocebus atys atys*) from the Tai Forest, Cote d'Ivoire: implications for the origin of epidemic human immunodeficiency virus type 2. J Virol 2005;79:12515–27.
[58] Van Heuverswyn F, Li Y, Bailes E, Neel C, Lafay B, Keele BF, et al. Genetic diversity and phylogeographic clustering of SIVcpzPtt in wild chimpanzees in Cameroon. Virology 2007;368:155–71.
[59] Aghokeng AF, Liu W, Bibollet-Ruche F, Loul S, Mpoudi-Ngole E, Laurent C, et al. Widely varying SIV prevalence rates in naturally infected primate species from Cameroon. Virology 2006;345:174–89.
[60] Ndongmo CB, Switzer WM, Pau CP, Zeh C, Schaefer A, Pieniazek D, et al. New multiple antigenic peptide-based enzyme immunoassay for detection of simian immunodeficiency virus infection in nonhuman primates and humans. J Clin Microbiol 2004;42: 5161–9.
[61] Simon F, Souquiere S, Damond F, Kfutwah A, Makuwa M, Leroy E, et al. Synthetic peptide strategy for the detection of and discrimination among highly divergent primate lentiviruses. AIDS Res Hum Retroviruses 2001;17:937–52.
[62] Vignali DA. Multiplexed particle-based flow cytometric assays. J Immunol Methods 2000;243:243–55.
[63] Lowenstine LJ, Lerche NW, Yee JL, Uyeda A, Jennings MB, Munn RJ, et al. Evidence for a lentiviral etiology in an epizootic of immune deficiency and lymphoma in stump-tailed macaques (*Macaca arctoides*). J Med Primatol 1992;21:1–14.
[64] Santiago ML, Bibollet-Ruche F, Gross-Camp N, Majewski AC, Masozera M, Munanura I, et al. Noninvasive detection of Simian immunodeficiency virus infection in a wild-living L'Hoest's monkey (*Cercopithecus lhoesti*). AIDS Res Hum Retroviruses 2003;19: 1163–6.
[65] Santiago ML, Lukasik M, Kamenya S, Li Y, Bibollet-Ruche F, Bailes E, et al. Foci of endemic simian immunodeficiency virus infection in wild-living eastern chimpanzees (*Pan troglodytes schweinfurthii*). J Virol 2003;77:7545–62.
[66] Santiago ML, Rodenburg CM, Kamenya S, Bibollet-Ruche F, Gao F, Bailes E, et al. SIVcpz in wild chimpanzees. Science 2002;295:465.

[67] Locatelli S, Roeder AD, Bruford MW, Noe R, Delaporte E, Peeters M. Lack of evidence of simian immunodeficiency virus infection among nonhuman primates in Tai National Park, Cote d'Ivoire: limitations of noninvasive methods and SIV diagnostic tools for studies of primate retroviruses. Int J Primatol 2011;32:288–307.

[68] Etienne L, Locatelli S, Ayouba A, Esteban A, Butel C, Liegeois F, et al. Noninvasive follow-up of simian immunodeficiency virus infection in wild-living nonhabituated western lowland gorillas in Cameroon. J Virol 2012;86:9760–72.

[69] Keele BF, Van Heuverswyn F, Li Y, Bailes E, Takehisa J, Santiago ML, et al. Chimpanzee reservoirs of pandemic and nonpandemic HIV-1. Science 2006;313:523–6.

[70] Li Y, Ndjango JB, Learn GH, Ramirez MA, Keele BF, Bibollet-Ruche F, et al. Eastern chimpanzees, but not bonobos, represent a simian immunodeficiency virus reservoir. J Virol 2012;86:10776–91.

[71] Leendertz SA, Locatelli S, Boesch C, Kucherer C, Formenty P, Liegeois F, et al. No evidence for transmission of SIVwrc from western red colobus monkeys (*Piliocolobus badius badius*) to wild West African chimpanzees (*Pan troglodytes verus*) despite high exposure through hunting. BMC Microbiol 2011;11:24.

[72] Prince AM, Brotman B, Lee DH, Andrus L, Valinsky J, Marx P. Lack of evidence for HIV type 1-related SIVcpz infection in captive and wild chimpanzees (*Pan troglodytes verus*) in West Africa. AIDS Res Hum Retroviruses 2002;18:657–60.

[73] Switzer WM, Parekh B, Shanmugam V, Bhullar V, Phillips S, Ely JJ, et al. The epidemiology of simian immunodeficiency virus infection in a large number of wild- and captive-born chimpanzees: evidence for a recent introduction following chimpanzee divergence. AIDS Res Hum Retroviruses 2005;21:335–42.

[74] Worobey M, Santiago ML, Keele BF, Ndjango JB, Joy JB, Labama BL, et al. Origin of AIDS: contaminated polio vaccine theory refuted. Nature 2004;428:820.

[75] Plantier JC, Leoz M, Dickerson JE, De Oliveira F, Cordonnier F, Lemee V, et al. A new human immunodeficiency virus derived from gorillas. Nat Med 2009;15:871–2.

[76] Gorillas in south-west Cameroon are the reservoir of HIV-1 group P ancestors. In: D'arc M, Ayouba A, Esteban A, Etienne L, Locatelli S, Delaporte E, et al., editors. 20th conference on retroviruses and opportunistic infections; 2013. Atlanta, GA.

[77] Sharp PM, Hahn BH. Origins of HIV and the AIDS pandemic. Cold Spring Harb Perspectives Med 2011;1:a006841.

[78] Apetrei C, Metzger MJ, Robinson D, Ling B, Telfer PT, Reed P, et al. Detection and partial characterization of new simian immunodeficiency virus (SIVsm) strains from bush meat samples from rural Sierra Leone. J Virol 2005;79:2631–6.

[79] Chen Z, Telfier P, Gettie A, Reed P, Zhang L, Ho DD, et al. Genetic characterization of new West African simian immunodeficiency virus SIVsm: geographic clustering of household-derived SIV strains with human immunodeficiency virus type 2 subtypes and genetically diverse viruses from a single feral sooty mangabey troop. J Virol 1996;70:3617–27.

[80] Ayouba A, Akoua-Koffi C, Adjogoua EV, Calvignac S, Esteban A, Locatelli S, et al. Identification of a new HIV-2 lineage in rural Côte d'Ivoire: an additional cross-species transmission from SIVsmm from sooty mangabeys to humans. In: 19th conference on retroviruses and opportunistic infections; 2012. Seattle, WA.

[81] Fa JE, Juste J, Delval JP, Castroviejo J. Impact of market hunting on mammal species in Equatorial-Guinea. Conserv Biol 1995;9:1107–15.

[82] Fa JE, Peres CA, Meeuwig J. Bushmeat exploitation in tropical forests: an intercontinental comparison. Conserv Biol 2002;16:232–7.

[83] Hart JA. From subsistence to market - case-study of Mbuti net hunters. Hum Ecol 1978;6:325–53.

[84] Wilkie DS, Carpenter JF. Bushmeat hunting in the Congo Basin: an assessment of impacts and options for mitigation. Biodivers Conservation 1999;8:27–955.

[85] Aghokeng AF, Ayouba A, Ahuka S, Liegoies F, Mbala P, Muyembe JJ, et al. Genetic diversity of simian lentivirus in wild De Brazza's monkeys (*Cercopithecus neglectus*) in Equatorial Africa. J Gen Virol 2010;91:1810–6.
[86] Leendertz SA, Junglen S, Hedemann C, Goffe A, Calvignac S, Boesch C, et al. High prevalence, coinfection rate, and genetic diversity of retroviruses in wild red colobus monkeys (*Piliocolobus badius badius*) in Tai National Park, Cote d'Ivoire. J Virol 2010;84:7427–36.
[87] Goldberg TL, Sintasath DM, Chapman CA, Cameron KM, Karesh WB, Tang S, et al. Coinfection of Ugandan red colobus (Procolobus [*Piliocolobus*] *rufomitratus tephrosceles*) with novel, divergent delta-, lenti-, and spumaretroviruses. J Virol 2009;83: 11318–29.
[88] Cooper R, Feistner A, Evans S, Tsujimoto H, Hayami M. A lack of evidence of sexual transmission of a simian immunodeficiency agent in a semifree-ranging group of mandrills. AIDS 1989;3:764.
[89] Estaquier J, Peeters M, Bedjabaga L, Honore C, Bussi P, Dixson A, et al. Prevalence and transmission of simian immunodeficiency virus and simian T-cell leukemia virus in a semi-free-range breeding colony of mandrills in Gabon. AIDS 1991;5:1385–6.
[90] Keele BF, Holland Jones J, Terio K, Estes JD, Rudicell RS, Wilson ML, et al. Increased mortality and AIDS-like immunopathology in wild chimpanzees infected with SIVcpz. Nature 2009;460:515–9.
[91] Otsyula MG, Gettie A, Suleman M, Tarara R, Mohamed I, Marx P. Apparent lack of vertical transmission of simian immunodeficiency virus (SIV) in naturally infected African green monkeys, Cercopithecus aethiops. Ann Trop Med Parasitol 1995;89:573–6.
[92] Phillips-Conroy JE, Jolly CJ, Petros B, Allan JS, Desrosiers RC. Sexual transmission of SIVagm in wild grivet monkeys. J Med Primatol 1994;23:1–7.
[93] Nerrienet E, Amouretti X, Muller-Trutwin MC, Poaty-Mavoungou V, Bedjebaga I, Nguyen HT, et al. Phylogenetic analysis of SIV and STLV type I in mandrills (*Mandrillus sphinx*): indications that intracolony transmissions are predominantly the result of male-to-male aggressive contacts. AIDS Res Hum Retroviruses 1998;14: 785–96.
[94] Ruprecht RM, Baba TW, Liska V, Ray NB, Martin LN, Murphey-Corb M, et al. Oral transmission of primate lentiviruses. J Infect Dis 1999;179(Suppl. 3):S408–12.
[95] Chenine AL, Siddappa NB, Kramer VG, Sciaranghella G, Rasmussen RA, Lee SJ, et al. Relative transmissibility of an R5 clade C simian-human immunodeficiency virus across different mucosae in macaques parallels the relative risks of sexual HIV-1 transmission in humans via different routes. J Infect Dis 2010;201:1155–63.
[96] Ma D, Jasinska AJ, Feyertag F, Wijewardana V, Kristoff J, He T, et al. International Vervet Research Consortium: Factors associated with siman immunodeficiency virus transmission in a natural African nonhuman primate host in the wild. J Virol 2014;88:5687–5705.
[97] Rey-Cuille MA, Berthier JL, Bomsel-Demontoy MC, Chaduc Y, Montagnier L, Hovanessian AG, et al. Simian immunodeficiency virus replicates to high levels in sooty mangabeys without inducing disease. J Virol 1998;72:3872–86.
[98] Corbet S, Muller-Trutwin MC, Versmisse P, Delarue S, Ayouba A, Lewis J, et al. env sequences of simian immunodeficiency viruses from chimpanzees in Cameroon are strongly related to those of human immunodeficiency virus group N from the same geographic area. J Virol 2000;74:529–34.
[99] Amedee AM, Rychert J, Lacour N, Fresh L, Ratterree M. Viral and immunological factors associated with breast milk transmission of SIV in rhesus macaques. Retrovirology 2004;1:17.
[100] Pandrea I, Parrish NF, Raehtz K, Gaufin T, Barbian HJ, Ma D, et al. Mucosal SIV transmission in African green monkeys: susceptibility to infection is proportional to target cell availability at mucosal sites. J Virol 2012;86:4158–68.

[101] Pandrea I, Onanga R, Souquiere S, Mouinga-Ondéme A, Bourry O, Makuwa M, et al. Paucity of CD4+CCR5+ T-cells may prevent breastfeeding transmission of SIV in natural non-human primate hosts. J Virol 2008;82:5501–9.
[102] Hahn BH, Shaw GM, De Cock KM, Sharp PM. AIDS as a zoonosis: scientific and public health implications. Science 2000;287:607–14.
[103] Pandrea I, Apetrei C. Where the wild things are: pathogenesis of SIV infection in African nonhuman primate hosts. Curr HIV/AIDS Rep 2010;7:28–36.
[104] Pandrea I, Sodora DL, Silvestri G, Apetrei C. Into the wild: simian immunodeficiency virus (SIV) infection in natural hosts. Trends Immunol 2008;29:419–28.
[105] Silvestri G, Paiardini M, Pandrea I, Lederman MM, Sodora DL. Understanding the benign nature of SIV infection in natural hosts. J Clin Invest 2007;117:3148–54.
[106] VandeWoude S, Apetrei C. Going wild: Lessons from T-lymphotropic naturally occurring lentiviruses. Clin Microbiol Rev 2006;19:728–62.
[107] Ling B, Apetrei C, Pandrea I, Veazey RS, Lackner AA, Gormus B, et al. Classic AIDS in a sooty mangabey after an 18-year natural infection. J Virol 2004;78:8902–8.
[108] Pandrea I, Onanga R, Rouquet P, Bourry O, Ngari P, Wickings EJ, et al. Chronic SIV infection ultimately causes immunodeficiency in African non-human primates. AIDS 2001;15:2461–2.
[109] Traina-Dorge V, Blanchard J, Martin L, Murphey-Corb M. Immunodeficiency and lymphoproliferative disease in an African green monkey dually infected with SIV and STLV-I. AIDS Res Hum Retroviruses 1992;8:97–100.
[110] Apetrei C, Gormus B, Pandrea I, Metzger M, ten Haaft P, Martin LN, et al. Direct inoculation of simian immunodeficiency virus from sooty mangabeys in black mangabeys (*Lophocebus aterrimus*): first evidence of AIDS in a heterologous African species and different pathologic outcomes of experimental infection. J Virol 2004;78: 11506–18.
[111] Pandrea I, Silvestri G, Apetrei C. AIDS in African nonhuman primate hosts of SIVs: a new paradigm of SIV infection. Curr HIV Res 2009;6:57–72.
[112] Peeters M, Courgnaud V, Abela B, Auzel P, Pourrut X, Bibollet-Ruche F, et al. Risk to human health from a plethora of simian immunodeficiency viruses in primate bushmeat. Emerg Infect Dis 2002;8:451–7.
[113] Rutjens E, Balla-Jhagjhoorsingh S, Verschoor E, Bogers W, Koopman G, Heeney J. Lentivirus infections and mechanisms of disease resistance in chimpanzees. Front Biosci 2003;8:d1134–45.
[114] ten Haaft P, Murthy K, Salas M, McClure H, Dubbes R, Koornstra W, et al. Differences in early virus loads with different phenotypic variants of HIV-1 and SIV(cpz) in chimpanzees. AIDS 2001;15:2085–92.
[115] Rudicell RS, Holland Jones J, Wroblewski EE, Learn GH, Li Y, Robertson JD, et al. Impact of simian immunodeficiency virus infection on chimpanzee population dynamics. PLoS Pathog 2010;6(9).
[116] Etienne L, Nerrienet E, LeBreton M, Bibila GT, Foupouapouognigni Y, Rousset D, et al. Characterization of a new simian immunodeficiency virus strain in a naturally infected *Pan troglodytes troglodytes* chimpanzee with AIDS related symptoms. Retrovirology 2011;8:4.
[117] Goldstein S, Ourmanov I, Brown CR, Beer BE, Elkins WR, Plishka R, et al. Wide range of viral load in healthy african green monkeys naturally infected with simian immunodeficiency virus. J Virol 2000;74:11744–53.
[118] Pandrea I, Onanga R, Kornfeld C, Rouquet P, Bourry O, Clifford S, et al. High levels of SIVmnd-1 replication in chronically infected Mandrillus sphinx. Virology 2003;317:119–27.
[119] Chen Z, Zhou P, Ho DD, Landau NR, Marx PA. Genetically divergent strains of simian immunodeficiency virus use CCR5 as a coreceptor for entry. J Virol 1997;71: 2705–14.

[120] Zhang Y, Lou B, Lal RB, Gettie A, Marx PA, Moore JP. Use of inhibitors to evaluate coreceptor usage by simian and simian/human immunodeficiency viruses and human immunodeficiency virus type 2 in primary cells. J Virol 2000;74:6893–910.
[121] Moore JP, Kitchen SG, Pugach P, Zack JA. The CCR5 and CXCR4 coreceptors–central to understanding the transmission and pathogenesis of human immunodeficiency virus type 1 infection. AIDS Res Hum Retroviruses 2004;20:111–26.
[122] Gautam R, Carter AC, Katz N, Butler IF, Barnes M, Hasegawa A, et al. In vitro characterization of primary SIVsmm isolates belonging to different lineages. In vitro growth on rhesus macaque cells is not predictive for in vivo replication in rhesus macaques. Virology 2007;362:257–70.
[123] Owen SM, Masciotra S, Novembre F, Yee J, Switzer WM, Ostyula M, et al. Simian immunodeficiency viruses of diverse origin can use CXCR4 as a coreceptor for entry into human cells. J Virol 2000;74:5702–8.
[124] Pandrea I, Kornfeld C, Ploquin MJ-I, Apetrei C, Faye A, Rouquet P, et al. Impact of viral factors on very early in vivo replication profiles in SIVagm-infected African green monkeys. J Virol 2005;79:6249–59.
[125] Schols D, De Clercq E. The simian immunodeficiency virus mnd(GB-1) strain uses CXCR4, not CCR5, as coreceptor for entry in human cells. J Gen Virol 1998;79:2203–5.
[126] Chen Z, Kwon D, Jin Z, Monard S, Telfer P, Jones MS, et al. Natural infection of a homozygous delta24 CCR5 red-capped mangabey with an R2b-tropic simian immunodeficiency virus. J Exp Med 1998;188:2057–65.
[127] Riddick NE, Hermann EA, Loftin LM, Elliott ST, Wey WC, Cervasi B, et al. A novel CCR5 mutation common in sooty mangabeys reveals SIVsmm infection of CCR5-null natural hosts and efficient alternative coreceptor use in vivo. PLoS Pathog 2010;6.
[128] Pandrea I, Apetrei C, Gordon S, Barbercheck J, Dufour J, Bohm R, et al. Paucity of $CD4^+CCR5^+$ T cells is a typical feature of natural SIV hosts. Blood 2007;109:1069–76.
[129] Paiardini M, Cervasi B, Reyes-Aviles E, Micci L, Ortiz AM, Chahroudi A, et al. Reduced CCR5 up-regulation upon activation limits virus replication in central-memory $CD4^+$ T cells of SIV-infected sooty mangabeys. Nat Med 2011;17:830–6.
[130] Sodora DL, Allan J, Apetrei C, Brenchley JM, Douek DC, Else J, et al. Towards an AIDS vaccine: Lessons from natural SIV infections of African nonhuman primate hosts. Nat Med 2009;15:861–5.
[131] Mellors JW, Rinaldo Jr CR, Gupta P, White RM, Todd JA, Kingsley LA. Prognosis in HIV-1 infection predicted by the quantity of virus in plasma. Science 1996;272:1167–70.
[132] Apetrei C, Gautam R, Sumpter B, Carter AC, Gaufin T, Staprans SI, et al. Virus-subtype specific features of natural SIVsmm infection in sooty mangabeys. J Virol 2007;81:7913–23.
[133] Diop OM, Gueye A, Dias-Tavares M, Kornfeld C, Faye A, Ave P, et al. High levels of viral replication during primary simian immunodeficiency virus SIVagm infection are rapidly and strongly controlled in African green monkeys. J Virol 2000;74(16):7538–47.
[134] Pandrea I, Apetrei C, Dufour J, Dillon N, Barbercheck J, Metzger M, et al. Simian immunodeficiency virus (SIV) SIVagm.sab infection of Caribbean African green monkeys: new model of the study of SIV pathogenesis in natural hosts. J Virol 2006;80:4858–67.
[135] Pandrea I, Gaufin T, Brenchley JM, Gautam R, Monjure C, Gautam A, et al. Experimentally-induced immune activation in natural hosts of SIV induces significant increases in viral replication and $CD4^+$ T cell depletion. J Immunol 2008;181:6687–91.
[136] Pandrea I, Gautam R, Ribeiro R, Brenchley JM, Butler IF, Pattison M, et al. Acute loss of intestinal $CD4^+$ T cells is not predictive of SIV virulence. J Immunol 2007;179:3035–46.
[137] Pandrea I, Silvestri G, Onanga R, Veazey RS, Marx PA, Hirsch VM, et al. Simian immunodeficiency viruses replication dynamics in African non-human primate hosts: common patterns and species-specific differences. J Med Primatol 2006;35:194–201.
[138] Silvestri G. Naturally SIV-infected sooty mangabeys: are we closer to understanding why they do not develop AIDS? J Med Primatol 2005;34:243–52.

[139] Silvestri G, Sodora DL, Koup RA, Paiardini M, O'Neil SP, McClure HM, et al. Nonpathogenic SIV infection of sooty mangabeys is characterized by limited bystander immunopathology despite chronic high-level viremia. Immunity 2003;18:441–52.

[140] Klatt NR, Villinger F, Bostik P, Gordon SN, Pereira L, Engram JC, et al. Availability of activated CD4+ T cells dictates the level of viremia in naturally SIV-infected sooty mangabeys. J Clin Invest 2008;118:2039–49.

[141] Goldstein S, Brown CR, Ourmanov I, Pandrea I, Buckler-White A, Erb C, et al. Comparison of simian immunodeficiency virus SIVagmVer replication and CD4+ T-cell dynamics in vervet and sabaeus African green monkeys. J Virol 2006;80:4868–77.

[142] Broussard SR, Staprans SI, White R, Whitehead EM, Feinberg MB, Allan JS. Simian immunodeficiency virus replicates to high levels in naturally infected African green monkeys without inducing immunologic or neurologic disease. J Virol 2001;75:2262–75.

[143] Fiebig EW, Wright DJ, Rawal BD, Garrett PE, Schumacher RT, Peddada L, et al. Dynamics of HIV viremia and antibody seroconversion in plasma donors: implications for diagnosis and staging of primary HIV infection. AIDS 2003;17:1871–9.

[144] Watkins DI, Burton DR, Kallas EG, Moore JP, Koff WC. Nonhuman primate models and the failure of the Merck HIV-1 vaccine in humans. Nat Med 2008;14:617–21.

[145] Keele BF, Giorgi EE, Salazar-Gonzalez JF, Decker JM, Pham KT, Salazar MG, et al. Identification and characterization of transmitted and early founder virus envelopes in primary HIV-1 infection. Proc Natl Acad Sci USA 2008;105:7552–7.

[146] Keele BF, Li H, Learn GH, Hraber P, Giorgi EE, Grayson T, et al. Low dose rectal inoculation of rhesus macaques by SIVsmE660 or SIVmac251 recapitulates human mucosal infection by HIV-1. J Exp Med 2009;206:1117–34.

[147] Brenchley JM, Price DA, Douek DC. HIV disease: fallout from a mucosal catastrophe? Nat Immunol 2006;7:235–9.

[148] Grossman Z, Meier-Schellersheim M, Paul WE, Picker LJ. Pathogenesis of HIV infection: what the virus spares is as important as what it destroys. Nat Med 2006;12:289–95.

[149] Haase AT. Perils at mucosal front lines for HIV and SIV and their hosts. Nat Rev Immunol 2005;5:783–92.

[150] Picker LJ, Hagen SI, Lum R, Reed-Inderbitzin EF, Daly LM, Sylwester AW, et al. Insufficient production and tissue delivery of CD4+ memory T cells in rapidly progressive simian immunodeficiency virus infection. J Exp Med 2004;200:1299–314.

[151] Picker LJ, Watkins DI. HIV pathogenesis: the first cut is the deepest. Nat Immunol 2005;6:430–2.

[152] Okoye A, Meier-Schellersheim M, Brenchley JM, Hagen SI, Walker JM, Rohankhedkar M, et al. Progressive CD4+ central memory T cell decline results in CD4+ effector memory insufficiency and overt disease in chronic SIV infection. J Exp Med 2007;204:2171–85.

[153] Douek DC, Betts MR, Hill BJ, Little SJ, Lempicki R, Metcalf JA, et al. Evidence for increased T cell turnover and decreased thymic output in HIV infection. J Immunol 2001;167:6663–8.

[154] Douek DC, McFarland RD, Keiser PH, Gage EA, Massey JM, Haynes BF, et al. Changes in thymic function with age and during the treatment of HIV infection. Nature 1998;396:690–5.

[155] Hellerstein MK, Hoh RA, Hanley MB, Cesar D, Lee D, Neese RA, et al. Subpopulations of long-lived and short-lived T cells in advanced HIV-1 infection. J Clin Invest 2003;112:956–66.

[156] Roederer M. T-cell dynamics of immunodeficiency. Nat Med 1995;1:621–2.

[157] Bosinger SE, Li Q, Gordon SN, Klatt NR, Duan L, Xu L, et al. Global genomic analysis reveals rapid control of a robust innate response in SIV-infected sooty mangabeys. J Clin Invest 2009;119:3556–72.

[158] Jacquelin B, Mayau V, Targat B, Liovat AS, Kunkel D, Petitjean G, et al. Nonpathogenic SIV infection of African green monkeys induces a strong but rapidly controlled type I IFN response. J Clin Invest 2009;119:3544–55.

[159] Harris LD, Tabb B, Sodora DL, Paiardini M, Klatt NR, Douek DC, et al. Downregulation of robust acute type I interferon responses distinguishes nonpathogenic simian immunodeficiency virus (SIV) infection of natural hosts from pathogenic SIV infection of rhesus macaques. J Virol 2010;84:7886–91.
[160] Estes JD, Gordon SN, Zeng M, Chahroudi AM, Dunham RM, Staprans SI, et al. Early resolution of acute immune activation and induction of PD-1 in SIV-infected sooty mangabeys distinguishes nonpathogenic from pathogenic infection in rhesus macaques. J Immunol 2008;180(10):6798–807.
[161] Gordon S, Klatt NR, Milush JM, Engram J, Dunham RM, Paiardini M, et al. Severe depletion of mucosal CD4$^+$ T cells in AIDS-free SIV-infected sooty mangabeys. J Immunol 2007;179:3026–34.
[162] Brenchley JM, Price DA, Schacker TW, Asher TE, Silvestri G, Rao S, et al. Microbial translocation is a cause of systemic immune activation in chronic HIV infection. Nat Med 2006;12:1365–71.

CHAPTER

4

Properties of Human and Simian Immunodeficiency Viruses

Daniel Sauter, Frank Kirchhoff

Institute of Molecular Virology, Ulm University Medical Center, Ulm, Germany

INTRODUCTION

Since the discovery that the acquired immune deficiency syndrome (AIDS) is caused by a lentivirus more than 30 years ago, it has become clear that HIV and AIDS are the result of relatively recent zoonotic transmissions of simian immunodeficiency viruses (SIVs) from African apes and monkeys to humans [1]. To date SIVs have been detected in more than 40 different non-human primate (NHP) species in the wild [2]. Thus, African NHPs represent a huge reservoir of lentiviruses that have the potential to cross species barriers. However, to our current knowledge only three species, i.e., chimpanzees, gorillas, and sooty mangabeys (SMs), have transmitted their viruses to humans (Figure 4.1) [3–6]. HIV-1 group M (*major*), the main causative agent of AIDS, and the rare group N viruses originated from SIVs infecting the central subspecies of chimpanzees (cpz*Ptt*; *Pan troglodytes troglodytes*) [4]. In contrast, the closest relatives of HIV-1 group O and P strains have been detected in gorillas (gor; *Gorilla gorilla gorilla*), who themselves most likely originally acquired their virus from chimpanzees (Figure 4.1) [5,7]. Natural infection with SIVcpz has also been observed in the eastern subspecies of chimpanzees (cpz*Pts*; *Pan troglodytes schweinfurthii*), but this virus has never been detected in humans. Notably, it has recently been shown that SIVcpz "naturally" infected chimpanzees develop AIDS-like disease and have a reduced lifespan in the wild [8], whereas it is currently unknown whether or not gorillas do also develop naturally acquired SIV-induced disease.

Natural Hosts of SIV.
http://dx.doi.org/10.1016/B978-0-12-404734-1.00004-8

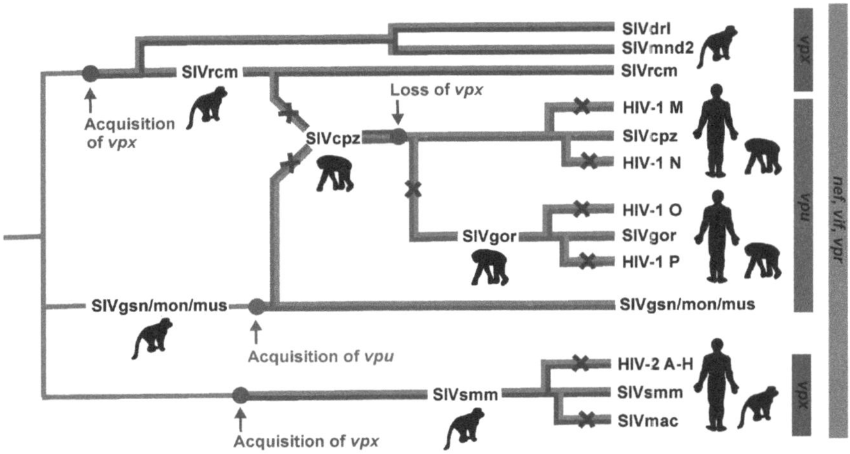

FIGURE 4.1 Gain and loss of accessory genes during primate lentiviral evolution. The cross-species transmission and recombination events that led to the emergence of *vpu* and *vpx* encoding primate lentiviruses are shown. Orange and blue indicate the presence of a *vpu* or *vpx* gene, respectively. All primate lentiviruses encode the accessory genes *vif*, *nef*, and *vpr* (green). Red crosses highlight cross-species transmissions. (See color plate at the back of the book.)

The second human immunodeficiency virus (HIV-2) originated from the sooty mangabey (smm; *Cercocebus atys*) [6]. This monkey species is infected with SIVsmm in the wild at high rates and does not develop immunodeficiency despite high levels of virus replication [9]. Another well-adapted host of SIV are African green monkeys (agm), which are widespread in Africa and also show a high seroprevalence [10]. Whereas SIVagm has never been detected in humans, SIVsmm was independently transmitted on at least nine different occasions resulting in HIV-2 groups A to I [11]. Only two of these transmissions (HIV-2 A and B), as well as a recombinant form thereof, however, have spread significantly in the human population [11]. SIVsmm is also infectious for Asian rhesus macaques (mac) that are not a natural host of SIV [6]. Macaques experimentally infected with select SIVsmm isolates may rapidly develop simian AIDS and are thus commonly used for studies on the pathogenesis of AIDS [12], whereas naturally SIV-infected sooty mangabeys and African green monkeys are examined to understand the mechanisms that allow some primate hosts to tolerate lentiviral infections [13,14]. As discussed in recent reviews, a variety of host and viral properties seems to determine the different clinical outcomes of these primate lentiviral infections [15–18]. The goal of this chapter is to provide a brief overview of the properties of primate lentiviruses and to discuss differences in outcomes conferred by the presence or function of the accessory viral genes.

COMMON FEATURES OF PRIMATE LENTIVIRUSES

Primate lentiviruses can be divided into various approximately equidistant major lineages, such as (1) SIVcpz/SIVgor/HIV-1, infecting chimpanzees, gorillas, and humans; (2) SIVsmm/HIV-2, found in sooty mangabeys and humans; (3) SIVagm, from various African green monkey species; (4) SIVsyk, infecting Sykes monkeys; (5) the SIVlho lineage, which is found in mandrills, l'Hoest, and sun-tailed monkeys; and (6) SIVcol, infecting Colobus monkeys [2]. Furthermore, some SIVs remain to be fully characterized [2]. Although these primate lentiviruses are highly divergent at the sequence level, many structural, molecular and biological features are well conserved. Primate lentiviral particles generally have a diameter of about 120 to 150 nm and are surrounded by a cell-derived lipid membrane containing the viral glycoproteins together with several cellular factors. The viral envelope (Env) glycoprotein is a trimer that interacts with CD4 as the primary receptor and one or several G-protein-coupled receptors (GPCRs) as coreceptor. To our current knowledge, both HIVs and the majority of SIVs most often use CCR5 as coreceptor for viral entry. In addition, HIV-1 strains obtained from late-stage AIDS patients frequently utilize CXCR4, whereas some SIVs use other alternative coreceptors, such as BOB/GPR15 or Bonzo/STRL-33/CXCR6, with high efficiency [19]. Recent findings suggest that these latter coreceptors may allow efficient viral replication even in sooty mangabeys lacking functional CCR5 expression [20]. The matrix protein forms a layer underneath the viral envelope. The viral genome is surrounded by a cone-shaped capsid and consists of two copies of positive single-stranded RNAs that are associated with the nucleocapsid protein, a tRNA primer, and enzymes required for reverse transcription and integration. The viral genome encompasses about 10,000 nucleotides and contains 8 or 9 genes encoding about 15 different proteins (Figure 4.2). The *gag*, *pol*, and *env* genes are present in all retroviruses and encode structural (Env and Gag) and enzymatic proteins (reverse transcriptase, integrase, and protease). Furthermore, all primate lentiviruses contain *tat* and *rev* genes encoding regulatory proteins, which are essential for efficient viral transcription and export of single or unspliced viral RNAs from the nucleus into the cytoplasm, as well as at least three additional accessory genes, i.e. *vif*, *vpr*, and *nef*. In contrast, two other accessory genes, *vpu* and *vpx*, are only found in some primate lentiviruses (Figures 4.1 and 4.2).

The importance of the *vif* (viral infectivity factor) gene is already evident from the fact that it is generally conserved in primate lentiviruses and also found in the genomes of prosimian, ovine-caprine, bovine, and feline lentiviruses [21,22]. Vif counteracts the host restriction factor APOBEC3G (apolipoprotein B mRNA-editing enzyme, catalytic polypeptide-like 3G) a cytidine deaminase, which would otherwise be incorporated into lentiviral virions to induce lethal G-to-A hypermutations by deaminating cytidine

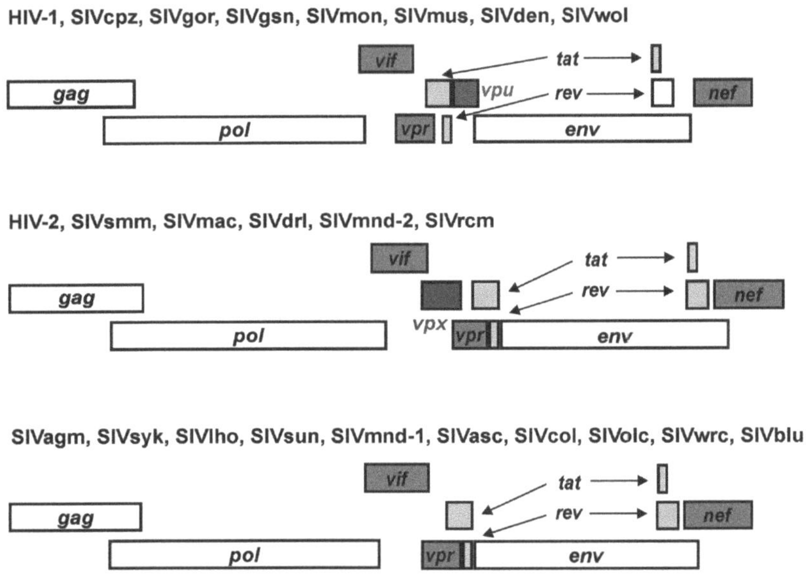

FIGURE 4.2 **Genomic organization of primate lentiviruses.** The *vpu* gene is only found in HIV-1 and its closest SIV counterparts, and the *vpx* gene is encoded by HIV-2 and some SIVs infecting the *Papionini* tribe of monkeys. ORFs encoding regulatory proteins Tat and Rev are shown in light gray, and the accessory genes *vif*, *vpr*, and *nef* in dark gray.

to uridine and by inhibiting reverse transcription [23–25]. Vif degrades APOBEC3G by inducing its poly-ubiquitination and proteasomal degradation [26]. The high selection pressure and ongoing genetic conflict between these two proteins is underscored by the finding that African green monkey APOBEC3G is counteracted by SIVagm in a subspecies-specific manner [27]. In contrast to Vif, the main function of Vpr (viral protein R) is still unclear. Multiple activities, such as cell-cycle arrest in the G2 phase, activation of proviral transcription, and induction of cell death, have been reported for this accessory protein. G2 cell-cycle arrest involves the interaction of Vpr with the cullin 4A-DDB1 complex via DCAF-1 (initially named VprBP) and is species-specific [28,29]. Furthermore, it has been shown that some SIV Vprs, not those encoded by HIV-1 and its direct SIVcpz and SIVgor precursors, antagonize the recently identified antiviral factor SAMHD1 (Figure 4.3) [30]. This may, however, require further study, because it has been reported that HIV-1 Vpr facilitates infection of macrophages [35], which would also be in agreement with antagonism of SAMHD1 or another as-yet-unknown restriction factor expressed in this cell type. The third accessory factor encoded by all primate lentiviruses is the Nef protein. Although originally named "negative factor," it has since become clear that

Vpr/Vpx	Vpr induces		Vpx induces	
	SAMHD1 degrad.	Cell cycle arrest	SAMHD1 degrad.	Cell cycle arrest
	dNTP↓	G2 → M	dNTP↓	G2 → M
HIV-1 M	✗	✓		
HIV-1 N, O, P	n.d.	n.d.		
SIVcpz	✗	n.d.		
SIVgor	n.d.	n.d.		
SIVgsn/mon	n.d.	n.d.		
SIVmus	✓	n.d.		
SIVrcm	✗	n.d.	✓	n.d.
HIV-2	✗	✓	✓	✗
SIVsmm	n.d.	✓	n.d.	✗
SIVmac	✗	✓	✓	✗
SIVagm	✓	✓		
SIVdeb	✓	n.d.		
SIVsyk	n.d.	✓		
SIVmnd-1	✗	n.d.		
SIVmnd-2	✗	n.d.	✓	n.d.
SIVdrl	n.d.	n.d.	n.d.	n.d.
SIVolc	✗	n.d.		

FIGURE 4.3 **Functional properties of primate lentiviral Vpx and Vpr proteins.** Check marks indicate the ability of Vpr or Vpx to counteract SAMHD1 or to induce cell-cycle arrest. A cross indicates lack of function. Table elements are shown in gray if the respective lentivirus does not encode a *vpx* gene. Note that the data are preliminary because only a very limited number of alleles have been analyzed for some of the lentiviruses and most data were generated in cells of human origin. (n.d., not determined.) [29–34].

Nef is actually required for efficient viral replication in vivo and accelerates disease progression in HIV-1-infected humans and in macaques experimentally infected with SIVmac [36–38]. It is not entirely clear whether Nef may enhance viral pathogenicity directly or only because it is associated with several orders of magnitude increased viral loads. Nef is myristoylated at its N-terminus and varies in size between 24 and 27 kDa in HIV-1 and between 22 and 37 kDa in HIV-2 and SIVs. Nef is expressed at high levels throughout the viral replication cycle and exerts multiple functions (Figure 4.4): it manipulates infected $CD4^+$ T cells by down-modulating CD4, MHC-I, and (less efficiently) CD28 and CXCR4 from the cell surface. Furthermore,

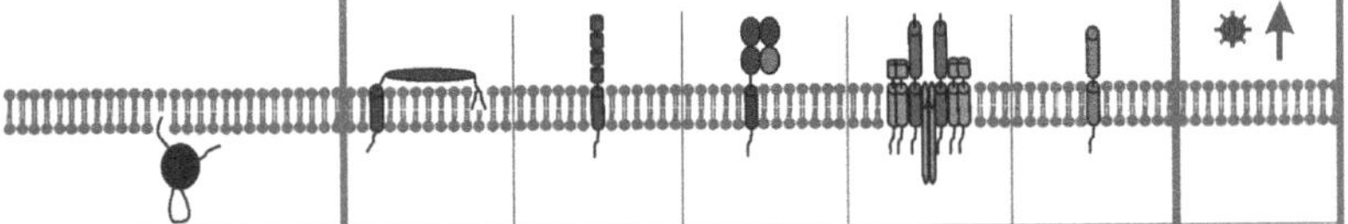

Nef	Nef displays activity against					
	Tetherin	CD4	MHC-I	CD3	CD28	Infectivity
HIV-1 M, N, O, P	✗	✓	✓	✗	✓	✓
SIVcpz	✓	✓	✓	✗	✓	✓
SIVgor	✓	✓	✓	✗	✓	✓
SIVgsn/mon/mus	✗	✓	✓	✗	✓	✓
SIVrcm	✓	✓	✓	✓	✗	✓
HIV-2 (A)	✗	✓	✓	✓	✓	✓
SIVsmm/mac	✓	✓	✓	✓	✓	✓
SIVagm	✓	✓	✓	✓	✓	✓
SIVdeb	n.d.	✓	✓	✓	✓	✓
SIVsyk	✓	✓	✓	✓	✓	✓
SIVblu	n.d.	✓	✓	✓	✓	✓
SIVsun	n.d.	✓	✓	✓	✓	✓
SIVwrc	✓	✓	✓	✓	✓	✓
SIVolc	✓	✓	✓	✗	✓	✓
SIVlho	✓	✓	✓	✓	✓	✓
SIVden	✗	✓	✓	✓	✓	✓
SIVasc	n.d.	✓	✓	✓	✓	✗

FIGURE 4.4 **Functional properties of primate lentiviral Nef proteins.** Check marks indicate the ability of Nef to counteract tetherin, to increase virion infectivity, or to decrease the surface expression levels of CD4, MHC-I, CD3, or CD28. A cross indicates lack of function. Note that the data are preliminary because only a very limited number of *nef* alleles have been analyzed for some of the lentiviruses and most data were generated in cells of human origin. (n.d., not determined.) [39–45].

many Nef proteins including those encoded by SIVcpz and SIVgor (the direct precursors of HIV-1) counteract the restriction factor tetherin (BST-2, CD317) that tethers nascent virions to the cell surface [39–42,46,47]. Nef has no catalytic activity and frequently seems to act as an adaptor that interacts with the cytoplasmic tails of these receptors and recruits them to the endocytic machinery [48–50]. These Nef functions may converge to modulate the functionality of virally infected T cells and to reduce their recognition and elimination by the immune system, thereby expanding the time period of virus production. Furthermore, Nef increases virion infectivity and may prevent superinfection [51,52]. For HIV-1 Nef, it has also been reported that it interacts with various cellular kinases and promotes the induction of cellular transcription factors, such as NF-AT, NF-κB, and AP-1, that promote transcription via the LTR promoter and thus viral replication [53–59]. Most Nef functions are highly conserved but fundamental differences do also exist: in contrast to *nef* alleles of HIV-1 and some *vpu*-containing SIVs, the great majority of primate lentiviral Nefs down-modulate TCR-CD3 from the cell surface, thereby attenuating the responsiveness of T cells to stimulation (Figure 4.4) [42,43]. Possible consequences of the differences in Nef function will be discussed in the following section.

UNIQUE FEATURES OF HIV-1 AND HIV-2 RELATED PRIMATE LENTIVIRUSES

The presence of a *vpu* gene distinguishes HIV-1 and its closest SIV counterparts from most other primate lentiviruses [60]. Vpu (viral protein U) is a small 16-kDa integral membrane protein that is co-expressed with the viral envelope glycoprotein (Env) during the late stage of the viral replication cycle [61]. The Vpu proteins of most primate lentiviruses degrade CD4 (Figure 4.5) [41]. For HIV-1 Vpu, it has been reported that this function involves the recruitment of a ubiquitin ligase complex to the cytoplasmic tail of CD4 to mediate poly-ubiquitinylation and subsequent proteasomal degradation [68,69]. CD4 degradation may promote virion release, reduce superinfection, and prevent incorporation of inactive gp120/CD4 complexes in progeny virions. However, most of these effects have only been observed under artificial conditions and need to be verified in primary cells. Secondly, the Vpu proteins of pandemic HIV-1 group M strains counteract tetherin, an interferon-induced host restriction factor that inhibits virion release [46,47]. Vpu interacts with the transmembrane domain of tetherin and targets the restriction factor to the trans-Golgi network or to early endosomes for proteasomal or lysosomal degradation by a mechanism that seems to involve β-TrCP [70–72]. Tetherin antagonism is conserved in Vpus of SIVs derived from *Cercopithecus* monkeys but not in SIVcpz and SIVgor, which instead use Nef to counteract the tetherin orthologues from their respective hosts [39–41,62]. After zoonotic transmission of the latter SIVs from apes to humans, only pandemic group M viruses evolved a fully functional Vpu that degrades CD4 and counteracts tetherin (Figure 4.5) [41,73]. Thus, effective tetherin antagonism may have been a prerequisite for the effective spread of the AIDS pandemic [74]. Recent evidence suggests that Vpus perform additional activities and may impair the function of natural killer (NK) and iNKT cells by down-modulation of NTB-A, PVR, and CD1d, respectively [63–65]. However, these effects are relatively modest compared to tetherin antagonism and CD4 degradation, and their relevance for viral immune evasion in vivo needs further study.

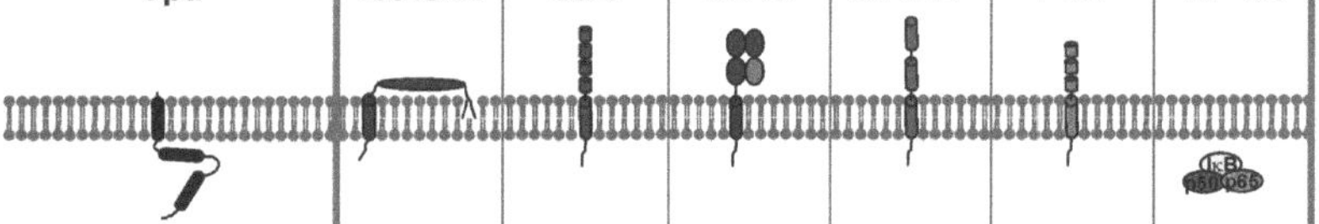

Vpu	Vpu displays activity against					
	Tetherin	CD4	CD1d	NTB-A	PVR	NF-κB
HIV-1 M	✓	✓	✓	✓	✓	✓
HIV-1 N	✓	✗	✗	✗	n.d.	n.d.
HIV-1 O,P	✗	✓	n.d.	n.d.	n.d.	n.d.
SIVcpz/gor	✗	✓	n.d.	n.d.	n.d.	n.d.
SIVgsn/mon/mus	✓	✓	n.d.	n.d.	n.d.	n.d.
SIVden/wol	✓	✓	n.d.	n.d.	n.d.	n.d.

FIGURE 4.5 **Functional properties of primate lentiviral Vpu proteins.** Check marks indicate the ability of Vpu to counteract tetherin, to inhibit the activation of NF-κB, or to decrease the surface expression levels of CD4, CD1d, NTB-A, or PVR. A cross indicates lack of function. Note that the data are preliminary because only a very limited number of *vpu* alleles have been analyzed for some of the lentiviruses and most data were generated in cells of human origin. SIVwol Vpu has not been functionally characterized but is closely related to that of SIVden. (n.d., not determined.) [39,41,42,46,47,62–67].

It will be interesting to further clarify whether the acquisition of a *vpu* gene promoted the evolution of primate lentiviruses with increased virulence [75]. As mentioned above, SIVcpz is associated with increased mortality rates in naturally infected chimpanzees [8]. Furthermore, *vpu* expressing SIVs infecting greater spot-nosed and mustached monkeys seem to show a low prevalence in the wild (~2–4%) compared to the 50–90% of "regular" SIV infections [76], which is a clear indicator for suboptimal virus-host adaptation. Vpu-mediated down-modulation of tetherin potentially prevents an inhibitory signal via ILT7 to plasmacytoid dendritic cells (pDCs) [77]. This lack of negative feedback signaling may induce IFN production by pDCs and thereby increase harmful immune activation in the host. On the other hand, however, it has recently been shown that Vpu reduces tetherin-mediated viral immune sensing and the induction of NF-κB dependent immune mediators [66,78–83]. Finally, Vpu has apparently facilitated the loss of Nef-mediated down-modulation of TCR-CD3 [43]. How all these viral properties converge to affect the levels of infection-associated inflammation and the clinical outcome of infection remains to be determined.

The *viral protein X* (Vpx) originated from a duplication of the *vpr* gene and has so far been shown to be encoded by HIV-2, SIVsmm, and SIVs infecting red-capped mangabeys, drills, or mandrills [84–86]. It seems that the two best-established functions of Vpr are segregated in *vpx*-expressing primate lentiviruses: Vpr induces cell-cycle arrest and Vpx facilitates infection of macrophages (Figure 4.3) [87,88]. Recently, it has been shown that Vpx (and some Vpr proteins) induce the degradation of SAMHD1, a dGTP-dependent nuclease that may reduce the levels of available dNTPs beyond those required for effective reverse transcription in myeloid cells [31,32]. Vpx binds DCAF-1 to interact with the cullin 4A-DDB1 complex, and this interaction seems to be required for its ability to degrade SAMHD1 and promote macrophage infection [89,90]. Notably, studies in the SIV/macaque model suggest that both accessory proteins may compensate for one another, because only combined deletions in both accessory genes resulted in a severely attenuated viral phenotype [38,87]. Interestingly, the *vpx* gene was deleted during the emergence of SIVcpz, which resulted in a novel unique *vif* gene that is able to antagonize ape APOBEC3G [85]. Thus, the loss of *vpx* may have facilitated the adaptation to the new chimpanzee host and the subsequent transmission to humans. Further studies to clarify whether infection of macrophages, dendritic cells, and resting T cells is particularly common in infections with *vpx*-containing viruses, such as SIVsmm or HIV-2, and whether the presence of this accessory gene may affect the viral coreceptor tropism to allow effective infection of these cell types, will be interesting.

PRIMATE LENTIVIRAL CORECEPTOR TROPISM AND PATHOGENESIS

Primate lentiviruses generally use CD4 as primary receptor and a seven-transmembrane G-protein-coupled receptor as a coreceptor to infect their target cells. CD4 is mainly expressed by T-helper cells and (at lower levels) by macrophages and dendritic cells. The main coreceptor for HIV-1 is CCR5, although virus variants capable of utilizing CXCR4 can be detected in about half of all AIDS patients [91]. Notably, a homozygous deletion of 32 bp of the *ccr5* gene (CCR5-Δ32), which is found in about 1% of the Caucasian population, is associated with a reduced risk of the acquisition of HIV-1 infection via the sexual route [92,93]. CCR5 also seems to be the major coreceptor of most SIVs and HIV-2, although promiscuous use of alternative coreceptors, such as CCR1, CCR2b, CCR3, CCR8, GPR1, GPR15/Bob, STRL-33/Bonzo, and ChemR23, seems to be quite common [19]. Importantly, SIVs can replicate efficiently in some natural host without using CCR5 or CXCR4. For example, SIVrcm uses CCR2b as the major entry co-factor [94]. Furthermore, lack of functional CCR5 expression in a subset of sooty mangabeys does not reduce their risk for infection and is only associated with moderately reduced viral loads, because SIVsmm may efficiently use CXCR6, GPR15, and GPR1 as alternative coreceptors [95].

Altogether, alternative coreceptor usage seems quite common among primate lentiviruses, and this clearly warrants further investigation.

The R5 to X4 coreceptor switch of HIV-1 in humans is not obligatory for the development of AIDS but associated with accelerated disease progression and a very poor prognosis in the absence of antiretroviral therapy [91]. In striking contrast, the very rare emergence of CXCR4-tropic SIVsmm strains in sooty mangabeys was associated with an almost complete loss of $CD4^+$ T cells but no disease during more than a decade of follow-up [95]. Recent data suggest that CD4/CD8 double-negative T cells that are resistant to SIV infection and may functionally compensate for the lack of $CD4^+$ helper T cells, may allow CD4-low sooty mangabeys to avoid the development of immunodeficiency [96–98]. CXCR4 can clearly function as an effective coreceptor of SIV and it is currently largely unclear why CXCR4-tropic viruses emerge much more frequently in HIV-1 than in SIV infection. Possible reasons are that intact CD4 homeostasis and thus high levels of activated memory $CCR5^+$ T cells are maintained throughout the course of natural SIV infection. Furthermore, it is tempting to speculate that the inability of Nef to down-modulate TCR-CD3 may be advantageous for HIV-1 replication in naive $CXCR4^+$ T cells, because it may allow their activation to levels that are sufficient for effective viral replication.

CONCLUSIONS AND PERSPECTIVES

Lentiviruses have apparently been infecting primates for millions of years and in some cases achieved a well-balanced relationship with their respective hosts, although it is becoming evident that not all natural SIV infections are nonpathogenic. Although important progress has been made, our understanding of the biological properties and evolution of primate lentiviruses is far from being complete. It is evident, for example, that some SIVs, such as SIVagm, can replicate to high levels and reach high seroprevalence rates in the wild although they lack both *vpu* and *vpx*. Thus, why was it advantageous for some viruses to acquire and maintain *vpu* and *vpx* genes? How do these accessory genes affect the cell tropism and replicative capacity of today's primate lentiviruses in vivo? Does the ability to counteract the human orthologues of host restriction factors explain why only 3 of 40 SIV-infected NHPs have transmitted their virus to humans, and does the presence of additional accessory genes, i.e., *vpu* and *vpx*, increase the fitness of primate lentiviruses for cross-species transmission? How do the differences in the repertoire and function of the viral accessory genes affect the predictive value of SIV/NHP models for studies on the pathogenesis of HIV-1 and AIDS vaccines? Finally, we are only beginning to understand the role of host restriction factors and their antagonists in viral immune sensing and in the efficacy of the antiviral immune response. A better knowledge about this may facilitate the development of more effective therapeutic and preventive strategies.

Acknowledgments

We thank Dré van der Merwe for critical reading of the manuscript. We apologize to the authors of many interesting studies that could not be cited due to space limitations. The authors are supported by grants from the Deutsche Forschungsgemeinschaft and the European Research Council.

References

[1] Gao F, Bailes E, Robertson DL, Chen Y, Rodenburg CM, Michael SF, et al. Origin of HIV-1 in the chimpanzee *Pan troglodytes troglodytes*. Nature 1999;397:436–41.
[2] Peeters M, Courgnaud V. HIV databases review article. http://www.hiv.lanl.gov/content/sequence/HIV/REVIEWS/PEETERS2002/Peeters2002.html [accessed 31.07.13].
[3] Plantier JC, Leoz M, Dickerson JE, De Oliveira F, Cordonnier F, Lemée V, et al. A new human immunodeficiency virus derived from gorillas. Nat Med 2009;15:871–2.
[4] Keele BF, Van Heuverswyn F, Li Y, Bailes E, Takehisa J, Santiago ML, et al. Chimpanzee reservoirs of pandemic and nonpandemic HIV-1. Science 2006;313:523–6.
[5] Van Heuverswyn F, Li Y, Neel C, Bailes E, Keele BF, Liu W, et al. Human immunodeficiency viruses: SIV infection in wild gorillas. Nature 2006;444:164.
[6] Hirsch VM, Olmsted RA, Murphey-Corb M, Purcell RH, Johnson PR. An African primate lentivirus (SIVsm) closely related to HIV-2. Nature 1989;339:389–92.
[7] Takehisa J, Kraus MH, Ayouba A, Bailes E, Van Heuverswyn F, Decker JM, et al. Origin and biology of simian immunodeficiency virus in wild-living western gorillas. J Virol 2009;83:1635–48.
[8] Keele BF, Jones JH, Terio KA, Estes JD, Rudicell RS, Wilson ML, et al. Increased mortality and AIDS-like immunopathology in wild chimpanzees infected with SIVcpz. Nature 2009;460:515–9.
[9] Santiago ML, Range F, Keele BF, Li Y, Bailes E, Bibollet-Ruche F, et al. Simian immunodeficiency virus infection in free-ranging sooty mangabeys (*Cercocebus atys atys*) from the Taï Forest, Côte d'Ivoire: implications for the origin of epidemic human immunodeficiency virus type 2. J Virol 2005;79:12515–27.
[10] Ma D, Jasinska A, Kristoff J, Grobler JP, Turner T, Jung Y, et al. SIVagm infection in wild African green monkeys from South Africa: epidemiology, natural history, and evolutionary considerations. PLoS Pathog 2013;9:e1003011.
[11] Damond F, Worobey M, Campa P, Farfara I, Colin G, Matheron S, et al. Identification of a highly divergent HIV type 2 and proposal for a change in HIV type 2 classification. AIDS Res Hum Retroviruses 2004;20:666–72.
[12] Chakrabarti L, Guyader M, Alizon M, Daniel MD, Desrosiers RC, Tiollais P, et al. Sequence of simian immunodeficiency virus from macaque and its relationship to other human and simian retroviruses. Nature 1987;328:543–7.
[13] Honjo S, Narita T, Kobayashi R, Hiyaoka A, Fujimoto K, Takasaka M, et al. Experimental infection of African green monkeys and cynomolgus monkeys with a SIVAGM strain isolated from a healthy African green monkey. J Med Primatol 1990;19:9–20.
[14] Kaur A, Grant RM, Means RE, McClure H, Feinberg M, Johnson RP. Diverse host responses and outcomes following simian immunodeficiency virus SIVmac239 infection in sooty mangabeys and rhesus macaques. J Virol 1998;72:9597–611.
[15] Chahroudi A, Bosinger SE, Vanderford TH, Paiardini M, Silvestri G. Natural SIV hosts: showing AIDS the door. Science 2012;335:1188–93.

[16] Brenchley JM, Paiardini M. Immunodeficiency lentiviral infections in natural and non-natural hosts. Blood 2011;118:847–54.
[17] Bosinger SE, Jacquelin B, Benecke A, Silvestri G, Müller-Trutwin M. Systems biology of natural simian immunodeficiency virus infections. Curr Opin HIV AIDS 2012;7:71–8.
[18] Liovat A-S, Jacquelin B, Ploquin MJ, Barré-Sinoussi F, Müller-Trutwin MC. African non human primates infected by SIV – why don't they get sick? Lessons from studies on the early phase of non-pathogenic SIV infection. Curr HIV Res 2009;7:39–50.
[19] Sina ST, Ren W, Cheng-Mayer C. Coreceptor use in nonhuman primate models of HIV infection. J Transl Med 2011;9(Suppl. 1):S7.
[20] Elliott ST, Riddick NE, Francella N, Paiardini M, Vanderford TH, Li B, et al. Cloning and analysis of sooty mangabey alternative coreceptors that support simian immunodeficiency virus SIVsmm entry independently of CCR5. J Virol 2012;86:898–908.
[21] Gifford RJ, Katzourakis A, Tristem M, Pybus OG, Winters M, Shafer RW. A transitional endogenous lentivirus from the genome of a basal primate and implications for lentivirus evolution. Proc Natl Acad Sci USA 2008;105:20362–7.
[22] Compton AA, Emerman M. Convergence and divergence in the evolution of the APOBEC3G-Vif interaction reveal ancient origins of simian immunodeficiency viruses. PLoS Pathog 2013;9:e1003135.
[23] Zhang H, Yang B, Pomerantz RJ, Zhang C, Arunachalam SC, Gao L. The cytidine deaminase CEM15 induces hypermutation in newly synthesized HIV-1 DNA. Nature 2003;424:94–8.
[24] Mangeat B, Turelli P, Caron G, Friedli M, Perrin L, Trono D. Broad antiretroviral defence by human APOBEC3G through lethal editing of nascent reverse transcripts. Nature 2003;424:99–103.
[25] Newman EN, Holmes RK, Craig HM, Klein KC, Lingappa JR, Malim MH, et al. Antiviral function of APOBEC3G can be dissociated from cytidine deaminase activity. Curr Biol CB 2005;15:166–70.
[26] Yu X, Yu Y, Liu B, Luo K, Kong W, Mao P, et al. Induction of APOBEC3G ubiquitination and degradation by an HIV-1 Vif-Cul5-SCF complex. Science 2003;302:1056–60.
[27] Compton AA, Hirsch VM, Emerman M. The host restriction factor APOBEC3G and retroviral Vif protein coevolve due to ongoing genetic conflict. Cell Host Microbe 2012;11:91–8.
[28] Le Rouzic E, Belaïdouni N, Estrabaud E, Morel M, Rain JC, Transy C, et al. HIV1 Vpr arrests the cell cycle by recruiting DCAF1/VprBP, a receptor of the Cul4-DDB1 ubiquitin ligase. Cell Cycle Georget Tex 2007;6:182–8.
[29] Stivahtis GL, Soares MA, Vodicka MA, Hahn BH, Emerman M. Conservation and host specificity of Vpr-mediated cell cycle arrest suggest a fundamental role in primate lentivirus evolution and biology. J Virol 1997;71:4331–8.
[30] Lim ES, Fregoso OI, McCoy CO, Matsen FA, Malik HS, Emerman M. The ability of primate lentiviruses to degrade the monocyte restriction factor SAMHD1 preceded the birth of the viral accessory protein Vpx. Cell Host Microbe 2012;11:194–204.
[31] Laguette N, Sobhian B, Casartelli N, Ringeard M, Chable-Bessia C, Ségéral E, et al. SAMHD1 is the dendritic- and myeloid-cell-specific HIV-1 restriction factor counteracted by Vpx. Nature 2011;474:654–7.
[32] Hrecka K, Hao C, Gierszewska M, Swanson SK, Kesik-Brodacka M, Srivastava S, et al. Vpx relieves inhibition of HIV-1 infection of macrophages mediated by the SAMHD1 protein. Nature 2011;474:658–61.
[33] Planelles V, Jowett JB, Li QX, Xie Y, Hahn B, Chen IS. Vpr-induced cell cycle arrest is conserved among primate lentiviruses. J Virol 1996;70:2516–24.
[34] Philippon V, Matsuda Z, Essex M. Transactivation is a conserved function among primate lentivirus Vpr proteins but is not shared by Vpx. J Hum Virol 1999;2:167–74.
[35] Jacquot G, Le Rouzic E, David A, Mazzolini J, Bouchet J, Bouaziz S, et al. Localization of HIV-1 Vpr to the nuclear envelope: impact on Vpr functions and virus replication in macrophages. Retrovirology 2007;4:84.

[36] Kirchhoff F, Greenough TC, Brettler DB, Sullivan JL, Desrosiers RC. Brief report: absence of intact nef sequences in a long-term survivor with nonprogressive HIV-1 infection. N Engl J Med 1995;332:228–32.
[37] Mariani R, Kirchhoff F, Greenough TC, Sullivan JL, Desrosiers RC, Skowronski J. High frequency of defective nef alleles in a long-term survivor with nonprogressive human immunodeficiency virus type 1 infection. J Virol 1996;70:7752–64.
[38] Desrosiers RC, Lifson JD, Gibbs JS, Czajak SC, Howe AY, Arthur LO, et al. Identification of highly attenuated mutants of simian immunodeficiency virus. J Virol 1998;72:1431–7.
[39] Jia B, Serra-Moreno R, Neidermyer W, Rahmberg A, Mackey J, Fofana IB, et al. Species-specific activity of SIV Nef and HIV-1 Vpu in overcoming restriction by tetherin/BST2. PLoS Pathog 2009;5:e1000429.
[40] Zhang F, Wilson SJ, Landford WC, Virgen B, Gregory D, Johnson MC, et al. Nef proteins from simian immunodeficiency viruses are tetherin antagonists. Cell Host Microbe 2009;6:54–67.
[41] Sauter D, Schindler M, Specht A, Landford WN, Münch J, Kim KA, et al. Tetherin-driven adaptation of Vpu and Nef function and the evolution of pandemic and nonpandemic HIV-1 strains. Cell Host Microbe 2009;6:409–21.
[42] Schmökel J, Sauter D, Schindler M, Leendertz FH, Bailes E, Dazza MC, et al. The presence of a vpu gene and the lack of Nef-mediated downmodulation of T cell receptor-CD3 are not always linked in primate lentiviruses. J Virol 2011;85:742–52.
[43] Schindler M, Münch J, Kutsch O, Li H, Santiago ML, Bibollet-Ruche F, et al. Nef-mediated suppression of T cell activation was lost in a lentiviral lineage that gave rise to HIV-1. Cell 2006;125:1055–67.
[44] Münch J, Rajan D, Schindler M, Specht A, Rücker E, Novembre FJ, et al. Nef-mediated enhancement of virion infectivity and stimulation of viral replication are fundamental properties of primate lentiviruses. J Virol 2007;81:13852–64.
[45] Schmökel J, Li H, Bailes E, Schindler M, Silvestri G, Hahn BH, et al. Conservation of Nef function across highly diverse lineages of SIVsmm. Retrovirology 2009;6:36.
[46] Neil SJD, Zang T, Bieniasz PD. Tetherin inhibits retrovirus release and is antagonized by HIV-1 Vpu. Nature 2008;451:425–30.
[47] Van Damme N, Goff D, Katsura C, Jorgenson RL, Mitchell R, Johnson MC, et al. The interferon-induced protein BST-2 restricts HIV-1 release and is downregulated from the cell surface by the viral Vpu protein. Cell Host Microbe 2008;3:245–52.
[48] Schwartz O, Maréchal V, Le Gall S, Lemonnier F, Heard JM. Endocytosis of major histocompatibility complex class I molecules is induced by the HIV-1 Nef protein. Nat Med 1996;2:338–42.
[49] Aiken C, Konner J, Landau NR, Lenburg ME, Trono D. Nef induces CD4 endocytosis: requirement for a critical dileucine motif in the membrane-proximal CD4 cytoplasmic domain. Cell 1994;76:853–64.
[50] Serra-Moreno R, Zimmermann K, Stern LJ, Evans DT. Tetherin/BST-2 antagonism by Nef depends on a direct physical interaction between Nef and tetherin, and on clathrin-mediated endocytosis. PLoS Pathog 2013;9:e1003487.
[51] Chowers MY, Spina CA, Kwoh TJ, Fitch NJ, Richman DD, Guatelli JC. Optimal infectivity in vitro of human immunodeficiency virus type 1 requires an intact nef gene. J Virol 1994;68:2906–14.
[52] Benson RE, Sanfridson A, Ottinger JS, Doyle C, Cullen BR. Downregulation of cell-surface CD4 expression by simian immunodeficiency virus Nef prevents viral super infection. J Exp Med 1993;177:1561–6.
[53] Djordjevic JT, Schibeci SD, Stewart GJ, Williamson P. HIV type 1 Nef increases the association of T cell receptor (TCR)-signaling molecules with T cell rafts and promotes activation-induced raft fusion. AIDS Res Hum Retroviruses 2004;20: 547–55.

[54] Fenard D, Yonemoto W, de Noronha C, Cavrois M, Williams SA, Greene WC. Nef is physically recruited into the immunological synapse and potentiates T cell activation early after TCR engagement. J Immunol Balt Md 2005;175:6050–7.
[55] Fortin J-F, Barat C, Beauséjour Y, Barbeau B, Tremblay MJ. Hyper-responsiveness to stimulation of human immunodeficiency virus-infected CD4+ T cells requires Nef and Tat virus gene products and results from higher NFAT, NF-kappaB, and AP-1 induction. J Biol Chem 2004;279:39520–31.
[56] Manninen A, Renkema GH, Saksela K. Synergistic activation of NFAT by HIV-1 nef and the Ras/MAPK pathway. J Biol Chem 2000;275:16513–7.
[57] Schrager JA, Marsh JW. HIV-1 Nef increases T cell activation in a stimulus-dependent manner. Proc Natl Acad Sci USA 1999;96:8167–72.
[58] Simmons A, Aluvihare V, McMichael A. Nef triggers a transcriptional program in T cells imitating single-signal T cell activation and inducing HIV virulence mediators. Immunity 2001;14:763–77.
[59] Wang JK, Kiyokawa E, Verdin E, Trono D. The Nef protein of HIV-1 associates with rafts and primes T cells for activation. Proc Natl Acad Sci USA 2000;97:394–9.
[60] Cohen EA, Terwilliger EF, Sodroski JG, Haseltine WA. Identification of a protein encoded by the vpu gene of HIV-1. Nature 1988;334:532–4.
[61] Strebel K, Klimkait T, Martin MA. A novel gene of HIV-1, vpu, and its 16-kilodalton product. Science 1988;241:1221–3.
[62] Yang SJ, Lopez LA, Hauser H, Exline CM, Haworth KG, Cannon PM. Anti-tetherin activities in Vpu-expressing primate lentiviruses. Retrovirology 2010;7:13.
[63] Moll M, Andersson SK, Smed-Sörensen A, Sandberg JK. Inhibition of lipid antigen presentation in dendritic cells by HIV-1 Vpu interference with CD1d recycling from endosomal compartments. Blood 2010;116:1876–84.
[64] Shah AH, Sowrirajan B, Davis ZB, Ward JP, Campbell EM, Planelles V, et al. Degranulation of natural killer cells following interaction with HIV-1-infected cells is hindered by downmodulation of NTB-A by Vpu. Cell Host Microbe 2010;8:397–409.
[65] Matusali G, Potestà M, Santoni A, Cerboni C, Doria M. The human immunodeficiency virus type 1 Nef and Vpu proteins downregulate the natural killer cell-activating ligand PVR. J Virol 2012;86:4496–504.
[66] Galão RP, Le Tortorec A, Pickering S, Kueck T, Neil SJD. Innate sensing of HIV-1 assembly by tetherin induces NFκB-dependent proinflammatory responses. Cell Host Microbe 2012;12:633–44.
[67] Sauter D, Unterweger D, Vogl M, Usmani SM, Heigele A, Kluge SF, et al. Human tetherin exerts strong selection pressure on the HIV-1 group N Vpu protein. PLoS Pathog 2012;8. http://dx.doi.org/10.1371/journal.ppat.1003093.
[68] Margottin F, Bour SP, Durand H, Selig L, Benichou S, Richard V, et al. A novel human WD protein, h-beta TrCp, that interacts with HIV-1 Vpu connects CD4 to the ER degradation pathway through an F-box motif. Mol Cell 1998;1:565–74.
[69] Schubert U, Antón LC, Bacík I, Cox JH, Bour S, Bennink JR, et al. CD4 glycoprotein degradation induced by human immunodeficiency virus type 1 Vpu protein requires the function of proteasomes and the ubiquitin-conjugating pathway. J Virol 1998;72:2280–8.
[70] Mitchell RS, Katsura C, Skasko MA, Fitzpatrick K, Lau D, Ruiz A, et al. Vpu antagonizes BST-2-mediated restriction of HIV-1 release via beta-TrCP and endo-lysosomal trafficking. PLoS Pathog 2009;5:e1000450.
[71] Douglas JL, Viswanathan K, McCarroll MN, Gustin JK, Früh K, Moses AV. Vpu directs the degradation of the human immunodeficiency virus restriction factor BST-2/tetherin via a {beta}TrCP-dependent mechanism. J Virol 2009;83:7931–47.
[72] Mangeat B, Gers-Huber G, Lehmann M, Zufferey M, Luban J, Piguet V. HIV-1 Vpu neutralizes the antiviral factor tetherin/BST-2 by binding it and directing its beta-TrCP2-dependent degradation. PLoS Pathog 2009;5:e1000574.

[73] Sauter D, Hué S, Petit SJ, Plantier JC, Towers GJ, Kirchhoff F, et al. HIV-1 Group P is unable to antagonize human tetherin by Vpu, Env or Nef. Retrovirology 2011;8:103.
[74] Sauter D, Specht A, Kirchhoff F. Tetherin: holding on and letting go. Cell 2010;141:392–8.
[75] Kirchhoff F. Is the high virulence of HIV-1 an unfortunate coincidence of primate lentiviral evolution? Nat Rev Microbiol 2009;7:467–76.
[76] Aghokeng AF, Liu W, Bibollet-Ruche F, Loul S, Mpoudi-Ngole E, Laurent C, et al. Widely varying SIV prevalence rates in naturally infected primate species from Cameroon. Virology 2006;345:174–89.
[77] Cao W, Bover L, Cho M, Wen X, Hanabuchi S, Bao M, et al. Regulation of TLR7/9 responses in plasmacytoid dendritic cells by BST2 and ILT7 receptor interaction. J Exp Med 2009;206:1603–14.
[78] Tokarev A, Suarez M, Kwan W, Fitzpatrick K, Singh R, Guatelli J. Stimulation of NF-κB activity by the HIV restriction factor BST2. J Virol 2013;87:2046–57.
[79] Cocka LJ, Bates P. Identification of alternatively translated tetherin isoforms with differing antiviral and signaling activities. PLoS Pathog 2012;8:e1002931.
[80] Hotter D, Kirchhoff F, Sauter D. HIV-1 Vpu does not degrade interferon regulatory factor 3. J Virol 2013;87:7160–5.
[81] Leulier F, Marchal C, Miletich I, Limbourg-Bouchon B, Benarous R, Lemaitre B. Directed expression of the HIV-1 accessory protein Vpu in Drosophila fat-body cells inhibits Toll-dependent immune responses. EMBO Rep 2003;4:976–81.
[82] Akari H, Bour S, Kao S, Adachi A, Strebel K. The human immunodeficiency virus type 1 accessory protein Vpu induces apoptosis by suppressing the nuclear factor kappaB-dependent expression of antiapoptotic factors. J Exp Med 2001;194:1299–311.
[83] Bour S, Perrin C, Akari H, Strebel K. The human immunodeficiency virus type 1 Vpu protein inhibits NF-kappa B activation by interfering with beta TrCP-mediated degradation of Ikappa B. J Biol Chem 2001;276:15920–8.
[84] Tristem M, Marshall C, Karpas A, Petrik J, Hill F. Origin of vpx in lentiviruses. Nature 1990;347:341–2.
[85] Etienne L, Hahn BH, Sharp PM, Matsen FA, Emerman M. Gene loss and adaptation to hominids underlie the ancient origin of HIV-1. Cell Host Microbe 2013;14:85–92.
[86] Sharp PM, Bailes E, Stevenson M, Emerman M, Hahn BH. Gene acquisition in HIV and SIV. Nature 1996;383:586–7.
[87] Gibbs JS, Regier DA, Desrosiers RC. Construction and in vitro properties of SIVmac mutants with deletions in 'nonessential' genes. AIDS Res Hum Retroviruses 1994;10:607–16.
[88] Fletcher 3rd TM, Brichacek B, Sharova N, Newman MA, Stivahtis G, Sharp PM, et al. Nuclear import and cell cycle arrest functions of the HIV-1 Vpr protein are encoded by two separate genes in HIV-2/SIV(SM). EMBO J 1996;15:6155–65.
[89] Planelles V. Restricted access to myeloid cells explained. Viruses 2011;3(9):1624–33.
[90] Ahn J, Hao C, Yan J, DeLucia M, Mehrens J, Wang C, et al. HIV/simian immunodeficiency virus (SIV) accessory virulence factor Vpx loads the host cell restriction factor SAMHD1 onto the E3 ubiquitin ligase complex CRL4DCAF1. J Biol Chem 2012;287:12550–8.
[91] Xiao L, Rudolph DL, Owen SM, Spira TJ, Lal RB. Adaptation to promiscuous usage of CC and CXC-chemokine coreceptors in vivo correlates with HIV-1 disease progression. AIDS 1998;12:F137–43.
[92] Samson M, Libert F, Doranz BJ, Rucker J, Liesnard C, Farber CM, et al. Resistance to HIV-1 infection in caucasian individuals bearing mutant alleles of the CCR-5 chemokine receptor gene. Nature 1996;382:722–5.
[93] Liu R, Paxton WA, Choe S, Ceradini D, Martin SR, Horuk R, et al. Homozygous defect in HIV-1 coreceptor accounts for resistance of some multiply-exposed individuals to HIV-1 infection. Cell 1996;86:367–77.
[94] Chen Z, Kwon D, Jin Z, Monard S, Telfer P, Jones MS, et al. Natural infection of a homozygous delta24 CCR5 red-capped mangabey with an R2b-tropic simian immunodeficiency virus. J Exp Med 1998;188:2057–65.

[95] Riddick NE, Hermann EA, Loftin LM, Elliott ST, Wey WC, Cervasi B, et al. A novel CCR5 mutation common in sooty mangabeys reveals SIVsmm infection of CCR5-null natural hosts and efficient alternative coreceptor use in vivo. PLoS Pathog 2010;6:e1001064.
[96] Milush JM, Mir KD, Sundaravaradan V, Gordon SN, Engram J, Cano CA, et al. Lack of clinical AIDS in SIV-infected sooty mangabeys with significant $CD4^+$ T cell loss is associated with double-negative T cells. J Clin Invest 2011;121:1102–10.
[97] Sundaravaradan V, Saleem R, Micci L, Gasper MA, Ortiz AM, Else J, et al. Multifunctional double-negative T cells in sooty mangabeys mediate t-helper functions irrespective of SIV infection. PLoS Pathog 2013;9:e1003441.
[98] Sundaravaradan V, Mir KD, Sodora DL. Double-negative T cells during HIV/SIV infections: potential pinch hitters in the T-cell lineup. Curr Opin HIV AIDS 2012;7:164–71.

CHAPTER

5

Simian Immunodeficiency Virus Infection of Chimpanzees (*Pan troglodytes*)

Edward J.D. Greenwood, Fabian Schmidt*, Jonathan L. Heeney*

Department of Veterinary Medicine, University of Cambridge, Cambridge, UK

INTRODUCTION

Simian immunodeficiency virus (SIV) of chimpanzees (*Pan troglodytes*) (SIVcpz) is now well established as the origin of the human immunodeficiency virus (HIV)-1 pandemic [1,2]. Because of this, and because of the close relationship between humans and chimpanzees, study of SIVcpz and its relationship with the chimpanzee promises to be of the greatest relevance to our understanding of the pathogenic mechanisms of HIV-1 infection in humans. Molecular genetic analyses have revealed a great deal regarding the origins, age, and distribution of SIVcpz within wild chimpanzee populations. SIVcpz is a chimeric virus [3], the result of recombination events between SIVs of other monkey species that are prey of chimpanzees, and therefore has presumably arisen relatively recently compared to other primate SIVs.

However, compared with the extensive study of two of the natural hosts of SIV infection, African green monkeys and sooty mangabeys—by far the best-described models—we know relatively little about the outcome of SIVcpz infection of chimpanzees. A key study of wild habituated chimpanzees suggests that SIVcpz induces increased mortality and an acquired immune deficiency syndrome (AIDS)–like disease in these animals [4].

*These authors contributed equally to this work.

Natural Hosts of SIV.
http://dx.doi.org/10.1016/B978-0-12-404734-1.00005-X

Reports of animals in captivity are somewhat anecdotal and conflicting; some animals also seem to suffer ill effects from SIV infection, whereas others live for decades without developing any clinical signs [5–8]. Here, we review these studies, summarize the case for and against a pathogenic outcome of SIVcpz infection, and comment on the potential reasons that the relationship between SIVcpz and its host is different from that found in other well-described natural hosts of SIV. Finally, we discuss the outstanding work required to provide a fuller understanding of the outcome of SIVcpz infection of chimpanzees, given the limitations and difficulties involved in working with this species.

DISTRIBUTION OF SIVcpz IN THE WILD

There are four recognized subspecies of the common chimpanzee: *verus*, *ellioti*, *troglodytes*, and *schweinfurthii*. Despite sampling of all subspecies, SIVcpz has been identified in the wild only in the *troglodytes* and *schweinfurthii* subspecies [1,9–11]; in both, a wide but uneven distribution spanning most of the habitat is observed (Figure 5.1). While sampling of wild animals can never be completely exhaustive, there is ample evidence that SIVcpz is not present in the *verus* subspecies. The vast majority of animals housed in primate centers and zoos in the United States and Europe are of the *verus* subspecies [12], whereas the only SIV-infected animals found in captivity in these countries are of the *troglodytes* [2] and *schweinfurthii* subspecies [13].

Estimates for the prevalence of SIVcpz infection in both infected subspecies range from complete absence in certain geographical regions to a prevalence that exceeds a third of sampled animals [9,14]. Unlike some monkey species, apes are banned from the bush-meat trade, the source of many key studies of SIV prevalence in other species. Insights into the prevalence and distribution of SIVcpz among wild chimpanzee populations have, therefore, been dependent on noninvasive strategies for SIV detection based on fecal and urine sampling [15]. Importantly, meaningful comparisons with other species can be only made when similar methodology is used, and demographic variations due to sampling should not be ignored (e.g., bush meat represents predominantly adult males). Nevertheless, fecal sampling approaches have also been applied to wild living populations of sooty mangabeys in the Taï Forest, Ivory Coast, with the finding of the presence of SIV infection in approximately 60% of the animals sampled [16]. Using a trapping strategy, a similar prevalence was found in African green monkeys [17]. The prevalence of SIV infection among sooty mangabeys and African green monkeys far exceeds that of SIVcpz in chimpanzees noted to date. Furthermore, unlike SIVcpz, SIVagm and SIVsm infections of African green monkeys and sooty mangabeys, respectively, are characterized by a geographically more even distribution

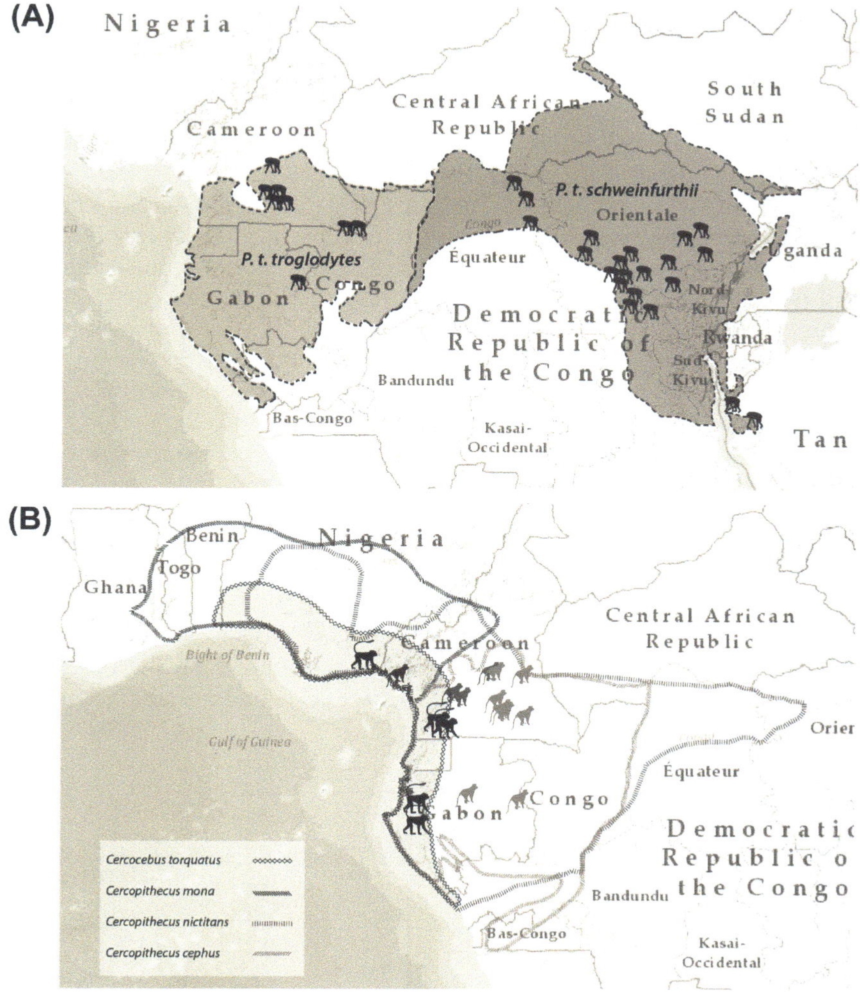

FIGURE 5.1 Ranges of the *Pan troglodytes troglodytes* (*P. t. troglodytes*) and *P. troglodytes schweinfurthii* (*P. t. schweinfurthii*) subspecies (A). Chimpanzee icons indicate communities found to have SIVcpz infection. (B) Ranges of the monkey species that may have contributed to the chimeric SIVcpz virus. Monkey icons indicate where SIV-infected animals have been identified. (See color plate at the back of the book.)

in the wild. For other wild living monkeys, especially arboreal species, data derived solely from bush-meat surveys are available, limiting a direct comparison of prevalence. With respect to an uneven geographic distribution, however, with regions where the virus is completely absent, chimpanzees show similarity only with a subset of guenons carrying an SIV related to part of the SIVcpz virus (see below) [18].

ORIGINS AND AGE OF SIVcpz

SIVcpz is a recombinant, or "chimeric," virus, with sequences contributed from two monkey species. Phylogenetic analyses comparing topologies of the three major genes, *gag*, *pol*, and *env*, revealed that the 5′ half of SIVcpz (*gag* and *pol*) is most closely related to SIV infection of red-capped mangabeys (*Cercocebus torquatus*) [3,19]. The virus contributing *env* and the accessory gene *vpu* was traced to one of three similar viruses infecting closely related guenon species—SIVgsn, SIVmon, and SIVmus of greater spot nosed guenons (*Cercopithecus nictitans*), mona guenons (*Cercopithecus mona*), and mustached guenons (*Cercopithecus cephus*), respectively [3,20]. Upon further investigation of the phylogenetic topology, the origin of the SIVcpz *nef* gene was attributed to the red-capped mangabey virus [21], despite its location at the 3′ half of the genome (Figure 5.2).

It is likely that these recombination events occurred after transmission of these SIVs to chimpanzees through their predation upon monkeys [3]. The geographic range of the red-capped mangabeys overlaps only with the range of Central African chimpanzees. It therefore is reasonable to assume that SIVcpz most likely originated within the geographic range of *Pan trogldytes troglodytes* (Figure 5.1). SIVcpz isolates from each of the *troglodytes* and *schweinfurthii* subspecies form monophyletic lineages in phylogenetic trees, with a high degree of divergence between these lineages (Figure 5.3), possibly indicative of a long period of segregation.

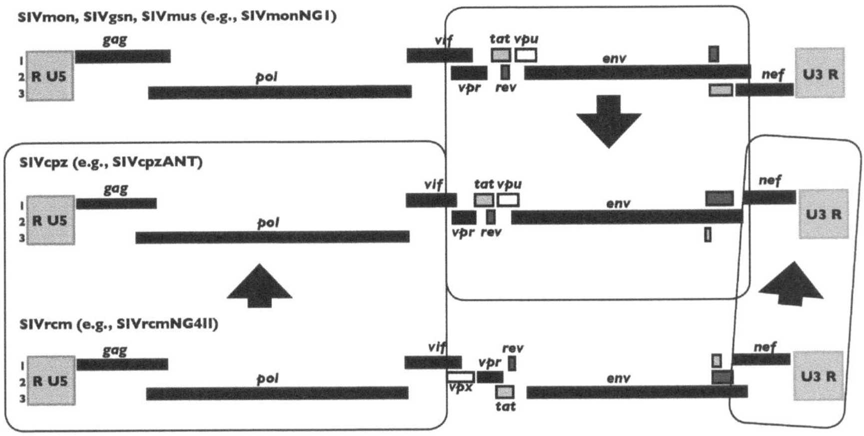

FIGURE 5.2 Chimeric nature of the SIV virus in chimpanzees (SIVcpz). Example genomes of SIV infection of mona guenons (SIVmon), SIVcpz, and SIV infection of red-capped mangabeys (SIVrcm) are displayed with genes displayed based on their reading frame (1, 2, or 3). Boxes and arrows indicate which elements of the two ancestral genomes contributed to SIVcpz.

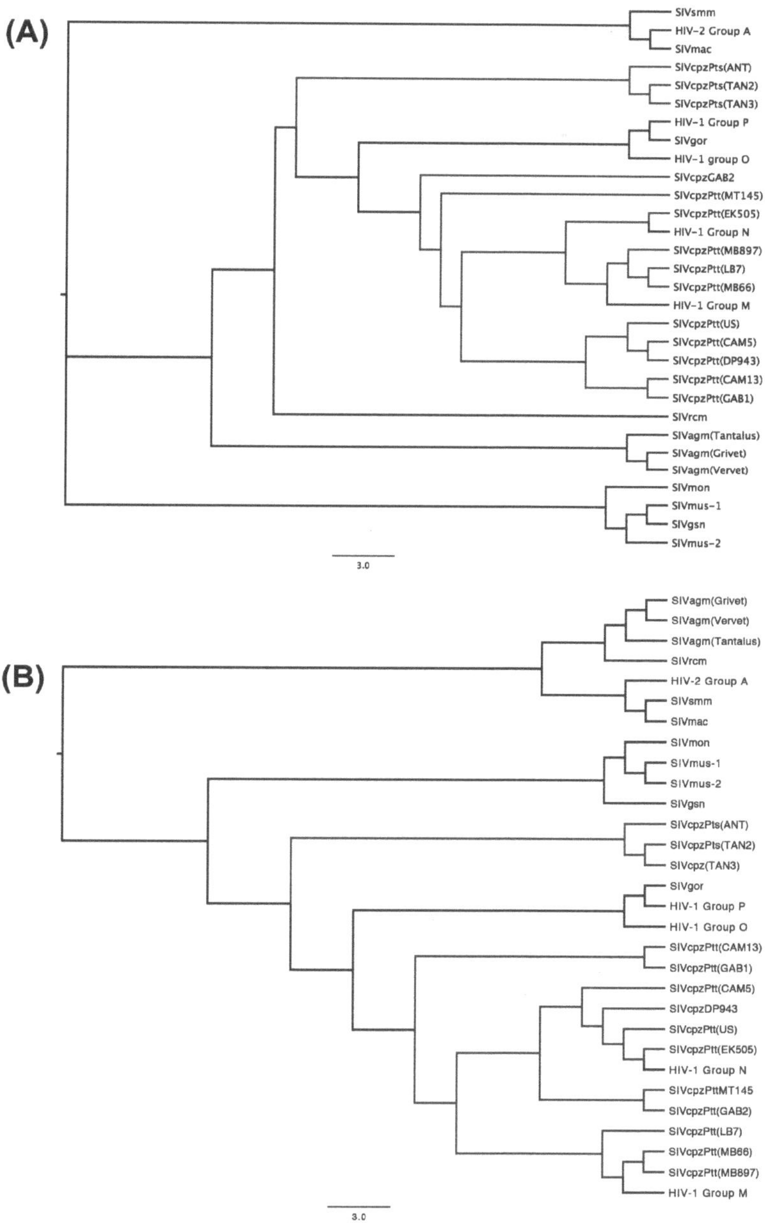

FIGURE 5.3 Phylogenetic trees of SIV virus in chimpanzees (SIVcpz) and other human and primate lentiviruses based on alignments of regions of the genes *pol* (A) and *gp41* (B).

The most recent estimate of the age of the African mainland SIV lineage, based on biogeographical features for molecular clock calibration, is that it is hundreds of thousands of years old; the SIVcpz lineage is estimated to be approximately 100,000 years old [17]. However, previous estimates using similar methods seem to underestimate the age of the SIV lineage [22]. In addition, an integrated lentivirus in the genome of the lemur is estimated to have integrated into the host genome at least four million years ago. Assuming an absence of an aerial vector for SIV, and based on the most recent time that Madagascar was accessible by land from the African mainland, the SIV lineage is at least 14 million years old [23,24]. If the true age of the primate lentivirus lineage is indeed millions of years old, the age of SIVcpz also would correspondingly be older than current molecular clock estimates.

Based on these estimates, it is conceivable that the distribution of SIVcpz among the chimpanzee subspecies is reflective of the time at which it arose, that is, that SIVcpz came into being before the split of the *schweinfurthii* from the *troglodytes* subspecies but after these subspecies split from *verus/ellioti*. Subsequent host-dependent evolution then led to the observed monophyletic SIV lineages of each subspecies, as has been proposed for the monophyletic lineages observed in African green monkey species [17,25]. The time to the most recent common ancestor of the *troglodytes* and *schweinfurthii* subspecies is 80,000 to 120,000 years [26,27]; these subspecies split from the *verus* and *ellioti* subspecies approximately 500,000 years ago. Thus, while the age of SIVcpz—and indeed all African primate lentiviruses—remains a matter of controversy, it can be concluded that while SIVcpz may have arisen more recently than other members of this lineage (which must be assumed because of its chimeric nature), SIVcpz has most likely been infecting chimpanzees for tens or hundreds of thousands of years.

OUTCOME OF SIVcpz INFECTION: EARLY ASSUMPTIONS

Consistent with a lengthy period of virus–host coevolution, and with well-described nonpathogenic SIV infections in African green monkeys and sooty mangabeys, most evidence before 2009 suggested that SIVcpz infection of chimpanzees was not pathogenic. This was based on an analysis of an extremely limited number of SIVcpz-infected animals held in primate centers in the United States and Europe and on the outcome of HIV-1 infection of chimpanzees. To the best of our knowledge, to date more than 200 chimpanzees have been experimentally infected with HIV-1 [28], but only a few progressed to disease [29,30]. A majority of these experimentally infected chimpanzees have shown no evidence of disease progression for more than 25 years.

The two most prominent SIVcpz-infected animals that are the basis of this assumption are "Marilyn" and "Noah," housed in the United States and The Netherlands, respectively. SIVcpz infection was detected retrospectively in Marilyn (subspecies *troglodytes*), after her death [2,6]. Marilyn was caught from the wild and, given the lack of other SIVcpz-infected animals in the center where she was housed, is presumed to have been infected before her capture in 1963 at the approximate age of 4 years, likely through vertical transmission at birth. Marilyn died in 1985 and therefore was infected for 22–26 years without the development of AIDS. The cause of death was apparently related to complications related to the birth of stillborn twins, and postmortem examination of tissues revealed only minor disruptions to the secondary lymphoid tissues (plasmacystosis and some degeneration of germinal centers) that are not consistent with the extensive changes seen in HIV-1-infected humans [6].

Noah (subspecies *schweinfurthii*) also was born in the wild and presumed to be infected vertically, certainly before age 2.5 [13]. Noah is alive to this day, in general good health, and is approximately 26 years old; his SIVcpz infection is assumed to be of the same duration [5]. The only notable clinical sign in Noah is a persistent and profound thrombocytopenia of unknown etiology. Since SIVcpz infection of Noah was discovered early in its lifetime, he was studied more extensively. Neither depleted $CD4^+$ T cells [31] nor generalized activation of peripheral blood T cells that characterizes pathogenic HIV/SIV infections in humans/macaques [32] were found after approximately 9 years of infection. However, moderate/marked follicular hyperplasia was identified within the secondary lymphoid tissues, although without the subsequent follicular involution that occurs in pathogenic infections [33].

AIDS IN WILD CHIMPANZEES?

Our understanding of the virus–host relationship in SIVcpz infection was profoundly altered by a study of wild chimpanzees of the *schweinfurthii* subspecies that were habituated to the presence of humans [4]. Over a 9-year period, SIVcpz infection status of the chimpanzees in this community was determined by noninvasive analysis of fecal and urine samples. Most importantly, SIVcpz infection was found to be associated with an increased risk of mortality compared to uninfected animals, conservatively estimated at 9.8-fold (95% confidence interval, 2.8–34.3) and less conservatively estimated at 15.6-fold (4.7–51.7). The authors of this important report suggest that this increased mortality is lower than the increased risk of mortality caused by HIV-1 infection of humans. However, when this result is compared with studies of the effect of HIV-1 infection in African populations (carried out for a similar duration), SIVcpz has a strikingly similar effect on mortality, with most studies finding a 10- to 16-fold increased risk of mortality (reviewed in [34]). It therefore seems that SIVcpz infection of chimpanzees may have a pathogenicity similar to that of HIV-1 infection in humans.

Supporting the observation of increased mortality in SIVcpz-infected chimpanzees, SIVcpz infection was also associated with a significant loss of splenic $CD4^+$ T cells and apparent increases in collagen deposition in three SIVcpz-infected animals that died during the course of the study [4]. While two of these deaths were related to trauma, one animal died following a period of profound weakness and lethargy; at necropsy, severe cachexia and numerous cysts relating to nematode worm infection were found. In addition, this animal had a profound general loss of lymphocytes, including all T cells and B cells, and substantial disruptions to the secondary lymphoid architecture. The authors therefore postulated that this animal died from an AIDS-like disease. While this case is of considerable interest, it should be noted that loss of lymphocytes and lymphoid structure is also consistent with autolysis of tissues that may occur after spending up to 18h in tropical conditions, the time between the death of this animal and its recovery for postmortem study (all cadavers were recovered between 6 and 18h after death, but unfortunately the authors did not specify the timeframe for each case). In addition, a later study of the same cohort demonstrated that moderate or numerous nematode granulomas are found in all animals in this community older than the age of 1 year for which postmortem study was possible. This includes one animal in which this parasitic infection was deemed to be the cause of death and was associated with weakness, diarrhea, and low body mass without concurrent SIVcpz infection [35]. Such nematode infections and clinical consequences seem to be common in this community and are not necessarily indicative of immunosuppression.

Limited further evidence that SIVcpz causes damage to its host was provided by a study of a single chimpanzee of the *troglodytes* subspecies housed in a primate center in Cameroon [7]. This animal, naturally infected before rescue (aged approximately 1.5 years) suffered from frequent ill health, including somewhat unusual infectious ailments such as oral candidiasis and recurrent *Balantidium coli* infection. This animal failed to achieve normal weight and suffered from thrombocytopenia. Shortly before this report was published, this animal lost further weight and had a suppressed peripheral blood $CD4^+$ count [7]. However, it should be noted that the $CD4^+$ cells cell count later recovered (S. Redrobe, personal communication). In addition, this animal has since been euthanized, at the approximate age of 10 years, because of ill health and suffering caused by the presence of an extensive lymphoma infiltrating multiple organs of the body, including several lesions of the face, jaw, and brain (S. Redrobe, personal communication). It can be speculated that SIVcpz infection may have been involved in the development of such a lymphoma; similar lesions often are found in HIV-1-infected patients [36], and such a lymphoma could be considered as an AIDS-defining condition in humans [37]. However, the other evidence of immunosuppression in this animal must be interpreted with caution, given the potential for confounding effects of such a neoplasm.

A low peripheral blood $CD4^+$ count was also reported for a single additional animal (from the troglodytes species) naturally infected with SIVcpz, aged 8 years, housed in Gabon [8]. This animal is assumed to have acquired SIV vertically and thus has been infected for 8 years without developing symptoms. That the findings of $CD4^+$ T-cell loss in both wild and captive SIVcpz-infected chimpanzees in Africa is not unique to these apparently pathogenic infections should be considered. Loss of peripheral blood $CD4^+$ T cells has been reported in a cohort of HIV-1-infected animals; no animal progressed to disease after >25 years of infection [31].

COMPARISON OF SIVCPZ INFECTION WITH OTHER MODELS

Based on these findings, it seems that the host–virus relationship in SIV infection of chimpanzees is radically different to the nonpathogenic SIV infection of other "natural host" species, in particular the well-studied models of SIV infection of African green monkeys and sooty mangabeys. However, comparison with these models is extremely problematic. Keele and colleagues [4] compare their survival analysis of wild habituated chimpanzees in Tanzania with the survival of SIV-positive and SIV-negative sooty mangabeys housed at the Yerkes National Primate Center in Atlanta, Georgia; no significant effect of SIV infection on mortality was found in sooty mangabeys. Considering that major causes of morbidity and mortality in the cohort of wild chimpanzees are intragroup aggression, trauma, and nematode infection—all of which are likely to be ameliorated or absent in captive research primate populations because of the management of behavior and high levels of veterinary care—the validity of such a comparison can be questioned. Unfortunately, the effects of SIV infection on long-term survival is not available for any other species or circumstance, a significant hindrance to drawing strong conclusions about the effect of SIVcpz infection of chimpanzees compared with other species. While African green monkeys in the wild also have been studied in a recent study of exceptional scale, unfortunately no measured parameters are shared between this study and that of wild chimpanzees [17]. For wild chimpanzees, only a survival analysis and examination of postmortem tissues are available [4], whereas in this recent study of African green monkeys, only serum was available for analysis, again preventing useful comparison [17].

Nevertheless, the finding that SIVcpz has a significant impact on chimpanzee survival that is comparable with the effect of HIV-1 on humans strongly suggests that the virus–host relationship between SIVcpz and chimpanzees differs from that found in the generally nonpathogenic infections of African green monkeys and sooty mangabeys. Shortly before the publication of the study of wild chimpanzees, it was proposed that the African lentivirus lineage, including SIVcpz, might only be hundreds of years old, and it was, therefore, perhaps not surprising that a nonpathogenic relationship did not develop in all species [38]. However, as outlined above, more recent publications indicate that the SIV lineage is ancient and that SIVcpz has also existed for a considerable amount of time. Therefore, an additional explanation as to why a nonpathogenic relationship has not developed in SIVcpz infection of chimpanzees, but is apparently possible in other species, is required.

INCREASED PATHOGENICITY OF THE SIVcpz/HIV-1 LINEAGE?

To address the issue of how pathogenic SIVcpz infection is able to persist in the chimpanzee population, with a relatively high prevalence in some communities and complete absence in others, mathematical modeling of the chimpanzee population and SIVcpz infection dynamics was applied. Intriguingly, in a model of the effect of SIVcpz on wild chimpanzee communities, Rudicell and colleagues [39] demonstrated that intercommunity migration exhibited a protective effect on connected host communities, enabling an overall positive population growth of those populations, even in the presence of pathogenic SIVs. Migration also leads to a stark reduction in chimpanzee populations that remain infected by SIVcpz, potentially explaining why SIVcpz is geographically absent in certain regions and has a very low prevalence in various investigated colonies. With plausible transmission rates in mind, that the SIV infection is pathogenic is a key requirement of the model that explains the persistently low prevalence with viral extinction from some communities.

Notably, among investigated SIVs, this pattern of low prevalence and uneven distribution is only found in two investigated SIVs that harbor *vpu*—SIVgsn and SIVmus—two viruses that are candidates for supplying most of the 5′ half of the SIVcpz genome [18]. It could therefore be speculated that this viral lineage, or the presence of *vpu*, results in greater pathogenicity to the host in all species in which it is found. Further evidence that this is the case is provided by the fact that HIV-1 shows much greater pathogenicity in humans than HIV-2 [34], which lacks *vpu* and has its origins in the apparently nonpathogenic infection of sooty mangabeys.

The *vpu* of HIV-1 inhibits the host restriction factor tetherin, a factor that is upregulated by type one interferon and prevents the release of viral particles unless disrupted [40]. However, while the *vpu* of SIVcpz is able to inhibit human tetherin, it is unable to interact with chimpanzee tetherin; the *nef* gene carries out this function for SIVcpz [41]. Notably, this is not unique to SIVcpz; all studied SIVs are able to counteract tetherin, and this activity is found in *vpu*, *nef*, or *env* [42–44]. One apparently unique function of HIV-1 *vpu* that has been identified is its ability to downregulate NTB-A, a molecule that promotes the degranulation of natural killer cells that, when present on the cell surface, promotes natural killers cells to eliminate infected target cells [45]. It is unknown whether SIVcpz is able to perform the same function in chimpanzees. The role of *vpu* in SIVcpz, and the reason it was maintained during the recombination events leading to its genesis, therefore remains cryptic.

However, it has been found that the presence of *vpu* has a striking effect on another viral gene, *nef*. The ability of *nef* to downregulate the T-cell receptor (CD3) in infected cells is conserved, as determined in the case of a majority of investigated SIVs and HIV-2. However, the *nef* gene has lost this ability in all SIVs that also contain a full-length *vpu* (HIV-1, SIVcpz, SIVmon, SIVmus, SIVgsn) [21]. The loss of this function in HIV-1 has been proposed to play a role in the chronic immune activation that characterizes progressive infection. A protective effect of this function has been demonstrated in both humans infected with HIV-2 and sooty mangabeys infected with SIV; in both cases *nef* is able to downmodulate CD3, but the efficiency of this activity is correlated with $CD4^+$ cell count [46,47]. It also is associated with activation of the $CD4^+$ subset, although this could be explained as a homeostatic response to T-cell loss. A correlation with the activation of $CD8^+$ cells has not yet been demonstrated, suggesting that the generalized immune activation found in HIV-1-infected humans likely has a different cause. However, while strong conclusions cannot be drawn on the basis of one animal, it should be noted that SIVcpzANT, the viral strain infecting the chimpanzee Noah, contains a *nef* that is not able to downmodulate CD3 (21, denoted "Nok5"). Despite this, depleted $CD4^+$ T cells or high levels of activated $CD4^+$ T cells are not found in this animal after nearly a decade of SIVcpz infection [31,32].

FURTHER CONFLICTS IN THE HISTORY OF SIV INFECTION OF CHIMPANZEES

Natural hosts of SIV possess a number of features that seem to be the result of convergent evolution to prevent the development of AIDS, many of which are evident even in uninfected animals, but few of these have been investigated in chimpanzees. A key hindrance is that the vast majority of chimpanzees housed in primate research centers in the United States and Europe are of the *verus* subspecies. One example of a resistance mechanism shared among multiple species is a low frequency of $CD4^+$ T cells that express CCR5 [48]. Curiously, chimpanzees of the *verus* subspecies share this feature, despite having apparently no history of SIV infection [48]. Several possibilities must therefore be considered; that SIV infection was not the driving factor for this apparent convergent evolution of this phenomenon; that the *verus* subspecies has previously been infected by an SIV that is no longer extant; and/or there is a CCR5 polymorphism that fails to be recognized by the anti-CCR5 reagent.

Interestingly, analysis of a number of genes indicates a "selective sweep" in the history of the chimpanzee, possibly due to retroviral infection, before chimpanzee subspeciation. De Groot et al. [49] demonstrated that chimpanzees have experienced a marked reduction in the major histocompatability complex class I repertoire, found in all subspecies and even in the bonobo (*Pan paniscus*), indicating that this selection occurred before the divergence of the two *Pan* species. The authors suggest that this is the result of an epidemic of SIV, or a similar retrovirus, because it seems that the major histocompatability complex class I alleles preferentially selected in this sweep are able to target conserved HIV-1 epitopes in a manner similar to human "elite controllers" of HIV-1 [50]. This logic is supported by the fact that the majority of experimentally HIV-1-infected chimpanzees also have extremely low viral loads, suggesting effective immune control. However, this theory is somewhat undermined by the fact that in all studied cases of SIVcpz infection, active viral replication is found, with peripheral blood viral loads only marginally lower than those found in other natural host species (4–5 log copies/ml) [7,8,51], suggesting that effective immune suppression of viral replication is not a feature of SIVcpz infection of chimpanzees. In addition, as noted, SIVcpz infection is not found in two of the four subspecies or in the bonobo. However, further evidence of an epidemic, potentially during the same time frame, is found in a selective sweep of the 5′ cis-regulatory region of CCR5 [52], with the extant modal haplotype being most similar to a human haplotype that promotes the least CCR5 expression and has been correlated with delayed progression to disease in HIV-1-infected patients of certain demographics [53]. In addition, chimpanzee Apobec3DE, an antiretroviral restriction factor, underwent rapid evolution through positive selection during the same period [54] and, unlike human Apobec3DE, has activity against HIV-1 and SIV in African green monkeys (although this in turn is countered by the viral gene *vif*).

While such an epidemic would predate current estimates of the age of the SIV lineage based on molecular clock estimates calibrated using biogeographical features [17,22], it bears restating that the presence of an endogenous lentivirus found in Malagasy prosimians suggests that SIVs have been circulating in Africa for more than 14 million years [24]. For as long as SIVs have been circulating in African primates, chimpanzees are likely to have been exposed through their predation upon monkeys. Nevertheless, it will likely never be possible to definitively determine whether SIV, a related virus, or a completely different infectious agent is responsible for these genetic phenomena shared by all chimpanzee subspecies.

PERSPECTIVES FOR FUTURE WORK AND CONCLUSIONS

Nearly 25 years have passed since the discovery of SIVcpz [55], and while the study of this virus, and the chimpanzee host, has been fruitful, many questions remain unanswered. Crucially, SIVcpz from the *troglodytes* subspecies has been identified as the source of pandemic HIV-1 and has been found to be the result of cross-species transmission and recombination of SIVs from prey monkeys. However, key elements of the virus–host relationship remain uncertain. First, comparisons with other well-described "natural host" models are hindered by a lack of relevant comparisons. Because of changes in ethical legislation, it is foreseeable that future insights on this animal model will have two possible sources. The first are retrospective follow-up studies of chimpanzees naturally and artificially infected with SIVcpz and housed in primate centers in the United States and Europe [56] would allow comparisons with the majority of investigations of the virus–host relationship in sooty mangabeys and African green monkeys in similar environments and demonstrate to what extent SIVcpz infection of chimpanzees mirrors that of other natural hosts.

A complementary approach would be the further study of other natural hosts in a wild environment using methods that are comparable to those used to study wild chimpanzees. This will allow the determination of whether the pathogenic outcome of SIVcpz of wild chimpanzees is truly unique to this species. Recent work by Ma et al. [17,57] has begun to address this question through studies of wild African green monkeys, with few significant findings associated with SIV infection. By contrast, our own recent study of mandrills infected with SIVmnd-1 in a "semi-wild" environment revealed a number of significant immune perturbations [58], including progressive loss of memory $CD4^+$ T cells, suggesting that SIV infection in the wild may have detrimental consequences in this species.

Finally, while the presence of the *vpu* gene may have influenced the virus–host relationship in SIVcpz infection, there are currently no described functions for *vpu* of SIVcpz that are not redundant with other viral genes. It seems likely that the function(s) for which it was preserved in the chimeric virus (apparently at the cost of the loss of the *vpx* gene) has not yet been elucidated. Given the evidence that this gene plays an important role in HIV-1 pathogenesis and seems to influence the virus–host relationship in all SIVs in which it is present, it is crucial that the role this gene plays in chimpanzee and other species is determined. In addition, further examination of the virus–host relationship in the SIVs harboring *vpu*, if possible, would greatly improve our understanding of the effect of this viral factor.

The study of SIVcpz has been a source of both optimism and pessimism in our attempts to combat HIV-1; our understanding of the outcome and age of SIVcpz infection have varied wildly. At first, the apparently nonpathogenic outcome, presumed to be the result of a long period of coevolution, suggested that it was possible for our closest relative to coexist "peacefully" with the virus that is the source of the HIV-1 pandemic. The discovery that SIVcpz causes AIDS in wild chimpanzees then was explained by a contemporaneous report that SIVcpz was only hundreds of years old—that they too were recent victims of the lentivirus epidemic and thus could not be expected to be resistant to disease. However, it now seems that SIVcpz is ancient, as initially thought, and yet a pathogenic outcome is observed. Specific features of the SIVcpz/HIV-1 lineage may have prevented a perfect virus–host relationship from developing in chimpanzees, a point that cannot be a cause for hope in our attempts to apply our understanding of natural host species to HIV-1 infection of humans.

Acknowledgments

We are grateful to Dr. Joel Wertheim for providing the sequence alignments used to generate Figure 5.3.

References

[1] Keele BF, Van Heuverswyn F, Li Y, Bailes E, Takehisa J, Santiago ML, et al. Chimpanzee reservoirs of pandemic and nonpandemic HIV-1. Science 2006;313(5786):523–6.
[2] Gao F, Bailes E, Robertson DL, Chen Y, Rodenburg CM, Michael SF, et al. Origin of HIV-1 in the chimpanzee *Pan troglodytes troglodytes*. Nature 1999;397(6718):436–41.
[3] Bailes E, Gao F, Bibollet-Ruche F, Courgnaud V, Peeters M, Marx PA, et al. Hybrid origin of SIV in chimpanzees. Science 2003;300(5626):1713.
[4] Keele BF, Jones JH, Terio KA, Estes JD, Rudicell RS, Wilson ML, et al. Increased mortality and AIDS-like immunopathology in wild chimpanzees infected with SIVcpz. Nature 2009;460(7254):515–9.
[5] Weiss RA, Heeney JL. Infectious diseases: an ill wind for wild chimps? Nature 2009;460(7254):470–1.
[6] Gilden RV, Arthur LO, Robey WG, Kelliher JC, Graham CE, Fischinger PJ. HTLV-III antibody in a breeding chimpanzee not experimentally exposed to the virus. Lancet 1986;1(8482):678–9.
[7] Etienne L, Nerrienet E, LeBreton M, Bibila GT, Foupouapouognigni Y, Rousset D, et al. Characterization of a new simian immunodeficiency virus strain in a naturally infected *Pan troglodytes troglodytes* chimpanzee with AIDS related symptoms. Retrovirology 2011;8:4.

[8] Souquiere S, Makuwa M, Salle B, Kazanji M. New strain of simian immunodeficiency virus identified in wild-born chimpanzees from central Africa. PLoS One 2012;7(9):e44298.
[9] Van Heuverswyn F, Li Y, Bailes E, Neel C, Lafay B, Keele BF, et al. Genetic diversity and phylogeographic clustering of SIVcpzPtt in wild chimpanzees in Cameroon. Virology 2007;368(1):155–71.
[10] Li Y, Ndjango JB, Learn GH, Ramirez MA, Keele BF, Bibollet-Ruche F, et al. Eastern chimpanzees, but not bonobos, represent a simian immunodeficiency virus reservoir. J Virol 2012;86(19):10776–91.
[11] Sharp PM, Hahn BH. Origins of HIV and the AIDS pandemic. Cold Spring Harb Perspect Med 2012;1(1):a006841.
[12] Switzer WM, Parekh B, Shanmugam V, Bhullar V, Phillips S, Ely JJ, et al. The epidemiology of simian immunodeficiency virus infection in a large number of wild- and captive-born chimpanzees: evidence for a recent introduction following chimpanzee divergence. AIDS Res Hum Retroviruses 2005;21(5):335–42.
[13] Peeters M, Fransen K, Delaporte E, Van den Haesevelde M, Gershy-Damet GM, Kestens L, et al. Isolation and characterization of a new chimpanzee lentivirus (simian immunodeficiency virus isolate cpz-ant) from a wild-captured chimpanzee. AIDS 1992;6(5):447–51.
[14] Rudicell RS, Piel AK, Stewart F, Moore DL, Learn GH, Li Y, et al. High prevalence of simian immunodeficiency virus infection in a community of savanna chimpanzees. J Virol 2011;85(19):9918–28.
[15] Santiago ML, Rodenburg CM, Kamenya S, Bibollet-Ruche F, Gao F, Bailes E, et al. SIVcpz in wild chimpanzees. Science 2002;295(5554):465.
[16] Santiago ML, Range F, Keele BF, Li Y, Bailes E, Bibollet-Ruche F, et al. Simian immunodeficiency virus infection in free-ranging sooty mangabeys (*Cercocebus atys atys*) from the Tai Forest, Cote d'Ivoire: implications for the origin of epidemic human immunodeficiency virus type 2. J Virol 2005;79(19):12515–27.
[17] Ma D, Jasinska A, Kristoff J, Grobler JP, Turner T, Jung Y, et al. SIVagm infection in wild African green monkeys from South Africa: epidemiology, natural history, and evolutionary considerations. PLoS Pathog 2013;9(1):e1003011.
[18] Aghokeng AF, Ayouba A, Mpoudi-Ngole E, Loul S, Liegeois F, Delaporte E, et al. Extensive survey on the prevalence and genetic diversity of SIVs in primate bushmeat provides insights into risks for potential new cross-species transmissions. Infect Genet Evol 2009;10(3):386–96.
[19] Beer BE, Foley BT, Kuiken CL, Tooze Z, Goeken RM, Brown CR, et al. Characterization of novel simian immunodeficiency viruses from red-capped mangabeys from Nigeria (SIVrcmNG409 and -NG411). J Virol 2001;75(24):12014–27.
[20] Courgnaud V, Salemi M, Pourrut X, Mpoudi-Ngole E, Abela B, Auzel P, et al. Characterization of a novel simian immunodeficiency virus with a vpu gene from greater spot-nosed monkeys (*Cercopithecus nictitans*) provides new insights into simian/human immunodeficiency virus phylogeny. J Virol 2002;76(16):8298–309.
[21] Schindler M, Munch J, Kutsch O, Li H, Santiago ML, Bibollet-Ruche F, et al. Nef-mediated suppression of T cell activation was lost in a lentiviral lineage that gave rise to HIV-1. Cell 2006;125(6):1055–67.
[22] Worobey M, Telfer P, Souquiere S, Hunter M, Coleman CA, Metzger MJ, et al. Island biogeography reveals the deep history of SIV. Science 2010;329(5998):1487.
[23] Gilbert C, Maxfield DG, Goodman SM, Feschotte C. Parallel germline infiltration of a lentivirus in two Malagasy lemurs. PLoS Genet 2009;5(3):e1000425.
[24] Gifford RJ, Katzourakis A, Tristem M, Pybus OG, Winters M, Shafer RW. A transitional endogenous lentivirus from the genome of a basal primate and implications for lentivirus evolution. Proc Natl Acad Sci USA 2008;105(51):20362–7.
[25] Sharp PM, Bailes E, Gao F, Beer BE, Hirsch VM, Hahn BH. Origins and evolution of AIDS viruses: estimating the time-scale. Biochem Soc Trans 2000;28(2):275–82.

[26] Bjork A, Liu W, Wertheim JO, Hahn BH, Worobey M. Evolutionary history of chimpanzees inferred from complete mitochondrial genomes. Mol Biol Evol 2011;28(1):615–23.
[27] Gonder MK, Locatelli S, Ghobrial L, Mitchell MW, Kujawski JT, Lankester FJ, et al. Evidence from Cameroon reveals differences in the genetic structure and histories of chimpanzee populations. Proc Natl Acad Sci USA 2011;108(12):4766–71.
[28] Committee on Long-Term Care of Chimpanzees IfLAR, Commission on Life Sciences, National Research Council. Chimpanzees in research: strategies for their ethical care, management, and use. Washington (DC): National Academy Press; 1997.
[29] Novembre FJ, Saucier M, Anderson DC, Klumpp SA, O'Neil SP, Brown II CR, et al. Development of AIDS in a chimpanzee infected with human immunodeficiency virus type 1. J Virol 1997;71(5):4086–91.
[30] Juompan LY, Hutchinson K, Montefiori DC, Nidtha S, Villinger F, Novembre FJ. Analysis of the immune responses in chimpanzees infected with HIV type 1 isolates. AIDS Res Hum Retroviruses 2008;24(4):573–86.
[31] Kestens L, Vingerhoets J, Peeters M, Vanham G, Vereecken C, Penne G, et al. Phenotypic and functional parameters of cellular immunity in a chimpanzee with a naturally acquired simian immunodeficiency virus infection. J Infect Dis 1995;172(4):957–63.
[32] Gougeon ML, Lecoeur H, Boudet F, Ledru E, Marzabal S, Boullier S, et al. Lack of chronic immune activation in HIV-infected chimpanzees correlates with the resistance of T cells to Fas/Apo-1 (CD95)-induced apoptosis and preservation of a T helper 1 phenotype. J Immunol 1997;158(6):2964–76.
[33] Koopman G, Haaksma AG, ten Velden J, Hack CE, Heeney JL. The relative resistance of HIV type 1-infected chimpanzees to AIDS correlates with the maintenance of follicular architecture and the absence of infiltration by $CD8^+$ cytotoxic T lymphocytes. AIDS Res Hum Retroviruses 1999;15(4):365–73.
[34] Jaffar S, Grant AD, Whitworth J, Smith PG, Whittle H. The natural history of HIV-1 and HIV-2 infections in adults in Africa: a literature review. Bull World Health Organ 2004;82(6):462–9.
[35] Terio KA, Kinsel MJ, Raphael J, Mlengeya T, Lipende I, Kirchhoff CA, et al. Pathologic lesions in chimpanzees (*Pan trogylodytes schweinfurthii*) from Gombe National Park, Tanzania, 2004–2010. J Zoo Wildl Med 2011;42(4):597–607.
[36] IARC Working Group on the Evaluation of Carcinogenic Risks to Humans. Human immunodeficiency viruses and human T-cell Lymphotropic viruses. Lyon, France, 1-18 June 1996. IARC Monogr Eval Carcinog Risks Hum 1996;67:1–424.
[37] 1993 revised classification system for HIV infection and expanded surveillance case definition for AIDS among adolescents and adults. MMWR Recomm Rep 1992;41(RR-17):1–19.
[38] Wertheim JO, Worobey M. Dating the age of the SIV lineages that gave rise to HIV-1 and HIV-2. PLoS Comput Biol 2009;5(5):e1000377.
[39] Rudicell RS, Holland Jones J, Wroblewski EE, Learn GH, Li Y, Robertson JD, et al. Impact of simian immunodeficiency virus infection on chimpanzee population dynamics. PLoS Pathog 2010;6(9):e1001116.
[40] Neil SJ, Zang T, Bieniasz PD. Tetherin inhibits retrovirus release and is antagonized by HIV-1 Vpu. Nature 2008;451(7177):425–30.
[41] Sauter D, Schindler M, Specht A, Landford WN, Munch J, Kim KA, et al. Tetherin-driven adaptation of Vpu and Nef function and the evolution of pandemic and nonpandemic HIV-1 strains. Cell Host Microbe 2009;6(5):409–21.
[42] Zhang F, Wilson SJ, Landford WC, Virgen B, Gregory D, Johnson MC, et al. Nef proteins from simian immunodeficiency viruses are tetherin antagonists. Cell Host Microbe 2009;6(1):54–67.

[43] Gupta RK, Mlcochova P, Pelchen-Matthews A, Petit SJ, Mattiuzzo G, Pillay D, et al. Simian immunodeficiency virus envelope glycoprotein counteracts tetherin/BST-2/CD317 by intracellular sequestration. Proc Natl Acad Sci USA 2009;106(49):20889–94.

[44] Le Tortorec A, Neil SJ. Antagonism to and intracellular sequestration of human tetherin by the human immunodeficiency virus type 2 envelope glycoprotein. J Virol 2009;83(22):11966–78.

[45] Shah AH, Sowrirajan B, Davis ZB, Ward JP, Campbell EM, Planelles V, et al. Degranulation of natural killer cells following interaction with HIV-1-infected cells is hindered by downmodulation of NTB-A by Vpu. Cell Host Microbe 2010;8(5):397–409.

[46] Schindler M, Schmokel J, Specht A, Li H, Munch J, Khalid M, et al. Inefficient Nef-mediated downmodulation of CD3 and MHC-I correlates with loss of $CD4^{+}T$ cells in natural SIV infection. PLoS Pathog 2008;4(7):e1000107.

[47] Khalid M, Yu H, Sauter D, Usmani SM, Schmokel J, Feldman J, et al. Efficient Nef-mediated downmodulation of TCR-CD3 and CD28 is associated with high $CD4^{+}$ T cell counts in viremic HIV-2 infection. J Virol 2012;86(9):4906–20.

[48] Pandrea I, Apetrei C, Gordon S, Barbercheck J, Dufour J, Bohm R, et al. Paucity of $CD4^{+}CCR5^{+}$ T cells is a typical feature of natural SIV hosts. Blood 2007;109(3):1069–76.

[49] de Groot NG, Otting N, Doxiadis GG, Balla-Jhagjhoorsingh SS, Heeney JL, van Rood JJ, et al. Evidence for an ancient selective sweep in the MHC class I gene repertoire of chimpanzees. Proc Natl Acad Sci USA 2002;99(18):11748–53.

[50] Balla-Jhagjhoorsingh SS, Koopman G, Mooij P, Haaksma TG, Teeuwsen VJ, Bontrop RE, et al. Conserved CTL epitopes shared between HIV-infected human long-term survivors and chimpanzees. J Immunol 1999;162(4):2308–14.

[51] ten Haaft P, Murthy K, Salas M, McClure H, Dubbes R, Koornstra W, et al. Differences in early virus loads with different phenotypic variants of HIV-1 and SIV(cpz) in chimpanzees. AIDS 2001;15(16):2085–92.

[52] Wooding S, Stone AC, Dunn DM, Mummidi S, Jorde LB, Weiss RK, et al. Contrasting effects of natural selection on human and chimpanzee CC chemokine receptor 5. Am J Hum Genet 2005;76(2):291–301.

[53] Gonzalez E, Bamshad M, Sato N, Mummidi S, Dhanda R, Catano G, et al. Race-specific HIV-1 disease-modifying effects associated with CCR5 haplotypes. Proc Natl Acad Sci USA 1999;96(21):12004–9.

[54] Duggal NK, Malik HS, Emerman M. The breadth of antiviral activity of Apobec3DE in chimpanzees has been driven by positive selection. J Virol 2011;85(21):11361–71.

[55] Peeters M, Honore C, Huet T, Bedjabaga L, Ossari S, Bussi P, et al. Isolation and partial characterization of an HIV-related virus occurring naturally in chimpanzees in Gabon. AIDS 1989;3(10):625–30.

[56] Heeney JL, Rutjens E, Verschoor EJ, Niphuis H, ten Haaft P, Rouse S, et al. Transmission of simian immunodeficiency virus SIVcpz and the evolution of infection in the presence and absence of concurrent human immunodeficiency virus type 1 infection in chimpanzees. J Virol 2006;80(14):7208–18.

[57] Ma D, Jasinska AJ, Feyertag F, Wijewardana V, Kristoff J, He T, et al. Factors associated with SIV transmission in a natural African nonhuman primate host in the wild. J Virol 2014;88(10):5687–705.

[58] Greenwood EJ, Schmidt F, Liegeois F, Kondova I, Herbert A, Ngoubangoye B, et al. Loss of memory $CD4^{+}$ T-cells in semi-wild mandrills (*Mandrillus sphinx*) naturally infected with species-specific simian immunodeficiency virus SIVmnd-1. J Gen Virol 2014;95(Pt. 1):201–12.

CHAPTER

6

Pathology of Sooty Mangabeys (*Cercocebus atys*)

Cynthia L. Courtney[1,2], *Anapatricia Garcia*[1,2], *Prachi Sharma*[1,2], *Daniel C. Anderson*[1], *Joyce K. Cohen*[1,3], *Elizabeth Strobert*[1], *Francois Villinger*[1,2], *Sanjeev Gumber*[1,2]

[1]Yerkes National Primate Research Center, Emory University, Atlanta, GA, USA; [2]Department of Pathology and Laboratory Medicine, Emory University School of Medicine, Atlanta, GA, USA; [3]Department of Psychiatry and Behavioral Sciences, Emory University School of Medicine, Atlanta, GA, USA

INTRODUCTION

Among the multiple species of African non-human primates (NHPs) that serve as natural hosts of SIV, the sooty mangabey monkeys (*Cercocebus atys*; SMs) and African green monkeys (*Cholorcebus sabeus*; AGMs) have been most extensively studied. SMs are native to West Africa and were initially imported for the purpose of research on leprosy, since they represent the only model of human lepromatous leprosy aside from armadillos [1–3]. Of relevance to this chapter, sooty mangabeys, although abundant in West Africa, are nevertheless considered a protected species by the US Fish and Wildlife Service (CITES Appendix II) and listed as a vulnerable species according to the IUCN [4,5], with the practical outcome that only non-invasive research is permitted, resulting in all captive SMs living out their lives, or should their clinical condition dictate it, being humanely euthanized. Studies testing the transmission of primate-adapted *Mycobacterium leprae* to macaque species led to the accidental transmission of a syndrome of immunodeficiency and the subsequent isolation of simian immunodeficiency virus (SIVsmm) from healthy SMs, first described at the Tulane

Natural Hosts of SIV.
http://dx.doi.org/10.1016/B978-0-12-404734-1.00006-1

National Primate Research Center [6,7]. In 1968, a captive group of 27 SMs was established at the Yerkes National Primate Research Center (YNPRC) in Atlanta, Georgia where it has expanded and continues to breed and remain under study [8,9]. This colony is the only captive breeding colony of SMs in the world, in which the presence of natural SIV infection was detected in 1986, although earlier serum samples (1981) confirmed the presence of antibodies to the virus in 82% of the colony [10].

Sooty mangabeys are of particular interest among natural simian hosts of SIV, since the mangabey isolate (SIVsmm) is considered to have spawned the closely related macaque SIV isolates (SIVmac, E660, mne, and derivatives), and to be the most likely source of transmission to humans and thus the origin of HIV-2 isolates infecting humans [11,12], via close contact and interactions with non-natural hosts. In fact, zoonotic transmission is likely still occurring in rural West Africa where humans and SMs intersect [13]. Similar to other African NHP natural hosts of SIV, the immune system of SMs has been the subject of intense investigation in comparison with disease-susceptible Asian macaque species, which as a whole develop disease very similar to HIV infections in humans [14]. One notable difference of African natural hosts from non-natural macaque hosts of SIV is the rapid resolution of proinflammatory responses upon acute infection with SIV and lack of chronic immune activation, suggesting more potent antiinflammatory control in these species [13–17]. Other differences in the innate immune system shown over the years between macaques, humans, and SMs suggest a definite tendency toward predominant Th2-biased T-cell responses [18,19], which correlates with their ability to develop lepromatous leprosy [1] as well as an inability to tolerize T cells upon signal one alone [20,21], suggesting lower dependence on second activating signals for T-cell activation. Finally, natural killer cells and natural killer T cells (NKs and NKTs), as part of the innate immune system, appear to be both present in markedly higher levels and exert a higher level of effector immunity in SMs compared to macaques [22–25].

While the respective importance of such species differences in the development of natural disease remains to be fully elucidated, they may in fact be related to some of the pathological manifestations described in this chapter. Thus the studies have focused to a large extent more heavily on comparative analysis of SIV+ and SIV− SMs than the other natural host species, primarily due to the unique historical perspective of this species and the availability of complete clinical follow-up available to the authors for this species. Thus, as explained below, beta islet destruction and amyloidosis comprise a large part of the disease in animals that consume a highly controlled diet, and thus the etiology of such disease manifestations has not been fully understood. The development of amyloidosis in particular contrasts with the mangabey's ability to limit immune inflammation induced by SIV, suggesting ongoing tissue remodeling that may be dissociated from the SIV infection but secondary to other ongoing immune processes.

A comprehensive review was performed of SMs ($n = 453$) that either died naturally or required euthanasia based on their clinical status at the YNPRC from 1986 to 2013. Table 6.1 presents the most common causes of mortality by frequency of occurrence and age groups at necropsy. The

TABLE 6.1 Most Common Causes of Mortality by Frequency of Occurrence and Age Groups in Sooty Mangabeys at YNPRC over 27-Year Period

Common Causes of Mortality	Neonate[a]	Infant/ Juvenile[b]	Adolescent[c]	Adult[d]	Elderly[e]	Total
Infectious	17	22	22	20	10	91
Neonatal death/ stillbirth	74	2				76
Trauma	28	15	20	7	5	75
Dehydration	21	14	2	1		38
Diabetes				11	27	38
Neoplasia			2	12	14	28
Diarrhea chronic		11	5	8	3	27
Other		4	4	9	7	24
Female reproductive disorders			1	5	10	16
Cardiovascular diseases				5	6	11
Intestinal disorders		1	4	2	3	10
Neurologic disorders		1	2	4	1	8
Degenerative disorders			1	2	4	7
Renal failure					2	2
Congenital anomalies	1	1				2
Total	**141**	**71**	**63**	**86**	**92**	**453**

[a] *<3 months of age.*
[b] *>0–3 years.*
[c] *3–10 years.*
[d] *>10–18 years.*
[e] *>18 years.*

host of pathologies reported in SMs aside from trauma appears to shift from infectious to degenerative conditions with advancing age; this is not very different from humans and rhesus monkeys [26]. The oldest sooty mangabey at YNPRC was 32 years old when she was euthanized for endometrial carcinoma, indicative of the long natural lifespan of this species in captivity.

TAXONOMY

The taxonomic classification of *C. atys* places them within the family of Cercopithecidae and subfamily of Cercopithecinae, with subspecies of *C. a. atys* and *C. a. lunulatus*. Common names for *C. a. atys* include sooty mangabey and white-crowned mangabey; and for *C. a. lunulatus*, white-collared mangabey and white-naped mangabey. While a subclassification of *C. atys* under *Cercocebus torquatus* has been proposed, consistent differences and a geographic separation between *C. torquatus* and *C. atys* populations lead to *C. atys* being considered a separate species, with the white-naped mangabey (*C. a. lunulatus*) as a subspecies [27,28]. Morphologically, mangabeys are diphyletic and classified into two genera: *Cercocebus* and *Lophocebus*. On the basis of skeletal and dental morphology, mandrills and drills are closely related to *Cercocebus* mangabeys (including SMs) while baboons and geladas are more closely related to *Lophocebus* mangabeys [28]. Similar findings were observed with Cercopithecoid phylogenies based on gamma-globin DNA sequences [29].

CLINICAL PATHOLOGY

A study was conducted on clinical pathology data collected at the YNPRC over a 20-year period [30]. Fifteen hematology analytes from 136 females and 96 males and 24 serum chemistry analytes from 57 females and 26 males were studied. SMs averages for several analytes differed markedly from accepted norms for the rhesus macaque, a laboratory standard for Old World primates [30]. For example, several parameters associated with erythrocytes and hemoglobin were higher in mangabeys, while platelet counts were markedly lower than in rhesus monkeys. SMs also differed substantially from rhesus macaque in several serum chemistry parameters, including higher average gamma-glutamyltransferase levels, and lower urea nitrogen levels. These differences were more striking in female than male SMs. Nevertheless, sex differences in hematology analytes were typical of primates, and included higher erythrocyte-associated values in males than females. Changes were observed with age in albumin, globulin, albumin/globulin ratio, calcium, and creatinine in sooty mangabey samples similar to those reported for the rhesus macaque [30].

BACTERIAL, MYCOTIC, AND PARASITIC INFECTIONS

Chronic Diarrhea

Chronic, recurring diarrhea that was nonresponsive to medical treatments was a common cause of morbidity and/or mortality in juvenile to adolescent SMs (1–4 years). Organisms cultured from feces included *Yersinia enterocolitica, pseudotuberculosis,* and/or *kristensenii; Shigella flexneri* type 4; and *Campylobacter coli* and/or *jejuni.* The most common finding associated with such chronic and recurring diarrhea cases was colitis, without small bowel involvement. Microscopically, colonic changes consisted of lymphoplasmacytic inflammation with crypt abscesses. The high incidence rate of recurring diarrhea in infants and juveniles is, however, not specific to mangabeys but a leading cause of animal morbidity requiring veterinary care in colonies of NHPs [31,32]. Colitis was also seen commonly in older SMs.

Streptococcus pneumoniae Meningitis

Streptococcus pneumoniae is a significant etiological agent for bacterial meningitis in NHPs. It is a disease of low morbidity and high mortality [33]. Suppurative meningitis due to *S. pneumoniae* has occurred sporadically in our SMs colony, with the exception of a cluster of 6 cases over a 3-month period in 2002. Over a series of 12 cases, the youngest death occurred in a one-year-old male and the oldest death in a 25-year-old female, with a median age of 10 years. The most common clinical findings include abnormal behavior, ataxia, decreased appetite, lethargy, and sudden death. Clinicopathologic changes include neutrophilia (inconsistently), cloudy cerebrospinal fluid (CSF), and positive cultures on CSF or blood. Gross findings in the brain included accumulation of a gray-yellowish exudate within sulci, congestion of the vasculature, or meningeal hemorrhage (Figure 6.1). Acute peritonitis with abscesses was also found in a 16-year-old female (SIV+) with streptococcal meningitis. Microscopic changes seen in most animals with this infection included extensive suppurative meningitis to meningoencephalitis with dense neutrophilic infiltrates in the meninges as well as neutrophils often seen perivascularly in the neuropil with numerous intralesional Gram-positive cocci. Suppurative inflammation was also seen in the lateral and third ventricles of the brain in one case. In addition, focal areas of hemorrhage were occasionally associated with acute inflammation in the cerebellum and occipital cortex. Congestion of the meningeal

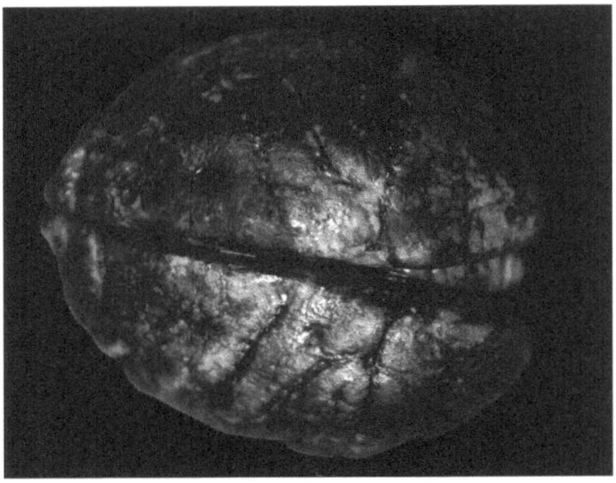

FIGURE 6.1 The meninges are diffusely congested and contain small amounts of pale purulent exudate due to streptococcal meningitis.

vasculature of the cerebrum/cerebellum and the presence of white, yellow, or gray-yellow purulent exudate filling the subarachnoid space has been reported in other NHP species [34,35]. Similarly, meningitis characterized by mild to severe fibrino-suppurative inflammation extending into the cerebral and cerebellar cortices has also been reported previously in NHPs [34–36]. Other common lesions due to *S. pneumoniae* infection in macaques include severe fibrino-purulent pneumonia, peritonitis, septicemia, and thromboembolic lymphadenitis [35,37].

Septicemia

Infections of SMs younger than 2 years of age, and occasionally older animals, included septicemia and/or pneumonia due to *Escherichia coli, Listeria monocytogenes, Klebsiella pneumoniae, Proteus vulgaris,* and/or *Campylobacter fetus subspecies fetus*. These organisms were isolated from the liver, blood, and/or spleen at the time of necropsy in over 25 cases. These findings were not specific to mangabeys but found in other NHP species as well [18]. In adult African green monkeys, multisystemic abscesses were detected due to *K. pneumoniae* infection [38].

Nontuberculous Mycobacterial (NTM) Infection

Colonization with NTM appears to be more common in New World primates usually in the absence of clinical disease [39]. The most commonly detected NTM isolates in New World primates were *Mycobacterium chelonae, Mycobacterium intracellulare, Mycobacterium nonchromogenicum, and Mycobacterium fortuitum* [39]. In our SM colony, only two mycobacterial cases have so far been diagnosed. The first case included peritoneal mycobacteriosis due to *M. fortuitum* in a 17-year-old female (SIV+) that presented with lethargy, watery diarrhea, and distended abdomen due to a firm, large peritoneal mass. Microscopically, chronic pyogranulomatous peritonitis, serositis, and granulomatous mesenteric lymphadenitis with intralesional acid-fast bacteria were seen. The second case was diagnosed as disseminated mycobacteriosis affecting the axillary and inguinal skin, oral mucosa, inguinal, axillary, cervical, periaortic lymph nodes, lingual tonsil, and liver due to *Mycobacterium avium* in a 14-year-old female (SIV+) that presented with chronic dermatitis. Microscopically, granulomatous inflammation was present in all affected organs. In contrast, other Old World primate species such as rhesus macaques are most commonly infected with disseminated *M. avium* complex during SIV infection [40]. It is of interest in this regard that among the various NHP species that were tested, only the SIV+ SMs were found to be susceptible to *M. leprae*.

Zygomycosis

Zygomycosis is a disseminated fungal infection that has been reported in NHPs [41–43]. Zygomycosis is caused by opportunistic fungi in the class Zygomycetes, which includes two orders: Mucorales and Entomophthorales. The order Mucorales has been implicated in NHP cases and includes multiple organisms such as Absidia and Mucor [44]. Organisms in the order Entomophthorales are less commonly observed, but in humans cause cutaneous infections [45]. Infection is usually acquired through trauma or inhalation, with several clinical forms of zygomycosis described in humans. A cutaneous form of zygomycosis has been reported in rhesus macaques following trauma [46]. Spontaneous zygomycosis was diagnosed in nine SMs with an age range of 4 to 24 years. Presenting clinical signs in these SMs were highly variable and included lethargy, anemia, weight loss, diarrhea, exophthalmos, and anorexia. Commonly observed gross lesions included gastric ulceration, peritonitis, periorbital granulomas, orchitis, hepatitis, pneumonia, nephritis, and lymphadenopathy. YNPRC reported a case of ruptured aortic aneurysm with zygomycetes infection (Figure 6.2) [43], and PCR analysis identified the organism as *Basidiobolus* spp [43]. Microscopically, granulomatous inflammation intermixed with nonseptate branching hyphae consistent with zygomycetes were present in all the cases. Immunosuppression caused by preexisting conditions such as diabetes, malnutrition, viral infection, and iatrogenic immunosuppression have been suggested as predisposing factors in humans [33,47]. Five of the animals in this report were naturally infected with SIV and four were seronegative for SIV. Although diabetes is commonly seen in older SMs, this condition did not appear to contribute to the susceptibility to this fungal infection in the YNPRC colony.

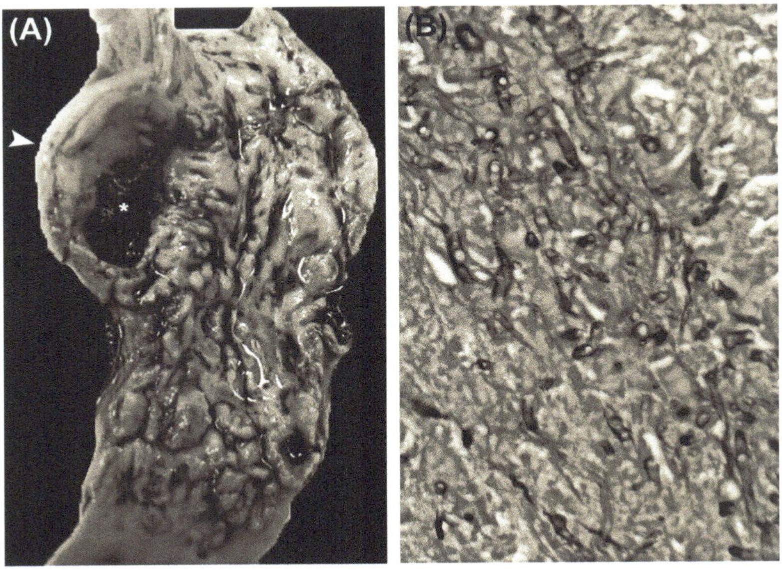

FIGURE 6.2 (A) The aorta contains an aneurysm (arrowhead), and the rupture site is covered by clotted blood (asterisk). The intimal surface has multiple, firm, pale nodules 0.25–0.5 cm in diameter. (B) Mats of fungal hyphae consistent with zygomycetes in an aortic intimal nodule. Gomori methenamine silver stain. ×600.

Amebiasis

Amebic lesions included necrohemorrhagic gastritis, hepatitis, and/or colitis with septicemia in seven animals: three cases occurred in animals <2 years of age (SIV−) and the remaining four cases occurred in animals >16 years (SIV+) of age.

Other common NHP parasites have been observed in case workups and include *Gongylonema* spp occasionally found in esophageal sections, intestinal strongyloidiasis, and colonic *Balantidium coli* infections; but in general the pathology occasioned by these parasites was not clinical and not the cause for death/euthanasia.

VIRAL DISEASES

Simian Immunodeficiency Virus

Similar to several other African NHPs, SMs have been known to harbor SIV as a naturally propagated infection in the wild and in captivity. Most animals have been shown to become infected as adults, suggesting primarily sexual transmission, although occasional juvenile monkeys have been identified as productively infected with SIV [48] and SIVsmm has been detected in breast milk (Chahroudi et al., submitted). While the virus has been shown to replicate in this species to levels similar to SIV in disease-susceptible macaques, or HIV in humans, SMs do not suffer clinical sequelae from this infection and chronic viral replication [14,49,50]. Only in very rare instances were symptoms described corresponding to disease progression [51], although several of the symptoms on which the diagnosis had been based, such as CD4 decline, are questionable based on the observation that mangabeys infected with SIV that has expanded its tropism to CXCR4 appear completely devoid of CD4+ T cells, without evidence of immune suppression [13]. Why mangabeys and other African NHPs are able to tolerate high SIV viremia remains incompletely understood, but the infection is associated with an absence of chronic immune inflammation in this species. In addition, recent data have suggested that while SIV infects primarily effector memory T cells but spares central memory T cells in this disease-resistant natural host, SIV predominantly infects central memory T cells in the disease-susceptible hosts like rhesus macaques. Thus, SIV potentially predominantly infects the fully differentiated and/or dispensable components of the immune system [16,52], which may contribute to the disease resistance in this species. Although monocyte-derived macrophages from SMs have been shown to support SIV replication in vitro, antibody-mediated depletion of CD4+ T cells in SIV-infected SMs demonstrated that the virus was dependent on the presence of activated CD4+ T cells for its replication [53], while rhesus macaques depleted of CD4+ T cells during acute infection readily replicated SIV in monocyte/macrophages [54], a finding that has being tied to pathogenesis [55]. However, a comprehensive retrospective analysis of the YNPRC SM colony mortality has led to the conclusion that SIV infection does not appear to shorten the mangabeys' lifespan, unlike a more recent report of HIV infection of chimpanzees in the wild [56].

Simian T-Cell Leukemia Virus

Simian T-lymphotropic virus-1 (STLV-1) serology is routinely conducted on samples from SMs at the YNPRC, and in one such study approximately 33% of the SMs were noted to be seropositive [57]. STLV-1 rarely induces pathology in SMs, and when present, the pathological findings include lymphoproliferative disease similar to acute lymphoblastic leukemia (ATL) in humans [57,58]. There have been four cases of malignant lymphoma and two cases of T-cell leukemia in SIV+ female mangabeys (ages 10 to 26; median 18 years of age). Coinfection with SIVsmm was present in four of the six cases. Neoplastic lymphocytes were confirmed by IHC as T cells in four cases. Phenotypic analysis of a recent case in an SIV+ and STLV+ 26-year-old female was confirmed as a CD3+/CD8+ T-cell lymphocytic leukemia/lymphoma. Although SIV or STLV-1 viruses could contribute to the development of such malignancies, at present there is no direct evidence for such an association.

METABOLIC DISORDERS

Diabetes Mellitus/Islet Amyloidosis

Spontaneous type 2 diabetes mellitus has been documented to occur at a high frequency in our population of SMs with an average lifespan of 18 years and with an age range of 7 to 28 years. The overall prevalence of diabetes in our colony was 11%, nearly twice the rate observed in other species of NHPs, which is typically less than 6% (Jones et al. submitted). Similar to humans, the initial clinical presentation of diabetes in NHPs can include polydipsia, polyuria, polyphagia, weight loss, and lethargy [59,60]. In SMs that died naturally or required euthanasia based on their clinical manifestation of diabetes, microscopic amyloid deposits were found in the pancreatic islets of Langerhans (Figure 6.3). Islet amyloidosis in the SMs is age-related, and similar to islet amyloidosis in other species. Spontaneous diabetes mellitus with islet amyloidosis has been reported in other NHP species including vervet monkeys [61–67].

Female Reproductive Disorders

Mortality due to endometriosis in SMs has been recorded to occur in females over 15 years of age (9 cases; 15–32 years of age). Clinical signs of endometriosis include a heavy menses, abnormal sexual swellings, pain during menstruation, anemia, weight loss, and lethargy. Cystic structures are frequently evident on ultrasound, and aspiration of

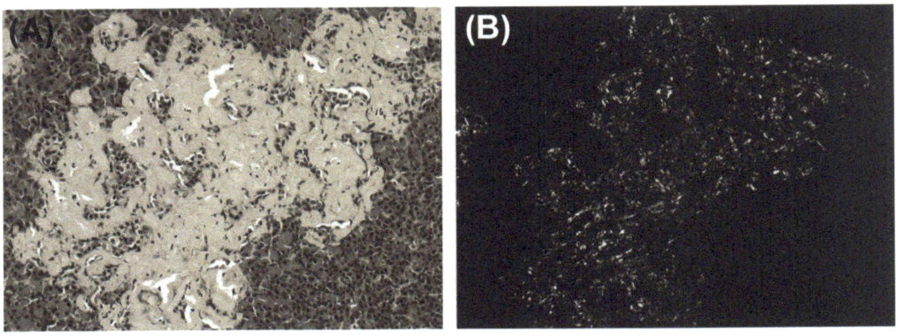

FIGURE 6.3 (A) Islets of Langerhans are replaced by abundant amorphous, homogenous, eosinophilic material (amyloid) during clinical diabetes mellitus. Hematoxylin and eosin (HE) stain. ×200. (B) Amyloid exhibits apple green birefringence with polarized light. Congo red stain. ×200. (See the color plate at the back of the book.)

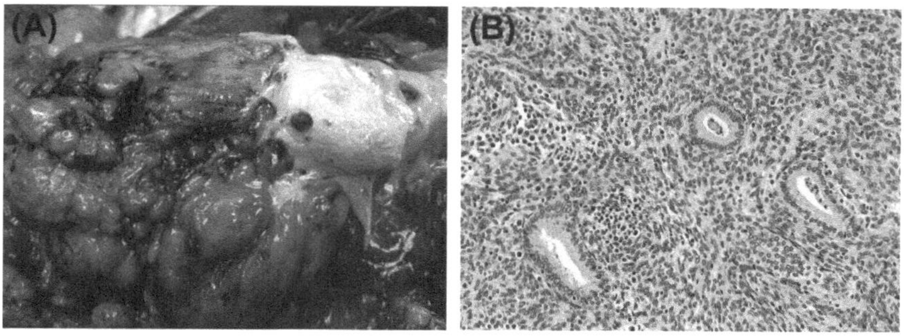

FIGURE 6.4 (A) Endometriosis: multifocal extensive fibrous adhesions and variably sized nodules adhered to serosal surface in the peritoneal cavity. (B) Multiple tortuous endometrial glands are surrounded by abundant, densely cellular endometrial stroma and small numbers of hemosiderin laden macrophages. HE stain. ×200.

cystic structures produces typical "chocolate-colored" fluid. Gross lesions included extensive adhesions, variably sized multiple solid nodules (Figure 6.4(A)), or cysts filled with red to brown fluid located predominately in the abdominal cavity. Typical microscopic findings included tortuous endometrial glands, uterine stroma, and scattered hemosiderin-laden macrophages outside the uterus in a majority of the cases (Figure 6.4(B)). However, one case with obstructive nephropathy resulting from endometrial tissue infiltration of the urinary bladder and blockage of the ureter that induced unilateral renal atrophy was present, as reported previously in humans [68]. Another case presented with hemoperitoneum and extensive fibrous adhesions throughout the abdomen, and the omentum presented as a mass surrounding the small intestines. On histologic examination, the omental mass was composed of reactive fibroplasia intermixed with fibrin, hemorrhage, and nests of endometrial glandular epithelium and sheets of polygonal to round cells with well-defined cell borders and moderate amounts of eosinophilic cytoplasm, as described previously for stromal decidualization of endometriosis in rhesus macaques [69,70].

AGING AND DEGENERATIVE DISORDERS

Osteoarthritis

Prevalence of degenerative joint disease with hind limb contracture, kyphosis, spondylosis, and/or scoliosis with atrophy of skeletal muscle has been noted to increase with advancing age in SMs (6–28 years). Clinical posterior paresis due to spinal cord compression secondary to severe kyphosis has resulted in euthanasia in the elderly population at YNPRC (Figure 6.5). Similar changes have also been observed in the aging macaque

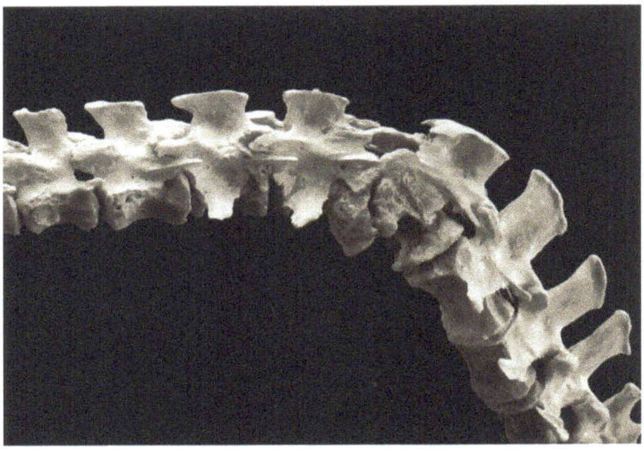

FIGURE 6.5 Spinal column showing degenerative osteoarthritis of lumbar vertebrae (L1-L2) resulting in marked kyphosis. *Courtesy of Dr. Elizabeth Uhl.*

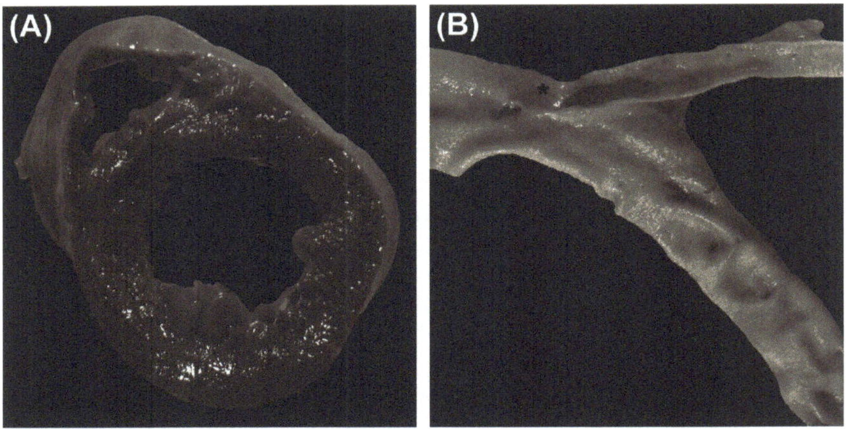

FIGURE 6.6 (A) Heart showing dilated cardiomyopathy with dilated left ventricle and thin left ventricular free wall. (B) There are variably sized pale yellow plaques on the intimal surface of the lower abdominal aorta and iliac arteries (aortoiliac bifurcation) due to atherosclerosis. (See the color plate at the back of the book.)

colonies both at the YNPRC as well as at the Wisconsin National Primate Research Center [26], and thus represent normal age-related symptoms.

Dilated Cardiomyopathy, Atherosclerosis, and Aortic Aneurysms

Eight cases of femoral artery aneurysm, aortic aneurysm, dilated cardiomyopathy (Figure 6.6(A)), and/or atherosclerosis (Figure 6.6(B)) have so far been recorded in male SMs >15years of age. Microscopically, myocardial fibrosis was frequently seen in cardiomyopathy cases. Cardiomyopathy has been reported as a major cause of death in a number of aged NHP

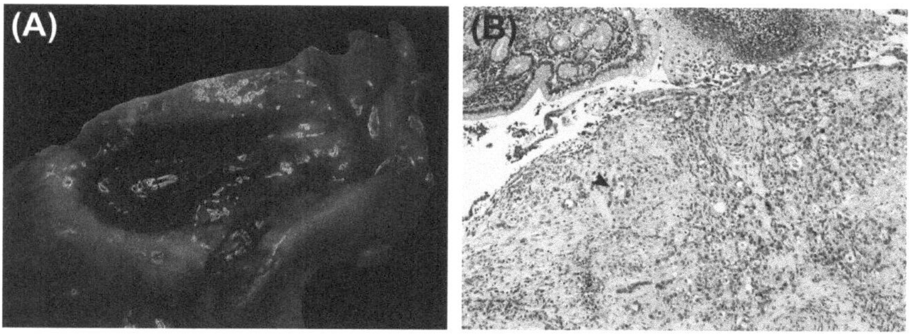

FIGURE 6.7 (A) The gastric mucosa is focally ulcerated and hemorrhagic, and the pyloric wall at the gastric–duodenal junction is severely thickened constricting the lumen. (B) The ulcerated pyloric mass is composed of neoplastic epithelial cells invading the submucosa and arranged in acini (arrowhead) intermixed with thick fibrous connective tissue (desmoplastic response). HE stain. ×100.

species [71,72]. Myocardial fibrosis has been associated with cardiac arrhythmias and/or sudden death in NHPs [73,74]. Myocardial fibrosis also carries a poor prognosis in humans [75]. Some NHP species have been used as models for atherosclerosis in humans by being fed a diet rich in fat and cholesterol [76]. Multiple predisposing risk factors have been reported and include genetics, old age, male sex, diet enriched in saturated fat, social stress, and ovariectomy [76]. Two cases of ruptured aortic aneurysms were also seen in SMs. Dissecting aortic aneurysm is another uncommon degenerative change described in NHPs. It is characterized by saccular dilatation of the aorta and often associated with a breach in the intimal lining leading to separation, degeneration of the muscular wall of the aorta, and forming a recess filled with clotted blood, fibrin, or degenerate cells with subsequent rupture [77].

NEOPLASMS

The frequency of spontaneously occurring tumors in SMs has rarely been reported in the literature [78–80]. A review of YNPRC cases found neoplasms present in 10 males and 22 females. The age of tumor-bearing animals ranged from 9 to 32 years. The majority of neoplasms were recorded as malignant tumors and were the cause of morbidity and mortality. Most of the tumors were seen in the gastrointestinal tract (eight gastric carcinomas (Figure 6.7), two ileocecal adenocarcinomas, two pancreatic carcinomas); followed by the female genital system (uterine polyp, uterine adenoma, uterine leiomyomas (Figure 6.8), ovarian teratoma, uterine cervical carcinoma, and uterine leiomyosarcoma); and the hematopoietic system (two T-cell leukemias and four lymphosarcomas, SIV+). Other neoplasms included squamous cell

FIGURE 6.8 The uterus contains three pedunculated nodules consistent with leiomyomas or fibroids.

carcinomas (nasal cavity, skin/mammary, stomach), cranial nerve sheath tumor, and spindle cell tumor of the thyroid gland. Spontaneous gastric carcinomas were overrepresented in SMs, with eight cases observed since the initial case was identified in 2002. An association of such gastric carcinomas with *Helicobacter pylori* was ruled out in these cases. The animals ranged in age from 10 to 30 years (median age 15 years) [80]. Interestingly, only a few cases of ileocecal adenocarcinomas were diagnosed compared to the high incidence reported in rhesus macaques [81,82].

MISCELLANEOUS DISORDERS

Trauma

Trauma due to fighting and bite wounds, and mortality due to secondary tetanus infections, were factors associated with maintenance in social primate colonies in indoor–outdoor compounds. The most common diagnoses in SMs dying before reaching 1 year of age were related to perinatal death (stillbirth, in utero death, dystocia) or inadequate maternal care (inanition/dehydration) and/or social colony interactions. Trauma included skull fractures, subdural hemorrhage, multiple lacerations, bite wounds, or hemoperitoneum secondary to fights within the social groups, or the natural change of the alpha male within a given group, leading to aggression against neonates and infants. In young adults, trauma associated with fight wounds and trauma-induced secondary tetanus infections were common causes of morbidity/mortality.

Acknowledgments

This work was supported by the Yerkes base grant P51OD11132. We also thank the animal care, veterinary department, colony management and pathology staff of the Yerkes National Primate Research Center for their excellent care of the animals and technical support for this study.

References

[1] Gormus BJ, Xu K, Baskin GB, Martin LN, Bohm RP, Blanchard JL, et al. Experimental leprosy in monkeys. I. Sooty mangabey monkeys: transmission, susceptibility, clinical and pathological findings. Lepr Rev 1995;66(2):96–104.

[2] Gormus BJ, Xu K, Cho SN, Baskin GB, Bohm RP, Martin LN, et al. Experimental leprosy in monkeys. II. Longitudinal serological observations in sooty mangabey monkeys. Lepr Rev 1995;66(2):105–25.

[3] Hamilton HK, Levis WR, Martiniuk F, Cabrera A, Wolf J. The role of the armadillo and sooty mangabey monkey in human leprosy. Int J Dermatol 2008;47(6):545–50.

[4] Oates JF, Gippoliti S, Groves CP. *Cercocebus atys*. In: IUCN 2013. IUCN red list of threatened species. Version 2013.2; 2008. http://www.iucnredlist.org.

[5] UNEP-WCMC. CITES species database; 2012. http://www.unep-wcmc.org/checklist-of-cites-species_499.html.

[6] Gormus BJ, Martin LN, Baskin GB. A brief history of the discovery of natural simian immunodeficiency virus (SIV) infections in captive sooty mangabey monkeys. Front Biosci 2004;9:216–24.

[7] Murphey-Corb M, Martin LN, Rangan SR, Baskin GB, Gormus BJ, Wolf RH, et al. Isolation of an HTLV-III-related retrovirus from macaques with simian AIDS and its possible origin in asymptomatic mangabeys. Nature 1986;321(6068):435–7.

[8] Bernstein IS. Agonistic behavior during group formation and intergroup interactions in sooty mangabeys (*Cercocebus atys*). In: Kummer H, editor. Proceedings of the third international congress of primatology. Zurich: Basel (CH):S. Karger; 1971. pp. 66–70.

[9] Gust DA, Gordon TP. Male age and reproductive-behavior in sooty mangabeys, *Cercocebus torquatus atys*. Anim Behav 1991;41:277–83.

[10] Fultz PN, McClure HM, Anderson DC, Swenson RB, Anand R, Srinivasan A. Isolation of a T-lymphotropic retrovirus from naturally infected sooty mangabey monkeys (*Cercocebus atys*). Proc Natl Acad Sci USA 1986;83(14):5286–90.

[11] Apetrei C, Kaur A, Lerche NW, Metzger M, Pandrea I, Hardcastle J, et al. Molecular epidemiology of simian immunodeficiency virus SIVsm in U.S. primate centers unravels the origin of SIVmac and SIVstm. J Virol 2005;79(14):8991–9005.

[12] Hirsch VM, Olmsted RA, Murphey-Corb M, Purcell RH, Johnson PR. An African primate lentivirus (SIVsm) closely related to HIV-2. Nature 1989;339(6223):389–92.

[13] Sodora DL, Allan JS, Apetrei C, Brenchley JM, Douek DC, Else JG, et al. Toward an AIDS vaccine: lessons from natural simian immunodeficiency virus infections of African nonhuman primate hosts. Nat Med 2009;15(8):861–5.

[14] Silvestri G, Sodora DL, Koup RA, Paiardini M, O'Neil SP, McClure HM, et al. Nonpathogenic SIV infection of sooty mangabeys is characterized by limited bystander immunopathology despite chronic high-level viremia. Immunity 2003;18(3):441–52.

[15] Estes JD, Gordon SN, Zeng M, Chahroudi AM, Dunham RM, Staprans SI, et al. Early resolution of acute immune activation and induction of PD-1 in SIV-infected sooty mangabeys distinguishes nonpathogenic from pathogenic infection in rhesus macaques. J Immunol 2008;180(10):6798–807.

[16] Paiardini M, Cervasi B, Reyes-Aviles E, Micci L, Ortiz AM, Chahroudi A, et al. Low levels of SIV infection in sooty mangabey central memory CD T cells are associated with limited CCR5 expression. Nat Med 2011;17(7):830–6.
[17] Silvestri G, Fedanov A, Germon S, Kozyr N, Kaiser WJ, Garber DA, et al. Divergent host responses during primary simian immunodeficiency virus SIVsm infection of natural sooty mangabey and nonnatural rhesus macaque hosts. J Virol 2005;79(7):4043–54.
[18] Ansari AA, Mayne A, Hunt D, Sundstrom JB, Villinger F. TH1/TH2 subset analysis. I. Establishment of criteria for subset identification in PBMC samples from nonhuman primates. J Med Primatol 1994;23(2–3):102–7.
[19] Bostik P, Dodd GL, Villinger F, Mayne AE, Ansari AA. Dysregulation of the polo-like kinase pathway in CD4+ T cells is characteristic of pathogenic simian immunodeficiency virus infection. J Virol 2004;78(3):1464–72.
[20] Bostik P, Mayne AE, Villinger F, Greenberg KP, Powell JD, Ansari AA. Relative resistance in the development of T cell anergy in CD4+ T cells from simian immunodeficiency virus disease-resistant sooty mangabeys. J Immunol 2001;166(1):506–16.
[21] Bostik P, Noble ES, Mayne AE, Gargano L, Villinger F, Ansari AA. Central memory CD4 T cells are the predominant cell subset resistant to anergy in SIV disease resistant sooty mangabeys. AIDS 2006;20(2):181–8.
[22] Mattapallil JJ, Douek DC, Hill B, Nishimura Y, Martin M, Roederer M. Massive infection and loss of memory CD4+ T cells in multiple tissues during acute SIV infection. Nature 2005;434(7037):1093–7.
[23] Pereira LE, Ansari AA. A case for innate immune effector mechanisms as contributors to disease resistance in SIV-infected sooty mangabeys. Curr HIV Res 2009;7(1):12–22.
[24] Pereira LE, Johnson RP, Ansari AA. Sooty mangabeys and rhesus macaques exhibit significant divergent natural killer cell responses during both acute and chronic phases of SIV infection. Cell Immunol 2008;254(1):10–9.
[25] Powell JD, McClure HM, Anderson D, Fultz PN, Sell KW, Ahmed-Ansari A. Phenotypic and functional differences in NK and LAK cells in the peripheral blood of sooty mangabeys and rhesus macaques. Cell Immunol 1989;124(1):107–18.
[26] Uno H. Age-related pathology and biosenescent markers in captive rhesus macaques. Age 1997;20(1):1–13.
[27] Booth AH. The Cercopithecidae of the gold and ivory coasts: geographic and systematic observations. J Nat Hist 1956;9(103):476–80.
[28] Grove C. Order primates. In: Wilson DE, Reeder DM, editors. Mammal species of the world: a taxonomic and geographic reference. 3rd ed. Baltimore (MD): Johns Hopkins U Pr; 2005. pp. 84–111.
[29] Page SL, Chiu C, Goodman M. Molecular phylogeny of old world monkeys (Cercopithecidae) as inferred from gamma-globin DNA sequences. Mol Phylogenet Evol 1999;13(2):348–59.
[30] Sharma P, Chennareddi L, Greene-Hartsfield EZ, Villinger F, Cohen JK, Herndon J. Hematology and serum chemistry values of sooty mangabeys (*Cercocebus atys*): comparison with rhesus monkeys. J Med Primatol 2014;43:78–88.
[31] Elmore DB, Anderson JH, Hird DW, Sanders KD, Lerche NW. Diarrhea rates and risk factors for developing chronic diarrhea in infant and juvenile rhesus monkeys. Lab Anim Sci 1992;42(4):356–9.
[32] Russell RG, Rosenkranz SL, Lee LA, Howard H, DiGiacomo RF, Bronsdon MA, et al. Epidemiology and etiology of diarrhea in colony-born *Macaca nemestrina*. Lab Anim Sci 1987;37(3):309–16.
[33] Simmons J, Gibson S. Bacterial and mycotic diseases of nonhuman primates. In: Abee CR, Mansfiled K, Tardif S, Morris T, editors. Nonhuman primates in biomedical research: diseases. Elsevier; 2012. pp. 105–72.
[34] Gilbert SG, Reuhl KR, Wong JH, Rice DC. Fatal pneumococcal meningitis in a colony-born monkey (*Macaca fascicularis*). J Med Primatol 1987;16(5):333–8.

[35] Fox JG, Wikse SE. Bacterial meningoencephalitis in rhesus monkeys: clinical and pathological features. Lab Anim Sci 1971;21(4):558–63.
[36] Graczyk TK, Cranfield MR, Kempske SE, Eckhaus MA. Fulminant streptococcus pneumoniae meningitis in a lion-tailed macaque (*Macaca silenus*) without detected signs. J Wildl Dis 1995;31(1):75–8.
[37] Kaufmann AF, Quist KD. Pneumococcal meningitis and peritonitis in rhesus monkeys. J Am Vet Med Assoc 1969;155(7):1158–62.
[38] Twenhafel NA, Whitehouse CA, Stevens EL, Hottel HE, Foster CD, Gamble S, et al. Multisystemic abscesses in African green monkeys (*Chlorocebus aethiops*) with invasive Klebsiella pneumoniae–identification of the hypermucoviscosity phenotype. Vet Pathol 2008;45(2):226–31.
[39] Alfonso R, Romero RE, Diaz A, Calderon MN, Urdaneta G, Arce J, et al. Isolation and identification of mycobacteria in new world primates maintained in captivity. Vet Microbiol 2004;98(3–4):285–95.
[40] Mansfield KG, Lackner AA. Simian immunodeficiency virus-inoculated macaques acquire *Mycobacterium avium* from potable water during AIDS. J Infect Dis 1997;175(1):184–7.
[41] Hessler JR, Woodard JC, Beattie RJ, Moreland AF. Mucormycosis in a rhesus monkey. J Am Vet Med Assoc 1967;151(7):909–13.
[42] Martin JE, Kroe DJ, Bostrom RE, Johnson DJ, Whitney Jr RA. Rhino-orbital phycomycosis in a rhesus monkey (*Macaca mulatta*). J Am Vet Med Assoc 1969;155(7):1253–7.
[43] Sharma P, Cohen JK, Lockhart SR, Hurst SF, Drew CP. Ruptured mycotic aortic aneurysm in a sooty mangabey (*Cercocebus atys*). Comp Med 2011;61(6):532–7.
[44] Chandler FW, Kaplan W, Ajello L. Color atlas and text of the histopathology of mycotic diseases. Chicago (IL): Year Book Medical Publishing; 1980.
[45] Mantadakis E, Samonis G. Clinical presentation of zygomycosis. Clin Microbiol Infect 2009;15(Suppl. 5):15–20.
[46] Baskin GB, Chandler FW, Watson EA. Cutaneous zygomycosis in rhesus monkeys (*Macaca mulatta*). Vet Pathol 1984;21(1):125–8.
[47] Lehrer RJ, Howard DH, Sypherd PS, Edwards Jr JE, Segal GP, Winston DJ. Mucormycosis. Ann Intern Med 1980;93:93–108.
[48] Jehuda-Cohen T, Villinger F, Powell JD, McClure HM, Sell KW, Ansari AA. Presence of SIV antibodies in the sera of infants born to SIV-seronegative monkeys. J Acquired Immune Defic Syndr 1991;4(2):204–5.
[49] Rey-Cuille MA, Berthier JL, Bomsel-Demontoy MC, Chaduc Y, Montagnier L, Hovanessian AG, et al. Simian immunodeficiency virus replicates to high levels in sooty mangabeys without inducing disease. J Virol 1998;72(5):3872–86.
[50] Jehuda-Cohen T, Powell JD, Villinger F, McClure HM, Sell KW, Ahmed-Ansari A. Comparison of SIV/SMM replication in CD4+ T cell and monocyte/macrophage cultures from rhesus macaques and sooty mangabeys. J Med Primatol 1990;19(3–4):251–67.
[51] Ling B, Apetrei C, Pandrea I, Veazey RS, Lackner AA, Gormus B, et al. Classic AIDS in a sooty mangabey after an 18-year natural infection. J Virol 2004;78(16):8902–8.
[52] Brenchley JM, Vinton C, Tabb B, Hao XP, Connick E, Paiardini M, et al. Differential infection patterns of CD4+ T cells and lymphoid tissue viral burden distinguish progressive and nonprogressive lentiviral infections. Blood 2012;120(20):4172–81.
[53] Klatt NR, Villinger F, Bostik P, Gordon SN, Pereira L, Engram JC, et al. Availability of activated CD4+ T cells dictates the level of viremia in naturally SIV-infected sooty mangabeys. J Clin Invest 2008;118(6):2039–49.
[54] Ortiz AM, Klatt NR, Li B, Yi Y, Tabb B, Hao XP, et al. Depletion of CD4(+) T cells abrogates post-peak decline of viremia in SIV-infected rhesus macaques. J Clin Invest 2011;121(11):4433–45.
[55] Kuroda MJ. Macrophages: do they impact AIDS progression more than CD4 T cells? J Leukocyte Biol 2010;87(4):569–73.

[56] Keele BF, Jones JH, Terio KA, Estes JD, Rudicell RS, Wilson ML, et al. Increased mortality and AIDS-like immunopathology in wild chimpanzees infected with SIVcpz. Nature 2009;460(7254):515–9.

[57] Fultz PN, Gordon TP, Anderson DC, McClure HM. Prevalence of natural infection with simian immunodeficiency virus and simian T-cell leukemia virus type I in a breeding colony of sooty mangabey monkeys. AIDS 1990;4(7):619–25.

[58] Wachtman L, Mansfield K. Viral diseases of nonhuman primates. In: Abee CR, Mansfiled K, Tardif S, Morris T, editors. Nonhuman primates in biomedical research: diseases. Elsevier; 2012. pp. 1–105.

[59] Ange-van Heugten KD, Burns R, Verstegen MW, Jansen WL, Ferket PR, van Heugten E. Evaluation of diabetes determinants in woolly monkeys (*Lagothrix lagotricha*). J Anim Physiol Anim Nutr (Berl) 2007;91(11–12):481–91.

[60] Howard Jr CF. Nonhuman primates as models for the study of human diabetes mellitus. Diabetes 1982;31(Suppl. 1 Pt 2):37–42.

[61] Blanchard JL, Baskin GB, Watson EA. Generalized amyloidosis in rhesus monkeys. Vet Pathol 1986;23(4):425–30.

[62] Howard Jr CF. Correlations of hemoglobin A1c and metabolic status in nondiabetic, borderline diabetic, and diabetic *Macaca nigra*. Diabetes 1982;31(12):1105–8.

[63] Ohagi S, Nishi M, Bell GI, Ensinck JW, Steiner DF. Sequences of islet amyloid polypeptide precursors of an old world monkey, the pig-tailed macaque (*Macaca nemestrina*), and the dog (*Canis familiaris*). Diabetologia 1991;34(8):555–8.

[64] Wagner JD, Carlson CS, O'Brien TD, Anthony MS, Bullock BC, Cefalu WT. Diabetes mellitus and islet amyloidosis in cynomolgus monkeys. Lab Anim Sci 1996;46(1):36–41.

[65] Hubbard GB, Steele KE, Davis 3rd KJ, Leland MM. Spontaneous pancreatic islet amyloidosis in 40 baboons. J Med Primatol 2002;31(2):84–90.

[66] Howard Jr CF, Palotay JL. Spontaneous diabetes mellitus in *Macaca cyclopis* and *Mandrillus leucophaeus*: case reports. Lab Anim Sci 1975;25(2):191–6.

[67] Cann JA, Kavanagh K, Jorgensen MJ, Mohanan S, Howard TD, Gray SB, et al. Clinicopathologic characterization of naturally occurring diabetes mellitus in vervet monkeys. Vet Pathol 2010;47(4):713–8.

[68] Ponticelli C, Graziani G, Montanari E. Ureteral endometriosis: a rare and underdiagnosed cause of kidney dysfunction. Nephron Clin Pract 2010;114(2):c89–93.

[69] Schiffer SP, Cary CJ, Peter GK, Cohen BJ. Hemoperitoneum associated with endometriosis in a rhesus monkey. J Am Vet Med Assoc 1984;185(11):1375–7.

[70] Usborne AL, Bolton IB, Slukvin I. Stromal decidualization of endometriosis in the rhesus macaque (*Macaca mulatta*): a case report. Comp Med 2002;52(2):167–70.

[71] David JM, Dick Jr EJ, Hubbard GB. Spontaneous pathology of the common marmoset (*Callithrix jacchus*) and tamarins (*Saguinus oedipus, Saguinus mystax*). J Med Primatol 2009;38(5):347–59.

[72] Zabka TS, Irwin M, Albassam MA. Spontaneous cardiomyopathy in cynomolgus monkeys (*Macaca fascicularis*). Toxicol Pathol 2009;37(6):814–8.

[73] Lammey ML, Baskin GB, Gigliotti AP, Lee DR, Ely JJ, Sleeper MM. Interstitial myocardial fibrosis in a captive chimpanzee (*Pan troglodytes*) population. Comp Med 2008;58(4):389–94.

[74] Vidal JD, Drobatz LS, Holliday DF, Geiger LE, Thomas HC. Spontaneous findings in the heart of Mauritian-origin cynomolgus macaques (*Macaca fascicularis*). Toxicol Pathol 2010;38(2):297–302.

[75] Arteaga E, de Araujo AQ, Bernstein M, Ramires FJ, Ianni BM, Fernandes F, et al. Prognostic value of the collagen volume fraction in hypertrophic cardiomyopathy. Arq Bras Cardiol 2009;92(3):210–4. 6-20.

[76] Clarkson TB. Nonhuman primate models of atherosclerosis. Lab Anim Sci 1998;48(6):569–72.

[77] Bunton TE. Aortic dissection. In: Jones TC, Mohr U, Hunt RD, editors. Nonhuman primates II. Berlin: Springer Verlag; 1993. pp. 118–22.
[78] Clark JD, Olsen RE. Hepatoma in a mangabey (*Cercocebus atys*). Vet Pathol 1973;10(2):89–93.
[79] Morales PR, Garcia A, Stroud FR, Paul KS. Spontaneous cutaneous squamous cell carcinoma in a sooty mangabey (*Cercocebus atys*): a case report. J Am Assoc Lab Anim Sci 2006;45(2):37–40.
[80] Sharma P, Cohen JK, Paul KS, Courtney CL, Johnson ZP, Anderson DC. Spontaneous gastric carcinomas in sooty mangabeys (*Cercocebus atys*). Comp Med 2011;61(6):527–31.
[81] Simmons HA, Mattison JA. The incidence of spontaneous neoplasia in two populations of captive rhesus macaques (*Macaca mulatta*). Antioxid Redox Signal 2011;14(2):221–7.
[82] Rodriguez NA, Garcia KD, Fortman JD, Hewett TA, Bunte RM, Bennett BT. Clinical and histopathological evaluation of 13 cases of adenocarcinoma in aged rhesus macaques (*Macaca mulatta*). J Med Primatol 2002;31(2):74–83.

CHAPTER

7

Gastrointestinal Immunity in Natural Hosts of Simian Immunodeficiency Virus

Molly R. Perkins, Jason M. Brenchley

Program in Tissue Immunity and Repair, Laboratory of Molecular Microbiology, NIAID, NIH, Bethesda, MD, USA

INTRODUCTION

One of the key immunological characteristics that distinguishes progressive from nonprogressive human immunodeficiency virus (HIV)/simian immunodeficiency virus (SIV) infections is chronic immune activation [1]. A multifaceted activation of both the adaptive and innate arms of the immune system is a hallmark of progressive HIV/SIV infection [2]. Indeed, the degree of immune activation is the best predictor of the time until the development of acquired immune deficiency syndrome (AIDS) during progressive infection [3]. However, in natural hosts with nonprogressive SIV infection, this immune activation is not observed; the absence of immune activation is thought to be of critical importance for the lack of disease progression in these species [4–6]. One important driver of immune activation in progressive HIV/SIV infection is reasoned to be the translocation of microbial products from a dysfunctional gastrointestinal (GI) tract into the systemic circulation because these can directly stimulate the immune system [7]. Importantly, microbial translocation is not observed in natural hosts with chronic SIV infection [8], and the containment of these microbial products may help explain the absence of immune activation in these species.

Therefore, the physiology of the GI tract acts as a barrier between the microbes and microbial products in the lumen of the intestine and systemic circulation of the host and seems to be a key factor in the nonprogressive

Natural Hosts of SIV.
http://dx.doi.org/10.1016/B978-0-12-404734-1.00007-3

nature of SIV infection of natural hosts. This chapter discusses the current understanding of the physiology of the immunological and structural barriers of the GI tract in SIV-infected natural host species and how these relate to the defects seen in progressive infections of non-natural host species.

CD4 T-CELL DEPLETION OCCURS IN THE GI TRACT OF NATURAL AND NON-NATURAL HOSTS

One of the most commonly used metrics to follow the clinical course of disease in HIV-infected individuals is the number of $CD4^+$ T cells per microliter of blood. Loss of $CD4^+$ T cells is a hallmark of progressive HIV/SIV infections, and $CD4^+$ T cell counts <200 cells/μl blood are coincident with the onset of opportunistic infections. This loss of $CD4^+$ T cells in peripheral blood is generally quite slow; HIV-infected humans and SIV-infected rhesus macaques (RM) lose approximately 40 $CD4^+$ T cells/μl/year [9,10]. In contrast, $CD4^+$ T-cell depletion in the GI tract is severe and rapid in these non-natural host species [11–14]. The GI tract has a large surface area and contains the majority of the $CD4^+$ T cells in the body; therefore the nearly complete depletion of GI tract $CD4^+$ T cells during the acute phase of HIV/SIV infection reflects a massive loss of this cell lineage early in infection.

Depletion of GI tract $CD4^+$ T cells also occurs in nonprogressive SIV infections of natural hosts, even though these animals do not tend to lose $CD4^+$ T cells from peripheral blood [15]. However, the degree to which $CD4^+$ T cells are depleted from the GI tracts of acutely infected natural hosts is not as extensive as that observed in acutely infected non-natural hosts. Furthermore, while depletion of GI tract $CD4^+$ T cells continues throughout disease progression in HIV/SIV-infected non-natural hosts, levels of $CD4^+$ T cells in the GI tract remain stable in sooty mangabeys (SMs) or even increase in African green monkeys (AGMs) after the acute phase of infection [16–19]. Notably, in African green monkeys the loss of $CD4^+$ T cells in the GI tract may be of little biological significance given that adult AGMs, even those who are not infected with SIV, typically have very low amounts of $CD4^+$ T cells within the GI tract and systemically [17].

One factor that may explain the differences in $CD4^+$ T-cell depletion in the GI tracts of AGMs and SMs compared to humans, RMs, and pig-tail macaques (PTMs) is the expression of CCR5 on $CD4^+$ T cells. SMs in particular, as well as AGMs, have far fewer $CCR5^+$ $CD4^+$ T cells in the GI tract than are observed in non-natural hosts [20]. This pattern of decreased expression of CCR5 by $CD4^+$ T cells is also observed in the blood of multiple natural host species [20]. The potentially adaptive nature of this phenomenon is highlighted by the fact that CCR5 expression is not lower in CD8 T cells in the same animals [20]. This lower expression of the coreceptor for HIV/SIV on $CD4^+$ T cells may limit viral infection and the depletion of these cells in natural hosts [21,22].

MUCOSAL ACTIVATION IS CONTROLLED AFTER ACUTE INFECTION IN NATURAL HOSTS

Antiviral immune responses lead to systemic and local immune activation during acute infection of natural and non-natural hosts of SIV. The downregulation of this response during the transition to the chronic phase of infection is a hallmark of natural host species [23]; non-natural hosts experience continued immune activation throughout the chronic phase of HIV/SIV infection and into AIDS, and immune activation is the strongest predictor of disease progression [3].

Most studies describing the ability of natural hosts to control the vigorous type I interferon responses after the acute phase of SIV infection have not probed the dynamics of immune activation in the mucosal tissues [24–27]. However, it is clear from the limited studies that have examined mucosal tissues that a similar pattern is observed in the GI tract. Gene expression profiling in the colon of AGMs and PTMs has identified a high level of activation of genes involved in the type I interferon response in both species at day 10 after infection. AGMs have higher activation of these genes than PTMs at day 10, but by day 45 the response is attenuated in AGMs, in contrast to PTMs, who experience sustained expression of genes involved in inflammation [28].

Natural hosts also seem to control T-cell activation in the GI tract after acute infection. In addition to the signature of attenuated innate immune activation at day 45 in the study described above, there was also less mucosal T-cell activation in the same animals, characterized by lower frequencies of $Ki67^+$ CD8 T cells in the jejunum, ileum, and colon [19].

Further support for the ability of natural hosts to limit chronic immune activation in the mucosa is the finding that SMs have small amounts of plasmacytoid dendritic cells (pDCs) in the colorectum during chronic infection, in contrast to RMs, in whom pDCs are recruited to the colorectum early in SIV infection and remain at elevated amounts through AIDS, when they are found at high numbers in the ileum, colon, and rectum [29]. Interestingly, upon antiretroviral therapy the frequency of colorectal pDCs decreases in RMs [29]. Taken together, these data suggest that recruitment of pDCs in RMs is driven by viral replication, but SMs do not recruit pDCs to the colorectum during chronic SIV infection despite ongoing viral replication. This may explain the decreased type I interferon signaling observed in natural hosts with chronic SIV infection.

INTERLEUKIN-17-PRODUCING CELLS ARE PREFERENTIALLY LOST IN THE GI TRACT DURING PATHOGENIC SIV INFECTION BUT ARE MAINTAINED IN NATURAL HOSTS

T helper 17 (Th17) cells are a subset of $CD4^+$ helper T cells that express interleukin (IL)-17 and IL-22 after T-cell receptor stimulation. Within the GI tract, these cells help maintain both the immunological and structural barriers that separate the luminal contents from the host tissue [30]. In healthy individuals, the highest frequencies of Th17 cells are found in the GI tract compared to all other anatomic sites studied [31], and the vast majority of IL-17-producing T cells circulating in the blood of RMs express GI tract homing receptors [32], consistent with their essential role in GI tract homeostasis.

During progressive HIV/SIV infection, IL-17-producing cells are preferentially depleted during the severe loss of $CD4^+$ T-cell in the GI tract. In RMs, PTMs, and humans the frequency of IL-17-producing $CD4^+$ T cells within the GI tract decreases after HIV/SIV infection [19,31,33,34]. Moreover, the degree to which these cells are depleted within the GI tract of progressively SIV-infected RMs correlates with damage to the structural barrier of the GI tract, with consequent microbial translocation that can drive immune activation [34]. In contrast, natural hosts maintain healthy frequencies of IL-17-producing cells within the GI tract after SIV infection [19,31,33,34], and this may explain how natural hosts are able to avoid microbial translocation during chronic SIV infection.

At least two mechanisms that might explain the preferential depletion of IL-17-producing cells have been proposed, and these are not mutually exclusive. $CD103^+$ dendritic cells (DCs) are antigen-presenting cells (APCs) that can induce T-cell expression of the transcription factor RORγt, which is responsible for the development and maintenance of IL-17-producing cells. In RMs, $CD103^+$ DCs express genes for the promotion of $IL\text{-}17^+$ cells and can directly promote a Th17 phenotype in vitro [34]. These $CD103^+$ DCs are depleted after SIV infection of RMs, and the frequency of these cells correlates with the frequency of all subsets of IL-17-producing lymphocytes [34]. The low frequencies of $CD103^+$ DCs might be attributed to the infiltration of proinflammatory myeloid cells expressing the tryptophan metabolizing enzyme indoleamine 2,3-dioxygenase [34,35]. These data suggest that an altered landscape of GI tract APCs might influence the functional profile of local lymphocytes in non-natural hosts. In contrast, natural hosts seem to maintain APCs in the GI tract after SIV infection; $CD103^+$ DCs are preserved in the GI tracts of SIV-infected SMs [34].

A second mechanism that has been proposed to underlie the low numbers of IL-17-producing lymphocytes in the GI tracts of non-natural hosts is the loss of IL-21-producing $CD4^+$ T cells [32]. Preferential loss of these cells occurs in the rectum of intrarectally infected RMs but is not observed in SMs [32]. The importance of the IL-21-producing $CD4^+$ T cells for the maintenance of IL-17-producing lymphocytes has been demonstrated by in vivo treatment of RMs with an IL-21 fusion protein, which results in an increase in Th17 cells in the peripheral blood [32].

Although the mechanisms that underlie the preferential loss of Th17 cells in non-natural hosts are still under investigation, it is clear that the maintenance of IL-17- and IL-22-producing lymphocytes in the GI tract of natural hosts may contribute to the nonprogressive nature of their infection; these cells aid in the repair of the GI tract barrier and prevent chronic microbial translocation.

NATURAL HOSTS PRODUCE $CD4^-$ T CELLS THAT CAN PERFORM FUNCTIONS OF $CD4^+$ T CELLS BUT ARE RESISTANT TO SIV INFECTION

Although IL-17-producing lymphocytes are not preferentially depleted in natural hosts [19,31,33,34], the overall loss of mucosal $CD4^+$ T cells including Th17 cells would be expected to hinder the ability of these animals to maintain the integrity of the GI tract after SIV infection. However, natural hosts have populations of T cells that do not express CD4 and are capable of performing effector functions normally associated with $CD4^+$ T cells. Importantly, these $CD4^-$ T cells are resistant to infection with SIV and therefore to infection-mediated depletion. The evolution of molecular mechanisms in natural hosts to generate a population of T cells that lack the receptor for SIV but preserve the immunological functions of $CD4^+$ T cells is thought to underlie the nonprogressive nature of SIV infection in these species.

An early indication that CD4 functions are being performed by $CD4^-$ T cells in natural hosts was the finding that AGMs have a large population of $CD4^-CD8\alpha^{dim}$ T cells, whose antigen-specific response is major histocompatibility complex class II–restricted, much like the major histocompatibility complex class II–restricted responses of classic CD4 T cells [36]. These cells can produce effector proteins normally made by CD4 T cells, including FoxP3, IL-17, IL-2, and CD40 ligand. Since the $CD4^-CD8\alpha^{dim}$ T cells outnumber the $CD4^+$ T cells in adult AGMs, there are as many or more $CD4^-$ T cells performing these functions as $CD4^+$ T cells [36]. In vitro and in vivo data suggest that these cells are created through the downregulation of CD4 by $CD4^+$ T cells [36]. Importantly, this mechanism seems to be the result of conserved or convergent evolution. Patas monkeys can also generate a functional population of $CD4^-$ T cells through the downregulation of CD4 [37]. Furthermore, many species that are natural hosts of SIV have $CD4^-$ T cells that can perform functions normally associated with classic $CD4^+$ T cells [38].

SMs also seem to have a population of $CD4^-$ T cells capable of maintaining immune function in the absence of $CD4^+$ T cells. A small number of SMs that experience profound CD4 T-cell depletion after SIV infection have been described [18,39]. These animals do not progress to AIDS despite this depletion, and they maintain a large population of $CD4^-CD8^-$ T cells that are not depleted after SIV infection [18,39]. In fact, in the rectum the frequency of these double-negative (DN) T cells increases almost threefold after infection, with DN T cells comprising more than 70% of rectal T cells at day 180 after infection [39]. That the DN T cells are able to aid in immunological processes that typically require $CD4^+$ T cells is indicated by the ability of these SMs to maintain mucosal integrity (and thereby avoid microbial translocation) and to mount robust antibody responses to influenza vaccination [39].

The finding that mucosal $CD4^+$ T cells in natural hosts are primarily CCR5-negative suggests that limiting CCR5 expression may be an additional mechanism to preserve functional T cells in natural hosts [20,21]. It is clear that the frequency of target cells is correlated with susceptibility to infection [22]. However, it is not clear that this mechanism of preservation is the main determinant of nonprogression, given the existence of SMs with extremely low numbers of CD4 T cells that still do not progress to AIDS [18,39]. Instead, in these animals the DN T cells are maintained or expanded and seem to be the providers of key functions [38,39]. Whether these DN T cells are important in SMs that do not experience such extreme CD4 T-cell depletion, as seems to be the case in AGMs even in the absence of SIV infection, remains to be determined.

MUCOSAL B CELLS IN NATURAL HOSTS

Mucosal B cells in natural hosts have not been well studied thus far. Treatment of SIV-infected AGMs with rituximab, an antibody that recognizes CD20, results in complete B-cell depletion in the proximal jejunum of the GI tract and other anatomic sites [40]. This depletion does not alter T-cell activation or proliferation within the GI tract. Moreover, depletion of B cells in AGMs with chronic SIV infection does not result in the elevation of plasma proinflammatory cytokines, and the animals show no signs of progressive disease [40]. Therefore, it is not clear whether GI tract B cells are important to nonprogression in this natural host species.

Given that antibody responses in non-natural hosts are negatively affected by HIV/SIV infection [41–43], there are indications that antibody responses in SIV-infected natural hosts may be superior to those in non-natural hosts [39], possibly because in natural hosts there is better maintenance of T-cell help from $CD4^-$ T cells. Studies have not yet examined antibody responses in the GI tract mucosa of natural host species. However, antibody secretion in breast milk also relies on immunoglobulin A (IgA)-positive (IgA^+) B cells, and as such, breast milk antibodies may reflect those in other mucosal tissues. The composition of the breast milk antibody differs dramatically between progressively SIV-infected RMs and nonprogressively SIV-infected AGMs. Specifically, AGMs have equivalent levels of IgA and total immunoglobulin in breast milk and blood plasma, whereas RMs have approximately 10-fold less IgA in breast milk than in blood plasma and approximately 100-fold less total immunoglobulin content in milk than in plasma. The IgA antibodies in the breast milk of natural hosts do not neutralize autologous SIV, so this effect seems to be unrelated to SIV-specific immunity and may reflect preserved B-cell functions in natural hosts [44]. Further study of mucosal IgA may shed light on differences between natural and non-natural hosts of SIV.

γδ T CELLS

T cells that express the γδ T-cell receptor, which recognizes microbial or stress-induced antigens, represent a minority of T cells in the blood but constitute a significant proportion of intraepithelial lymphocytes in the GI tract. γδ T cells belong to one of two classes based on differential expression of either TCR Vδ1 or Vδ2. In healthy humans, RMs, SMs, and AGMs, peripheral blood γδ T cells predominantly comprise Vδ2 T cells, whereas γδ T cells in the GI tract are predominantly Vδ1 T cells [45]. However, in HIV/SIV-infected humans and RMs, Vδ1 T cells are expanded and become dysfunctional; this phenomenon does not occur in SIV-infected AGMs [45] and SMs [46]. Given that γδ T cells respond to bacterial antigens, the observed γδ T-cell dysfunctions in HIV/SIV-infected non-natural hosts might result from microbial translocation and chronic antigenemia. Whether the preserved γδ T-cell functionality in SIV-infected natural hosts is important for their clinical well-being is currently unknown.

NATURAL HOSTS MAINTAIN THE STRUCTURAL INTEGRITY OF THE GI TRACT DURING SIV INFECTION AND THEREBY AVOID THE VICIOUS CYCLE OF MICROBIAL TRANSLOCATION, IMMUNE ACTIVATION, AND ENTERIC DYSREGULATION

Depletion of $CD4^+$ T cells in the GI tract of non-natural hosts during progressive infection results in insult to both the structural and immunological barriers that separate the luminal contents of the intestine from the host. In non-natural hosts, the damage to the GI tract barrier allows the translocation of microbial products into the systemic circulation, and these products can then stimulate the immune system and contribute to the chronic immune activation that is a hallmark of progressive HIV/SIV infection [8]. However, this vicious cycle does not occur in natural host species, in whom microbial translocation occurs transiently during early infection but does not persist into the chronic phase [7,16].

The contrast in the barrier function of the GI tract is stark when comparing natural and non-natural hosts infected with HIV/SIV (Table 7.1). In SMs there is no evidence of structural damage to the GI tract or microbial translocation during the chronic phase of infection, whereas in RMs there is clear evidence of both [47]. Furthermore, the extent of damage to the epithelial barrier of the GI tract correlates with the extent of microbial translocation in SIV-infected RMs, but this is not observed in SIV-infected SMs [47]. Even in SIV-infected SMs with severe CD4+ T-cell depletion, the mucosal integrity is maintained, as evidenced by expression of the tight junction protein occludin, and microbial translocation is not observed, perhaps as a result of IL-17 and IL-22 production by the DN T cells observed in these animals [39]. Indeed, although the precise mechanisms leading to maintenance of the immunological and structural barriers of the GI tract in SIV-infected natural hosts are still unclear, it has been proposed that the presence of $CD4^-$ T cells throughout the GI tract allows maintenance of immunity to extracellular pathogens, IL-17 and IL-22 production, and other functions of $CD4^+$ T cells and thereby limits the harm caused by CD4 T-cell depletion in the GI tract.

TABLE 7.1 Characteristics of Gastrointestinal (GI) Tract Changes in Natural and Non-natural Hosts after HIV/SIV Infection

	Humans	Asian Macaques	Sooty Mangabeys	African Green Monkeys
Plasma viral load	++	+++	+++	++
GI tract CD4 depletion	+++	+++	++	+
GI tract Th17 preferential depletion	Yes	Yes	No	No
Acute GI immune activation	Yes	Yes	Yes	Yes
Chronic GI immune activation	Yes	Yes	No	No
Functional GI tract $CD4^-$ T cells	No	No	Yes	Yes
Structural integrity of GI tract	No	No	Yes	Unknown
Microbial translocation during the chronic phase	Yes	Yes	No	No
Outgrowth of enteric viruses	Unknown	Yes	Unknown	No

Further evidence that natural hosts avoid immune activation by limiting microbial translocation is demonstrated by experimental administration of the microbial product lipopolysaccharide (LPS) to SIV-infected AGMs to mimic microbial translocation. LPS administration to natural SIV-infected hosts led to increased viral load, simulating the expected effect of microbial translocation if it were to occur in natural hosts. However, such LPS administration did not lead to any evidence of disease [48].

Finally, additional evidence that mucosal immunity is maintained in natural hosts comes from comparative genomics of the fecal microbiome in multiple cohorts of SIV-infected AGMs and RMs [49]. After SIV infection of RMs, there is a marked expansion of enteric viruses that is not observed after SIV infection of AGMs [49]. This finding both supports the assertion that antiviral mucosal immunity in natural hosts limits changes to the microbiome and suggests an additional mechanism of GI tract damage that is restricted by natural hosts. It has been proposed that enteric viruses can damage the intestine and that the maintenance of a limited enteric virome may be a mechanism by which AGMs successfully preserve the mucosal barrier during nonprogressive SIV infection [49].

CONCLUSIONS

One observation about the information presented here is that the prototype AGMs and SMs that are natural hosts seem to share some, but not all, immunological characteristics within their GI tracts. Although the primary goal driving natural host research has been to characterize differences between nonprogressive infection of natural hosts and progressive infection of non-natural hosts, the potential differences between the natural host species themselves should not be discounted. Virus–host coevolution has driven the adaptive differences in mucosal immunity observed in species of natural SIV hosts. The mechanisms that allow each species of natural host to avoid pathogenic disease may differ among species, as the virus exerts a continued selective pressure on the host after species divergence. Indeed, driving species to diverge is a hallmark of a host–pathogen evolutionary "arms race." This process has been described for host restriction factors against retroviruses and the viral proteins that counteract them, and both divergent and convergent adaptations are observed in natural hosts of SIV. For this reason, it is expected that the mechanisms by which AGMs and SMs avoid progressive SIV will not be identical. In particular, while both species are thought to restrict the target cells of SIV to maintain CD4 effector functions in the GI tract and systemically, SMs seem to limit CCR5 expression as a primary mechanism to protect particular subsets of memory CD4 T cells from depletion and maintain a smaller population of DN T cells that cannot become infected by SIV. AGMs, on the other hand, limit CCR5 expression to a lesser degree and form a numerically dominant population of CD4$^-$CD8α^{dim} T cells even in the absence of SIV infection. Both mechanisms are thought to preserve CD4$^+$ T-cell function during SIV infection to maintain the immunological and structural barriers of the GI tract and limit the downstream pathogenic effects of CD4$^+$ T-cell depletion. The complex evolutionary history of SIV infection in each of these species is itself an additional tool to identify common and unique features in each natural host species that may bear on our understanding of HIV and SIV pathogenesis and future therapeutic approaches.

References

[1] Silvestri G, Sodora DL, Koup RA, Paiardini M, O'Neil SP, McClure HM, et al. Nonpathogenic SIV infection of sooty mangabeys is characterized by limited bystander immunopathology despite chronic high-level viremia. Immunity 2003;18(3):441–52.
[2] Hunt PW. Role of immune activation in HIV pathogenesis. Curr HIV/AIDS Rep 2007;4(1):42–7.
[3] Giorgi JV, Hultin LE, McKeating JA, Johnson TD, Owens B, Jacobson LP, et al. Shorter survival in advanced human immunodeficiency virus type 1 infection is more closely associated with T lymphocyte activation than with plasma virus burden or virus chemokine coreceptor usage. J Infect Dis 1999;179(4):859–70.
[4] Brenchley JM, Paiardini M. Immunodeficiency lentiviral infections in natural and non-natural hosts. Blood 2011;118(4):847–54.
[5] Silvestri G, Paiardini M, Pandrea I, Lederman MM, Sodora DL. Understanding the benign nature of SIV infection in natural hosts. J Clin Invest 2007;117(11):3148–54.
[6] Silvestri G. Naturally SIV-infected sooty mangabeys: are we closer to understanding why they do not develop AIDS? J Med Primatol 2005;34(5–6):243–52.
[7] Brenchley JM, Price DA, Schacker TW, Asher TE, Silvestri G, Rao S, et al. Microbial translocation is a cause of systemic immune activation in chronic HIV infection. Nat Med 2006;12(12):1365–71.
[8] Marchetti G, Tincati C, Silvestri G. Microbial translocation in the pathogenesis of HIV infection and AIDS. Clin Microbiol Rev 2013;26(1):2–18.
[9] Moore RD, Keruly J, Richman DD, Creagh-Kirk T, Chaisson RE. Natural history of advanced HIV disease in patients treated with zidovudine. The Zidovudine Epidemiology Study Group. AIDS 1992;6(7):671–7.
[10] Mellors JW, Munoz A, Giorgi JV, Margolick JB, Tassoni CJ, Gupta P, et al. Plasma viral load and CD4+ lymphocytes as prognostic markers of HIV-1 infection. Ann Intern Med 1997;126(12):946–54.
[11] Brenchley JM, Schacker TW, Ruff LE, Price DA, Taylor JH, Beilman GJ, et al. CD4+ T cell depletion during all stages of HIV disease occurs predominantly in the gastrointestinal tract. J Exp Med 2004;200(6):749–59.
[12] Mehandru S, Poles MA, Tenner-Racz K, Horowitz A, Hurley A, Hogan C, et al. Primary HIV-1 infection is associated with preferential depletion of CD4+ T lymphocytes from effector sites in the gastrointestinal tract. J Exp Med 2004;200(6):761–70.
[13] Veazey RS, DeMaria M, Chalifoux LV, Shvetz DE, Pauley DR, Knight HL, et al. Gastrointestinal tract as a major site of CD4+ T cell depletion and viral replication in SIV infection. Science 1998;280(5362):427–31.
[14] Guadalupe M, Reay E, Sankaran S, Prindiville T, Flamm J, McNeil A, et al. Severe CD4+ T-cell depletion in gut lymphoid tissue during primary human immunodeficiency virus type 1 infection and substantial delay in restoration following highly active antiretroviral therapy. J Virol 2003;77(21):11708–17.

[15] Paiardini M, Pandrea I, Apetrei C, Silvestri G. Lessons learned from the natural hosts of HIV-related viruses. Annu Rev Med 2009;60:485–95.
[16] Gordon SN, Klatt NR, Bosinger SE, Brenchley JM, Milush JM, Engram JC, et al. Severe depletion of mucosal CD4+ T cells in AIDS-free simian immunodeficiency virus-infected sooty mangabeys. J Immunol 2007;179(5):3026–34.
[17] Pandrea IV, Gautam R, Ribeiro RM, Brenchley JM, Butler IF, Pattison M, et al. Acute loss of intestinal CD4+ T cells is not predictive of simian immunodeficiency virus virulence. J Immunol 2007;179(5):3035–46.
[18] Milush JM, Reeves JD, Gordon SN, Zhou D, Muthukumar A, Kosub DA, et al. Virally induced CD4+ T cell depletion is not sufficient to induce AIDS in a natural host. J Immunol 2007;179(5):3047–56.
[19] Favre D, Lederer S, Kanwar B, Ma ZM, Proll S, Kasakow Z, et al. Critical loss of the balance between Th17 and T regulatory cell populations in pathogenic SIV infection. PLoS Pathog 2009;5(2):e1000295.
[20] Pandrea I, Apetrei C, Gordon S, Barbercheck J, Dufour J, Bohm R, et al. Paucity of CD4+CCR5+ T cells is a typical feature of natural SIV hosts. Blood 2007;109(3):1069–76.
[21] Paiardini M, Cervasi B, Reyes-Aviles E, Micci L, Ortiz AM, Chahroudi A, et al. Low levels of SIV infection in sooty mangabey central memory CD(4)(+) T cells are associated with limited CCR5 expression. Nat Med 2011;17(7):830–6.
[22] Pandrea I, Parrish NF, Raehtz K, Gaufin T, Barbian HJ, Ma D, et al. Mucosal simian immunodeficiency virus transmission in African green monkeys: susceptibility to infection is proportional to target cell availability at mucosal sites. J Virol 2012;86(8):4158–68.
[23] Mir KD, Gasper MA, Sundaravaradan V, Sodora DL. SIV infection in natural hosts: resolution of immune activation during the acute-to-chronic transition phase. Microbes Infect/Institut Pasteur 2011;13(1):14–24.
[24] Bosinger SE, Li Q, Gordon SN, Klatt NR, Duan L, Xu L, et al. Global genomic analysis reveals rapid control of a robust innate response in SIV-infected sooty mangabeys. J Clin Invest 2009;119(12):3556–72.
[25] Estes JD, Gordon SN, Zeng M, Chahroudi AM, Dunham RM, Staprans SI, et al. Early resolution of acute immune activation and induction of PD-1 in SIV-infected sooty mangabeys distinguishes nonpathogenic from pathogenic infection in rhesus macaques. J Immunol 2008;180(10):6798–807.
[26] Jacquelin B, Mayau V, Targat B, Liovat AS, Kunkel D, Petitjean G, et al. Nonpathogenic SIV infection of African green monkeys induces a strong but rapidly controlled type I IFN response. J Clin Invest 2009;119(12):3544–55.
[27] Kim N, Dabrowska A, Jenner RG, Aldovini A. Human and simian immunodeficiency virus-mediated upregulation of the apoptotic factor TRAIL occurs in antigen-presenting cells from AIDS-susceptible but not from AIDS-resistant species. J Virol 2007;81(14):7584–97.
[28] Lederer S, Favre D, Walters KA, Proll S, Kanwar B, Kasakow Z, et al. Transcriptional profiling in pathogenic and non-pathogenic SIV infections reveals significant distinctions in kinetics and tissue compartmentalization. PLoS Pathog 2009;5(2):e1000296.
[29] Kwa S, Kannanganat S, Nigam P, Siddiqui M, Shetty RD, Armstrong W, et al. Plasmacytoid dendritic cells are recruited to the colorectum and contribute to immune activation during pathogenic SIV infection in rhesus macaques. Blood 2011;118(10):2763–73.
[30] Sonnenberg GF, Fouser LA, Artis D. Border patrol: regulation of immunity, inflammation and tissue homeostasis at barrier surfaces by IL-22. Nat Immunol 2011;12(5):383–90.
[31] Brenchley JM, Paiardini M, Knox KS, Asher AI, Cervasi B, Asher TE, et al. Differential Th17 CD4 T-cell depletion in pathogenic and nonpathogenic lentiviral infections. Blood 2008;112(7):2826–35.
[32] Micci L, Cervasi B, Ende ZS, Iriele RI, Reyes-Aviles E, Vinton C, et al. Paucity of IL-21-producing CD4(+) T cells is associated with Th17 cell depletion in SIV infection of rhesus macaques. Blood 2012;120(19):3925–35.

[33] Nigam P, Kwa S, Velu V, Amara RR. Loss of IL-17-producing CD8 T cells during late chronic stage of pathogenic simian immunodeficiency virus infection. J Immunol 2011;186(2):745–53.
[34] Klatt NR, Estes JD, Sun X, Ortiz AM, Barber JS, Harris LD, et al. Loss of mucosal CD103+ DCs and IL-17+ and IL-22+ lymphocytes is associated with mucosal damage in SIV infection. Mucosal Immunol 2012;5(6):646–57.
[35] Favre D, Mold J, Hunt PW, Kanwar B, Loke P, Seu L, et al. Tryptophan catabolism by indoleamine 2,3-dioxygenase 1 alters the balance of TH17 to regulatory T cells in HIV disease. Sci Transl Med 2010;2(32):32r–6.
[36] Beaumier CM, Harris LD, Goldstein S, Klatt NR, Whitted S, McGinty J, et al. CD4 downregulation by memory CD4+ T cells in vivo renders African green monkeys resistant to progressive SIVagm infection. Nat Med 2009;15(8):879–85.
[37] Apetrei C, Gaufin T, Gautam R, Vinton C, Hirsch V, Lewis M, et al. Pattern of SIVagm infection in patas monkeys suggests that host adaptation to simian immunodeficiency virus infection may result in resistance to infection and virus extinction. J Infect Dis 2010;202(Suppl. 3):S371–6.
[38] Vinton C, Klatt NR, Harris LD, Briant JA, Sanders-Beer BE, Herbert R, et al. CD4-like immunological function by CD4– T cells in multiple natural hosts of simian immunodeficiency virus. J Virol 2011;85(17):8702–8.
[39] Milush JM, Mir KD, Sundaravaradan V, Gordon SN, Engram J, Cano CA, et al. Lack of clinical AIDS in SIV-infected sooty mangabeys with significant CD4+ T cell loss is associated with double-negative T cells. J Clin Invest 2011;121(3):1102–10.
[40] Gaufin T, Pattison M, Gautam R, Stoulig C, Dufour J, MacFarland J, et al. Effect of B-cell depletion on viral replication and clinical outcome of simian immunodeficiency virus infection in a natural host. J Virol 2009;83(20):10347–57.
[41] Cagigi A, Nilsson A, Pensieroso S, Chiodi F. Dysfunctional B-cell responses during HIV-1 infection: implication for influenza vaccination and highly active antiretroviral therapy. Lancet Infect Dis 2010;10(7):499–503.
[42] Moir S, Fauci AS. B cells in HIV infection and disease. Nat Rev Immunol 2009;9(4):235–45.
[43] Klatt NR, Vinton CL, Lynch RM, Canary LA, Ho J, Darrah PA, et al. SIV infection of rhesus macaques results in dysfunctional T- and B-cell responses to neo and recall Leishmania major vaccination. Blood 2011;118(22):5803–12.
[44] Wilks AB, Perry JR, Ehlinger EP, Zahn RC, White R, Gauduin MC, et al. High cell-free virus load and robust autologous humoral immune responses in breast milk of simian immunodeficiency virus-infected african green monkeys. J Virol 2011;85(18):9517–26.
[45] Harris LD, Klatt NR, Vinton C, Briant JA, Tabb B, Ladell K, et al. Mechanisms underlying gammadelta T-cell subset perturbations in SIV-infected Asian rhesus macaques. Blood 2010;116(20):4148–57.
[46] Kosub DA, Lehrman G, Milush JM, Zhou D, Chacko E, Leone A, et al. Gamma/Delta T-cell functional responses differ after pathogenic human immunodeficiency virus and nonpathogenic simian immunodeficiency virus infections. J Virol 2008;82(3):1155–65.
[47] Estes JD, Harris LD, Klatt NR, Tabb B, Pittaluga S, Paiardini M, et al. Damaged intestinal epithelial integrity linked to microbial translocation in pathogenic simian immunodeficiency virus infections. PLoS Pathog 2010;6(8):e1001052.
[48] Pandrea I, Gaufin T, Brenchley JM, Gautam R, Monjure C, Gautam A, et al. Cutting edge: experimentally induced immune activation in natural hosts of simian immunodeficiency virus induces significant increases in viral replication and CD4+ T cell depletion. J Immunol 2008;181(10):6687–91.
[49] Handley SA, Thackray LB, Zhao G, Presti R, Miller AD, Droit L, et al. Pathogenic simian immunodeficiency virus infection is associated with expansion of the enteric virome. Cell 2012;151(2):253–66.

CHAPTER

8

Innate Immunity in Simian Immunodeficiency Virus Infection

Roger Keith Reeves[1], *Steven E. Bosinger*[2]

[1]Division of Immunology, New England Primate Research Center, Harvard Medical School, Southborough, MA, USA; [2]Division of Microbiology and Immunology, Emory Vaccine Center, Yerkes National Primate Research Center, Atlanta, GA, USA

OVERVIEW

Since the discovery of Toll-like receptors (TLR) in the late 1990s, an explosion of research describing the interaction of HIV with the innate immune system has emerged in the literature. Most of this work has fallen into two broad categories: (1) elucidating the cellular machinery and signaling pathways responsible for recognition of HIV-derived pathogen-associated molecular patterns (PAMPs) and (2) examining the fate of innate cells and their contribution to the control of virus in HIV-infected individuals. More recently, the concept has emerged that unmitigated stimulation of the innate immune system by HIV PAMPs may be a significant contributor to the systemic immune activation associated with HIV disease progression. Finally, a number of studies have attempted to exploit the potency of TLR ligands by using them as adjuvants to enhance adaptive immunity in a number of preclinical vaccine candidates or as potential microbicides.

The simian immunodeficiency virus (SIV)/macaque model has been invaluable in addressing several of these questions. The primary focus of this chapter is on summarizing current concepts of the interplay of innate immunity and HIV infection that have been shaped by work

Natural Hosts of SIV.
http://dx.doi.org/10.1016/B978-0-12-404734-1.00008-5

done in the SIV/macaque model. The chapter will be divided into five parts:

1. Non-human primate (NHP) models of SIV infection
2. Innate immune sensors of HIV and SIV
3. SIV infection and cellular innate immunity
4. Physical and mucosal immune barriers in SIV infection
5. Innate immunity as a driver of immunopathogenesis in HIV/SIV infection
6. In vivo administration of innate ligands to NHPs.

Several topics peripheral to innate immunity will not be covered in detail within this chapter, because they are beyond its scope or are directly addressed in an adjoining chapter, and we will not directly describe the considerable research on adaptive immune responses in SIV infection or major histocompatibility (MHC) alleles associated with viral control, or discuss in detail HIV/SIV restriction factors.

NHP MODELS OF SIV INFECTION

Characteristics of the SIV/Macaque Model

Experimental inoculation of Asian macaque species (*Macaca mulatta*—rhesus macaque (RM), *Macaca fasicularis*—cynomolgus monkey, and *Macaca nemestrina*—pig-tailed macaques) with SIVmac or SIVsmm has resulted in the most relevant NHP models of in vivo HIV-1 infection [1]. The course of disease progression after SIV infection in macaques has proven to have a high degree of variability, and depends on the SIV isolates and molecular clones. Disease severity can range from very low or none, as in the case of cloned SIVmac142, to rapid and pronounced, as observed for SIVmac239 or SIVmac251 infections, which causes death in 25% of animals by 3 months postinfection [2,3].

The clinical disease associated with SIVmac infection of Asian macaques is strikingly similar to that observed in HIV-1-infected humans and can be classified into three stages: acute, post-acute asymptomatic and AIDS [2]. During the acute phase, occurring in the first 2–3 weeks after inoculation, a massive viremia develops, accompanied by large decreases in the frequency and absolute numbers of $CD4^+$ T lymphocytes, both in the peripheral circulation and in the gastrointestinal (GI) tissues [4]. Other clinical signs during the acute phase, such as fever, lymphadenopathy, rash, and malaise, are characteristic of simian AIDS [3]. SIV infection generally enters a clinically asymptomatic phase, during which the viremia is controlled and maintained at a stable set point and $CD4^+$ T lymphocyte levels undergo a slow but continual decline [2]. During the later AIDS stage, there is severe depletion of $CD4^+$ T cells and concomitant occurrence of opportunistic infections. Protozoal infections such as *Pneumocystis jiroveci* are usually the first to develop; however, viral (typically cytomegalovirus, adenovirus, and simian virus 40) and bacterial infections (*Mycobacterium avium*) are also prevalent [3]. Immunopathologies include hyperplasia within lymphoid follicles early in infection, switching to lymphoid depletion in advanced disease [3,5]. Thymic atrophy and inflammation of the lungs are also common [6–8]. Most pathogenic SIVmac viruses, such as SIVmac239, cause death from AIDS-like symptoms at approximately 1 year postinfection [9]. The similarities between human and simian AIDS suggest that the macaque model is highly relevant. The high efficiency of infection after challenge also makes it a useful model for vaccine efficacy studies.

Nonpathogenic/Natural Hosts of SIV Infection

Unlike SIV infection of Asian macaque species, infection of most African NHPs with their corresponding SIV does not induce an overtly appreciable disease. The notable exception is chimpanzees, in which pathology consistent with AIDS has been reported in experimental infection with HIV-1 [10] and in SIVcpz infections in the wild [11]. The best characterized natural host species are sooty mangabeys (SMs) and African Green monkeys (AGMs) [12]. SIV infection of natural hosts shares phenotypic similarities with pathogenic infection: a high level of plasma viremia (~10^5 copies/ml), tropism for $CD4^+$ T cells, and rapid but transient depletion of mucosal $CD4^+$ T cells during acute infection. Despite these similarities, in addition to remaining AIDS-free, SMs and AGMs lack many of the classical clinical features of pathogenic HIV/SIV: They maintain stable numbers of peripheral blood $CD4^+$ T cells and do not display bystander lymphocyte activation. The mechanisms by which SMs and AGMs remain AIDS-free have been extensively studied in recent years, and several hypotheses have emerged, as reviewed in Ref. [12]. One key difference appears to be tropism in key subsets of $CD4^+$ T cells: The $CD4^+$ T cells from SMs have a reduced expression of the SIV co-receptor CCR5 on central memory cells (Tcm) and $CD4^+$ Tcm cells harbor lower levels of SIV in vivo relative to their effector counterparts (Tem) [13]. A strong, inverse correlation between disease progression and preservation of the $CD4^+$ Tcm pool has been demonstrated in SIV-infected macaques; the ability of SMs to shunt the bulk of SIV replication away from this subset likely is linked to their ability to maintain stable bulk $CD4^+$ T-cell numbers. The precise immunological mechanisms by which natural host species remain AIDS-free are under intense investigation; however, they remain a crucial model system to dissect the immunological aspects of SIV infection and differentiate aspects of normal host response to a viral infection from a dysregulated, pathogenic immune activation driven by SIV. As we discuss in detail subsequently, in recent years, several studies have directly compared the innate immune response to SIV among natural pathogenic hosts, and a clear picture of the differences among species has begun to emerge.

INNATE IMMUNE SENSORS OF HIV AND SIV

A number of innate receptors have been demonstrated to mediate recognition of HIV. The best characterized are the TLRs, which recognize viral-derived ssRNA and viral DNA. Most of the current knowledge of innate recognition has come from in vitro studies of HIV-1; however, in a few examples, these studies have been extended to primate studies, typically with the following aims: (1) to empirically test innate stimuli for its adjuvant activity with candidate anti-HIV/SIV vaccines, or (2) to understand the contribution of innate stimuli to pathogenic immune activation. A thorough discussion of the innate immune receptors shown to recognize HIV can be found in Iwasaki [14].

Toll-like Receptors

Toll genes and their corresponding proteins, termed TLRs, were originally described for their antifungal properties in the *Drosophila* species [15]. Eleven mammalian homologs of TLRs that recognize different components of microbes, called PAMPs, have been identified to date [16,17]. Different leukocytes express specific profiles of TLRs; in humans, the plasmacytoid dendritic cells (pDCs) express only TLR7 and TLR9 [18], both of which, along with TLR3 and -8, belong to a TLR subfamily that detects PAMPS intracellularly [19–21]. Toll-like receptor-7 binds single-stranded viral RNA, whereas TLR9 binds hypo-methylated cytosine-phosphate-guanine (CpG) dinucleotides that are enriched in bacterial DNA but are less prevalent in eukaryotic DNA [20,22]. Signaling occurs when the ligands are endocytosed and transported to lysosomal compartments, where they bind their cognate receptors, TLR7 or TLR9 [23,24]. The ligand–receptor complex recruits the global TLR adaptor molecule MyD88, which initiates a signaling cascade of transcription factors including, nuclear factor-κB, activator protein-1, c-Jun N-terminal kinases, and extracellular signal-regulated kinases [19,24,25].

The best-studied TLR recognizing HIV and SIV is TLR7, which is expressed endosomally and recognizes ssRNA [26]. It is expressed at high levels in pDCs, and signaling via TLR7 induces high expression of interferon-alpha (IFN-α) [27]. Whereas HIV and SIV induce high levels of IFN-α by pDCs, these cells are only productively infected at low levels [28]. However, nonproductive internalization of HIV by pDCs is sufficient to induce IFN-α secretion [29,30], as noted with aldothiothreitol-inactivated SIV, which is capable of inducing high levels of IFN-α induction in RMs and SMs [31]. HIV-infected CD4+ T cells are more potent inducers of IFN production in pDCs than purified virion [32], and autophagy is required for efficient IFN production, which suggests that infected CD4+ T cells, rather than HIV itself, may be the primary stimuli in vivo. Upon stimulation, TLR7 interacts with the MYD88 adaptor molecule and activates phosphorylation of IRF7, which translocates to the nucleus and *trans*-activates type I IFN transcription [33]. During viral infection, IRF7 associates with TLR7 in early endosomes and initiates primarily IFN production. However feedback mechanisms induce maturation of the endosomes, consequently the IFN produced leads to an up-regulation of surface co-stimulatory molecules. In contrast to other viruses and TLR7 ligands, activation of TLR7 by HIV causes the TLR7/IRF7 signaling complexes to remain sequestered in early endosomes, with prolonged production of IFN-α [34].

Emerging Innate Receptors in the Recognition of HIV

Toll-like receptors that recognize HIV are expressed within endosomes, and require virion fusion and delivery to the endosome to become activated. However, until recently, it was unclear how cytosolic recognition of HIV occurs. HIV-1 virions harbor two positive-sense viral RNA genomes. The RIG-I family of helicases has been demonstrated to recognize viral RNA in the cytoplasm, and recently it has been reported that HIV genomic RNA activates RIG-I, the prototypical member of this family [35,36]. Activation of RIG-I in virally infected cells may be masked by HIV's ability to target the IRF3 transcription factor for degradation [35,37,38] and suppress IFN-beta expression. The lattice of HIV capsid also interacts with the restriction factor TRIM5 and enhances innate signaling [39]. Finally, an endogenous receptor for HIV DNA intermediates that arise during reverse transcription of the HIV genome has been hypothesized [14]. Activation of this receptor is masked by the cellular expression of TREX1, a host exonuclease that digests ssDNA [40] and prevents HIV DNA from accumulating to high enough levels to stimulate the as-yet undescribed sensor [41]. With the exception of TRIM5, as described subsequently, the role of these factors in infection at the organism level in either humans or NHPs has not yet been described.

Restriction Factors in SIV Vaccine Studies

Restriction factors are intracellular proteins that act directly to block viral infection at the intracellular level. Although they have primarily been studied for their activity against HIV and SIV in vitro, recent genetic data demonstrated that they can affect viral load and transmission in SIV infection. The most well-studied restriction factors against HIV are TRIM5a [42], APOBEC3G [43,44], Tetherin/BST-2 [45,46], and SAMHD1 [47,48], each of which has been able to demonstrate highly effective blockade against HIV infection in vitro. These restriction factors have been hypothesized to be important in safeguarding against transmission of lentiviruses between species [49]. TRIM5α is the best studied of the lentiviral restriction factors. Human TRIM5 has low activity against HIV; however, TRIM5 in RMs demonstrates much higher activity in vitro against SIV [42]. Furthermore, TRIM5 in macaques is highly polymorphic, and macaques inheriting restrictive alleles of TRIM5 have lower viral load [50,51] and are more resistant to low-dose challenge by the intrarectal [52] or penile [53] routes. Importantly, these studies have employed SIVsmm strains of SIV; other groups have shown that TRIM5 genotypes have no effect on viral replication or transmission of SIVmac strains [54]. Owing to their considerable influence on viral load and transmission, the prevalence of restrictive or permissive TRIM5 genotypes is an important consideration in SIV vaccine studies.

SIV INFECTION AND CELLULAR INNATE IMMUNITY

Dendritic Cells

The primary function of dendritic cells (DCs) is to act as potent or professional antigen-presenting cells that activate the adaptive immune response. Resident DCs at sites of infection express a multitude of innate pathogen recognition receptors (PRRs) described previously at the plasma membrane, cytoplasm, and endosome, with which they recognize PAMP-containing structures in foreign microorganisms such as viruses. Upon recognition, DCs become activated and traffic to secondary lymphoid organs, where they present foreign antigen via MHC class I and II surface molecules to $CD4^+$ and $CD8^+$ T cells. In this manner, DCs represent the interface between innate and adaptive immunity. Because of their primary importance in modulating T-cell responses, modulation of DC response by vaccination and adjuvants has been a subject of interest in the development of HIV vaccines; we will describe methodologies targeting the innate machinery in the subsequent section. In addition, over the past decade, it has been demonstrated that HIV and SIV are also capable of hijacking sentinel DCs and using them to disseminate from primary sites of infection to secondary lymphoid organs, which will be discussed later in more detail.

Dendritic Cell Subsets

Unlike lymphocytes, DCs are classified by both their cell-surface markers and anatomic location. In blood, the best-studied human myeloid DC subsets are typically defined as $CD11c^+$ and $CD1c/BDCA1^+$ Lin-cells, sometimes referred to as conventional or classical DCs (cDCs). Phenotypically, these cells have been demonstrated to be similar in RMs and SMs [55]. However, in RMs, several lineage markers are present on cDCs that are absent on their human counterparts, including CD16, CD56, and CD11b [56,57]. More recently, novel blood DC subsets have been described in humans: $CD141^+/BDCA3^+$ DCs, capable of efficient cross-presentation and representing the functional analog of mouse CD8α DCs [58]. However, similar subsets of these cells have not been reported as yet in NHPs. A third subset of nonmyeloid blood DC, the pDC (discussed subsequently), differs significantly from myeloid DCs in that its primary function appears to be the secretion of high amounts of IFN-α, rather than antigen presentation. Within the skin and mucosa, Langerhans cells (expressing langerin) are prevalent in the epidermal layers as two dermal DC subsets: $CD1a^+$ or $CD14^+$. More recently, gut-resident $CD103^+$ DCs have been described in RMs and humans [59] that promote tolerance by induction of T-regulatory cell response and maintain homeostasis of the gut epithelial barrier (described in detail later). Of interest, this subset is also notable for being depleted from the mucosa during SIV infection [60].

Highjacking of DCs for Viral Entry and Dissemination

Early studies in vaginal transmission models of SIV demonstrated that SIV enters the mucosa and nonproductively infects resident DCs [61] and LCs [62]. Within 24 h of infection, however, the HIV-infected DCs are detectable in the draining LNs [61]. HIV-1 binds to a C-type lectin receptor, DC-SIGN (CD209), on the surface of mucosal DCs, and potently enhances *trans*-infection of $CD4^+$ T cells. Similar activities have been demonstrated for DC-SIGN + DCs in RMs, chimpanzees, and AGMs in vitro [63–65].

Plasmacytoid DCs

pDCs are a unique subset of DCs that are present at low frequencies in peripheral blood. Unlike most DCs, the primary function of pDCs does not seem to be antigen presentation, but production of massive levels of the antiviral cytokine IFN-α [66]. pDCs express high levels of the interleukin (IL)-3Rα chain, CD123 [67–70], as well as BDCA-2 and -4 [71]. Although the function of BDCA-2 is still being defined, BDCA-4 is identical to neuropilin-1, a neuronal receptor for axon guidance factors and a receptor on endothelial and tumor cells for vascular endothelial growth factor (VEGF-A) [71–73]. Isolated pDCs are characteristically smooth round cells, whereas mDCs have prominent dendrites [74]. These morphological differences suggest that pDCs in blood are generally more immature, a finding supported by low cell surface expression of the T-cell co-stimulatory molecules CD40 and CD80/CD86 [75]. pDCs were first recognized in human lymph nodes as cells with plasma cell morphology but devoid of B cell and plasma cell markers [76]. They were originally referred to as plasmacytoid T cells because of their close association with T cells in lymph nodes [77], but the nomenclature was later changed to plasmacytoid monocytes when they were found to share some common markers with myelomonocytic cells [74]. Several research groups recognized concurrently that stimulating peripheral blood mononuclear cell (PBMC) cultures with viruses or virus-infected cells resulted in high concentrations of IFN-α accumulation in the supernatants, but only a small number of cells appeared to be responsible for the bulk of IFN-α production [78–80]. These cells, dubbed natural interferon-producing cells (nIPCs), lacked T, B, NK, and monocytic markers, but expressed MHC-II molecules, and were clearly distinct from classical peripheral blood DCs [78,80,81]. After nearly a decade of independent study, both Siegal et al. [82] and Cella et al. [83] determined that plasmacytoid monocytes and nIPCs in peripheral blood and secondary lymphoid tissues were the same cell type, whereupon the designation pDCs was adopted.

In response to a wide range of viruses and bacteria, pDCs produce up to 10 pg of IFN-α/activated cell, which is 10-fold greater than the amount of IFN-α produced by activated monocytes [84]. Although IFN-α1 is predominant, human pDCs secrete most of the 13 IFN-α subtypes, as well as moderate levels of IFN-β [85,86]. Interferon regulatory factors (IRFs) control expression of both IFN-α and -β, with IRF-5 and -7 controlling IFN-α secretion in response to virus infections. pDCs express most of the IRF genes, and in particular constitutively express high levels of IRF-5 and IRF-7, which can be detected by the presence at both the mRNA and protein levels [86]. The high level of constitutive IRF-7 expression in pDCs is thought to contribute to rapid IFN-α synthesis in response to virus infections [87]. pDCs are adept at recognizing and becoming activated in response to a variety of viruses including hepatitis C virus (HCV), human T-lymphotropic virus (HTLV), and herpes simplex virus 2 (HSV-2). pDCs have been directly linked to resolution of disease in respiratory syncytial virus, Dengue fever virus, and HSV-2 [88–90], but are also depleted in a number of diseases including HTLV, HCV, severe acute respiratory syndrome (SARS) coronavirus (SARS-CoV), and HIV [91–97].

pDCs have been well studied during HIV infection. Human pDCs stimulated in vitro with infectious or inactivated HIV, or even with recombinant gp120, secrete IFN-α, and up-regulate CD40, CD83, and CD86 [98–102]. Compared with uninfected controls, circulating pDCs in HIV-infected patients express higher levels of CD86 on the cell surface [103]. Although both TLR7- and TLR9-dependent mechanisms of HIV-mediated pDC activation in vitro have been described, activation in vivo primarily depends on TLR7 pathways and contributes to the overall immune activation observed in advanced HIV disease [104–106].

However, in chronically infected HIV patients, the frequencies and absolute numbers of circulating pDCs are reduced, as is pDC-dependent IFN-α production ex vivo; these reductions generally correlate with increasing viral loads and decreasing numbers of $CD4^+$ T cells [103,107–110]. Downregulated CCR7 expression on pDCs in HIV patients is also related to viral load [111]. After the administration of highly active antiretroviral activity, circulating $CD4^+$ T-cell numbers usually increase and plasma viremia declines; however, pDC numbers are only partially restored and IFN-α secretion remains defective [112,113]. Furthermore, pDCs isolated from HIV patients were impaired in their ability to stimulate allogeneic T cells, and not only harbored proviral DNA, but also transferred virus to $CD4^+$ T cells in co-culture [100,114,115]. HIV-1 isolates that used both R5 and X4 infected human pDCs productively in vitro, although R5-tropic viruses tended to replicate more efficiently [116–118]. In whole PBMC cultures infected with R5-using strains, pDCs had higher proviral loads than $CD4^+$ T cells, which suggests that pDCs are preferentially infected [119].

Similar to what has been observed in HIV, pDC depletion has been demonstrated during acute and chronic SIV infection of RMs, pig-tailed, and cynomolgus macaques [56,120,121], but not nonpathogenic host species of SIV such as SMs and AGMs [122–124]. Interestingly, studies of acute pathogenic SIV infection in macaques have demonstrated a transient increase of pDCs in peripheral blood after rapid egress from the bone marrow, followed by depletion of circulating pDCs and accumulation of apoptotic pDCs in lymph nodes [57,120]. Furthermore, recent evidence from multiple laboratories has shown that pDCs are not necessarily depleted during pathogenic SIV infection, but rather, accumulate in large numbers in the GI tract [124–126]. Of note, this phenomenon appears to be absent in nonpathogenic SIV infections.

Natural Killer Cells

Natural killer (NK) cells are generally thought of as the primary effector cells of the innate immune system. Although NK cells are prototypically thought of as cytotoxic cells, in which they are among the first cells to intersect pathogens and eliminate neoplastic cells, they also have more recently been shown to have significant homeostatic and regulatory functions [127–132]. In humans, two subsets of NK cells predominate [129]. The dominant subset in peripheral blood is $CD16^+CD56^{dim}$, which primarily mediates cytotoxic activity and secretes relatively little IFN-γ. In contrast, a distinct subset of $CD16^{-/low}CD56^{hi}$ NK cells displays little cytotoxic activity but secretes relatively large amounts of IFN-γ. The $CD56^{hi}$ subset also expresses CCR7 and CD62L and is the dominant population found in lymph nodes and in tonsillar tissues [133]. Mucosal NK cells have also been identified throughout the digestive tract, from the labial and oropharyngeal mucosae to the large and small intestines [132,134–136]. Mucosal NK cells are typically CD56$^+$ and can be found both within the lamina propria and localized to intraepithelial tissues. Multiple mucosal NK cell subpopulations described in both humans and macaque models reveal significant diversity, but generally express low levels of cytotoxic enzymes such as granzyme and perforin and robust secretion of IFN-γ and tumor necrosis factor-α (TNF-α) [132,137–139]. NK cells have also been found throughout the female reproductive tract and have a major role in placental implantation and modulation of pregnancy [127–132].

Because of the significant interest in NK cell biology in the context of tumor biology and infectious diseases, including HIV/SIV, accurate definitions of NK cells in animal models, including NHPs, are of critical importance. Recent studies have accurately identified macaque NK cells in the peripheral blood as being CD3$^-$CD8αα$^+$NKG2A$^+$ cells, and then further divided them into three distinct subpopulations: a major NK cell population that is CD16$^+$CD56$^{-/dim}$ and two minor populations, CD16$^{-/dim}$CD56^{hi} and CD16$^-$CD56$^-$. A subsequent publication by another group validated this definition of macaque NK cells and demonstrated that it applies to SMs [140]. The CD56/CD16 functional profiles found in human NK cells seem to be analogous in NHPs, but interestingly, whereas CD16$^+$ NK cells predominate in peripheral blood, most NK cells in tissues are CD56$^+$ or DN.

NK cells have long been shown to inhibit HIV replication in in vitro cultures with β-chemokine-dependent inhibition and NKG2D ligand-mediated killing as a dominant mechanism [141]. Interestingly, acute HIV disease is associated with expansion of the CD56^{dim}CD16$^+$ NK cell subsets in humans, which suggests a virus-driven increase in cytolytic activity [142], and long-term nonprogressors have increased NK cell cytotoxicity compared with viremic individuals [143]. With the advent of more precise molecular profiling, it has recently been shown that KIR3DS1 and its putative ligand HLA-B Bw4-80I are also associated with slower disease progression, and the interaction of the two molecules leads to inhibition of HIV-1 replication in vitro [144].

A number of studies suggest that NK cells can help mediate control of SIV infection. NK cells have been shown to readily lyse SIV-infected T cells or cells pulsed with SIV [145]. Acute infection with SIVmac251 induces activation of macaque NK cells, reflected by the up-regulation of CD69 expression and increased lysis of susceptible K562 cells [146]. Longitudinal studies following SIV infection demonstrate that early NK cell responses can lead to long-term virus control in RMs, as well as in the natural host species, SMs and AGMs [140,147]. Mathematical models suggest that NK cells can modulate SIV replication and disease outcome in SMs, but not in RMs [148]. During SIV infection, NK cells in the mucosa up-regulate cytotoxic functions, and an overall increase in mucosal homing of NK cells is evident [138,139,149]. In pig-tailed macaques infected with SIV, an inverse association between the magnitude of the NK cell response and the probability of development of neuro-AIDS was observed [145]. Macaque studies have thus far yielded conflicting results on the role of Killer cell immunoglobulin-like receptors (KIRs) in controlling SIV infection, with some studies suggesting positive control in the presence of cognate ligands [150,151]; others have suggested that KIRs are related to loss of control of SIV replication [152]. Although imperfect, macaque models have allowed for partial depletion of NK cells in vivo. Depletion of NK cells using a JAK3 inhibitor resulted in transient increases in viral loads, which suggests that there may be some role for NK cells in regulating viral control [153]. However, anti-CD16 depletion [154,155] showed no effect on early infection, but these studies may have been limited by inadequate depletion in tissues and a lack of effect on non-CD16 NK cell subpopulations.

Although NK cells generally exert their functions, either cytolytic or regulatory, directly on target cells, they can also cooperate with the adaptive immune system. Antibody-dependent cell-mediated cytotoxicity (ADCC) responses develop within the first few weeks postinfection [156] and are thought typically to be exerted by NK cells, but can also involve neutrophils. ADCC activity has been associated with delayed progression to AIDS in SIV-infected macaques [157], involved in the efficacy of live attenuated SIV vaccines [158], and with reducing viral loads in various nonreplicating vaccine modalities [159–161], including the RV144 HIV vaccine trial [162,163]. Anti-HIV ADCC antibodies have also been observed in highly exposed seronegative individuals [164], and are associated with the control of virus replication in HIV elite controllers and slow progressors [165,166].

Th17 Cells in HIV/SIV Infection

Another prominent $CD4^+$ subset in the mucosa that is profoundly affected by HIV and SIV infection are Th17 cells [167,168]. The Th17 subset represents an independent lineage from other Th cell populations in both function (high production of IL17A, IL17F IL22, IL21, and IL26) and differentiation (induced by the ROR$\gamma\tau$ transcription factor). Although they are rare in the periphery, Th17 cells are highly enriched in the lamina propria (LP) [168,169]. Th17 cells function to coordinate innate immunity against bacterial and fungal properties by regulating granulopoiesis, recruitment of neutrophils, and induction of antimicrobial peptides [170]. Th17 lymphocytes are also integral to maintenance of an intact epithelial barrier via secretion of IL22[170]. Interleukin-22 acts on epithelial and other nonhematopoietic cells to promote differentiation and proliferation, and triggers production of β-defensins, S100 family proteins, RegIIIγ, and other antimicrobial compounds [167,170,171]. During HIV and pathogenic SIV infection [172,173], Th17 are depleted from the mucosa but are maintained in natural hosts such as SMs. HIV-infected long-term non-progressors also maintain higher Th17 levels than patients with progressive disease. The loss of Th17 cells and other IL17-producing cell types such as innate lymphoid cells (ILCs) [60], and $CD8^+$ IL17 cells [174], also occurs during pathogenic infection. In 2006, a seminal work by Brenchley and Douek and colleagues showed that during pathogenic HIV and SIV infections, a significant amount of LPS and other microbial byproducts could be detected in the plasma [175]. This process, which they termed microbial translocation (MT) (described in greater detail subsequently), was argued to be a natural sequelae of the massive $CD4^+$ T-cell depletion and disruption of lymphoid architecture observed in the mucosa, and could be a significant source of the systemic immune activation observed in HIV and SIV infection. Later, Raffatellu et al. [172] demonstrated that the loss of Th17 cells was associated with translocation of *Salmonella* in SIV-infected macaques, providing strong evidence linking Th17 loss with breakdown of the epithelial barrier, a role that seems logical considering their role in IL22 production [172]. Providing another link between the loss of Th17 cells and SIV pathogenesis was the demonstration that whereas Th17 are depleted from SIV-infected macaques, they are maintained in infected, nonpathogenic species such as SMs [173].

The mechanism by which Th17 cells are depleted is an area of active investigation. Although it has been reported that Th17 can be efficiently infected in vitro [176], they are not preferentially infected in vivo [173]. Furthermore, that other cells capable of secreting IL17 that do not bear CD4 are also depleted from the mucosa during SIV infection suggests a mechanism independent of direct infection of IL17-producing cells. One likely mechanism may be the loss of IL21-producing cells, because the depletion of $IL21^+CD4^+$ T cells correlates with the loss of Th17 cells [177], and exogenously given IL21 attenuates the loss of Th17 cells during SIV infection [178]. An alternative, overlapping mechanism, may be due to the dysregulated metabolism by activated DCs and APCs expressing the indoleamine 2,3-dioxygenase (IDO1) enzyme responsible for the catabolism of tryptophan. IDO1 activity and its catabolites inversely correlate with Th17 levels in HIV-infected individuals [179].

Innate Lymphoid Cells

Recently, a subpopulation of mucosae-restricted cells that bear features similar to both NK cells, Th17 and Th22 cells, has been identified in mice, humans, and RMs [139,180–186]. These cells have been termed ILCs and are partially identifiable by the high expression of NKp44; they are predominately found in mucosa-associated lymphoid tissues, including all components of the GI tract, including gut-draining lymph nodes, tonsils, and lungs. ILCs express high levels of the Th17 cell transcription factor RORγt and share a phylogenetic lineage with lymphoid tissue-inducing cells (LTis) [187], but are generally non-cytotoxic and secrete large quantities of IL17, IL22, IL26, and CCL6. Unfortunately, due to limited access to mucosal tissues in humans, ILCs have been problematic to study and their functional niche in primates is unclear. This caveat has been partially alleviated by the more recent description of this cell type in nonhuman primates by multiple groups [139,188,189]. Interestingly, during SIV infection ILCs are massively depleted in number and have a significantly altered functional repertoire. Furthermore, loss of ILCs negatively correlates with breaches in the gut epithelia during chronic SIV infection, suggesting they may be necessary for the maintenance of gut homeostasis [60]. The link is likely to be functionally related given the prominent roles IL-17 and IL-22 play in the regulation of gut integrity. Although the mechanism of ILC suppression during SIV infection is not entirely clear, it appears to be partially linked to increases in soluble inflammatory mediators in the GI tract and an accumulation of indoleamine 2,3-dioxygenase 1 (IDO1) catabolites [139].

PHYSICAL AND MUCOSAL IMMUNE BARRIERS IN SIV INFECTION

Innate Physical Nonimmune Barriers in HIV/SIV Infection

In general, the rate of HIV/SIV transmission in the absence of vaccines is relatively low, partially because of low virus infectivity and sufficient innate immune responses. However, although most transmission events occur at vaginal or rectal mucosal surfaces, the role that physical barriers have in these tissues in blocking transmission has been traditionally ignored. Empirical evidence suggests that physical barriers may limit many transmission events because intravenous infection (where such barriers are obviously absent) has significantly higher rates of transmission compared with mucosal HIV infection [190]. Furthermore, intravenous (IV) infections characteristically have greater numbers of viral variants, with as many as 16 founder viruses that have been observed in samples from HIV-1–infected intravenous drug users. By comparison, vaginal transmission is typically characterized by one to three founder populations and rectal transmission, both experimentally and in cohorts of men who have sex with men (MSM), has a greater number of founder virus populations than vaginal transmission, but fewer than intravenous transmission [191–194]. Finally, although founder virus populations have yet to be enumerated in cases of oral HIV transmission, this route of infection is generally thought to be inefficient. These data detailing the relative success rates of virus transmission by various routes offer clues about the substantial role that physical barriers have in preventing HIV infection.

During mucosal transmission of HIV/SIV, infection likely involves one of four mechanisms: (1) direct infection of epithelial cells; (2) transcytosis through epithelial cells; (3) transmigration of infected cells; or (4) direct virus entry through breaches in the epithelial layer. All of these methods feature a role for the mucosal epithelia. Interestingly, HIV has evolved to disrupt mucosal epithelia even in the absence of infection [195], which further suggests that these physical barriers put selection pressures on the virus. Thus, we can ask which characteristics of the various epithelia may be involved in delaying spread of virus or blocking infection entirely. Most HIV transmissions occur across the vaginal epithelia, but transmission rates are higher across the rectal mucosa. The mean thickness of human epithelia in rectum is 24.6 mm, compared with vaginal epithelia, with a mean thickness of 215.5 mm, a 10-fold difference [196]; and the rectal mucosa is far more vulnerable to epithelial breaches during intercourse. Furthermore, most of the vaginal epithelia are stratified squamous arranged in layers compared with single-layer columnar epithelium in the rectal mucosa. However, the thick vaginal epithelial layer thins during the luteal phase of the menstrual cycle as a result of increased progesterone levels, and elevated infection rates are observed during this period. In fact, experimentally induced thinning of the vaginal epithelia in macaque models using Depo-Provera treatment is commonly used to increase infection efficacy [197].

Intestinal epithelia are generally thought of as simply providing a physical barrier; however, they have an active role in mucosal immunity by production of a broad array of mucins, inflammatory cytokines, and chemokines [198], including β-chemokines capable of blocking HIV/SIV infection directly. Increased elasticity of vaginal mucus and a low pH in the follicular phase are also thought to prevent transmission, and during the luteal phase, mucus thinning and increase in the pH correlate with increased transmission rates [197]. Cervicovaginal mucus inhibits HIV diffusion and penetration, and provides another physical barrier to infection [199].

Finally, oral HIV/SIV infection occurs at low rates during mother-to-child transmission [200], but it is debatable whether sexual oral transmission occurs at a detectable frequency. Regardless, oral transmission is inefficient compared with other routes of transmission. This may be partly because of particularly strong physical barriers in the oral mucosa, including the thickest epithelial layer among mucosal surfaces, with a mean thickness of 263 mm [196]. The oral mucosa is also bathed in saliva, which contains multiple salivary proteases and amylases that are known to inhibit virus infectivity, as well as mucins that can physically trap virus particles [201].

Mucosal Immunity

The epithelial lining that includes luminal mucus layers provides a considerable physical and biochemical barrier to microbes. The chemical barriers include secretory immunoglobulin A (IgA), defensins, RegIII proteins, and other soluble factors [202]. Underlying and within the epithelial lining is a complex mucosal immune system composed of its own specialized structures and diverse immunological subsets [203]. The mucosal immune system faces a unique challenge in the constant barrage of foreign antigens represented by the microbiota present in the intestinal lumen. The mucosal compartment is a vibrantly dynamic system in which each of the three major components (the epithelial barrier, the mucosal immune structure including the cells that form the structure, and commensal microbiota) all actively contribute to the development and homeostasis of the system as a whole. Perturbations to one part of the system, as occurs during HIV/SIV infection, can cause significant disruption of the others [171].

Structurally, the mucosal immune system has several features analogous to the classical immune system. Activation of adaptive responses is initiated in inductive sites such as mesenteric LNs, Peyer patches (PP), and isolated lymphoid follicles (ILFs) [203]. A cardinal feature of the PP and ILF that distinguishes them from nonmucosal LNs is the encapsulation by specialized follicle-associated epithelium that contains pinocytotic M cells, which are responsible for sampling the antigenic content of the intestinal lumen. After presentation to T and B cells, they migrate to mucosal effector sites such as the LP and the basal layer underlying the gut epithelium, and within the intra-epithelium lymphocyte compartment (IEL).

The mucosal immune system is also enriched in several unique cellular subsets that are found only minimally in the periphery. Immunoglobulin A–producing B cells and plasma cells reside in the LP and produce polymeric IgA, which are transported to the intestinal lumen and contribute to the soluble immunity provided by the extra-epithelial mucus layer. Within the IEL, most T cells are $CD8^+$ as either CD8αβ heterodimers or CD8αα homodimers [169,203,204]. The IEL $CD8^+$ compartment is split into a TCRαβ subset that resembles conventional $CD8^+$ T cells, but is enriched for cells expressing TCRγδ chains [203]. The frequency of γδ T cells is about 4% of the T-cell population in the peripheral blood, but can comprise up to 40–50% of the mucosal T cells [171]. Gamma-delta T cells express TLR1 and TLR2 and produce chemokines that recruit other effector cells during bacterial infections and can make IL17 [205]. In humans and NHPs, these cells express the TCR Vδ1 or Vδ2, and the Vδ1 population expands, but their ability to produce IL17 is impaired [205,206].

In contrast to the IEL, most T cells in the lamina propria express CD4 [169]. Lamina propria–resident $CD4^+$ T cells are predominantly activated and express CCR5, which makes them efficient targets for

HIV/SIV infection [167,207]. Whereas the $CD4^+$ compartment is heterogeneous, there are three predominant populations: Th1, Th17, and $CD25^+$ Treg cells [169]. The fate of mucosal T cells during HIV and SIV infection has been well studied (reviewed in Ref. [208]) and they undergo a rapid and massive depletion [209–211] that is largely refractory to reconstitution during antiretroviral (ART) administration.

INNATE IMMUNITY AS A DRIVER OF IMMUNOPATHOGENESIS IN HIV/SIV INFECTION

Persistent Activation of Bacterial PRRs Induced by Microbial Translocation

As described earlier, disease progression in HIV infection has been demonstrated to be intimately linked to persistent activation of the immune system manifested as elevated rates of T-cell turnover [212], expression of activation markers on T and B lymphocytes [213], and elevated plasma levels of proinflammatory cytokines [214], among others. The cause of immune activation is likely multifactorial, and one of the widely accepted hypotheses is that the persistence of virus provides unabated stimuli for activation of both the innate and adaptive immune systems [122]. However, immune activation is not completely eradicated in HIV-infected patients who have effectively suppressed virus after the institution of ART [215]. The residual immune activation in virologic responders has been argued to result from low levels of viral replication, but it could also be caused by alternative mechanisms of immune activation that are independent of viral replication [216]. In the late 1990s and early 2000s, several groups demonstrated that HIV and SIV infection caused a rapid and near total depletion of $CD4^+$ T cells in the GI tract [209–211,217,218]. These findings led Brenchley and Douek to propose a model in which the initial insult to the mucosal lymphocyte compartment during primary infection established a permanent perturbation to the host immune system [208]. Shortly thereafter, the same group demonstrated that elevated levels of bacterial structural components such as lipopolysaccharide (LPS) could be detected in the plasma of HIV-infected patients and SIV-infected RMs in the absence of overt bacteremia, and correlated with lymphocyte markers of immune activation [175]. These findings provided a mechanism by which immunological damage inflicted during early infection could translate into chronic immune activation. These views led to coining of the term "microbial translocation," and the model postulates that depletion of gut $CD4^+$ T cells results in disruption of the epithelial barrier structures of the GI tract that allows for penetrance of microbial byproducts to enter the systemic circulation, where they provide continual stimuli for innate pattern recognition receptors [167,219]. Evidence of elevated microbial components in plasma in HIV infection has since been replicated by several groups [220–223] Data from the NHP model provided direct visualization of LPS in breaches of the mucosal barrier in SIV-infected RMs [224], and the absence of MT in natural host species has helped link MT to pathogenesis [175].

Studies of MT in HIV infection have involved the detection and quantitation of plasma levels of LPS or sCD14 using appropriate assay systems; however, other models have demonstrated that translocation of several microbial byproducts capable of activating multiple PRRs also occurs, including peptidoglycan, lipoteichoic acid, flagellin, and CpG motifs in bacterial DNA. Recently, expansion of several viral species has been demonstrated to occur in the mucosa of SIV-infected macaques, but it is not yet known whether these viruses can be detected systemically [225]. The translocation of these products correlates with elevated levels of the pro-inflammatory cytokines IL6, IL8, and IFN-α. Moreover, IFN-α expression co-localizes with bacterial proteins in the GI tract of SIV-infected RMs. In this manner, microbial translocation may be an alternate mechanism of pathogenic stimulation of viral and bacterial innate pathways during HIV infection.

Plasmacytoid Dendritic Cell Activation and Chronic Production of Type I Interferon

Most chronic viral infections are efficiently controlled by the immune system and remain largely latent except for minor bursts of viral replication [226]. HIV and SIV differ from most chronic viral infections in that in the absence of ART therapy, HIV and SIV replicate at high levels (~10^5 RNA copies/ml plasma) in most subjects for life. The persistent presence of viral particles provides perpetual stimuli for both innate pattern recognition receptors and antigen-specific cells, and is considered one of the primary sources for the chronic immune activation observed in pathogenic HIV/SIV infection. Unabated antigen stimulation as a cause of immune activation and $CD4^+$ T-cell depletion have been extensively described [227]. In recent years, more attention has been directed toward the effects that chronic virus replication has on the innate system. In particular, the role of activated pDCs and subsequent IFN-α production as a factor in driving pathogenesis has garnered significant interest [228]. As described earlier, exposure of HIV particles to pDCs initiates high levels of IFN-α and inflammatory cytokine production via the TLR7 and TLR9 receptors and the activation of the transcription factor IRF7. Several lines of in vitro evidence have implicated pDCs and/or IFN-α production in HIV pathogenesis: (1) pDCs activated by viral particles induce apoptosis of $CD4^+$ T-cell through the up-regulation of TNF-related apoptosis inducing ligand (TRAIL/TNFSF10) [229,230], and

(2) HIV-derived RNA treatment of pDCs stimulates the production of IDO, which skews T-cell maturation toward Treg development, which typically suppresses antiviral immunity. An overview of models of the contribution of pDCs to chronic immune activation is summarized in Figure 8.1. In pathogenically infected rhesus, cynomolgus, or pig-tailed macaques, SIV infection initiates and sustains the chronic induction of interferon-stimulated gene (ISG) expression, which lasts indefinitely [231–235]. Similarly, persistent ISG expression is detected in HIV-infected humans [236,237]. In contrast, natural hosts of SIV that avoid AIDS, such as SMs and AGMs, demonstrate a transient ISG induction that disappears during chronic infection [232–235]. pDCs have been reported to be depleted from the blood of SIV-infected RMs [238]; subsequent studies demonstrated that they relocate to the mucosa in large numbers rapidly after infection [124,125,239] and co-localize with strong ISG expression.

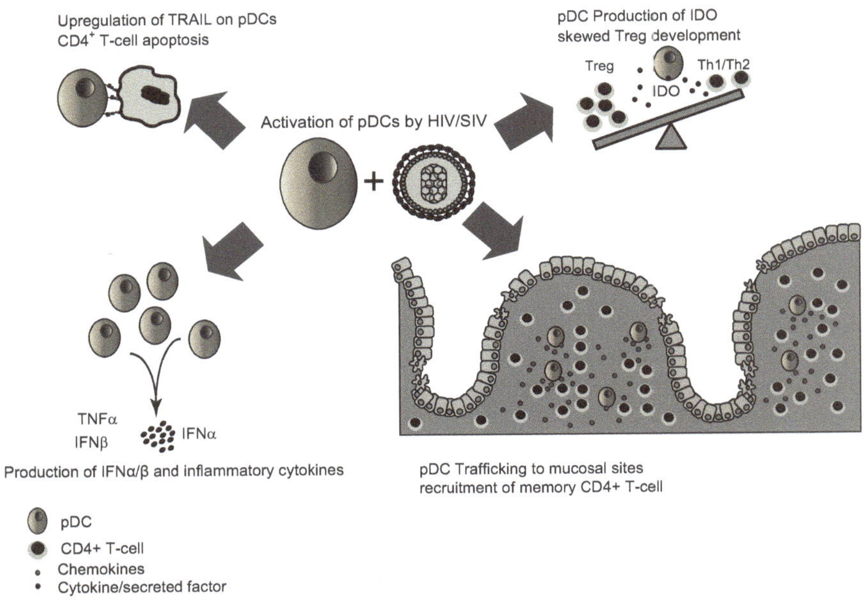

FIGURE 8.1 Current hypotheses of pDC-driven immune activation in HIV/SIV infection. Activation of pDCs by HIV/SIV leads to their activation and secretion of high levels of IFNα and -β and other inflammatory cytokines. Plasmacytoid dendritic cell activation by HIV or TLR7/8 ligands up-regulates surface expression of the TRAIL molecule, and TRAIL+pDCs are capable of driving apoptosis in CD4$^+$ T-cell lines by TRAIL-dependent pathways. Activated pDCs produce high levels of IDO that influence CD4$^+$ T-cell maturation toward a Treg phenotype and inhibit classical CD4$^+$ T-cell responses. Finally, in early SIV infection, pDCs migrate to the vaginal and rectal mucosa, where they are associated with expression of ISGs and recruitment of CD4$^+$ T cells by chemokines such as MIP3α. (See color plate at the back of the book.)

Plasmacytoid dendritic cells do not accumulate in the gut of SIV-infected SMs [124]. The absence of long-term ISG expression and pDC re-localization in species that resist SIV-disease suggests that aberrant activation of pDC contributes to pathogenesis in HIV/SIV-infected humans and macaques, respectively. The molecular mechanisms underlying the divergence between natural hosts and pathogenic species are currently unknown: A previous study reported that IFN-α production by SM pDCs was defective because of species-specific mutations in IRF7 [55]; however, those observations were later demonstrated to be largely the result of technical errors and sequencing artifact [232]. Further elucidation of differences in the innate signaling pathways between pathogenic and disease-resistant model species will likely provide valuable insight into the molecular basis of pathogenesis.

IN VIVO MANIPULATION OF THE INNATE IMMUNE SYSTEM IN NHPs

Over recent years, a number of studies have attempted to manipulate the innate system in NHPs, to affect SIV transmission rates during experimental vaccine studies or alter disease progression. The goals of these studies have largely fallen into two categories: (1) to understand the role of innate signaling as a cause of chronic pathogenic immune activation in SIV infection, and (2) to test the efficacy of TLR ligands in enhancing protection from SIV challenge.

In vivo Pathogenesis Studies

In what are now considered classic studies, immune activation has been demonstrated to be a stronger predictor of mortality and disease progression in HIV infection than CD4 counts and/or other markers [240]. These have been complemented by comparative studies of SIV infection in pathogenic or nonpathogenic infection [241]. Rhesus and pig-tailed macaques, which both ultimately develop disease during SIV infection, show widespread innate signaling and interferon responses well into chronic infection; in contrast, nonpathogenic species such as AGMs and SMs rapidly control this response to pre-infection levels.

Two primary pathways have been implicated as the most likely candidates for pathogenic innate signaling: (1) bacterial ligands and associated TLR receptors (TLR2 and TLR4), and (2) unabated stimulation of innate receptors for viral PAMPs (TLR3, TLR7, TLR9, and TLR9) provided by persistent replication of HIV/SIV. The rationale for bacterial signaling stems from a seminal study in which HIV-infected patients and SIV-infected macaques were demonstrated to have elevated plasma levels of

LPS owing to increased infiltration of micro-particles from gut-resident bacteria [175]. This phenomena, referred to as microbial translocation, is thought to be the natural sequelae of the massive depletion of mucosal CD4+ T cells [218], Th17 cells [173], and CD103+ mucosal dendritic cells [60], and subsequent disruption to mucosal structural integrity [224]. In a logical progression from the microbial translocation model, two studies administered LPS to nonpathogenic SIV+AGMs to determine the potential for inducing immune activation in a manner similar to that seen in macaques. A single bolus injection of LPS resulted in elevation of activation markers on CD4+ T cells that lasted several days [242]. A follow-up study using multiple doses showed that exogenous LPS could also induce elevated levels of coagulation markers similar to those observed in pathogenic SIV and HIV infections [243].

Other facets of the innate system that have been considered a likely candidate for driving host immune activation are viral PRRs engaged by the persistent stimuli provided by HIV/SIV replication. The persistent interferon signaling that is observed in humans and pathogenic hosts, but not nonpathogenic hosts, provided some of the rationale for this hypothesis [232–235,244]. Prolonged administration of viral TLRs in mice results in lymphopenia and degradation of lymphoid architecture [245]. To test the role of innate signaling and resultant chronic interferon in pathogenesis, Vanderford and colleagues [246] administered recombinant IFN-α2 to SIV-infected SMs that were otherwise clinically benign. Although the IFN-α elicited transient increases in activated CD8+ T cells, this activation was not sustained, and no CD4+ T-cell depletion was observed in blood, lymph node, or mucosa. Recently, Kader and colleagues performed the reciprocal experiment, attempting to block TLR7 and TLR9 signaling using a synthetic antagonist in SIV-infected RMs [247]. No decrease was observed in CD4+ or CD8+ T-cell activation in animals receiving blockade compared with saline controls. However, blockade was also unable to inhibit the expression of interferon-stimulated genes in vivo, which made it difficult to interpret the link between ISGs and T-cell activation.

As yet, modulation of the innate system in vivo has been unable to unequivocally reverse SIV-related immune activation in macaques. Attempts to break the nonpathogenic phenotype maintained by natural hosts using in vivo manipulation of the innate system have resulted in relatively modest increases in activation but have not induced widespread dysregulation of the immune system observed in pathogenic hosts. This likely reflects the considerable technical difficulty in performing in vivo interventions in NHPs, and that the mechanisms of immunopathology (or nonpathogenesis) are multifactorial and difficult to target with a single modality. Nevertheless, these studies provide a technical framework for future studies.

Harnessing of the Innate System as Adjuvants in HIV Vaccines

Rationale of NHPs in SIV Challenge Studies

Many of the most effective vaccines are live attenuated strains of the pathogen. To date, live attenuated SIV strains have demonstrated by far the highest efficacy of experimental vaccines against SIV in their ability to protect from acquisition in challenge studies. Although safety issues preclude the use of live attenuated HIV as a human vaccine, the superior protection afforded by whole virus has fueled inquiry aimed at defining the underlying immunological mechanisms responsible. In contrast, subunit vaccines targeting individual viral particles or combinations of them are typically much poorer immunogens on their own. Improvement in vaccination efficacy has been shown to be enhanced through the use of adjuvants, which for the better part of the century have chiefly been accomplished with the use of formulations of aluminum salts or oil–water emulsions [248]. However, over the past decade, the study of vaccinology as a science has largely shifted from an empiric science to that of reductionist immunology and continues to evolve, now employing high-throughput "omics" approaches [249]. Whereas the concept has long been observed that nonself signals provided by live attenuated organisms provide a robust activation signal, the demonstration that individual TLR ligands could polarize T-cell responses toward Th1 or Th2 activity [250] or target specific DC subtypes [251,252] suggests that manipulation of the innate immune system could result in fine-tuning vaccine responses. Since the early 2000s, much progress has been made in understanding the effect of adjuvants composed of bacterial and viral TLR ligands and how they affect vaccine responses; the section that follows reviews data derived from TLR adjuvants used in NHP/SIV vaccine studies.

Non-human Primates and Adjuvant Development for HIV Vaccines

RMs have been the workhorse animal model for testing and evaluating candidate HIV vaccines, primarily owing to their ability to be infected by SIV and experience disease in a manner highly parallel to HIV infection of humans [1,253]. However, the use of NHPs in vaccine preclinical development is due to more than just their ability to be infected by SIV. Most adjuvants involve stimulation of TLRs or other PRRs in DCs, yet there are significant differences in the biology of DC populations between rodents and primates, as well the cellular distribution of TLRs [248]. Compared with mice, NHPs share a much higher degree of similarity with humans in both their DC populations, and are much more concordant in their cellular expression of PRRs and response to ligands [254]. In addition, the typical administration of vaccines in mice is via intraperitoneal or intravenous routes, rather than subcutaneous or intramuscular routes as in humans and primates, which leads to adjuvants/vaccines that are being presented to differing subsets of DCs [248]. For these reasons, adjuvant immunogenicity data derived in RMs offer a more direct translation to human studies, and subsequently increased use of NHPs in studies of non-SIV vaccines in recent years.

Early work in mice demonstrated that injection of either the TLR7/8 agonist R848 or TLR9 ligand CpG oligodeoxynucleotides were capable of eliciting high levels of cytokine production and phenotypic maturation of conventional DCs, but that R848 was a poor adjuvant for raising antigen-specific $CD4^+$ Th1 and $CD8^+$ T-cell activity toward an HIV-Gag protein; however, these responses could be improved by direct conjugation of the Gag antigen to a TLR7/8 analogue [255,256]. A parallel study was performed in RMs and demonstrated that a TLR7/8 ligand-Gag protein conjugate significantly enhanced Gag-specific $CD4^+$ and $CD8^+$ T cells cytokine production (IL2, TNF-α, and IFN-γ) and included enhancement of antibody avidity [257]. Combination of TLR7/8 or TLR9 ligand-Gag conjugates with additional oil-based adjuvant, Montanide ISA 51, also led to enhanced cytokine release and a higher degree of polyfunctional cytokine responses in Gag-specific CD4 and CD8 compared with Gag + Montanide alone, and these enhancements were also observed after boost with adenoviral vectors expressing Gag. Direct comparison of TLR ligand adjuvants using the Gag-protein prime/Ad-Gag boost system demonstrated that Gag + TLR3 ligand followed by Gag + TLR7/8 and Gag + TLR9 ligands were most effective at inducing Gag-specific $CD4^+$ and $CD8^+$ T cells [258]. Although none of the ligands was able to augment resistance to SIV challenge, animals receiving TLR adjuvants had reduced viremia after infection [258]. Kwissa and colleagues observed a similar effect for TLR9 and TLR7/8 adjuvants for improving responses to a vaccine regimen composed of DNA prime and modified vaccinia virus Ankara boost [259].

The ability of innate adjuvants to enhance the immunogenicity of vaccines is becoming well established. However, basic experiments in mice demonstrate that the combinations of TLR ligands together in a single adjuvant formulation could have synergistic effects and improve immunogenicity even further [248]. This principle was driven home dramatically by studies of the yellow fever vaccine, one of the most efficacious vaccines developed, which was demonstrated to activate TLR2, TLR7, TLR8, and TLR9 [260]. Following up on these insights, Kasturi and colleagues developed a candidate composite TLR adjuvant of nanoparticles that approximate viruses in terms of size and carry both a TLR4 ligand (MPL) and TLR7 ligand (R837 in mice and R848 in monkeys) that enabled long-lived antibodies against influenza in both mice and RMs [261]. This adjuvant regimen is currently in preclinical studies to test its efficacy against SIV challenge.

Innate immune system activating ligands have demonstrated a robust ability to enhance the immunogenicity of vaccines, and their application will continue to be intensely studied. However, there have been instances in which innate stimuli have been demonstrated to be detrimental to host immune responses against SIV challenge, owing to their ability to recruit potential target cells to the site of infection. Intravaginal application of either CpG oligodeoxynucleotides (TLR9 ligand) or imiquimod (TLR7/8 ligand) induced a proinflammatory milieu and cellular infiltrate of $CD4^+$ T cells and other immune cells in the cervicovaginal mucosa. Subsequent high-dose challenge with SIV did not lead to protection; in fact, animals receiving TLR9 or TLR7 ligands developed higher set-point viremia compared with control animals [262]. Conversely, intravaginal treatment of monkeys with glycerol monolaurate, an antimicrobial that suppresses production of inflammatory cytokines and chemokines, suppressed production of CCL20/MIP3A and recruitment of $CD4^+$ T cells; and most remarkably, GML treatment led to protection from SIV transmission when exposed to an intravaginal high-dose challenge [263]. These studies demonstrate that although innate immune system activating ligands have a robust capacity to enhance the immunogenicity, immune functionality, and persistence of responses induced by a vaccine, they also have the potential to increase target cell recruitment and aid infection. Although the innate immune responses induced by adjuvants are expected to wane shortly after vaccination and not affect target cell recruitment, there is a paucity of data on the long-term effects of adjuvants.

CONCLUSIONS

The early 2000s saw an explosion in our understanding of the innate immune system at the cellular and molecular levels. The description of the innate system provided a theoretical framework for understanding how HIV and SIV cause chronic immune activation. In more recent years, several groups have translated murine studies using innate immune system activating ligands as adjuvants to empirically demonstrate that modulation of the innate immune system can enhance vaccine immunogenicity in primates and, in some cases, improved protective efficacy against SIV challenge. Future research efforts aimed at optimizing TLR combinations and defining the effect on target cells will be needed to maximize the protective potential of vaccines and move toward clinical trials. Comparative studies of SIV infection between species that develop AIDS and natural hosts have been informative in highlighting differences in the innate immune system that are important for pathogenesis. Finally, a handful of studies have begun to modulate the innate system in vivo in monkey models, and these will become increasingly important for the development of therapeutics capable of ameliorating disease.

References

[1] Evans DT, Silvestri G. Nonhuman primate models in AIDS research. Curr Opin HIV AIDS 2013;8(4):255–61.

[2] Hirsch VM, Johnson PR. Pathogenic diversity of simian immunodeficiency viruses. Virus Res 1994;32(2):183–203.

[3] Simon MA, Brodie SJ, Sasseville VG, Chalifoux LV, Desrosiers RC, Ringler DJ. Immunopathogenesis of SIVmac. Virus Res 1994;32(2):227–51.

[4] Letvin NL, Daniel MD, Sehgal PK, Desrosiers RC, Hunt RD, Waldron LM, et al. Induction of AIDS-like diseases in macaque monkeys with T-cell tropic retrovirus STLV-III. Science 1985;230:71–3.

[5] Ringler DJ, Wyand MS, Walsh DG, MacKey JJ, Chalifoux LV, Popovic M, et al. Cellular localization of simian immunodeficiency virus in lymphoid tissues. I. Immunohistochemistry and electron microscopy. Am J Pathol 1989;134(2):373–83.

[6] Baskin GB, Murphey-Corb M, Martin LN, Davison-Fairburn B, Hu FS, Kuebler D. Thymus in simian immunodeficiency virus-infected rhesus monkeys. Lab Invest 1991;65:400–7.

[7] Baskin GB, Murphey-Corb M, Martin LN, Soike KF, Hu FS, Kuebler D. Lentivirus-induced pulmonary lesions in rhesus monkeys (*Macaca mulatta*) infected with simian immunodeficiency virus. Vet Pathol 1991;28(6):506–13.

[8] Kestler III HW, Ringler DJ, Mori K Panicali DL, Sehgal PK, Daniel MD, et al. Importance of the nef gene for maintenance of high virus loads and for development of AIDS. Cell 1991;65:651–62.

[9] King NW, Chalifoux LV, Ringler DJ, Wyand MS, Sehgal PK, Daniel MD, et al. Comparative biology of natural and experimental SIVmac infection in macaque monkeys: a review. J Med Primatol 1990;19(2):109–18.

[10] Novembre FJ, Saucier M, Anderson DC, Klumpp SA, O'Neil SP, Brown 2nd CR, et al. Development of AIDS in a chimpanzee infected with human immunodeficiency virus type 1. J Virol 1997;71:4086–91.

[11] Keele BF, Jones JH, Terio KA, Estes JD, Rudicell RS, Wilson ML, et al. Increased mortality and AIDS-like immunopathology in wild chimpanzees infected with SIVcpz. Nature 2009;460(7254):515–9.

[12] Chahroudi A, Bosinger SE, Vanderford TH, Paiardini M, Silvestri G. Natural SIV hosts: showing AIDS the door. Science 2012;335(6073):1188–93.

[13] Paiardini M, Cervasi B, Reyes-Aviles E, Micci L, Ortiz AM, Chahroudi A, et al. Low levels of SIV infection in sooty mangabey central memory CD(4)(+) T cells are associated with limited CCR5 expression. Nat Med 2011;17(7):830–6.

[14] Iwasaki A. Innate immune recognition of HIV-1. Immunity 2012;37(3):389–98.

[15] Lemaitre B, Nicolas E, Michaut L, Reichhart JM, Hoffmann JA. The dorsoventral regulatory gene cassette spatzle/Toll/cactus controls the potent antifungal response in *Drosophila* adults. Cell 1996;86:973–83.

[16] Rock FL, Hardiman G, Timans JC, Kastelein RA, Bazan JF. A family of human receptors structurally related to *Drosophila* Toll. Proc Natl Acad Sci USA 1998;95:588–93.

[17] Creagh EM, O'Neill LAJ. TLRs, NLRs, and RLRs: a trinity of pathogen sensors that co-operate in innate immunity. Trends Immunol 2006;27:352–7.

[18] Kadowaki N, Ho S, Antonenko S, Malefyt RW, Kastelein RA, Bazan F, et al. Subsets of human dendritic cell precursors express different toll-like receptors and respond to different microbial antigens. J Exp Med 2001;194:863–9.

[19] Ahmad-Nejad P, Hacker H, Rutz M, Bauer S, Vabulas RM, Wagner H. Bacterial CpG-DNA and lipopolysaccharides activate Toll-like receptors at distinct cellular compartments. Eur J Immunol 2002;32:1958–68.

[20] Diebold SS, Kaisho T, Hemmi H, Akira S, Reis e Sousa C. Innate antiviral responses by means of TLR7-mediated recognition of single-stranded RNA. Science 2004;303: 1529–31.
[21] Nishiya T, DeFranco AL. Ligand-regulated chimeric receptor approach reveals distinctive subcellular localization and signaling properties of the Toll-like receptors. J Biol Chem 2004;279:19008–17.
[22] Hemmi H, Takeuchi O, Kawai T, Kaisho T, Sato S, Sanjo H, et al. A Toll-like receptor recognizes bacterial DNA. Nature 2000;408:740–5.
[23] Lee HK, Lund JM, Ramanathan B, Mizushima N, Iwasaki A. Autophagy-dependent viral recognition by plasmacytoid dendritic cells. Science 2007;315:1398–401.
[24] Latz E, Schoenemeyer A, Visintin A, Fitzgerald KA, Monks BG, Knetter CF, et al. TLR9 signals after translocating from the ER to CpG DNA in the lysosome. Nat Immunol 2004;5:190–8.
[25] Wagner H. Toll meets bacterial CpG-DNA. Immunity 2001;14:499–502.
[26] Diebold SS, Kaisho T, Hemmi H, Akira S, Reis e Sousa C. Innate antiviral responses by means of TLR7-mediated recognition of single-stranded RNA. Science 2004;303(5663):1529–31.
[27] Gilliet M, Cao W, Liu YJ. Plasmacytoid dendritic cells: sensing nucleic acids in viral infection and autoimmune diseases. Nat Rev Immunol 2008;8(8):594–606.
[28] Lore K, Smed-Sorensen A, Vasudevan J, Mascola JR, Koup RA. Myeloid and plasmacytoid dendritic cells transfer HIV-1 preferentially to antigen-specific CD4+ T cells. J Exp Med 2005;201(12):2023–33.
[29] Fitzgerald-Bocarsly P, Jacobs ES. Plasmacytoid dendritic cells in HIV infection: striking a delicate balance. J Leukoc Biol 2010;87(4):609–20.
[30] Beignon AS, McKenna K, Skoberne M, Manches O, DaSilva I, Kavanagh DG, et al. Endocytosis of HIV-1 activates plasmacytoid dendritic cells via Toll-like receptor-viral RNA interactions. J Clin Invest 2005;115(11):3265–75.
[31] Bosinger SE, Johnson ZP, Folkner KA, Patel N, Hashempour T, Jochems SP, et al. Intact type I interferon production and IRF7 function in sooty mangabeys. PLoS Pathog 2013;9(8):e1003597.
[32] Lepelley A, Louis S, Sourisseau M, Law HKW, Pothlichet J, Schilte C , et al. Innate sensing of HIV-infected cells. PLoS Pathog 2011;7(2):e1001284.
[33] Ravasi T, Suzuki H, Cannistraci CV, Katayama S, Bajic VB, Tan K, et al. An atlas of combinatorial transcriptional regulation in mouse and man. Cell 2010;140(5):744–52.
[34] O'Brien M, Manches O, Sabado RL, Baranda SJ, Wang Y, Marie I, et al. Spatiotemporal trafficking of HIV in human plasmacytoid dendritic cells defines a persistently IFN-alpha-producing and partially matured phenotype. J Clin Invest 2011;121(3):1088–101.
[35] Solis M, Nakhaei P, Jalalirad M, Lacoste J, Douville R, Arguello M, et al. RIG-I-mediated antiviral signaling is inhibited in HIV-1 infection by a protease-mediated sequestration of RIG-I. J Virol 2011;85(3):1224–36.
[36] Berg RK, Melchjorsen J, Rintahaka J, Diget E, Søby S, Horan KA, et al. Genomic HIV RNA induces innate immune responses through RIG-I-dependent sensing of secondary-structured RNA. PLoS One 2012;7(1):e29291.
[37] Doehle BP, Hladik F, McNevin JP, McElrath MJ, Gale Jr M. Human immunodeficiency virus type 1 mediates global disruption of innate antiviral signaling and immune defenses within infected cells. J Virol 2009;83(20):10395–405.
[38] Okumura A, Alce T, Lubyova B, Ezelle H, Strebel K, Pitha PM. HIV-1 accessory proteins VPR and Vif modulate antiviral response by targeting IRF-3 for degradation. Virology 2008;373(1):85–97.
[39] Pertel T, Hausmann S, Morger D, Züger S, Guerra J, Lascano J, et al. TRIM5 is an innate immune sensor for the retrovirus capsid lattice. Nature 2011;472(7343):361–5.

[40] Stetson DB, Ko JS, Heidmann T, Medzhitov R. Trex1 prevents cell-intrinsic initiation of autoimmunity. Cell 2008;134(4):587–98.
[41] Yan N, Regalado-Magdos AD, Stiggelbout B, Lee-Kirsch MA, Lieberman J. The cytosolic exonuclease TREX1 inhibits the innate immune response to human immunodeficiency virus type 1. Nat Immunol 2010;11(11):1005–13.
[42] Stremlau M, Owens CM, Perron MJ, Kiessling M, Autissier P, Sodroski J. The cytoplasmic body component TRIM5a restricts HIV-1 infection in Old World monkeys. Nature 2004;427:848–53.
[43] Sheehy AM, Gaddis NC, Choi JD, Malim MH. Isolation of a human gene that inhibits HIV-1 infection and is suppressed by the viral Vif protein. Nature 2002;418(6898): 646–50.
[44] Zhang H, Yang B, Pomerantz RJ, Zhang C, Arunachalam SC, Gao L. The cytidine deaminase CEM15 induces hypermutation in newly synthesized HIV-1 DNA. Nature 2003;424(6944):94–8.
[45] Neil SJ, Zang T, Bieniasz PD. Tetherin inhibits retrovirus release and is antagonized by HIV-1 Vpu. Nature 2008;451(7177):425–30.
[46] Van Damme N, Goff D, Katsura C, Jorgenson RL, Mitchell R, Johnson MC, et al. The interferon-induced protein BST-2 restricts HIV-1 release and is downregulated from the cell surface by the viral Vpu protein. Cell Host Microbe 2008;3(4):245–52.
[47] Laguette N, Sobhian B, Casartelli N, Ringeard M, Chable-Bessia C, Ségéral E, et al. SAMHD1 is the dendritic- and myeloid-cell-specific HIV-1 restriction factor counteracted by Vpx. Nature 2011;474(7353):654–7.
[48] Hrecka K, Hao C, Gierszewska M, Swanson SK, Kesik-Brodacka M, Srivastava M, et al. Vpx relieves inhibition of HIV-1 infection of macrophages mediated by the SAMHD1 protein. Nature 2011;474(7353):658–61.
[49] Newman RM, Hall L, Kirmaier A, Pozzi LA, Pery E, Farzan M, et al. Evolution of a TRIM5-CypA splice isoform in old world monkeys. PLoS Pathog 2008;4(2):e1000003.
[50] Lim SY, Rogers T, Chan T, Whitney JB, Kim J, Sodroski J, et al. TRIM5alpha modulates immunodeficiency virus control in rhesus monkeys. PLoS Pathog 2010;6(1):e1000738.
[51] Wu F, Kirmaier A, Goeken R, Ourmanov I, Hall L, Morgan JS, et al. TRIM5 alpha drives SIVsmm evolution in rhesus macaques. PLoS Pathog 2013;9(8):e1003577.
[52] Letvin NL, Rao SS, Montefiori DC, Seaman MS, Sun Y, Lim SY, et al. Immune and genetic correlates of vaccine Protection against mucosal infection by SIV in monkeys. Sci Transl Med 2011;3(81):81ra36.
[53] Yeh WW, Rao SS, Lim SY, Zhang J, Hraber PT, Brassard LM, et al. The TRIM5 gene modulates penile mucosal acquisition of simian immunodeficiency virus in rhesus monkeys. J Virol 2011;85(19):10389–98.
[54] Fenizia C, Keele BF, Nichols D, Cornara S, Binello N, Vaccari M, et al. TRIM5alpha does not affect simian immunodeficiency virus SIV(mac251) replication in vaccinated or unvaccinated Indian rhesus macaques following intrarectal challenge exposure. J Virol 2011;85(23):12399–409.
[55] Mandl JN, Barry AP, Vanderford TH, Kozyr N, Chavan R, Klucking S, et al. Divergent TLR7 and TLR9 signaling and type I interferon production distinguish pathogenic and nonpathogenic AIDS virus infections. Nat Med 2008;14(10):1077–87.
[56] Brown KN, Trichel A, Barratt-Boyes SM. Parallel loss of myeloid and plasmacytoid dendritic cells from blood and lymphoid tissue in simian AIDS. J Immunol 2007;178(11):6958–67.
[57] Brown KN, Wijewardana V, Liu X, Barratt-Boyes SM. Rapid influx and death of plasmacytoid dendritic cells in lymph nodes mediate depletion in acute simian immunodeficiency virus infection. PLoS Pathog 2009;5(5):e1000413.
[58] Ziegler-Heitbrock L, Ancuta P, Crowe S, Dalod M, Grau V, Hart DN, et al. Nomenclature of monocytes and dendritic cells in blood. Blood 2010;116(16):e74–80.

[59] Jaensson E, Uronen-Hansson H, Pabst O, Eksteen B, Tian J, Coombes JL, et al. Small intestinal CD103+ dendritic cells display unique functional properties that are conserved between mice and humans. J Exp Med 2008;205(9):2139–49.

[60] Klatt NR, Estes JD, Sun X, Ortiz AM, Barber JS, Harris LD, et al. Loss of mucosal CD103+ DCs and IL-17+ and IL-22+ lymphocytes is associated with mucosal damage in SIV infection. Mucosal Immunol 2012;5(6):646–57.

[61] Hu J, Gardner MB, Miller CJ. Simian immunodeficiency virus rapidly penetrates the cervicovaginal mucosa after intravaginal inoculation and infects intraepithelial dendritic cells. J Virol 2000;74(13):6087–95.

[62] Miller CJ, Hu J. T cell-tropic simian immunodeficiency virus (SIV) and simian-human immunodeficiency viruses are readily transmitted by vaginal inoculation of rhesus macaques, and Langerhans' cells of the female genital tract are infected with SIV. J Infect Dis 1999;179(Suppl. 3):S413–7.

[63] Geijtenbeek TB, Koopman G, van Duijnhoven GC, van Vliet SJ, van Schijndel AC, Engering A, et al. Rhesus macaque and chimpanzee DC-SIGN act as HIV/SIV gp120 trans-receptors, similar to human DC-SIGN. Immunol Lett 2001;79(1–2):101–7.

[64] Yu Kimata MT, Cella M, Biggins JE, Rorex C, White R, Hicks S, et al. Capture and transfer of simian immunodeficiency virus by macaque dendritic cells is enhanced by DC-SIGN. J Virol 2002;76(23):11827–36.

[65] Ploquin MJ, Diop OM, Sol-Foulon N, Mortara L, Faye A, Soares MA, et al. DC-SIGN from African green monkeys is expressed in lymph nodes and mediates infection in trans of simian immunodeficiency virus SIVagm. J Virol 2004;78(2):798–810.

[66] Liu YJ. IPC: professional type 1 interferon-producing cells and plasmacytoid dendritic cell precursors. Annu Rev Immunol 2005;23:275–306.

[67] Summers KL, Hock BD, McKenzie JL, Hart DNJ. Phenotypic characterization of five dendritic cell subsets in human tonsils. Am J Pathol 2001;159:285–95.

[68] MacDonald KP, Munster DJ, Clark GJ, Dzionek A, Schmitz J, Hart DN. Characterization of human blood dendritic cell subsets. Blood 2002;100:4512–20.

[69] Li W, Liu Y-J. Development of dendritic-cell lineages. Immunity 2007;26:741–50.

[70] Colonna M, Trinchieri G, Liu YJ. Plasmacytoid dendritic cells in immunity. Nat Immunol 2004;5:1219–26.

[71] Dzionek A, Fuchs A, Schmidt P, Cremer S, Zysk M, Miltenyi S, et al. BDCA-2, BDCA-3, and BDCA-4: three markers for distinct subsets of dendritic cells in human peripheral blood. J Immunol 2000;165:6037–46.

[72] Grage-Griebenow E, Loseke S, Kauth M, Gehlhar K, Zawatzky R, Bufe A. Anti-BDCA-4 (neuropilin-1) antibody can suppress virus-induced IFN-alpha production of plasmacytoid dendritic cells. Immunol Cell Biol 2007;85:383–90.

[73] Dzionek A, Sohma Y, Nagafune J, Cella M, Colonna M, Facchetti F, et al. BDCA-2, a novel plasmacytoid dendritic cell-specific type II C-type lectin, mediates antigen capture and is a potent inhibitor of interferon alpha-beta induction. J Exp Med 2001;194:1823–34.

[74] Facchetti F, de Wolf-Peeters C, Mason DY, Pulford K, van den Oord JJ, Desmet VJ. Plasmacytoid T cells. Immunohistochemical evidence for their monocyte/macrophage origin. Am J Pathol 1988;133:15–21.

[75] O'Doherty U, Peng M, Gezelter S, Swiggard WJ, Betjies M, Bhardwaj N, et al. Human blood contains two subsets of dendritic cells, one immunologically mature and the other immature. Immunology 1994;82:487–93.

[76] Lennert K, Remmele W. Karyometric research on lymph node cells in man. I. Germinoblasts, lymphoblasts, and lymphocytes. Acta Haematol 1958;19:99–113.

[77] Vollenweider R, Lennert K. Plasmacytoid T-cell clusters in non-specific lymphadenitis. Virchows Arch B Cell Pathol Incl Mol Pathol 1983;44:1–14.

[78] Perussia B, Fanning V, Trinchieri G. A leukocyte subset bearing HLA-DR antigens is responsible for in vitro alpha interferon production in response to viruses. Nat Immun Cell Growth Regul 1985;4:120–37.

[79] Ronnblom L, Ramstedt U, Alm GV. Properties of human natural interferon-producing cells stimulated by tumor cell lines. Eur J Immunol 1983;13:471–6.

[80] Abb J, Abb H, Deinhardt F. Phenotype of human alpha-interferon producing leucocytes identified by monoclonal antibodies. Clin Exp Immunol 1983;52:179–84.

[81] Chehimi J, Starr SE, Kawashima H, Miller DS, Trinchieri G, Perussia B, et al. Dendritic cells and IFN-alpha-producing cells are two functionally distinct non-B, non-monocytic HLA-DR+ cell subsets in human peripheral blood. Immunology 1989;68: 486–90.

[82] Siegal FP, Kadowaki N, Shodell M, Fitzgerald-Bocarsly PA, Shah K, Ho S, et al. The nature of the principal type 1 interferon-producing cells in human blood. Science 1999;284:1835–7.

[83] Cella M, Jarrossay D, Facchetti F, Alebardi O, Nakajima H, Lanzavecchia A, et al. Plasmacytoid monocytes migrate to inflamed lymph nodes and produce large amounts of type I interferon. Nat Med 1999;5:919–23.

[84] Fitzgerald-Bocarsly P, Feldman M, Mendelsohn M, Curl S, Lopez C. Human mononuclear cells which produce interferon-alpha during NK(HSV-FS) assays are HLA-DR positive cells distinct from cytolytic natural killer effectors. J Leukoc Biol 1988;43:323–34.

[85] Coccia EM, Severa M, Giacomini E, Monneron D, Remoli ME, Julkunen I, et al. Viral infection and Toll-like receptor agonists induce a differential expression of type I and lambda interferons in human plasmacytoid and monocyte-derived dendritic cells. Eur J Immunol 2004;34:796–805.

[86] Izaguirre A, Barnes BJ, Amrute S, Yeow WS, Megjugorac N, Dai J, et al. Comparative analysis of IRF and IFN-alpha expression in human plasmacytoid and monocyte-derived dendritic cells. J Leukoc Biol 2003;74:1125–38.

[87] Kerkmann M, Rothenfusser S, Hornung V, Towarowski A, Wagner M, Sarris A, et al. Activation with CpG-A and CpG-B oligonucleotides reveals two distinct regulatory pathways of type I IFN synthesis in human plasmacytoid dendritic cells. J Immunol 2003;170:4465–74.

[88] Gill MA, Palucka K, Barton T, Ghaffar F, Jafri H, Banchereau J, Ramilo O. Mobilization of plasmacytoid and myeloid dendritic cells to mucosal sites in children with respiratory syncytial virus and other viral respiratory infections. J Infect Dis 2005;191:1105–15.

[89] Lund JM, Linehan MM, Iijima N, Iwasaki A. Plasmacytoid dendritic cells provide innate immune protection against mucosal viral infection in situ. J Immunol 2006;177: 7510–4.

[90] Pichyangkul S, Endy TP, Kalayanrooj S, Nisalak A, Yongvanitchit K, Green S, et al. A blunted blood plasmacytoid dendritic cell response to an acute systemic viral infection is associated with increased disease severity. J Immunol 2003;171:5571–8.

[91] Hishizawa M, Imada K, Kitawaki T, Ueda M, Kadowaki N, Uchiyama T. Depletion and impaired interferon-alpha-producing capacity of blood plasmacytoid dendritic cells in human T-cell leukaemia virus type 1-infected individuals. Br J Haematol 2004;125:568–75.

[92] Zhang Z, Xu D, Li Y, Jin L, Shi M, Wang M, et al. Longitudinal alteration of circulating dendritic cell subsets and its correlation with steroid treatment in patients with severe acute respiratory syndrome. Clin Immunol 2005;116:225–35.

[93] Duan XZ, Wang M, Li HW, Zhuang H, Xu D, Wang FS. Decreased frequency and function of circulating plasmocytoid dendritic cells (pDC) in hepatitis B virus infected humans. J Clin Immunol 2004;24:637–46.

[94] Goutagny N, Vieux C, Decullier E, Ligeoix B, Epstein A, Trepo C, et al. Quantification and functional analysis of plasmacytoid dendritic cells in patients with chronic hepatitis C infection. J Infect Dis 2004;189:1646–55.

[95] Anthony DD, Yonkers NL, Post AB, Asaad R, Heinzel FP, Lederman MM, et al. Selective impairments in dendritic cell-associated function distinguish hepatitis C virus and HIV infection. J Immunol 2004;172:4907–16.
[96] Dolganiuc A, Chang S, Kodys K, Mandrekar P, Bakis G, Cormier M, et al. Hepatitis C virus (HCV) core protein-induced, monocyte-mediated mechanisms of reduced IFN-apha and plasmacytoid dendritic cell loss in chronic HCV infection. J Immunol 2007;177:6758–68.
[97] Yonkers NL, Rodriguez B, Milkovich KA, Asaad R, Lederman MM, Heeger PS, et al. TLR ligand-dependent activation of naive CD4 T cells by plasmacytoid dendritic cells is impaired in hepatitis C virus infection. J Immunol 2007;178:4436–44.
[98] Del Corno M, Gauzzi MC, Penna G, Belardelli F, Adorini L, Gessani S. Human immunodeficiency virus type 1 gp120 and other activation stimuli are highly effective in triggering alpha interferon and CC chemokine production in circulating plasmacytoid but not myeloid dendritic cells. J Virol 2005;79:12597–601.
[99] Ferbas JJ, Toso JF, Logar AJ, Navaratil JS, Rinaldo Jr CR. CD4+ blood dendritic cells are potent producers of IFN-alpha in response to in vitro HIV-1 infection. J Immunol 1994;152:4649–62.
[100] Fong L, Mengozzi M, Abbey NW, Herndier BG, Engleman EG. Productive infection of plasmacytoid dendritic cells with human immunodeficiency virus type 1 is triggered by CD40 ligation. J Virol 2002;76:11033–41.
[101] Yonezawa A, Morita R, Takaori-Kondo A, Kadowaki N, Kitawaki T, Hori T, et al. Natural alpha interferon-producing cells respond to human immunodeficiency virus type 1 with alpha interferon production and maturation into dendritic cells. J Virol 2003;77:3777–84.
[102] Schmidt B, Ashlock BM, Foster H, Fujimura SH, Levy JA. HIV-infected cells are major inducers of plasmacytoid dendritic cell interferon production, maturation, and migration. Virology 2005;343:256–66.
[103] Barron MA, Blyveis N, Palmer BE, MaWhinney S, Wilson CC. Influence of plasma viremia on defects in number and immunophenotype of blood dendritic cell subsets in human immunodeficiency virus 1-infected individuals. J Infect Dis 2003;187:26–37.
[104] Meier A, Alter G, Frahm N, Sidhu H, Li B, Bagchi A, et al. MyD88-dependent immune activation mediated by human immunodeficiency virus type 1-encoded toll-like receptor ligands. J Virol 2007;81:8180–91.
[105] Martinelli E, Cicala C, Van Ryk D, Goode DJ, Macleod K, Arthos J, et al. HIV-1 gp120 inhibits TLR9-mediated activation and IFN-(alpha) secretion in plasmacytoid dendritic cells. Proc Natl Acad Sci USA 2007;104:3396–401.
[106] Beignon AS, McKenna K, Skoberne M, Manches O, DaSilva I, Kavanagh DG, et al. Enocytosis of HIV-1 activates plasmacytoid dendritic cells via Toll-like receptor-viral RNA interactions. J Clin Invest 2005.
[107] Feldman S, Stein D, Amrute S, Denny T, Garcia Z, Kloser P, et al. Decreased interferon-alpha production in HIV-infected patients correlates with numerical and functional deficiencies in circulating type 2 dendritic cell precursors. Clin Immunol 2001;101:201–10.
[108] Pacanowski J, Kahi S, Baillet M, Lebon P, Deveau C, Goujard C, et al. Reduced blood CD123+ (lymphoid) and CD11c+ (myeloid) dendritic cell numbers in primary HIV-1 infection. Blood 2001;98:3016–21.
[109] Soumelis V, Scott I, Gheyas G, Bouhour D, Cozon G, Cotte L, et al. Depletion of circulating natural type 1 interferon-producing cells in HIV-infected AIDS patients. Blood 2001;98:906–12.
[110] Donaghy H, Pozniak A, Gazzard B, Qazi N, Gilmour J, Gotch F, et al. Loss of blood CD11c(+) myeloid and CD11c(-) plasmacytoid dendritic cells in patients with HIV-1 infection correlates with HIV-1 RNA virus load. Blood 2001;98:2574–6.

[111] Desai S, Chaparro A, Liu H, Haslett P, Arheart K, Scott G, et al. Impaired CCR7 expression on plasmacytoid dendritic cells of HIV-infected children and adolescents with immunologic and virologic failure. J Acquir Immune Defic Syndr 2007;45(5): 501–7.
[112] Chehimi J, Campbell DE, Azzoni L, Bacheller D, Papasavvas E, Jerandi G, et al. Persistent decreases in blood plasmacytoid dendritic cell number and function despite effective highly active antiretroviral therapy and increased blood myeloid dendritic cells in HIV-infected individuals. J Immunol 2002;168:4796–801.
[113] Zhang Z, Fu J, Zhao Q, He Y, Jin L, Zhang H, et al. Differential restoration of myeloid and plasmacytoid dendritic cells in HIV-1 infected children after treatment with highly active antiretroviral therapy. J Immunol 2006;176:5644–51.
[114] Schmidt B, Scott I, Whitmore RG, Foster H, Fujimura S, Schmitz J, et al. Low-level HIV infection of plasmacytoid dendritic cells: onset of cytopathic effects and cell death after PDC maturation. Virology 2004;329:280–8.
[115] Donaghy H, Gazzard B, Gotch F, Patterson S. Dysfunction and infection of freshly isolated blood myeloid and plasmacytoid dendritic cells in patients infected with HIV-1. Blood 2003;101:4505–11.
[116] Lore K, Smed-Sorensen A, Vasudevan J, Mascola JR, Koup RA. Myeloid and plasmacytoid dendritic cells transfer HIV-1 preferentially to antigen specific CD4+ T cells. J Exp Med 2005;201:2023–33.
[117] Schmitt N, Nugeyre MT, Scott-Algara D, Cumont MC, Barre-Sinoussi F, Pancino G, et al. Differential susceptibility of human thymic dendritic cell subsets to X4 and R5 HIV-1 infection. AIDS 2006;20:533–42.
[118] Patterson S, Rae A, Hockey N, Gilmour J, Gotch F. Plasmacytoid dendritic cells are highly susceptible to human immunodeficiency virus type 1 infection and release infectious virus. J Virol 2001;75:6710–3.
[119] Cameron PU, Handley AJ, Baylis DC, Solomon AE, Bernard N, Purcell DF, et al. Preferential infection of dendritic cells during human immunodeficiency virus type 1 infection of blood leukocytes. J Virol 2007;81:2297–306.
[120] Reeves RK, Fultz PN. Disparate effects of acute and chronic infection with SIVmac239 or SHIV-89.6P on macaque plasmacytoid dendritic cells. Virology 2007;365(2): 356–68.
[121] Malleret B, Karisson I, Maneglier B, Brochard P, Delache B, Andrieu T, et al. Effect of SIVmac infection on plasmacytoid and CD1c+ myeloid dendritic cells in cynomolgus macaques. Immunology 2008;124:223–33.
[122] Bosinger SE, Sodora DL, Silvestri G. Generalized immune activation and innate immune responses in simian immunodeficiency virus infection. Curr Opin HIV AIDS 2011;6(5):411–8.
[123] Diop OM, Ploquin MJ, Mortara L, Faye A, Jacquelin B, Kunkel D, et al. Plasmacytoid dendritic cell dynamics and alpha interferon production during Simian immunodeficiency virus infection with a nonpathogenic outcome. J Virol 2008;82(11):5145–52.
[124] Kwa S, Kannanganat S, Nigam P, Siddiqui M, Shetty RD, Armstrong W, et al. Plasmacytoid dendritic cells are recruited to the colorectum and contribute to immune activation during pathogenic SIV infection in rhesus macaques. Blood 2011;118(10):2763–73.
[125] Reeves RK, Evans TI, Gillis J, Wong FE, Kang G, Li Q, et al. SIV infection induces accumulation of plasmacytoid dendritic cells in the gut mucosa. J Infect Dis 2012;206(9):1462–8.
[126] Li H, Gillis J, Johnson RP, Reeves RK. Multi-functional plasmacytoid dendritic cells redistribute to gut tissues during simian immunodeficiency virus infection. Immunology 2013;140(2):244–9.
[127] Lanier LL. Evolutionary struggles between NK cells and viruses. Nat Rev Immunol 2008;8(4):259–68.

[128] Caligiuri MA. Human natural killer cells. Blood 2008;112(3):461–9.
[129] Cooper MA, Fehniger TA, Caligiuri MA. The biology of human natural killer-cell subsets. Trends Immunol 2001;22(11):633–40.
[130] El Costa H, Casemayou A, Aguerre-Girr M, Rabot M, Berrebi A, Parant O, et al. Critical and differential roles of NKp46- and NKp30-activating receptors expressed by uterine NK cells in early pregnancy. J Immunol 2008;181(5):3009–17.
[131] Tiemessen CT, Shalekoff S, Meddows-Taylor S, Schramm DB, Papathanasopoulos MA, Gray GE, et al. Cutting Edge: unusual NK cell responses to HIV-1 peptides are associated with protection against maternal-infant transmission of HIV-1. J Immunol 2009;182(10):5914–8.
[132] Pang G, Buret A, Batey RT, Chen QY, Couch L, Cripps A, et al. Morphological, phenotypic and functional characteristics of a pure population of CD56+ CD16- CD3- large granular lymphocytes generated from human duodenal mucosa. Immunol 1993;79(3):498–505.
[133] Campbell JJ, Qin S, Unutmaz D, Soler D, Murphy KE, Hodge MR, et al. Unique subpopulations of CD56+ NK and NK-T peripheral blood lymphocytes identified by chemokine receptor expression repertoire. J Immunol 2001;166(11):6477–82.
[134] de Repentigny L, Lewandowski D, Jolicoeur P. Immunopathogenesis of oropharyngeal candidiasis in human immunodeficiency virus infection. Clin Microbiol Rev 2004;17(4):729–59.
[135] Challacombe SJ, Naglik JR. The effects of HIV infection on oral mucosal immunity. Adv Dent Res 2006;19(1):29–35.
[136] Zancope E, Costa NL, Junqueira-Kipnis AP, Valadares MC, Silva TA, Leles CR, et al. Differential infiltration of CD8+ and NK cells in lip and oral cavity squamous cell carcinoma. J Oral Pathol Med 2010;39(2):162–7.
[137] Chinen H, Matsuoka K, Sato T, Kamada N, Okamoto S, Hisamatsu T, et al. Lamina propria c-kit+ immune precursors reside in human adult intestine and differentiate into natural killer cells. Gastroenterology 2007;133(2):559–73.
[138] Reeves RK, Gillis J, Wong FE, Yu Y, Connole M, Johnson RP. CD16- natural killer cells: enrichment in mucosal and secondary lymphoid tissues and altered function during chronic SIV infection. Blood 2010;115(22):4439–46.
[139] Reeves RK, Rajakumar PA, Evans TI, Connole M, Gillis J, Wong FE, et al. Gut inflammation and indoleamine deoxygenase inhibit IL-17 production and promote cytotoxic potential in NKp44+ mucosal NK cells during SIV infection. Blood 2011;118(12):3321–30.
[140] Pereira LE, Johnson RP, Ansari AA. Sooty mangabeys and rhesus macaques exhibit significant divergent natural killer cell responses during both acute and chronic phases of SIV infection. Cell Immunol 2008;254(1):10–9.
[141] Ward J, Bonaparte M, Sacks J, Guterman J, Fogli M, Mavilio D, et al. HIV modulates the expression of ligands important in triggering natural killer cell cytotoxic responses on infected primary T-cell blasts. Blood 2007;110(4):1207–14.
[142] Alter G, Teigen N, Davis BT, Addo MM, Suscovich TJ, Waring MT, et al. Sequential deregulation of NK cell subset distribution and function starting in acute HIV-1 infection. Blood 2005;106(10):3366–9.
[143] O'Connor GM, Holmes A, Mulcahy F, Gardiner CM. Natural killer cells from long-term non-progressor HIV patients are characterized by altered phenotype and function. Clin Immunol 2007;124(3):277–83.
[144] Alter G, Martin MP, Teigen N, Carr WH, Suscovich TJ, Schneidewind A, et al. Differential natural killer cell-mediated inhibition of HIV-1 replication based on distinct KIR/HLA subtypes. J Exp Med 2007;204(12):3027–36.
[145] Shieh TM, Carter DL, Blosser RL, Mankowski JL, Zink MC, Clements JE. Functional analyses of natural killer cells in macaques infected with neurovirulent simian immunodeficiency virus. J Neurovirol 2001;7(1):11–24.

[146] Giavedoni LD, Velasquillo MC, Parodi LM, Hubbard GB, Hodara VL. Cytokine expression, natural killer cell activation, and phenotypic changes in lymphoid cells from rhesus macaques during acute infection with pathogenic simian immunodeficiency virus. J Virol 2000;74(4):1648–57.
[147] Holznagel E, Norley S, Holzammer S, Coulibaly C, Kurth R. Immunological changes in simian immunodeficiency virus (SIV(agm))-infected African green monkeys (AGM): expanded cytotoxic T lymphocyte, natural killer and B cell subsets in the natural host of SIV(agm). J Gen Virol 2002;83:631–40.
[148] Kouyos RD, Gordon SN, Staprans SI, Silvestri G, Regoes RR. Similar impact of CD8+ T cell responses on early virus dynamics during SIV infections of rhesus macaques and sooty mangabeys. PLoS Comput Biol 2010;6(8).
[149] Reeves RK, Evans TI, Gillis J, Johnson RP. Simian immunodeficiency virus infection induces expansion of alpha4beta7+ and cytotoxic CD56+ NK cells. J Virol 2010;84(17):8959–63.
[150] Colantonio AD, Bimber BN, Neidermyer Jr WJ, Reeves RK, Alter G, Altfeld M, et al. KIR polymorphisms modulate peptide-dependent binding to an MHC class I ligand with a Bw6 motif. PLoS Pathog 2011;7(3):e1001316.
[151] Hellmann I, Lim SY, Gelman RS, Letvin NL. Association of activating KIR copy number variation of NK cells with containment of SIV replication in rhesus monkeys. PLoS Pathog 2011;7(12):e1002436.
[152] Bostik P, Kobkitjaroen J, Tang W, Villinger F, Pereira LE, Little DM, et al. Decreased NK cell frequency and function is associated with increased risk of KIR3DL allele polymorphism in simian immunodeficiency virus-infected rhesus macaques with high viral loads. J Immunol 2009;182(6):3638–49.
[153] Takahashi Y, Mayne AE, Khowawisetsut L, Pattanapanyasat K, Little D, Villinger F, et al. In vivo administration of a JAK3 inhibitor to chronically SIV infected rhesus macaques leads to NK cell depletion associated with transient modest increase in viral loads. PLoS One 2013;8(7):e70992.
[154] Choi EI, Reimann KA, Letvin NL. In vivo natural killer cell depletion during primary simian immunodeficiency virus infection in rhesus monkeys. J Virol 2008;82(13):6758–61.
[155] Choi EI, Wang R, Peterson L, Letvin NL, Reimann KA. Use of an anti-CD16 antibody for in vivo depletion of natural killer cells in rhesus macaques. Immunology 2008;124(2):215–22.
[156] Sun Y, Asmal M, Lane S, Permar SR, Schmidt SD, Mascola JR, et al. Antibody-dependent cell-mediated cytotoxicity in simian immunodeficiency virus-infected rhesus monkeys. J Virol 2011;85(14):6906–12.
[157] Banks ND, Kinsey N, Clements J, Hildreth JE. Sustained antibody-dependent cell-mediated cytotoxicity (ADCC) in SIV-infected macaques correlates with delayed progression to AIDS. AIDS Res Hum Retroviruses 2002;18(16):1197–205.
[158] Alpert MD, Harvey JD, Lauer WA, Reeves RK, Piatak Jr M, Carville A, et al. ADCC develops over time during persistent infection with live-attenuated SIV and is associated with complete protection against SIV(mac)251 challenge. PLoS Pathog 2012;8(8):e1002890.
[159] Hidajat R, Xiao P, Zhou Q, Venzon D, Summers LE, Kalyanaraman VS, et al. Correlation of vaccine-elicited systemic and mucosal nonneutralizing antibody activities with reduced acute viremia following intrarectal simian immunodeficiency virus SIVmac251 challenge of rhesus macaques. J Virol 2009;83(2):791–801.
[160] Ferrari G, Pollara J, Kozink D, Harms T, Drinker M, Freel S, et al. An HIV-1 gp120 envelope human monoclonal antibody that recognizes a C1 conformational epitope mediates potent antibody-dependent cellular cytotoxicity (ADCC) activity and defines a common ADCC epitope in human HIV-1 serum. J Virol 2011;85(14):7029–36.

[161] Barouch DH, Liu J, Li H, Maxfield LF, Abbink P, Lynch DM, et al. Vaccine protection against acquisition of neutralization-resistant SIV challenges in rhesus monkeys. Nature 2012;482(7383):89–93.
[162] Bonsignori M, Pollara J, Moody MA, Alpert MD, Chen X, Hwang KK, et al. Antibody-dependent cellular cytotoxicity-mediating antibodies from an HIV-1 vaccine efficacy trial target multiple epitopes and preferentially use the VH1 gene family. J Virol 2012;86(21):11521–32.
[163] Haynes BF, Gilbert PB, McElrath MJ, Zolla-Pazner S, Tomaras GD, Alam SM, et al. Immune-correlates analysis of an HIV-1 vaccine efficacy trial. N Engl J Med 2012;366(14):1275–86.
[164] Scott-Algara D, Truong LX, Versmisse P, David A, Luong TT, Nguyen NV, et al. Cutting edge: increased NK cell activity in HIV-1-exposed but uninfected Vietnamese intravascular drug users. J Immunol 2003;171(11):5663–7.
[165] Lambotte O, Ferrari G, Moog C, Yates NL, Liao HX, Parks RJ, et al. Heterogeneous neutralizing antibody and antibody-dependent cell cytotoxicity responses in HIV-1 elite controllers. AIDS 2009;23(8):897–906.
[166] Wren LH, Chung AW, Isitman G, Kelleher AD, Parsons MS, Amin J, et al. Specific antibody-dependent cellular cytotoxicity responses associated with slow progression of HIV infection. Immunology 2013;138(2):116–23.
[167] Brenchley JM. Mucosal immunity in human and simian immunodeficiency lentivirus infections. Mucosal Immunol 2013;6(4):657–65.
[168] Bixler SL, Mattapallil JJ. Loss and dysregulation of Th17 cells during HIV infection. Clin Dev Immunol 2013;2013:852418.
[169] Shale M, Schiering C, Powrie F. CD4(+) T-cell subsets in intestinal inflammation. Immunol Rev 2013;252(1):164–82.
[170] Rutz S, Eidenschenk C, Ouyang W. IL-22, not simply a Th17 cytokine. Immunol Rev 2013;252(1):116–32.
[171] Honda K, Littman DR. The microbiome in infectious disease and inflammation. Annu Rev Immunol 2012;30:759–95.
[172] Raffatellu M, Santos RL, Verhoeven DE, George MD, Wilson RP, Winter SE, et al. Simian immunodeficiency virus-induced mucosal interleukin-17 deficiency promotes *Salmonella* dissemination from the gut. Nat Med 2008;14(4):421–8.
[173] Brenchley JM, Paiardini M, Knox KS, Asher AI, Cervasi B, Asher TE, et al. Differential Th17 CD4 T-cell depletion in pathogenic and nonpathogenic lentiviral infections. Blood 2008;112(7):2826–35.
[174] Nigam P, Kwa S, Velu V, Amara RR. Loss of IL-17-producing CD8 T cells during late chronic stage of pathogenic simian immunodeficiency virus infection. J Immunol 2011;186(2):745–53.
[175] Brenchley JM, Price DA, Schacker TW, Asher TE, Silvestri G, Rao S, et al. Microbial translocation is a cause of systemic immune activation in chronic HIV infection. Nat Med 2006;12(12):1365–71.
[176] El Hed A, Khaitan A, Kozhaya L, Manel N, Daskalakis D, Borkowsky W, et al. Susceptibility of human Th17 cells to human immunodeficiency virus and their perturbation during infection. J Infect Dis 2010;201(6):843–54.
[177] Micci L, Cervasi B, Ende ZS, Iriele RI, Reyes-Aviles E, Vinton C, et al. Paucity of IL-21-producing CD4(+) T cells is associated with Th17 cell depletion in SIV infection of rhesus macaques. Blood 2012;120(19):3925–35.
[178] Pallikkuth S, Micci L, Ende ZS, Iriele RI, Cervasi B, Lawson B, et al. Maintenance of intestinal Th17 cells and reduced microbial translocation in SIV-infected rhesus macaques treated with interleukin (IL)-21. PLoS Pathog 2013;9(7):e1003471.
[179] Favre D, Mold J, Hunt PW, Kanwar B, Loke P, Seu L, et al. Tryptophan catabolism by indoleamine 2,3-dioxygenase 1 alters the balance of TH17 to regulatory T cells in HIV disease. Sci Transl Med 2010;2(32):32ra6.

[180] Cella M, Fuchs A, Vermi W, Facchetti F, Otero K, Lennerz JK, et al. A human natural killer cell subset provides an innate source of IL-22 for mucosal immunity. Nature 2009;457(7230):722–5.
[181] Wills-Karp M, Finkelman FD. Innate lymphoid cells wield a double-edged sword. Nat Immunol 2011;12(11):1025–7.
[182] Hughes T, Becknell B, McClory S, Briercheck E, Freud AG, Zhang X, et al. Stage 3 immature human natural killer cells found in secondary lymphoid tissue constitutively and selectively express the TH 17 cytokine interleukin-22. Blood 2009;113(17):4008–10.
[183] Cella M, Otero K, Colonna M. Expansion of human NK-22 cells with IL-7, IL-2, and IL-1beta reveals intrinsic functional plasticity. Proc Natl Acad Sci USA 2010;107(24):10961–6.
[184] Colonna M. Interleukin-22-producing natural killer cells and lymphoid tissue inducer-like cells in mucosal immunity. Immunity 2009;31(1):15–23.
[185] Hughes T, Becknell B, Freud AG, McClory S, Briercheck E, Yu J, et al. Interleukin-1beta selectively expands and sustains interleukin-22+ immature human natural killer cells in secondary lymphoid tissue. Immunity 2010;32(6):803–14.
[186] Satoh-Takayama N, Vosshenrich CA, Lesjean-Pottier S, Sawa S, Lochner M, Rattis F, et al. Microbial flora drives interleukin 22 production in intestinal NKp46+ cells that provide innate mucosal immune defense. Immunity 2008;29(6):958–70.
[187] Crellin NK, Trifari S, Kaplan CD, Cupedo T, Spits H. Human NKp44+IL-22+ cells and LTi-like cells constitute a stable RORC+ lineage distinct from conventional natural killer cells. J Exp Med 2010;207(2):281–90.
[188] Klatt NR, Canary LA, Sun X, Vinton CL, Funderburg NT, Morcock DR, et al. Probiotic/prebiotic supplementation of antiretrovirals improves gastrointestinal immunity in SIV-infected macaques. J Clin Invest 2013;123(2):903–7.
[189] Xu H, Wang X, Liu DX, Moroney-Rasmussen T, Lackner AA, Veazey RS. IL-17-producing innate lymphoid cells are restricted to mucosal tissues and are depleted in SIV-infected macaques. Mucosal Immunol 2012;5(6):658–69.
[190] Cohen MS. Preventing sexual transmission of HIV. Clin Infect Dis 2007;45(Suppl. 4): S287–92.
[191] Keele BF, Giorgi EE, Salazar-Gonzalez JF, Decker JM, Pham KT, Salazar MG, et al. Identification and characterization of transmitted and early founder virus envelopes in primary HIV-1 infection. Proc Natl Acad Sci USA 2008;105(21):7552–7.
[192] Boeras DI, Hraber PT, Hurlston M, Evans-Strickfaden T, Bhattacharya T, Giorgi EE, et al. Role of donor genital tract HIV-1 diversity in the transmission bottleneck. Proc Natl Acad Sci USA 2011;108(46):E1156–63.
[193] Keele BF. Identifying and characterizing recently transmitted viruses. Curr Opin Hiv AIDS 2010;5(4):327–34.
[194] Li H, Bar KJ, Wang S, Decker JM, Chen Y, Sun C, et al. High multiplicity infection by HIV-1 in men who have sex with men. PLoS Pathog 2010;6(5):e1000890.
[195] Nazli A, Chan O, Dobson-Belaire WN, Ouellet M, Tremblay MJ, Gray-Owen SD, et al. Exposure to HIV-1 directly impairs mucosal epithelial barrier integrity allowing microbial translocation. PLoS Pathog 2010;6(4):e1000852.
[196] Hussain LA, Lehner T. Comparative investigation of Langerhans' cells and potential receptors for HIV in oral, genitourinary and rectal epithelia. Immunology 1995;85(3):475–84.
[197] Vishwanathan SA, Livingston L, McNicholl JM, Kersh EN. Progesterone and the use of macaque models to study vaginal HIV infection. In: Berhardt LV, editor. Advances in Medicine and biology. NOVA Publishers; 2012. pp. 161–78.
[198] Stadnyk AW. Intestinal epithelial cells as a source of inflammatory cytokines and chemokines. Can J gastroenterol 2002;16(4):241–6.
[199] Shukair SA, Allen SA, Cianci GC, Stieh DJ, Anderson MR, Baig SM, et al. Human cervicovaginal mucus contains an activity that hinders HIV-1 movement. Mucosal Immunol 2013;6(2):427–34.

[200] Wood LF, Chahroudi A, Chen HL, Jaspan HB, Sodora DL. The oral mucosa immune environment and oral transmission of HIV/SIV. Immunol Rev 2013;254(1):34–53.
[201] Shine N, Konopka K, Duzgunes N. The anti-HIV-1 activity associated with saliva. J Dent Res 1997;76(2):634–40.
[202] Hooper LV, Littman DR, Macpherson AJ. Interactions between the microbiota and the immune system. Science 2012;336(6086):1268–73.
[203] Cauley LS, Lefrancois L. Guarding the perimeter: protection of the mucosa by tissue-resident memory T cells. Mucosal Immunol 2013;6(1):14–23.
[204] Kamada N, Seo SU, Chen GY, Nunez G. Role of the gut microbiota in immunity and inflammatory disease. Nat Rev Immunol 2013;13(5):321–35.
[205] Kosub DA, Lehrman G, Milush JM, Zhou D, Chacko E, Leone A, et al. Gamma/Delta T-cell functional responses differ after pathogenic human immunodeficiency virus and nonpathogenic simian immunodeficiency virus infections. J Virol 2008;82(3):1155–65.
[206] Harris LD, Klatt NR, Vinton C, Briant JA, Tabb B, Ladell K, et al. Mechanisms underlying gammadelta T-cell subset perturbations in SIV-infected Asian rhesus macaques. Blood 2010;116(20):4148–57.
[207] Xu H, Wang X, Veazey RS. Mucosal immunology of HIV infection. Immunol Rev 2013;254(1):10–33.
[208] Brenchley JM, Price DA, Douek DC. HIV disease: fallout from a mucosal catastrophe? Nat Immunol 2006;7(3):235–9.
[209] Veazey RS, DeMaria M, Chalifoux LV, Shvetz DE, Pauley DR, Knight HL, et al. Gastrointestinal tract as a major site of CD4 T lymphocyte depletion and viral replication in SIV infection. Science 1998;280:427–31.
[210] Mehandru S, Poles MA, Tenner-Racz K, Horowitz A, Hurley A, Hogan C, et al. Primary HIV-1 infection is associated with preferential depletion of CD4+ T lymphocytes from effector sites in the gastrointestinal tract. J Exp Med 2004;200(6):761–70.
[211] Brenchley JM, Schacker TW, Ruff LE, Price DA, Taylor JH, Beilman GJ, et al. CD4+ T cell depletion during all stages of HIV disease occurs predominantly in the gastrointestinal tract. J Exp Med 2004;200(6):749–59.
[212] Hellerstein M, Hanley MB, Cesar D, Siler S, Papageorgopoulos C, Wieder E, et al. Directly measured kinetics of circulating T lymphocytes in normal and HIV-1-infected humans. Nat Med 1999;5(1):83–9.
[213] Hazenberg MD, Stuart JW, Otto SA, Borleffs JC, Boucher CA, de Boer RJ, et al. T-cell division in human immunodeficiency virus (HIV)-1 infection is mainly due to immune activation: a longitudinal analysis in patients before and during highly active antiretroviral therapy (HAART). Blood 2000;95(1):249–55.
[214] Kuller LH, Tracy R, Belloso W, De Wit S, Drummond F, Lane HC, et al. Inflammatory and coagulation biomarkers and mortality in patients with HIV infection. PLoS Med 2008;5(10):e203.
[215] Hunt PW, Martin JN, Sinclair E, Bredt B, Hagos E, Lampiris H, et al. T cell activation is associated with lower CD4+ T cell gains in human immunodeficiency virus-infected patients with sustained viral suppression during antiretroviral therapy. J Infect Dis 2003;187(10):1534–43.
[216] Katlama C, Deeks SG, Autran B, Martinez-Picado J, van Lunzen J, Rouzioux C, et al. Barriers to a cure for HIV: new ways to target and eradicate HIV-1 reservoirs. Lancet 2013;381(9883):2109–17.
[217] Li Q, Duan L, Estes JD, Ma ZM, Rourke T, Wang Y, et al. Peak SIV replication in resting memory CD4+ T cells depletes gut lamina propria CD4+ T cells. Nature 2005;434(7037):1148–52.
[218] Mattapallil JJ, Douek DC, Hill BJ, Nishimura Y, Martin M, Roederer M. Massive infection and loss of memory CD4+ T cells in multiple tissues during acute SIV infection. Nature 2005;434(7037):1093–7.
[219] Sandler NG, Douek DC. Microbial translocation in HIV infection: causes, consequences and treatment opportunities. Nat Rev Microbiol 2012;10(9):655–66.

[220] Chege D, Sheth PM, Kain T, Kim CJ, Kovacs C, Loutfy M, et al. Sigmoid Th17 populations, the HIV latent reservoir, and microbial translocation in men on long-term antiretroviral therapy. AIDS 2011;25(6):741–9.

[221] Papasavvas E, Pistilli M, Reynolds G, Bucki R, Azzoni L, Chehimi J, et al. Delayed loss of control of plasma lipopolysaccharide levels after therapy interruption in chronically HIV-1-infected patients. AIDS 2009;23(3):369–75.

[222] Wallet MA, Rodriguez CA, Yin L, Saporta S, Chinratanapisit S, Hou W, et al. Microbial translocation induces persistent macrophage activation unrelated to HIV-1 levels or T-cell activation following therapy. AIDS 2010;24(9):1281–90.

[223] Burdo TH, Lo J, Abbara S, Wei J, DeLelys ME, Preffer F, et al. Soluble CD163, a novel marker of activated macrophages, is elevated and associated with noncalcified coronary plaque in HIV-infected patients. J Infect Dis 2011;204(8):1227–36.

[224] Estes JD, Harris LD, Klatt NR, Tabb B, Pittaluga S, Paiardini M, et al. Damaged intestinal epithelial integrity linked to microbial translocation in pathogenic simian immunodeficiency virus infections. PLoS Pathog 2010;6(8).

[225] Handley SA, Thackray LB, Zhao G, Presti R, Miller AD, Droit L, et al. Pathogenic simian immunodeficiency virus infection is associated with expansion of the enteric virome. Cell 2012;151(2):253–66.

[226] Virgin HW, Wherry EJ, Ahmed R. Redefining chronic viral infection. Cell 2009;138(1):30–50.

[227] Silvestri G, Feinberg MB. Turnover of lymphocytes and conceptual paradigms in HIV infection. J Clin Invest 2003;112(6):821–4.

[228] Herbeuval JP, Shearer GM. HIV-1 immunopathogenesis: how good interferon turns bad. Clin Immunol 2007;123(2):121–8.

[229] Hardy AW, Graham DR, Shearer GM, Herbeuval JP. HIV turns plasmacytoid dendritic cells (pDC) into TRAIL-expressing killer pDC and down-regulates HIV coreceptors by Toll-like receptor 7-induced IFN-alpha. Proc Natl Acad Sci USA 2007;104(44):17453–8.

[230] Stary G, Klein I, Kohlhofer S, Koszik F, Scherzer T, Müllauer L, et al. Plasmacytoid dendritic cells express TRAIL and induce CD4+ T-cell apoptosis in HIV-1 viremic patients. Blood 2009;114(18):3854–63.

[231] Bosinger SE, Hosiawa KA, Cameron MJ, Persad D, Ran L, Xu L, et al. Gene expression profiling of host response in models of acute HIV infection. J Immunol 2004;173(11):6858–63.

[232] Bosinger SE, Li Q, Gordon SN, Klatt NR, Duan L, Xu L, et al. Global genomic analysis reveals rapid control of a robust innate response in SIV-infected sooty mangabeys. J Clin Invest 2009;119(12):3556–72.

[233] Jacquelin B, Mayau V, Targat B, Liovat AS, Kunkel D, Petitjean G, et al. Nonpathogenic SIV infection of African green monkeys induces a strong but rapidly controlled type I IFN response. J Clin Invest 2009;119(12):3544–55.

[234] Lederer S, Favre D, Walters KA, Proll S, Kanwar B, Kasakow Z, et al. Transcriptional profiling in pathogenic and non-pathogenic SIV infections reveals significant distinctions in kinetics and tissue compartmentalization. PLoS Pathog 2009;5(2):e1000296.

[235] Favre D, Lederer S, Kanwar B, Ma ZM, Proll S, Kasakow Z, et al. Critical loss of the balance between Th17 and T regulatory cell populations in pathogenic SIV infection. PLoS Pathog 2009;5(2):e1000295.

[236] Hyrcza M, Kovacs C, Loutfy M, Halpenny R, Heisler L, Yang S, et al. Distinct transcriptional profiles in ex vivo CD4+ and CD8+ T cells are established early in human immunodeficiency virus type 1 infection and are characterized by a chronic interferon response as well as extensive transcriptional changes in CD8+ T cells. J Virol 2007;81(7):3477–86.

[237] Sedaghat AR, German J, Teslovich TM, Cofrancesco Jr J, Jie CC, Talbot Jr CC, et al. Chronic CD4+ T-cell activation and depletion in human immunodeficiency virus type 1 infection: type I interferon-mediated disruption of T-cell dynamics. J Virol 2008;82(4):1870–83.

[238] Barratt-Boyes SM, Wijewardana V, Brown KN. In acute pathogenic SIV infection plasmacytoid dendritic cells are depleted from blood and lymph nodes despite mobilization. J Med Primatol 2010;39(4):235–42.
[239] Hong HS, Rajakumar PA, Billingsley JM, Reeves RK, Johnson RP. No monkey business: why studying NK cells in non-human primates pays off. Front Immunol 2013;4:32.
[240] Giorgi JV, Ho H-N, Hirji K, Chou CC, Hultin LE, O'Rourke S, et al. CD8+ lymphocyte activation at human immunodeficiency virus Type 1 seroconversion: development of HLA-DR+ CD38- CD+ cells is associated with subsequent stable CD4+ cell levels. J Infect Dis 1994;170:775–81.
[241] Sodora DL, Allan JS, Apetrei C, Brenchley JM, Douek DC, Else JG, et al. Toward an AIDS vaccine: lessons from natural simian immunodeficiency virus infections of African nonhuman primate hosts. Nat Med 2009;15(8):861–5.
[242] Pandrea I, Gaufin T, Brenchley JM, Gautam R, Monjure C, Gautam A, et al. Cutting edge: experimentally induced immune activation in natural hosts of simian immunodeficiency virus induces significant increases in viral replication and CD4+ T cell depletion. J Immunol 2008;181(10):6687–91.
[243] Pandrea I, Cornell E, Wilson C, Ribeiro RM, Ma D, Kristoff J, et al. Coagulation biomarkers predict disease progression in SIV-infected nonhuman primates. Blood 2012;120(7):1357–66.
[244] Manches O, Bhardwaj N. Resolution of immune activation defines nonpathogenic SIV infection. J Clin Invest 2009;119(12):3512–5.
[245] Heikenwalder M, Polymenidou M, Junt T, Sigurdson C, Wagner H, Akira S, et al. Lymphoid follicle destruction and immunosuppression after repeated CpG oligodeoxynucleotide administration. Nat Med 2004;10(2):187–92.
[246] Vanderford TH, Slichter C, Rogers KA, Lawson BO, Obaede R, Else J, et al. Treatment of SIV-infected sooty mangabeys with a type-I IFN agonist results in decreased virus replication without inducing hyperimmune activation. Blood 2012;119(24):5750–7.
[247] Kader M, Smith AP, Guiducci C, Wonderlich ER, Normolle D, Watkins SC, et al. Blocking TLR7- and TLR9-mediated IFN-alpha production by plasmacytoid dendritic cells does not diminish immune activation in early SIV infection. PLoS Pathog 2013;9(7):e1003530.
[248] Coffman RL, Sher A, Seder RA. Vaccine adjuvants: putting innate immunity to work. Immunity 2010;33(4):492–503.
[249] Pulendran B, Oh JZ, Nakaya HI, Ravindran R, Kazmin DA. Immunity to viruses: learning from successful human vaccines. Immunol Rev 2013;255(1):243–55.
[250] Kapsenberg ML. Dendritic-cell control of pathogen-driven T-cell polarization. Nat Rev Immunol 2003;3(12):984–93.
[251] Loré K, Betts MR, Brenchley JM, Kuruppu J, Khojasteh S, Perfetto S, et al. Toll-like receptor ligands modulate dendritic cells to augment cytomegalovirus- and HIV-1-specific T cell responses. J Immunol 2003;171(8):4320–8.
[252] Gujer C, Sundling C, Seder RA, Karlsson Hedestam GB, Lore K. Human and rhesus plasmacytoid dendritic cell and B-cell responses to Toll-like receptor stimulation. Immunology 2011;134(3):257–69.
[253] Shedlock DJ, Silvestri G, Weiner DB. Monkeying around with HIV vaccines: using rhesus macaques to define 'gatekeepers' for clinical trials. Nat Rev Immunol 2009;9(10):717–28.
[254] Verthelyi D, Kenney RT, Seder RA, Gam AA, Friedag B, Klinman DM. CpG oligodeoxynucleotides as vaccine adjuvants in primates. J Immunol 2002;168(4):1659–63.
[255] Wille-Reece U, Flynn BJ, Loré K, Koup RA, Miles AP, Saul A, et al. Toll-like receptor agonists influence the magnitude and quality of memory T cell responses after prime-boost immunization in nonhuman primates. J Exp Med 2006;203(5):1249–58.
[256] Wille-Reece U, Wu CY, Flynn BJ, Kedl RM, Seder RA. Immunization with HIV-1 Gag protein conjugated to a TLR7/8 agonist results in the generation of HIV-1 Gag-specific Th1 and CD8+ T cell responses. J Immunol 2005;174(12):7676–83.

[257] Wille-Reece U, Flynn BJ, Loré K, Koup RA, Kedl RM, Mattapallil JJ, et al. HIV Gag protein conjugated to a Toll-like receptor 7/8 agonist improves the magnitude and quality of Th1 and CD8+ T cell responses in nonhuman primates. Proc Natl Acad Sci USA 2005;102(42):15190–4.
[258] Park H, Adamson L, Ha T, Mullen K, Hagen SI, Nogueron A, et al. Polyinosinic-polycytidylic acid is the most effective TLR adjuvant for SIV Gag protein-induced T cell responses in nonhuman primates. J Immunol 2013;190(8):4103–15.
[259] Kwissa M, Amara RR, Robinson HL, Moss B, Alkan S, Jabbar A, et al. Adjuvanting a DNA vaccine with a TLR9 ligand plus Flt3 ligand results in enhanced cellular immunity against the simian immunodeficiency virus. J Exp Med 2007;204(11):2733–46.
[260] Querec TD, Akondy RS, Lee EK, Cao W, Nakaya HI, Teuwen D, et al. Systems biology approach predicts immunogenicity of the yellow fever vaccine in humans. Nat Immunol 2009;10(1):116–25.
[261] Kasturi SP, Skountzou I, Albrecht RA, Koutsonanos D, Hua T, Nakaya HI, et al. Programming the magnitude and persistence of antibody responses with innate immunity. Nature 2011;470(7335):543–7.
[262] Wang Y, Abel K, Lantz K, Krieg AM, McChesney MB, Miller CJ. The Toll-like receptor 7 (TLR7) agonist, imiquimod, and the TLR9 agonist, CpG ODN, induce antiviral cytokines and chemokines but do not prevent vaginal transmission of simian immunodeficiency virus when applied intravaginally to rhesus macaques. J Virol 2005;79(22):14355–70.
[263] Li Q, Estes JD, Schlievert PM, Duan L, Brosnahan AJ, Southern PJ, et al. Glycerol monolaurate prevents mucosal SIV transmission. Nature 2009;458(7241):1034–8.

CHAPTER

9

Humoral Immune Responses in SIV Infection of Sooty Mangabeys: Implications for HIV-1 Pathogenesis and Vaccine Design

Cynthia A. Derdeyn

Department of Pathology and Laboratory Medicine, Emory University, Atlanta, GA, USA

A BRIEF HISTORY OF SIMIAN IMMUNODEFICIENCY VIRUS INFECTION OF SOOTY MANGABEYS

Each of more than 40 primate species in sub-Saharan Africa is known to harbor infection by a unique strain of simian immunodeficiency virus (SIV). With the exception of natural SIV infection of chimpanzees (SIVcpz) [1], these infections do not seem to cause overt disease, perhaps reflecting long-term co-evolution between the virus and the host toward reduced pathogenicity [2]. The prevalence of SIV in nonhuman primate species that serve as natural hosts for infection ranges from less than 1% to >50%, suggesting marked differences in virus transmission and host susceptibility [3–6]. Furthermore, natural prevalence of SIV also is influenced by demographic factors, such as age and sex, within a primate community [7]. Cross-species introduction of SIV into a nonnatural host, such as a different nonhuman primate species or humans, can result in a dead-end infection, the generation of novel recombinant viruses, or the beginning of a pandemic; the latter phenomenon is perhaps most notably recognized in the current group M human immunodeficiency virus (HIV)-1 pandemic, which originated

Natural Hosts of SIV.
http://dx.doi.org/10.1016/B978-0-12-404734-1.00009-7

via SIVcpz transmission into humans and affects more than 30 million individuals globally [2,8,9].

In 1986, a virus similar to HIV-1 was isolated from patients diagnosed with acquired immune deficiency syndrome (AIDS) in West Africa [10,11]. This virus (HIV-2) was distantly related to HIV-1 [12] but was genetically similar to the SIV isolated from a macaque (SIVmac) in captivity that died from an AIDS-like illness [13]. Further studies revealed that SIVmac was inadvertently generated via inoculation of infected material from sooty mangabeys infected with SIV (SIVsm) into macaques in captivity [14–18]. Shortly thereafter, SIVsm was identified as the viral agent associated with a naturally occurring, nonpathogenic infection of sooty mangabeys [19] and the ancestor of HIV-2 in humans [20]. Thus, SIVsm had adapted to replication in two nonnative host species, with transmission to humans resulting in the HIV-2 epidemic in West Africa. Interestingly, and in contrast to HIV-1, HIV-2 infection has remained largely restricted to individuals in West Africa, and the overall prevalence seems to be declining [21–23]. This lack of sustained human-to-human transmission could reflect the lower viral loads characteristic of HIV-2 infection; viral load is a strong predictor of HIV-1 transmission [22,24]. In fact, in individuals affected by both HIV-1 and HIV-2 with progressing disease, HIV-1 viral loads are higher than those of HIV-2 [25], and a larger proportion of HIV-1-infected individuals progress to AIDS [22,26]. Moreover, marked differences between the pathogenesis, immune responses, and transmissibility of HIV-1 and HIV-2 exist [22]. Over time, various strains of SIVsm have been adapted (through serial passaging) for replication in rhesus macaques in captivity, resulting in increased pathogenicity and a disease course that is similar to that of HIV-1 in humans but with more rapid kinetics [27–29]. Because of this, the SIVmac viruses (SIVmac239, SIVmac251) and other related macaque-adapted SIVs (SIVsmE660, SIVsmE543) now serve as the standards for investigations of pathogenesis, vaccine-mediated immune responses, and protection in macaques, especially as studies involving the experimental SIV infection of SIV-negative sooty mangabeys are currently prohibited [30].

PRESERVATION OF GERMINAL CENTER FUNCTION

Adaptive antibody responses normally develop in the B-cell follicles of germinal centers within secondary lymphoid tissue. This process is, however, disrupted in pathogenic HIV-1 and SIV infections by dissolution of lymph node architecture, follicular hyperplasia, and follicular dendritic cell (FDC) trapping of virions in immune complexes, all of which lead to unresolved immune activation and B-cell dysfunction [31,32]. The follicular helper subset of CD4 T cells (T_{FH}) also resides in germinal center B cell follicles, and these cells promote development of memory B cells, antibody affinity maturation through somatic hypermutation, and isotype switching [33]. Thus, the protection of T_{FH} cells, which are susceptible to SIV (and HIV-1) infection, and the restriction of viral burden in lymph nodes could be vital contributions to generating and regulating B-cell function and antibody responses during natural SIVsm infection.

Martinot et al. [34] recently investigated the germinal center microenvironment and viral burden in lymph nodes collected from sooty mangabeys and pigtail macaques experimentally infected with SIV during the acute/early infection phase. In sooty mangabeys, a rapid germinal center response was observed, and the burden of SIV+ cells in lymph nodes was similar to that in the pigtail macaques at early time points, but it subsequently declined. Even sooty mangabeys whose CD8 T cells had been experimentally depleted rarely had detectable levels of SIV+ cells in lymph nodes at the later time points. Within germinal centers, SIV+ cells were present in the pigtail macaques but not in sooty mangabeys, and FDC trapping of virions within B-cell follicles in the germinal center was also limited to lymph nodes from the pigtail macaques. Brenchley and colleagues [35] performed similar comparative studies of the lymph node architecture and viral burden during the chronic stage of infection in sooty mangabeys and rhesus macaques infected with SIV and humans infected with HIV-1. In the SIV-infected rhesus macaques, half of the SIV+ cells in the lymph nodes were located within B-cell follicles in the germinal center, compared to <10% in sooty mangabeys. In addition, more SIV+ T_{FH} cells were observed in the lymph nodes of macaques and humans than in sooty mangabeys. Overall, the proportion of SIV+ cells located within and outside of the B-cell follicles in HIV-1-infected humans was more similar to the macaques than the sooty mangabeys. FDC trapping of the virus was also increased in macaques and humans compared with sooty mangabeys. In addition, a recent study of pathogenic SIV infection of pigtail macaques also demonstrated that lymph node T_{FH} cells were infected with SIV, despite the fact that they did not express high levels of coreceptors associated with SIV entry, including CCR5 [36]. The lack of evidence for selective pressures in the SIV envelope (Env) across infected CD4 T-cell subsets in this study suggested that T_{FH} cells are most likely infected at an earlier stage of differentiation, when CCR5 is expressed at a level that supports infection. In tandem, these dual analyses demonstrate that the sooty mangabey uniquely exhibits a lower SIV+ burden in lymph nodes (particularly within the B-follicle regions in the germinal center), a lower frequency of SIV+ T_{FH} cells, and less FDC trapping of SIV virions. Collectively, these circumstances could depress immune activation and preserve the functional architecture of the lymph node in nonpathogenic SIVsm infection.

B-CELL RESPONSES IN SIVSM

The ability of sooty mangabeys to maintain normal follicular morphology and function during SIV infection could also help to regulate B-cell responses in this natural host despite ongoing viral replication. In the African green monkey (AGM), an experimental model of natural SIV infection, depletion of B cells does not result in increased viremia or accelerated disease progression [37,38]. This result suggests that antibody effector functions do not directly suppress viral replication. While this depletion experiment has not been performed in comparable studies using sooty mangabeys, no evidence that humoral immune responses suppress viral replication exists for this natural host either. The depletion findings, nevertheless, do not preclude B cells and antibodies from contributing to a nonprogressive disease state. Indeed, recent studies suggest that B cells and antibodies in SIVsm are qualitatively and quantitatively different from pathogenic SIV or HIV-1 infection.

Using BrdU labeling, Kaur et al. [39] demonstrated that the rate of peripheral B-cell turnover was similar between natural SIVsm infected and uninfected sooty mangabeys, although the absolute B-cell count was significantly lower in the infected animals. This finding suggests that SIVsm may affect the balance within the B-cell compartment without inducing a hyperactivated or dysfunctional state. During experimental infection of sooty mangabeys, Li et al. [40] reported a transient increase in peripheral B cell Ki67+ proliferation at day 30, which subsequently returned to near baseline levels by day 60, consistent with active resolution of early immune activation in the B-cell compartment. In the same study, the magnitude of autologous neutralizing antibody activity across naturally infected sooty mangabeys (discussed in more detail below) directly correlated with the number of peripheral CD4 T cells, reinforcing the concept that maintenance of certain CD4 T-cell subsets augments B-cell function. Kaur et al. [39] also identified a connection between the CD4 T-cell and B-cell compartments in natural SIVsm, in that the rate of B-cell turnover was inversely correlated with CD4 T-cell turnover. This finding, combined with a lack of correlation between B-cell and CD8 T-cell turnover, further distinguishes SIVsm from SIV-infected rhesus macaques, in which the turnover of B cells is directly correlated with both CD4 and CD8 T cells [41]. Perhaps activation and effector function within the B-cell compartment in natural SIVsm infections are linked with the preservation of CD4 T cells, which are actively regulated even though viral replication persists, and lead to "functional clearance" of the virus.

ANTIBODY RESPONSES IN SOOTY MANGABEYS

Our understanding of the antibody response to SIVsm is limited, hampered by the current lack of a sooty mangabey experimental model. An early study of natural SIVsm demonstrated that these animals were seropositive but displayed very low levels of neutralizing antibody activity against a heterologous SIVsm viral isolate [19]. In a comparative follow-up study, Fultz et al. [42] asked whether anti-SIV antibodies played a role in the disease progression of SIVsm-infected rhesus or pigtail macaques or the lack thereof in sooty mangabeys. Anti-SIVsm-binding antibody titers were higher in macaques than in sooty mangabeys, and the magnitude was positively associated with progression to disease in macaques. Neutralizing antibodies were not detected in the sooty mangabeys but were present at relatively high titers in 3 of 13 SIVsm-infected macaques; neutralizing capacity did not, however, correlate with disease progression. When sooty mangabeys and rhesus macaques were inoculated intravenously with SIVmac239 (an SIVsm strain adapted for replication in rhesus macaques) and evaluated longitudinally, Kaur et al. [43] found low viral loads and undetectable neutralizing activity in the three infected sooty mangabeys, whereas relatively potent neutralizing antibody activity was detected in two of three macaques. Taken together, these early studies suggested that high titer neutralizing antibody activity was not a common feature of SIVsm infection.

More recently, Li et al. [40] measured neutralizing antibody activity in sooty mangabeys against their autologous SIVsm Envs (that were cloned directly from each infected animal) using a sensitive method typically and ubiquitously used to gauge humoral responses against HIV-1. For each animal, this evaluation system uses pseudoviruses displaying the viral Envs that B cells would have encountered in vivo. Responses were evaluated in plasma collected during the early stages of experimentally induced infection and established infection of naturally infected sooty mangabeys; these neutralizing activities were then compared against autologous plasma and Env combinations collected from HIV-1-infected human subjects at similar time points. Because of the widespread resistance of the contemporaneous virus to autologous neutralizing antibodies in HIV-1 infection, all plasma samples used for neutralization assays were collected 6–9 months after the Envs were cloned. Autologous neutralizing activity was significantly lower in the sooty mangabeys in both early and established infection compared to HIV-1 infection. Thus, consistent with previous studies, the B-cell and/or neutralizing antibody response in SIVsm infection may be blunted to prevent the hyperimmune activation, dysregulation, and dysfunction of B cells that persists throughout HIV-1 infection [31,44]. Li et al. proposed that the quantitative contrast between SIVsm and HIV-1 infection could also stem from qualitative differences, in which antibodies targeted distinct regions of the SIVsm and HIV-1 Envs. This hypothesis is consistent with the poor cross-reactivity and low level of amino acid sequence conservation between SIVsm/HIV-2 and SIVcpz/HIV-1 Envs, which provide evidence for subtle conformational differences, and is supported by the recent identification of neutralizing antibody epitopes in SIVsm and HIV-2 Envs. These and related topics are discussed in more detail below and are presented in summary form in Table 9.1.

TABLE 9.1 Summary of Envelope Properties

Property	HIV-1	HIV-2	SIVmac	SIVsmE660	Natural SIVsm
Env lineage	SIVcpz[a]	SIVsm[b]	SIVsm	SIVsm	SIVsm
Coreceptor usage	Mainly restricted to CCR5 and CXCR4	Uses CCR5 and CXCR4; promiscuous for other coreceptors	CCR5, GPR15/BOB, STRL33/BONZO/CXCR6, GPR1	CCR5	Uses CCR5; promiscuous for other coreceptors but not CXCR4
Autologous nAb Titers	Rapid, high	Rapid, very high	Variable	Rapid, high	Slow, low
Susceptibility to heterologous nAb	Low	High	Low	High	Low/Moderate[l]
Defined nAb epitopes	MPER[c], CD4bs[d], N332 glycan patch[e], α2 helix[f], V1V2[g], base of V3[h]	V3[i], CD4i[j], CD4bs, V4[k]	V1V2, V4	V3, CD4i, CD4bs, V4	Unknown[l]
nAb escape variants	Rapid, efficient escape in most individuals	Inefficient escape, persistence of sensitive variants	Variable	Rapid and efficient escape	Unknown
Diversifying selection on Env	High across isolates	Conflicting reports across isolates	High during SIVmac infection	Low in E660 isolates; high during E660 infection	Low across isolates

[a] *cpz—chimpanzee.*
[b] *sm—sooty mangabey.*
[c] *MPER—membrane proximal external region of gp41.*
[d] *CD4bs—CD4 binding site on gp120.*
[e] *An array of asparagine-linked carbohydrates-centered on residue asparagine 332 based on HIV-1 HXB2 numbering.*
[f] *Structural region located on the gp120 outer domain.*
[g] *The first and second variable domains of gp120.*
[h] *A conformational region formed by regions that flank the third variable domain on gp120.*
[i] *The third variable domain on gp120.*
[j] *CD4-induced epitopes in gp120, containing the coreceptor binding domain.*
[k] *The fourth variable domain of gp120.*
[l] *C.A. Derdeyn, unpublished data.*

It is also possible that during natural infection of sooty mangabeys, SIVsm is initially neutralized, becomes resistant, and, in contrast to HIV-1, fails to initiate a de novo response against the escape variant. In addition, immunoglobulins from sooty mangabeys and humans could have different functional properties that influence the magnitude, targeting, and outcome of neutralizing antibody activity [45].

COMPARATIVE STUDIES OF HIV-1 AND SIVSM/HIV-2 ANTIBODY RESPONSES

The envelope (*env*) gene of HIV/SIV is the most variable of viral genes and is the principal target of the humoral immune response. Many studies have focused on neutralizing antibody activity against Env in HIV/SIV infection, reflecting the reputed importance of this antibody effector function in achieving vaccine-mediated protection against HIV-1 [46–48]. It is important to note that genetically diverse HIV-2 and SIVsm-derived Env sequences form a single phylogenetic lineage that is distinct from the HIV-1/SIVcpz group [49]; SIVsm and HIV-2 Envs are, therefore, defined by unique genetic and phenotypic properties. Indeed, Decker et al. [50] reported that HIV-1 and HIV-2 Envs share only about 40% amino acid identity. The amino acid identity among HIV-1, HIV-2, and SIVsm is about 55% in the well-conserved bridging sheet or coreceptor binding domain, whereas the overall level of conservation in gp120 rests at only about 25% (excluding regions that cannot be aligned) [50]. It is not surprising, then, that antibodies poorly cross-react between HIV-1 and HIV-2 Envs but strongly cross-react (and cross-neutralize) between HIV-2 and SIVsm Envs [50–54]. The genetic relatedness between HIV-2 and SIVsm Envs, paired with the apparent lower pathogenicity and less positive selection of HIV-2 in humans, suggests that at least some aspects of antibody responses and Env evolution in HIV-2 infection could be comparable to natural SIVsm.

The widespread use of relatively standardized methods in studies of neutralizing antibody responses in HIV-1 infection, and more recently in HIV-2 infection, provides a strong foundation for comparison. Monoclonal antibodies derived from individuals with chronic HIV-1 infection with an elite breadth of cross-neutralizing antibodies and recently infected individuals with very focused neutralizing antibody activity suggest that HIV-1-neutralizing antibodies frequently recognize four regions of the Env: the V1V2 domain, the base of the V3 domain/α2 helix/glycan patch region, the CD4 binding site (CD4bs) of gp120, and the membrane-proximal external region domain of gp41 [55–59] (Figure 9.1(A)). While the HIV-1 gp120 V3 domain itself is immunogenic and frequently elicits cross-reactive binding antibodies in natural infection, the HIV-1 Env effectively shields this domain, rendering it inaccessible on the virion-associated Env

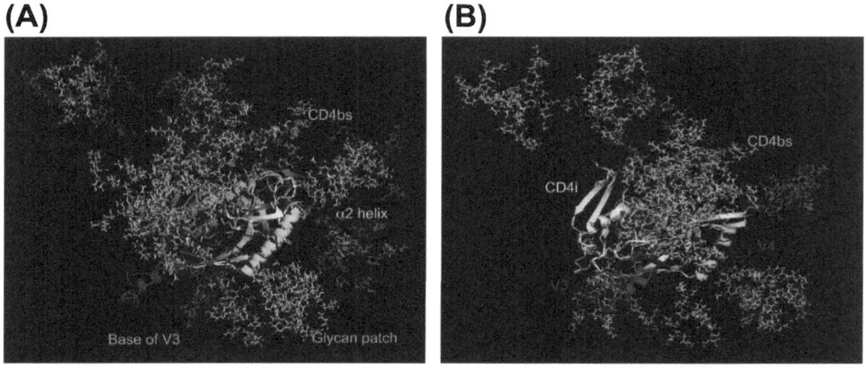

FIGURE 9.1 Neutralizing antibody targets on human immunodeficiency virus (HIV)-1 and HIV-2/simian immunodeficiency virus (SIV) in sooty mangabeys gp120. A homology-modeled, CD4-bound subtype A HIV-1 gp120 monomer, described previously [59] was used to display targets of neutralizing antibodies that have been identified in HIV-1 (A), and HIV-2 or SIV (B) infections. A ribbon diagram of the gp120 crystal structure is shown in gray in two different orientations in A and B to visualize the corresponding targets. Carbohydrates are represented as sticks and are colored by element. Gp120 structural domains are color-coded as follows: V1V2 is red, V3 is magenta, V4 is blue, α2 helix is cyan, the CD4-induced (CD4i) bridging sheet is yellow, and residues comprising the conformational CD4 binding site (CD4bs) are green. For orientation, the α2 helix (cyan) is located on the outer domain of gp120; the CD4i region (yellow) is located in the inner domain. (A) Neutralizing antibody targets identified for HIV-1 infection are V1V2, CD4bs, α2 helix, a high mannose glycan patch on the outer domain, and conformational epitopes formed at the base of the V3. (B) In HIV-2/SIV infection, the targets identified thus far are within V1V2, CD4bs, CD4i, V4, and V3. Bridging sheet residues and CD4 contact residues were identified according to Decker et al. [50] and Zhou et al. [60]. The web-based "glycoprot" tool (http://www.glycosciences.de/modeling/glyprot/php/main.php) was used to add carbohydrates to gp120. MacPyMol for Mac OS X was used to visualize the gp120 structure. (See the color plate at the back of the book.)

trimer [61–63]. Transferring the HIV-1 V3 domain into an HIV-2 Env backbone results in a dramatically increased sensitivity to V3-mediated neutralization by monoclonal and HIV-1+ plasma antibodies [62]. Thus, in the context of the HIV-2 Env, the V3 loop likely exists in a configuration on the virion that is accessible to neutralizing antibodies. Likewise, neutralizing antibodies are directed against the coreceptor binding domain during infection but generally do not neutralize HIV-1 Envs unless first triggered by soluble CD4 (called CD4-induced or CD4i epitopes) [50,63]. Interestingly, the bridging sheet is one of the most conserved regions among HIV-1, HIV-2, and genetically diverse SIVs, and antibodies against this domain become widely cross-neutralizing when the Envs are first triggered by soluble CD4 [50].

Few studies have been performed to map the targets of neutralizing or other antibody specificities in infected sooty mangabeys; nevertheless, a series of recent studies, some aided by cross-neutralizing monoclonal antibodies derived from HIV-2-infected subjects, have demonstrated that many HIV-2 and some SIVsm Envs are sensitive to neutralization by recognition of epitopes in the gp120 V3 and V4 domains, CD4i epitopes, and the CD4bs [52–54,64,65] (Figure 9.1(B)). These observations contrast with HIV-1 because, as discussed earlier, HIV-1 Envs are generally not sensitive to neutralization by V3 or CD4i antibodies, even though these antibodies are commonly found in HIV-1-infected individuals. Antibodies against the CD4bs are also commonly found in HIV-1-infected subjects but infrequently mediate neutralization of autologous patient-derived HIV-1 [66,67]. Neutralizing antibodies that target V4 epitopes in HIV-1 infection do not seem to be common, although mutations in or near V4 can affect neutralization targeted to other regions [68,69]. The V4 loop is located on the gp120 outer domain and could be more accessible to neutralizing antibodies in the HIV-2 Env because it is less glycosylated compared to the HIV-1 Env [52]. Interestingly, in addition to eliciting different antibody specificities than HIV-1 infection, HIV-2 infection seems to elicit antibodies with a greater breadth and potency of cross-neutralization [52,70,71]. A panel of monoclonal antibodies isolated from HIV-2-infected individuals described by Kong et al. [52] potently neutralized most of the 32 strains of HIV-2 that were tested. This study also discussed the interesting point that while HIV-1 and HIV-2 Envs both elicit antibodies against V3, CD4i epitopes, and the CD4bs during infection, these antibody specificities are generally only neutralizing in the context of the HIV-2 Env. The authors proposed that the HIV-2 Env is unexpectedly vulnerable to heterologous neutralization, unlike HIV-1. Differences in the high mannose glycosylation content of gp120 between SIVcpz/HIV-1 isolates and SIVsm/HIV-2 isolates may also influence their susceptibility to neutralization [72]. In consequence, the immunogenicity and/or antigenicity of HIV-1 and HIV-2 Envs differs (Figure 9.1(A) and (B)). The HIV-2 Env more readily exposes conserved structures, such as V3 and the coreceptor binding domain, that render the virus susceptible to neutralizing antibodies. One might then expect that SIVsm Envs also are antigenically distinct from HIV-1 Envs.

ANTIBODY NEUTRALIZATION OF SIVSM ENVS

The neutralization sensitivity of SIVsm-derived Envs to heterologous SIV+ plasma and HIV-2 monoclonal antibodies varies widely. For example, SIVmac239 tests as highly resistant, whereas SIVsmE660, a commonly used challenge virus for nonhuman primate vaccine studies, is extremely sensitive [54]. Lopker et al. [54] recently characterized the neutralization sensitivity of Envs from an SIVsmE660 quasispecies and found an overabundance of Env variants that were highly susceptible to neutralization by heterologous plasma. In fact, some of the SIVsmE660 Envs were neutralized by heterologous SIV-infected macaque serum or plasma samples with 50% inhibitory titers of 1:100,000; HIV-2 monoclonal antibodies targeting V3, CD4i epitopes, V4, and the CD4bs also potently neutralized these Envs. This type of extreme neutralization sensitivity is rare for patient-derived HIV-1 Envs that have not been subjected to in vitro cell culture [73]. These findings indicate that, in comparison to the majority of circulating HIV-1 viruses, SIVsmE660 exhibits substantially less masking (and/or glycosylation) of certain Env epitopes, making it more globally susceptible to neutralization. These studies collectively and indirectly suggest that the immunogenicity and/or antigenicity of SIVsm-derived Envs differ from commonly circulating HIV-1 Envs, which may lead to an overestimation (SIVsmE660) or an underestimation (SIVmac239) of vaccine-induced antibody protection.

The existence of neutralization-resistant, macaque-adapted SIVsm Envs suggests that, like the HIV-1 Env, the SIVsm Env also has the capacity to develop resistance against neutralizing antibodies. In fact, Wu et al. [74] recently demonstrated that SIVsmE660-derived molecular clones exhibited immune escape from neutralizing antibodies in rhesus macaques that was reminiscent of HIV-1 infection. The SIVsmE660 Envs elicited autologous neutralizing antibody responses that were detectable by 4 weeks after infection, which was slightly more rapid but close to the range of most HIV-1 infections [55,75–77]. Neutralization escape variants were generated and were associated with changes in the V1 and V4 domains of gp120 [74]. De novo antibodies were also generated against the escape variants in a pattern that recapitulates this process in HIV-1 infection [55,59,77–80]. Lauren et al. [81] also examined autologous neutralizing antibody responses against SIVsm in cynomolgus macaques. The authors found that each infected monkey developed a neutralizing antibody response against the inoculum virus; these responses developed earlier in the animals infected intravenously compared to those infected intrarectally. Neutralization escape variants were present by 3 months after infection in animals that rapidly progressed to disease but were found later in monkeys with slower progression rates. Consequently, in this pathogenic infection model, route of infection influenced neutralizing antibody responses and the development of resistance, and perhaps affected the rate of disease progression. Ultimately, the SIVsm Env can escape from neutralizing antibodies during pathogenic infection of two nonnatural hosts; however, whether cycles of viral escape from neutralizing antibodies occur in natural SIVsm remains to be determined.

BIOLOGICAL PROPERTIES OF HIV-1- AND SIVSM-DERIVED ENVS

To understand antibody responses in natural versus non-natural hosts, it makes sense to consider inherent similarities and differences between the target Env glycoproteins. All naturally occurring HIV-1, HIV-2, and SIVsm viruses use CD4 as their primary receptor for entry into host cells [82–85]. The gp120-CD4 interaction promotes conformational changes in the gp120 inner domain that lead to the formation of a bridging sheet or coreceptor binding domain. Exposure of the CD4bs on gp120 must be tightly regulated to prevent recognition by neutralizing antibodies and to inhibit any premature changes in conformation that could lead to inactivation of the Env spike. Furthermore, neutralizing antibodies seem to be a strong selective force that constrains HIV-1, HIV-2, and SIV to dependence on CD4 in vivo [50,86,87]. Subtle conformational differences do exist between HIV-2/SIVsm and HIV-1 Envs, despite their shared dependence on CD4 and common entry processes. For example, SIVsm-derived Envs (including HIV-2 Envs) are generally less dependent on CD4 and can use lower levels to enter target cells than HIV-1 Envs isolated from peripheral blood or lymph nodes, which, in contrast, depend significantly on CD4 [52,88–95]. SIV can also attain efficient CD4-independent utilization of the CCR5 coreceptor in vivo, following depletion of CD4 T cells in rhesus macaques experimentally infected with SIVmac251 [86,96]. It is important to note that at least nine phylogenetically distinct subtypes of naturally occurring SIVsm viral isolates have been identified, and thus the biological properties of their Envs may vary [14,97].

HIV-1 viruses are generally restricted to utilization of CCR5 or CXCR4, although usage of alternative coreceptors have been described in vitro [98–108]; see Peden and Farber [109] for a comprehensive review. While CCR5 is the main coreceptor for almost all SIVsm-derived Envs, including HIV-2, this lineage of Envs is more promiscuous than HIV-1 with regard to its coreceptor usage in vitro [110–120]. Coreceptors that seem to be most widely and efficiently utilized by SIVsm-derived Envs are GPR15 (BOB) and CXCR6 (BONZO, STRL33). More recently, Elliott et al. [121] cloned chemokine receptor genes from naturally infected sooty mangabeys and evaluated their ability to support in vitro entry of SIVsm Envs from this naturally infected host. The authors found that smCXCR6 supported efficient infection by all of the natural SIVsm Envs tested at levels comparable to smCCR5. Coreceptors smGPR15, smGPR1, and smAPJ supported lower levels of entry, whereas smCCR2b, smCCR3, smCCR4, smCCR8, and smCXCR4 were not used by most isolates. CXCR4 usage by SIVsm-derived Envs, excluding HIV-2, is rare [121,122] but was described in experimentally infected sooty mangabeys that experienced profound, virally induced CD4 depletion [123]. SIVsm Envs from these

CD4-depleted sooty mangabeys used smCXCR4, smCXCR6, and smCCR5 most efficiently, as well as smCCR3, smCCR8, smGPR1, smGPR15, and smAPJ to a lesser extent [121]. SIVsm Envs from naturally infected sooty mangabeys that do not express CCR5 because of a genetic mutation also use CXCR6, GPR15, and GPR1 [124].

A recent study suggests that the varied requirements for CD4 utilization between HIV-1- and SIVsm-derived Envs potentially arise from inherent differences in how their respective gp120 molecules modulate association of the Env trimer and CD4 binding [125]. Finzi et al. [125] described evidence for a structural difference between SIV and HIV-1 gp120 in the cavity that binds the Phe43 residue of CD4. SIVsm-derived Envs have a bulky tryptophan residue at position 375 in the CD4 binding cavity, whereas HIV-1 Envs carry a much smaller serine residue, which leaves a void that can be filled by Phe43 [60] (Figure 9.2). Because of this and other gp120 properties, SIV Envs are more likely to sample conformations resembling the CD4-bound state than HIV-1 gp120 [125]. Furthermore, when serine 375 is replaced with tryptophan in HIV-1, the gp120 molecule becomes more like SIV in its propensity to sample CD4-bound conformations [126]. This and other differences between SIVsm-derived and HIV-1 Envs could influence the efficiency of CD4 and coreceptor utilization, immunogenicity of the CD4bs, as well as susceptibility to CD4bs neutralizing antibodies.

Interestingly, the nature of the Phe43 cavity on gp120 could have been important in the adaptation of HIV-1 for efficient human-to-human spread. Finzi et al. [125] compared gp120 sequences from genetically diverse SIVs and HIVs and found that residue 375 is tryptophan in most monkey SIVs but has evolved to other large residues (Met, His, Tyr, and Phe) in ape SIVs and in group N and O HIV-1, which remain restricted to a small number of individuals. A serine at residue 375, by contrast, was favored in the evolution of pandemic group M HIV-1 from SIVcpz, and this serine is highly conserved across HIV-1 group M clades, illustrating its importance. Nevertheless, differences in CD4 utilization and structural characteristics of the CD4bs could stem from divergent selective pressures placed on Envs in natural versus nonnatural hosts. As described earlier, several strains of SIVsm have been adapted for replication and AIDS-like pathogenesis in the rhesus macaque, a non-natural host. However, little is known about the adaptive events in Envs that lead to the ability to establish infection, replicate, and induce disease progression in a different species. Demma et al. [127] and Vanderford et al. [128] investigated potential determinants of adaptation in the SIVsm Env when they juxtaposed the outcomes of parallel inoculation of an SIVsm quasispecies into sooty mangabeys and rhesus macaques. They found that the variants that established infection in rhesus macaques were enriched for specific changes within the V1V2 domain but that this type of selection was not favored upon SIVsm inoculation into its natural host. They hypothesized that Env adaptation could

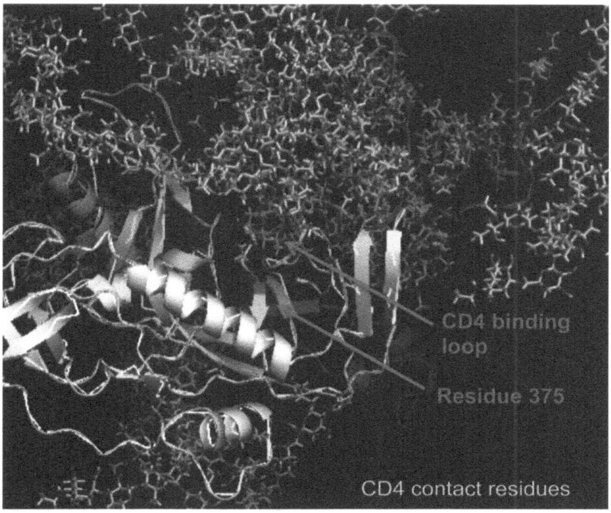

FIGURE 9.2 **The location of residue serine 375 and the CD4 binding loop on the HIV-1 gp120 monomer.** A homology-modeled, CD4-bound subtype A gp120 monomer, described previously [59], was used to display the location of the CD4 binding loop and residue Ser375 (based on the human immunodeficiency virus [HIV]-1 HXB2 numbering scheme), which in HIV-1 accommodates the Phe43 residue of the CD4 molecule. In simian immunodeficiency virus (SIV), this position is occupied by a bulky tryptophan residue [125]. The diagram is zoomed into this region of the gp120 inner domain. A ribbon diagram of the gp120 crystal structure is shown in gray. Surrounding carbohydrates are represented as sticks and are colored by element. Gp120 structural domains are color-coded as follows: V1V2 is red, V3 is magenta, V4 is blue, α2 helix is cyan, the CD4-induced bridging sheet is yellow, and residues comprising the conformational CD4 binding site (CD4bs) are green. Orange color and arrows indicate the CD4 binding loop and residue Ser375. Bridging sheet residues and CD4 contact residues were identified according to Decker et al. [50] and Zhou et al. [60]. The web-based "glycoprot" tool at (http://www/glycosciences.de/modeling/glyprot/php.main.php) was used to add carbohydrates to gp120. MacPyMol for Mac OS X was used to visualize the gp120 structure. (See the color plate at the back of the book.)

have involved interactions with the rhesus macaque CD4 molecule, as V1V2 can modulate this interaction. However, signatures of Env adaptation to the rhesus macaque outside of V1V2 were not assessed.

EVOLUTIONARY PATHWAYS OF SIVsm Envs

As discussed earlier, virtually all vaccine protection studies in nonhuman primates use SIVsm-derived Env immunogens and challenge viruses. Studies that are designed to evaluate antibody-mediated protection often use heterologous pairs of Envs, such as SIVmac239/251 and

SIVsmE660, in which the immunogen Env is genetically distinct from the challenge virus Env. The rationale for this approach is to approximate the divergence that exists between an HIV-1 Env immunogen and the genetically diverse HIV-1 isolates circulating within a particular geographic region or population. Fischer et al. [129] recently investigated the validity of this approach by comparing the "evolutionary pathways" of pandemic HIV-1 and SIVsm, including 20 genetically diverse viral isolates of SIVsm from naturally infected sooty mangabeys. While this study analyzed the sequence of the entire viral genome, here we focus the discussion on Envs. Overall, Fischer et al. found that the SIVsm Env gp120 exhibited more synonymous mutations, less evidence of positive selection, and lower diversity than the HIV-1 Env gp120. This is consistent with a previous study demonstrating that the ratio of nonsynonymous to synonymous substitutions in Envs was higher for pathogenic HIV-1 and SIV infections than for nonpathogenic SIVsm and SIVagm infections [130]. Evidence for lower levels of positive selection on Envs in HIV-2 versus HIV-1 infection also exists in some studies [70,130,131] but not others [132]. Fischer et al. concluded that the HIV-1 Env is adapted for rapid immune escape, whereas the SIVsm Env is likely subjected to less and perhaps qualitatively different immunologic pressure. Indeed, Zhang et al. [133] demonstrated that the shifting of N-linked glycosylation sites, a phenomenon associated with escape from neutralizing antibodies in both HIV-1 and SIV infection, is more pronounced in the SIVcpz/HIV-1 lineage than in the HIV-2/SIVsm lineage. In that same study, however, HIV-2 exhibited greater shifting of potential N-linked glycan addition sites than SIVsm, perhaps reflecting intensified immune pressure following transmission into a new host.

The distinct evolutionary pathways for SIVsm and HIV-1 Envs nicely parallel their distinct antibody elicitation profiles. High titer autologous neutralizing activity in HIV-1 infection drives cycles of viral escape and de novo neutralization and is reflected in the consistently increasing Env diversity [59,77–80]. This phenomenon contrasts with a single comparable study showing a lower magnitude of autologous neutralizing antibody responses in sooty mangabeys [40]. In that study, however, experiments did not address whether escape from neutralizing antibodies ensued. Demma et al. [134] demonstrated that the V1V2 domain of the SIVsm Env gp120 underwent length, glycosylation, and amino acid sequence variation within infected sooty mangabeys, whereas the Gag p27 protein in these animals experienced purifying selection. Thus, SIVsm Envs do undergo changes within hypervariable regions of gp120, as does HIV-1, but it is not known whether neutralizing antibodies are the selective force. Regardless, on a larger scale, HIV-1 has evolved more extensive genetic diversity in fewer than 100 years than SIVsm has over thousands of years [129]. Although the genetic diversity of SIVsm in primate centers within the United States has been suggested to "mirror" that of HIV-1 group M isolates [14], the genetic distance between SIVsm Envs on a phylogenetic tree may not be "functionally equivalent" to the same parameter between HIV-1 Envs. Evolution of the SIVsm Env likely projects in a direction opposite from HIV-1: toward long-term reduced pathogenicity and perhaps even reduced transmission [135]. For a comparison of biological properties of HIV-1, HIV-2, and SIVsm Envs relevant to pathogenesis and vaccine studies in macaques and natural infection of sooty mangabeys, see Table 9.1. A better understanding of the genetic and phenotypic properties of SIVsm Envs, and the antibody responses they elicit and to which they are susceptible, will inform more accurate models of HIV-1 vaccination and protection; this knowledge may also facilitate dissection of the underlying basis for reduced pathogenicity in natural host infections.

CONCLUSIONS

SIVsm has clearly followed a unique evolutionary trajectory in its natural sooty mangabey host that has shaped the immunology and course of this nonpathogenic infection. Deciphering the many similarities in antibody responses and Env evolution between SIVsm and HIV-2, and the marked differences from HIV-1, could lead to the development of tailored interventions for pathogenic HIV-1 and HIV-2 infections. While the T-cell compartment has generally been the focus of studies of immune activation and dysfunction in HIV-1 infection, comparative studies of SIVsm infection suggest that strategies to limit immune activation and dysfunction of B cells could also be beneficial. For instance, preservation of "normal" B-cell germinal center functions and T_{FH} responses may contribute to reduced pathogenesis. It is possible that high levels of viral antigen alone do not disrupt B-cell/germinal center function, as is observed in SIVsm infection, but that a persistent viral load combined with the unrelenting emergence of viral escape variants in HIV-1 infection is also detrimental. Thus far, it is apparent that SIVsm Envs share antigenic properties with HIV-2 Envs but are quite distinct from HIV-1 Envs in this regard (Table 9.1, Figure 9.1). Neutralizing antibody responses seem to differ both qualitatively and quantitatively between pathogenic and nonpathogenic SIVsm, HIV-2, and HIV-1 infections, suggesting that SIVsm-derived Env immunogens could elicit a different profile of neutralizing antibodies than HIV-1 Env immunogens. Therefore, SIVsm-based vaccination and challenge models could underestimate or overestimate the level of antibody-mediated protection that could be achieved against circulating HIV-1 strains, depending on the selected SIV strains. Furthermore, the concept of heterologous challenge after vaccination based on pairwise distance between the immunogen and challenge Envs could be somewhat misconstrued. Thus, for HIV-1 vaccine studies, a proper infrastructure should be developed to gauge the relevance and/or applicability of using SIVsm-derived Envs as diagnostic models, which could also facilitate more apt selection of vaccine immunogens. Together, studies designed to recognize the interconnections between SIVsm, HIV-1, and HIV-2 viral populations, with a focus on Env, and to unravel the differences in their respective immune responses and infection trajectories, stand to enhance global health in multiple regards.

Acknowledgments

I thank Dr Megan Murphy for critical reading and editing of the chapter and also acknowledge funding support from NIH: R01-AI58706, U19-AI96187, P51-RR000165.

References

[1] Keele BF, Jones JH, Terio KA, Estes JD, Rudicell RS, Wilson ML, et al. Increased mortality and AIDS-like immunopathology in wild chimpanzees infected with SIVcpz. Nature 2009;460:515–9.

[2] Apetrei C, Robertson DL, Marx PA. The history of SIVS and AIDS: epidemiology, phylogeny and biology of isolates from naturally SIV infected non-human primates (NHP) in Africa. Front Biosci 2004;9:225–54.

[3] Aghokeng AF, Ayouba A, Mpoudi-Ngole E, Loul S, Liegeois F, Delaporte E, et al. Extensive survey on the prevalence and genetic diversity of SIVs in primate bushmeat provides insights into risks for potential new cross-species transmissions. Infect Genet Evol 2010;10:386–96.

[4] Aghokeng AF, Liu W, Bibollet-Ruche F, Loul S, Mpoudi-Ngole E, Laurent C, et al. Widely varying SIV prevalence rates in naturally infected primate species from Cameroon. Virology 2006;345:174–89.

[5] Phillips-Conroy JE, Jolly CJ, Petros B, Allan JS, Desrosiers RC. Sexual transmission of SIVagm in wild grivet monkeys. J Med Primatol 1994;23:1–7.

[6] Santiago ML, Range F, Keele BF, Li Y, Bailes E, Bibollet-Ruche F, et al. Simian immunodeficiency virus infection in free-ranging sooty mangabeys (*Cercocebus atys atys*) from the Tai Forest, Cote d'Ivoire: implications for the origin of epidemic human immunodeficiency virus type 2. J Virol 2005;79:12515–27.

[7] Ma D, Jasinska A, Kristoff J, Grobler JP, Turner T, Jung Y, et al. SIVagm infection in wild African green monkeys from South Africa: epidemiology, natural history, and evolutionary considerations. PLoS Pathog 2013;9:e1003011.

[8] Gao F, Bailes E, Robertson DL, Chen Y, Rodenburg CM, Michael SF, et al. Origin of HIV-1 in the chimpanzee pan troglodytes troglodytes. Nature 1999;397:436–41.

[9] Keele BF, Van Heuverswyn F, Li Y, Bailes E, Takehisa J, Santiago ML, et al. Chimpanzee reservoirs of pandemic and nonpandemic HIV-1. Science 2006;313:523–6.

[10] Clavel F, Guyader M, Guetard D, Salle M, Montagnier L, Alizon M. Molecular cloning and polymorphism of the human immune deficiency virus type 2. Nature 1986; 324:691–5.

[11] Clavel F, Mansinho K, Chamaret S, Guetard D, Favier V, Nina J, et al. Human immunodeficiency virus type 2 infection associated with AIDS in West Africa. N Engl J Med 1987;316:1180–5.

[12] Guyader M, Emerman M, Sonigo P, Clavel F, Montagnier L, Alizon M. Genome organization and transactivation of the human immunodeficiency virus type 2. Nature 1987;326:662–9.

[13] Chakrabarti L, Guyader M, Alizon M, Daniel MD, Desrosiers RC, Tiollais P, et al. Sequence of simian immunodeficiency virus from macaque and its relationship to other human and simian retroviruses. Nature 1987;328:543–7.
[14] Apetrei C, Kaur A, Lerche NW, Metzger M, Pandrea I, Hardcastle J, et al. Molecular epidemiology of simian immunodeficiency virus SIVsm in U.S. primate centers unravels the origin of SIVmac and SIVstm. J Virol 2005;79:8991–9005.
[15] Apetrei C, Lerche NW, Pandrea I, Gormus B, Silvestri G, Kaur A, et al. Kuru experiments triggered the emergence of pathogenic SIVmac. AIDS 2006;20:317–21.
[16] Daniel MD, Letvin NL, King NW, Kannagi M, Sehgal PK, Hunt RD, et al. Isolation of T-cell tropic HTLV-III-like retrovirus from macaques. Science 1985;228: 1201–4.
[17] Mansfield KG, Lerch NW, Gardner MB, Lackner AA. Origins of simian immunodeficiency virus infection in macaques at the New England Regional Primate Research Center. J Med Primatol 1995;24:116–22.
[18] Murphey-Corb M, Martin LN, Rangan SR, Baskin GB, Gormus BJ, Wolf RH, et al. Isolation of an HTLV-III-related retrovirus from macaques with simian AIDS and its possible origin in asymptomatic mangabeys. Nature 1986;321:435–7.
[19] Fultz PN, McClure HM, Anderson DC, Swenson RB, Anand R, Srinivasan A. Isolation of a T-lymphotropic retrovirus from naturally infected sooty mangabey monkeys (*Cercocebus atys*). Proc Natl Acad Sci USA 1986;83:5286–90.
[20] Hirsch VM, Olmsted RA, Murphey-Corb M, Purcell RH, Johnson PR. An African primate lentivirus (SIVsm) closely related to HIV-2. Nature 1989;339:389–92.
[21] Hamel DJ, Sankale JL, Eisen G, Meloni ST, Mullins C, Gueye-Ndiaye A, et al. Twenty years of prospective molecular epidemiology in Senegal: changes in HIV diversity. AIDS Res Hum Retroviruses 2007;23:1189–96.
[22] Nyamweya S, Hegedus A, Jaye A, Rowland-Jones S, Flanagan KL, Macallan DC. Comparing HIV-1 and HIV-2 infection: lessons for viral immunopathogenesis. Rev Med Virol 2013;23:221–40.
[23] van der Loeff MF, Awasana AA, Sarge-Njie R, van der Sande M, Jaye A, Sabally S, et al. Sixteen years of HIV surveillance in a West African research clinic reveals divergent epidemic trends of HIV-1 and HIV-2. Int J Epidemiol 2006;35:1322–8.
[24] Fideli US, Allen SA, Musonda R, Trask S, Hahn BH, Weiss H, et al. Virologic and immunologic determinants of heterosexual transmission of human immunodeficiency virus type 1 in africa. AIDS Res Hum Retroviruses 2001;17:901–10.
[25] Raugi DN, Gottlieb GS, Sow PS, Toure M, Sall F, Gaye A, et al. HIV-1 outcompetes HIV-2 in dually infected senegalese subjects with low CD4 counts. AIDS 2013;27(15): 2441–50.
[26] Rowland-Jones SL, Whittle HC. Out of Africa: what can we learn from HIV-2 about protective immunity to HIV-1? Nat Immunol 2007;8:329–31.
[27] Johnson PR, Hirsch VM. SIV infection of macaques as a model for AIDS pathogenesis. Int Rev Immunol 1992;8:55–63.
[28] Kestler 3rd HW, Naidu YN, Kodama T, King NW, Daniel MD, Li Y, et al. Use of infectious molecular clones of simian immunodeficiency virus for pathogenesis studies. J Med Primatol 1989;18:305–9.
[29] Naidu YM, Kestler 3rd HW, Li Y, Butler CV, Silva DP, Schmidt DK, et al. Characterization of infectious molecular clones of simian immunodeficiency virus (SIVmac) and human immunodeficiency virus type 2: persistent infection of rhesus monkeys with molecularly cloned SIVmac. J Virol 1988;62:4691–6.
[30] Klatt NR, Silvestri G, Hirsch V. Nonpathogenic simian immunodeficiency virus infections. Cold Spring Harb Perspect Med 2012;2:a007153.
[31] Moir S, Fauci AS. Pathogenic mechanisms of B-lymphocyte dysfunction in HIV disease. J Allergy Clin Immunol 2008;122:12–9. quiz 20–11.

[32] Pantaleo G, Cohen O, Graziosi C, Vaccarezza M, Paolucci S, Demarest JF, et al. Immunopathogenesis of HIV infection. In: Vincent J, DeVita T, Hellman Samuel, Rosenberg Steven A, editors. AIDS: biology, diagnosis, treatment and prevention. 4th ed. Philadelphia (PA): Lippincott-Raven Publishers; 1997.
[33] Schmitt N, Ueno H. Human T follicular helper cells: development and subsets. Adv Exp Med Biol 2013;785:87–94.
[34] Martinot AJ, Meythaler M, Pozzi LA, Dalecki Boisvert K, Knight H, Walsh D, et al. Acute SIV infection in sooty mangabey monkeys is characterized by rapid virus clearance from lymph nodes and absence of productive infection in germinal centers. PLoS One 2013;8:e57785.
[35] Brenchley JM, Vinton C, Tabb B, Hao XP, Connick E, Paiardini M, et al. Differential infection patterns of CD4+ T cells and lymphoid tissue viral burden distinguish progressive and nonprogressive lentiviral infections. Blood 2012;120:4172–81.
[36] Xu Y, Weatherall C, Bailey M, Alcantara S, De Rose R, Estaquier J, et al. Simian immunodeficiency virus infects follicular helper CD4 T cells in lymphoid tissues during pathogenic infection of pigtail macaques. J Virol 2013;87:3760–73.
[37] Schmitz JE, Zahn RC, Brown CR, Rett MD, Li M, Tang H, et al. Inhibition of adaptive immune responses leads to a fatal clinical outcome in SIV-infected pigtailed macaques but not vervet African green monkeys. PLoS Pathog 2009;5:e1000691.
[38] Zahn RC, Rett MD, Li M, Tang H, Korioth-Schmitz B, Balachandran H, et al. Suppression of adaptive immune responses during primary SIV infection of sabaeus African green monkeys delays partial containment of viremia but does not induce disease. Blood 2010;115:3070–8.
[39] Kaur A, Di Mascio M, Barabasz A, Rosenzweig M, McClure HM, Perelson AS, et al. Dynamics of T- and B-lymphocyte turnover in a natural host of simian immunodeficiency virus. J Virol 2008;82:1084–93.
[40] Li B, Stefano-Cole K, Kuhrt DM, Gordon SN, Else JG, Mulenga J, et al. Nonpathogenic simian immunodeficiency virus infection of sooty mangabeys is not associated with high levels of autologous neutralizing antibodies. J Virol 2010;84:6248–53.
[41] De Boer RJ, Mohri H, Ho DD, Perelson AS. Turnover rates of B cells, T cells, and NK cells in simian immunodeficiency virus-infected and uninfected rhesus macaques. J Immunol 2003;170:2479–87.
[42] Fultz PN, Stricker RB, McClure HM, Anderson DC, Switzer WM, Horaist C. Humoral response to SIV/SMM infection in macaque and mangabey monkeys. J Acquir Immune Defic Syndr 1990;3:319–29.
[43] Kaur A, Grant RM, Means RE, McClure H, Feinberg M, Johnson RP. Diverse host responses and outcomes following simian immunodeficiency virus SIVmac239 infection in sooty mangabeys and rhesus macaques. J Virol 1998;72:9597–611.
[44] Boliar S, Murphy MK, Tran TC, Carnathan DG, Armstrong WS, Silvestri G, et al. B-Lymphocyte dysfunction in chronic HIV-1 infection does not prevent cross-clade neutralization breadth. J Virol 2012;86:8031–40.
[45] Scinicariello F, Masseoud F, Jayashankar L, Attanasio R. Sooty mangabey (*Cercocebus torquatus atys*) IGHG and IGHA genes. Immunogenetics 2006;58:955–65.
[46] Ahmed R, Burton DR. Viral vaccines: past successes and future challenges. Curr Opin Virol 2013;3(3):307–8.
[47] Burton DR, Ahmed R, Barouch DH, Butera ST, Crotty S, Godzik A, et al. A blueprint for HIV vaccine discovery. Cell Host Microbe 2012;12:396–407.
[48] Nabel GJ, Kwong PD, Mascola JR. Progress in the rational design of an AIDS vaccine. Philos Trans R Soc Lond B Biol Sci 2011;366:2759–65.
[49] Wertheim JO, Worobey M. Dating the age of the SIV lineages that gave rise to HIV-1 and HIV-2. PLoS Comput Biol 2009;5:e1000377.
[50] Decker JM, Bibollet-Ruche F, Wei XP, Wang S, Levy DN, Wang W, et al. Antigenic conservation and immunogenicity of the HIV coreceptor binding site. J Exp Med 2005;201:1–14.

[51] Bottiger B, Karlsson A, Andreasson PA, Naucler A, Costa CM, Norrby E, et al. Envelope cross-reactivity between human immunodeficiency virus types 1 and 2 detected by different serological methods: correlation between cross-neutralization and reactivity against the main neutralizing site. J Virol 1990;64:3492–9.
[52] Kong R, Li H, Bibollet-Ruche F, Decker JM, Zheng NN, Gottlieb GS, et al. Broad and potent neutralizing antibody responses elicited in natural HIV-2 infection. J Virol 2012;86:947–60.
[53] Kong R, Li H, Georgiev I, Changela A, Bibollet-Ruche F, Decker JM, et al. Epitope mapping of broadly neutralizing HIV-2 human monoclonal antibodies. J Virol 2012;86:12115–28.
[54] Lopker M, Easlick J, Sterrett S, Decker JM, Barbian H, Learn G, et al. Heterogeneity in neutralization sensitivity of viruses comprising the SIVsmE660 isolate and vaccine challenge stock. J Virol 2013;87(10):5477–92.
[55] Bar KJ, Tsao CY, Iyer SS, Decker JM, Yang Y, Bonsignori M, et al. Early low-titer neutralizing antibodies impede HIV-1 replication and select for virus escape. PLoS Pathog 2012;8:e1002721.
[56] Gray ES, Moody MA, Wibmer CK, Chen X, Marshall D, Amos J, et al. Isolation of a monoclonal antibody that targets the alpha-2 helix of gp120 and represents the initial autologous neutralizing-antibody response in an HIV-1 subtype C-infected individual. J Virol 2011;85:7719–29.
[57] Kwong PD, Mascola JR. Human antibodies that neutralize HIV-1: identification, structures, and B cell ontogenies. Immunity 2012;37:412–25.
[58] Lynch RM, Rong R, Boliar S, Sethi A, Li B, Mulenga J, et al. The B cell response is redundant and highly focused on V1V2 during early subtype C infection in a Zambian seroconverter. J Virol 2011;85:905–15.
[59] Murphy MK, Yue L, Pan R, Boliar S, Sethi A, Tian J, et al. Viral escape from neutralizing antibodies in early subtype a HIV-1 infection drives an increase in autologous neutralization breadth. PLoS Pathog 2013;9:e1003173.
[60] Zhou T, Xu L, Dey B, Hessell AJ, Van Ryk D, Xiang SH, et al. Structural definition of a conserved neutralization epitope on HIV-1 gp120. Nature 2007;445:732–7.
[61] Bou-Habib DC, Roderiquez G, Oravecz T, Berman PW, Lusso P, Norcross MA. Cryptic nature of envelope V3 region epitopes protects primary monocytotropic human immunodeficiency virus type 1 from antibody neutralization. J Virol 1994;68:6006–13.
[62] Davis KL, Bibollet-Ruche F, Li H, Decker JM, Kutsch O, Morris L, et al. Human immunodeficiency virus type 2 (HIV-2)/HIV-1 envelope chimeras detect high titers of broadly reactive HIV-1 V3-specific antibodies in human plasma. J Virol 2009;83:1240–59.
[63] Moore PL, Gray ES, Choge IA, Ranchobe N, Mlisana K, Karim SS, et al. The c3-v4 region is a major target of autologous neutralizing antibodies in Hiv-1 subtype C infection. J Virol 2008;82(4):1860–9.
[64] Kuwata T, Takaki K, Yoshimura K, Enomoto I, Wu F, Ourmanov I, et al. Conformational epitope consisting of the v3 and v4 loops as a target for potent and broad neutralization of simian immunodeficiency viruses. J Virol 2013;87:5424–36.
[65] Yeh WW, Brassard LM, Miller CA, Basavapathruni A, Zhang J, Rao SS, et al. Envelope variable region 4 is the first target of neutralizing antibodies in early simian immunodeficiency virus mac251 infection of rhesus monkeys. J Virol 2012;86:7052–9.
[66] Gray ES, Taylor N, Wycuff D, Moore PL, Tomaras GD, Wibmer CK, et al. Antibody specificities associated with neutralization breadth in plasma from human immunodeficiency virus type 1 subtype C-infected blood donors. J Virol 2009;83:8925–37.
[67] Lynch RM, Tran L, Louder MK, Schmidt SD, Cohen M, Dersimonian R, et al. The development of CD4 binding site antibodies during HIV-1 infection. J Virol 2012;86:7588–95.
[68] Moore PL, Gray ES, Choge IA, Ranchobe N, Mlisana K, Abdool Karim SS, et al. The c3-v4 region is a major target of autologous neutralizing antibodies in human immunodeficiency virus type 1 subtype C infection. J Virol 2008;82:1860–9.

[69] Tang H, Robinson JE, Gnanakaran S, Li M, Rosenberg ES, Perez LG, et al. Epitopes immediately below the base of the V3 loop of gp120 as targets for the initial autologous neutralizing antibody response in two HIV-1 subtype B-infected individuals. J Virol 2011;85:9286–99.
[70] de Silva TI, Aasa-Chapman M, Cotten M, Hue S, Robinson J, Bibollet-Ruche F, et al. Potent autologous and heterologous neutralizing antibody responses occur in HIV-2 infection across a broad range of infection outcomes. J Virol 2012;86:930–46.
[71] Ozkaya Sahin G, Holmgren B, da Silva Z, Nielsen J, Nowroozalizadeh S, Esbjornsson J, et al. Potent intratype neutralizing activity distinguishes human immunodeficiency virus type 2 (HIV-2) from HIV-1. J Virol 2012;86:961–71.
[72] Stansell E, Desrosiers RC. Fundamental difference in the content of high-mannose carbohydrate in the HIV-1 and HIV-2 lineages. J Virol 2010;84:8998–9009.
[73] Seaman MS, Janes H, Hawkins N, Grandpre LE, Devoy C, Giri A, et al. Tiered categorization of a diverse panel of HIV-1 Env pseudoviruses for assessment of neutralizing antibodies. J Virol 2010;84:1439–52.
[74] Wu F, Ourmanov I, Kuwata T, Goeken R, Brown CR, Buckler-White A, et al. Sequential evolution and escape from neutralization of simian immunodeficiency virus SIVsmE660 clones in rhesus macaques. J Virol 2012;86:8835–47.
[75] Gray ES, Moore PL, Choge IA, Decker JM, Bibollet-Ruche F, Li H, et al. Neutralizing antibody responses in acute human immunodeficiency virus type 1 subtype C infection. J Virol 2007;81:6187–96.
[76] Li B, Decker JM, Johnson RW, Bibollet-Ruche F, Wei X, Mulenga J, et al. Evidence for potent autologous neutralizing antibody titers and compact envelopes in early infection with subtype C human immunodeficiency virus type 1. J Virol 2006;80:5211–8.
[77] Wei X, Decker JM, Wang S, Hui H, Kappes JC, Wu X, et al. Antibody neutralization and escape by HIV-1. Nature 2003;422:307–12.
[78] Moore PL, Gray ES, Wibmer CK, Bhiman JN, Nonyane M, Sheward DJ, et al. Evolution of an HIV glycan-dependent broadly neutralizing antibody epitope through immune escape. Nat Med 2012;18:1688–92.
[79] Moore PL, Ranchobe N, Lambson BE, Gray ES, Cave E, Abrahams MR, et al. Limited neutralizing antibody specificities drive neutralization escape in early HIV-1 subtype C infection. PLoS Pathog 2009;5:e1000598.
[80] Rong R, Li B, Lynch RM, Haaland RE, Murphy MK, Mulenga J, et al. Escape from autologous neutralizing antibodies in acute/early subtype C HIV-1 infection requires multiple pathways. PLoS Pathog 2009;5:e1000594.
[81] Lauren A, Thorstensson R, Fenyo EM. Comparative studies on mucosal and intravenous transmission of simian immunodeficiency virus (SIVsm): the kinetics of evolution to neutralization resistance are related to progression rate of disease. J Gen Virol 2006;87:595–606.
[82] Clapham PR, Weber JN, Whitby D, McIntosh K, Dalgleish AG, Maddon PJ, et al. Soluble CD4 blocks the infectivity of diverse strains of HIV and SIV for T cells and monocytes but not for brain and muscle cells. Nature 1989;337:368–70.
[83] Dalgleish AG, Beverley PC, Clapham PR, Crawford DH, Greaves MF, Weiss RA. The CD4 (T4) antigen is an essential component of the receptor for the AIDS retrovirus. Nature 1984;312:763–7.
[84] Klatzmann D, Champagne E, Chamaret S, Gruest J, Guetard D, Hercend T, et al. T-lymphocyte T4 molecule behaves as the receptor for human retrovirus LAV. Nature 1984;312:767–8.
[85] Sattentau QJ, Clapham PR, Weiss RA, Beverley PC, Montagnier L, Alhalabi MF, et al. The human and simian immunodeficiency viruses HIV-1, HIV-2 and SIV interact with similar epitopes on their cellular receptor, the CD4 molecule. AIDS 1988;2:101–5.
[86] Francella N, Gwyn SE, Yi Y, Li B, Xiao P, Elliott ST, et al. CD4+ T cells support production of SIV Env antibodies that enforce CD4-dependent entry and shape tropism in vivo. J Virol 2013;87(17):9719–32.

[87] Thomas ER, Shotton C, Weiss RA, Clapham PR, McKnight A. CD4-dependent and CD4-independent HIV-2: consequences for neutralization. AIDS 2003;17:291–300.
[88] Edinger AL, Mankowski JL, Doranz BJ, Margulies BJ, Lee B, Rucker J, et al. CD4-independent, CCR5-dependent infection of brain capillary endothelial cells by a neurovirulent simian immunodeficiency virus strain. Proc Natl Acad Sci USA 1997;94:14742–7.
[89] Martin KA, Wyatt R, Farzan M, Choe H, Marcon L, Desjardins E, et al. CD4-independent binding of SIV gp120 to rhesus CCR5. Science 1997;278:1470–3.
[90] Mori K, Rosenzweig M, Desrosiers RC. Mechanisms for adaptation of simian immunodeficiency virus to replication in alveolar macrophages. J Virol 2000;74:10852–9.
[91] Pohlmann S, Davis C, Meister S, Leslie GJ, Otto C, Reeves JD, et al. Amino acid 324 in the simian immunodeficiency virus SIVmac V3 loop can confer CD4 independence and modulate the interaction with CCR5 and alternative coreceptors. J Virol 2004;78:3223–32.
[92] Puffer BA, Altamura LA, Pierson TC, Doms RW. Determinants within gp120 and gp41 contribute to CD4 independence of SIV Envs. Virology 2004;327:16–25.
[93] Puffer BA, Pohlmann S, Edinger AL, Carlin D, Sanchez MD, Reitter J, et al. CD4 independence of simian immunodeficiency virus Envs is associated with macrophage tropism, neutralization sensitivity, and attenuated pathogenicity. J Virol 2002; 76:2595–605.
[94] Reeves JD, Hibbitts S, Simmons G, McKnight A, Azevedo-Pereira JM, Moniz-Pereira J, et al. Primary human immunodeficiency virus type 2 (HIV-2) isolates infect CD4-negative cells via CCR5 and CXCR4: comparison with HIV-1 and simian immunodeficiency virus and relevance to cell tropism in vivo. J Virol 1999;73:7795–804.
[95] Schenten D, Marcon L, Karlsson GB, Parolin C, Kodama T, Gerard N, et al. Effects of soluble CD4 on simian immunodeficiency virus infection of CD4-positive and CD4-negative cells. J Virol 1999;73:5373–80.
[96] Ortiz AM, Klatt NR, Li B, Yi Y, Tabb B, Hao XP, et al. Depletion of CD4(+) T cells abrogates post-peak decline of viremia in SIV-infected rhesus macaques. J Clin Invest 2011;121:4433–45.
[97] Apetrei C, Gautam R, Sumpter B, Carter AC, Gaufin T, Staprans SI, et al. Virus subtype-specific features of natural simian immunodeficiency virus SIVsmm infection in sooty mangabeys. J Virol 2007;81:7913–23.
[98] Alkhatib G, Combadiere C, Broder CC, Feng Y, Kennedy PE, Murphy PM, et al. CC CKR5: a RANTES, MIP-1alpha, MIP-1beta receptor as a fusion cofactor for macrophage-tropic HIV-1. Science 1996;272:1955–8.
[99] Bazan HA, Alkhatib G, Broder CC, Berger EA. Patterns of CCR5, CXCR4, and CCR3 usage by envelope glycoproteins from human immunodeficiency virus type 1 primary isolates. J Virol 1998;72:4485–91.
[100] Berger EA, Murphy PM, Farber JM. Chemokine receptors as HIV-1 coreceptors: roles in viral entry, tropism, and disease. Annu Rev Immunol 1999;17:657–700.
[101] Choe H, Farzan M, Sun Y, Sullivan N, Rollins B, Ponath PD, et al. The beta-chemokine receptors CCR3 and CCR5 facilitate infection by primary HIV-1 isolates. Cell 1996;85:1135–48.
[102] Deng H, Liu R, Ellmeier W, Choe S, Unutmaz D, Burkhart M, et al. Identification of a major co-receptor for primary isolates of HIV-1. Nature 1996;381:661–6.
[103] Doranz BJ, Rucker J, Yi Y, Smyth RJ, Samson M, Peiper SC, et al. A dual-tropic primary HIV-1 isolate that uses fusin and the beta-chemokine receptors CKR-5, CKR-3, and CKR-2b as fusion cofactors. Cell 1996;85:1149–58.
[104] Dragic T, Litwin V, Allaway GP, Martin SR, Huang Y, Nagashima KA, et al. HIV-1 entry into CD4+ cells is mediated by the chemokine receptor CC-CKR-5. Nature 1996;381:667–73.
[105] Feng Y, Broder CC, Kennedy PE, Berger EA. HIV-1 entry cofactor: functional cDNA cloning of a seven- transmembrane, G protein-coupled receptor. Science 1996;272:872–7.

[106] Moore JP, Kitchen SG, Pugach P, Zack JA. The CCR5 and CXCR4 coreceptors–central to understanding the transmission and pathogenesis of human immunodeficiency virus type 1 infection. AIDS Res Hum Retroviruses 2004;20:111–26.
[107] Tscherning C, Alaeus A, Fredriksson R, Bjorndal A, Deng H, Littman DR, et al. Differences in chemokine coreceptor usage between genetic subtypes of HIV-1. Virology 1998;241:181–8.
[108] Zhang L, He T, Huang Y, Chen Z, Guo Y, Wu S, et al. Chemokine coreceptor usage by diverse primary isolates of human immunodeficiency virus type 1. J Virol 1998;72: 9307–12.
[109] Peden KW, Farber JM. Coreceptors for human immunodeficiency virus and simian immunodeficiency virus. Adv Pharmacol 2000;48:409–78.
[110] Blaak H, Boers PH, Gruters RA, Schuitemaker H, van der Ende ME, Osterhaus AD. CCR5, GPR15, and CXCR6 are major coreceptors of human immunodeficiency virus type 2 variants isolated from individuals with and without plasma viremia. J Virol 2005;79:1686–700.
[111] Bron R, Klasse PJ, Wilkinson D, Clapham PR, Pelchen-Matthews A, Power C, et al. Promiscuous use of CC and CXC chemokine receptors in cell-to-cell fusion mediated by a human immunodeficiency virus type 2 envelope protein. J Virol 1997;71:8405–15.
[112] Deng HK, Unutmaz D, KewalRamani VN, Littman DR. Expression cloning of new receptors used by simian and human immunodeficiency viruses. Nature 1997;388: 296–300.
[113] Edinger AL, Hoffman TL, Sharron M, Lee B, O'Dowd B, Doms RW. Use of GPR1, GPR15, and STRL33 as coreceptors by diverse human immunodeficiency virus type 1 and simian immunodeficiency virus envelope proteins. Virology 1998;249:367–78.
[114] Farzan M, Choe H, Martin K, Marcon L, Hofmann W, Karlsson G, et al. Two orphan seven-transmembrane segment receptors which are expressed in CD4-positive cells support simian immunodeficiency virus infection. J Exp Med 1997;186:405–11.
[115] Guillon C, van der Ende ME, Boers PH, Gruters RA, Schutten M, Osterhaus AD. Coreceptor usage of human immunodeficiency virus type 2 primary isolates and biological clones is broad and does not correlate with their syncytium-inducing capacities. J Virol 1998;72:6260–3.
[116] Marcon L, Choe H, Martin KA, Farzan M, Ponath PD, Wu L, et al. Utilization of C-C chemokine receptor 5 by the envelope glycoproteins of a pathogenic simian immunodeficiency virus, SIVmac239. J Virol 1997;71:2522–7.
[117] McKnight A, Dittmar MT, Moniz-Periera J, Ariyoshi K, Reeves JD, Hibbitts S, et al. A broad range of chemokine receptors are used by primary isolates of human immunodeficiency virus type 2 as coreceptors with CD4. J Virol 1998;72:4065–71.
[118] Morner A, Bjorndal A, Albert J, Kewalramani VN, Littman DR, Inoue R, et al. Primary human immunodeficiency virus type 2 (HIV-2) isolates, like HIV-1 isolates, frequently use CCR5 but show promiscuity in coreceptor usage. J Virol 1999;73:2343–9.
[119] Owen SM, Ellenberger D, Rayfield M, Wiktor S, Michel P, Grieco MH, et al. Genetically divergent strains of human immunodeficiency virus type 2 use multiple coreceptors for viral entry. J Virol 1998;72:5425–32.
[120] Zhang Y, Lou B, Lal RB, Gettie A, Marx PA, Moore JP. Use of inhibitors to evaluate coreceptor usage by simian and simian/human immunodeficiency viruses and human immunodeficiency virus type 2 in primary cells. J Virol 2000;74:6893–910.
[121] Elliott ST, Riddick NE, Francella N, Paiardini M, Vanderford TH, Li B, et al. Cloning and analysis of sooty mangabey alternative coreceptors that support simian immunodeficiency virus SIVsmm entry independently of CCR5. J Virol 2012;86:898–908.
[122] Chen Z, Gettie A, Ho DD, Marx PA. Primary SIVsm isolates use the CCR5 coreceptor from sooty mangabeys naturally infected in west Africa: a comparison of coreceptor usage of primary SIVsm, HIV-2, and SIVmac. Virology 1998;246:113–24.

[123] Milush JM, Reeves JD, Gordon SN, Zhou D, Muthukumar A, Kosub DA, et al. Virally induced CD4+ T cell depletion is not sufficient to induce AIDS in a natural host. J Immunol 2007;179:3047–56.

[124] Riddick NE, Hermann EA, Loftin LM, Elliott ST, Wey WC, Cervasi B, et al. A novel CCR5 mutation common in sooty mangabeys reveals SIVsmm infection of CCR5-null natural hosts and efficient alternative coreceptor use in vivo. PLoS Pathog 2010;6:e1001064.

[125] Finzi A, Pacheco B, Xiang SH, Pancera M, Herschhorn A, Wang L, et al. Lineage-specific differences between human and simian immunodeficiency virus regulation of gp120 trimer association and CD4 binding. J Virol 2012;86:8974–86.

[126] Xiang SH, Kwong PD, Gupta R, Rizzuto CD, Casper DJ, Wyatt R, et al. Mutagenic stabilization and/or disruption of a CD4-bound state reveals distinct conformations of the human immunodeficiency virus type 1 gp120 envelope glycoprotein. J Virol 2002;76:9888–99.

[127] Demma LJ, Logsdon Jr JM, Vanderford TH, Feinberg MB, Staprans SI. SIVsm quasispecies adaptation to a new simian host. PLoS Pathog 2005;1:e3.

[128] Vanderford TH, Demma LJ, Feinberg MB, Staprans SI, Logsdon Jr JM. Adaptation of a diverse simian immunodeficiency virus population to a new host is revealed through a systematic approach to identify amino acid sites under selection. Mol Biol Evol 2007;24:660–9.

[129] Fischer W, Apetrei C, Santiago ML, Li Y, Gautam R, Pandrea I, et al. Distinct evolutionary pressures underlie diversity in simian immunodeficiency virus and human immunodeficiency virus lineages. J Virol 2012;86:13217–31.

[130] Shpaer EG, Mullins JI. Rates of amino acid change in the envelope protein correlate with pathogenicity of primate lentiviruses. J Mol Evol 1993;37:57–65.

[131] Barroso H, Taveira N. Evidence for negative selective pressure in HIV-2 evolution in vivo. Infect Genet Evol 2005;5:239–46.

[132] Skar H, Borrego P, Wallstrom TC, Mild M, Marcelino JM, Barroso H, et al. HIV-2 genetic evolution in patients with advanced disease is faster than that in matched HIV-1 patients. J Virol 2010;84:7412–5.

[133] Zhang M, Gaschen B, Blay W, Foley B, Haigwood N, Kuiken C, et al. Tracking global patterns of N-linked glycosylation site variation in highly variable viral glycoproteins: HIV, SIV, and HCV envelopes and influenza hemagglutinin. Glycobiology 2004;14:1229–46.

[134] Demma LJ, Vanderford TH, Logsdon Jr JM, Feinberg MB, Staprans SI. Evolution of the uniquely adaptable lentiviral envelope in a natural reservoir host. Retrovirology 2006;3:19.

[135] Chahroudi A, Meeker T, Lawson B, Ratcliffe S, Else J, Silvestri G. Mother-to-infant transmission of simian immunodeficiency virus is rare in sooty mangabeys and is associated with low viremia. J Virol 2011;85:5757–63.

CHAPTER

10

Cellular Immune Responses in Natural and Non-natural Hosts of Simian Immunodeficiency Virus Infection

Namita Rout, Amitinder Kaur

Division of Immunology, New England Primate Research Center, Harvard Medical School, Southborough, MA, USA

INTRODUCTION

Simian immunodeficiency virus (SIV) infection is naturally prevalent in several African non-human primates and, to date, at least 40 different African monkey species harboring their own species-specific SIV have been identified [1]. Based largely on studies in sooty mangabeys (SM), African green monkeys (AGM), and mandrills, it appears that SIV infection in its natural host is generally asymptomatic and does not progress to AIDS, whereas experimental SIV infection in non-natural hosts such as macaques, is pathogenic and leads to AIDS [2,3]. Paradoxically, nonpathogenic SIV infection in natural hosts is associated with high levels of viral replication. Similar to pathogenic infection, viremia levels are maintained by productive infection in short-lived $CD4^+$ T lymphocytes and acute infection is associated with a profound initial decline in gut $CD4^+$ T cells and a transient decline in peripheral blood $CD4^+$ T cells. In striking contrast to pathogenic infection, ongoing chronic viral replication is associated with neither increased lymphocyte turnover nor aberrant immune activation, features that are pathognomonic of pathogenic infection [4]. Thus, absence of chronic immune activation is a key feature that distinguishes nonpathogenic SIV infection in natural hosts from pathogenic lentiviral infection in AIDS-susceptible HIV/SIV-infected humans and macaques.

Natural Hosts of SIV.
http://dx.doi.org/10.1016/B978-0-12-404734-1.00010-3

TABLE 10.1 T Lymphocyte Subsets Mediating Cellular Immunity in Natural and Non-natural Hosts of SIV Infection

Type of Immunity	T Lymphocyte Subsets	Natural Host	Non-natural Host	References
Innate immunity	T Regulatory cells	+	+	[47–49,55,56]
	Th17 cells	+	+	[55,56]
	Natural Killer T cells	+	+	[71,73]
	Double-negative (DN) T cells	+	?	[75–77]
	Mucosal associated invariant T cells	?	?	–
	Gamma delta T cells	+	+	[61,63]
Adaptive immunity	SIV-specific CD8$^+$ T	+	+	[11,19,25–27, 29–31,41]
	SIV-specific CD4$^+$ T	+	+	[26,27]
	SIV-specific DNT	±	?	[75,76]

Comparative studies of SIV infection in natural and non-natural hosts have highlighted differences in the host response and provided valuable insight into mechanisms underlying protection against AIDS in SIV-infected natural hosts. In this chapter, T lymphocyte responses in natural and non-natural hosts of SIV infection are reviewed with emphasis on both adaptive and innate cellular immune responses (Table 10.1). The role of adaptive cellular immune responses in immune control and prevention of immunodeficiency, and that of innate T lymphocyte subsets in modulating immune activation and preserving mucosal integrity are discussed.

CELLULAR IMMUNITY AND IMMUNE CONTROL IN NATURAL HOSTS OF SIV INFECTION

Studies in HIV-infected humans and SIV-infected macaques have provided overwhelming evidence for a central role of virus-specific cellular immunity in control of viral replication and protection against AIDS [5]. Natural hosts of SIV provide a new paradigm of protection. Unlike pathogenic lentiviral infection where prevention of disease progression is predicated on eradication or suppression of viremia, natural hosts remain disease free despite the presence of persistent viremia at levels that induce AIDS in SIV-infected macaques or HIV-infected humans. This has raised the question of whether or not cellular immunity has a role in conferring disease resistance to natural hosts.

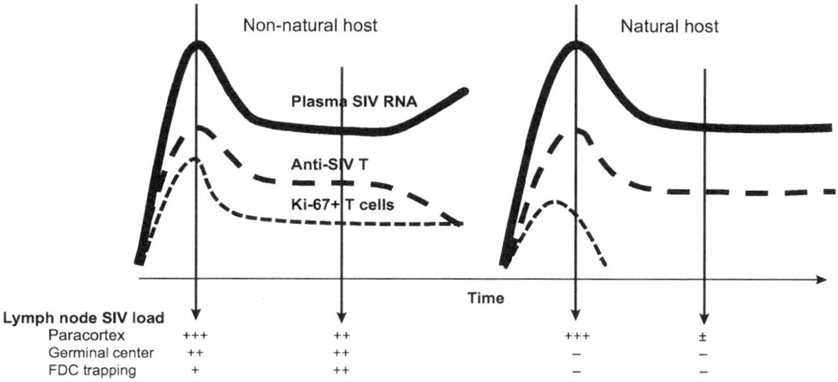

FIGURE 10.1 **Natural history of SIV infection in natural and non-natural hosts.** Line diagram showing similarities and differences in SIV load, cellular immunity and immune activation in acute and chronic SIV infection.

Despite ongoing viral replication, some aspects of viral control are different in SIV-infected natural and non-natural hosts (Figure 10.1). While natural hosts can display marked inter-animal variation in set-point plasma viremia levels without impacting the course of SIV infection, these levels remain stable over prolonged periods of time, generally through the lifetime of the host [6–8]. In this regard it is noteworthy that in two case reports of AIDS in a natural host, onset of disease was preceded by a significant increase in plasma viremia levels [9,10]. The sustained steady-state equilibrium in viremia manifested by asymptomatic natural hosts suggests that durable viral control, even if partial, is necessary to maintain nonpathogenic infection (Figure 10.1). The factors determining set-point viremia levels are likely multifactorial and include both immune control and target cell availability. Another point of difference between natural and non-natural hosts is the disparity between plasma viremia levels and lymphoid tissue-associated viral load in SIV-infected animals. Even with high plasma viremia levels, natural hosts rapidly clear productively infected cells from lymphoid tissue in acute SIV infection and SIV-infected cells are rarely detected in chronic infection [11–13]. Moreover, lymphoid tissue-associated SIV in natural hosts is largely confined to the paracortex with sparing of germinal center $CD4^+$ T cells and absence of virus trapping on follicular dendritic cells [12,14] (Figure 10.1). These observations suggest that total SIV burden is substantially different in natural and non-natural hosts and raise the possibility that an overall lower viral burden is central to maintenance of nonpathogenicity in natural hosts.

Questions pertinent to the role of cellular immunity in natural hosts include its requirement for (1) maintenance of set-point viral loads, (2) decline of peak viremia to set-point viral loads, (3) control of tissue-associated virus, (4) down modulation of acute immune activation and (5) prevention of chronic aberrant immune activation. While virus-specific cellular immunity is clearly required for viral suppression, its impact on immune activation is less clear. In this regard it is noteworthy that a recent study on PD-1 blockade in chronic SIV infection in rhesus macaques showed reduction in aberrant immune activation and microbial translocation, possibly due to enhanced microbial immunity against gut-residing pathogens [15,16].

T LYMPHOCYTES

The ability of natural hosts to withstand aberrant immune activation and increased lymphocyte turnover following SIV infection might suggest the presence of more quiescent T cells [17,18]. Surprisingly, studies in SIV-negative SM suggest the contrary (Table 10.2). SIV-negative SM were found to have a significantly higher frequency of circulating $CD8^+$

TABLE 10.2 Comparison of $CD4^+$ and $CD8^+$ T Lymphocyte-Mediated Immune Responses in Natural and Non-natural Hosts of SIV Infection

	Natural Host	Non-natural Host	References
T LYMPHOCYTES			
• Frequency of granzyme B-positive $CD8^+$ T	++	+	[11,19]
• $CD4^+$ T cell IL-2 production	++	+	[21]
• Resistance of $CD4^+$ T to in vitro TCR-stimulation induced anergy	Yes	No	[20,22]
IMMUNE RESPONSES IN ACUTE SIV INFECTION			
• Magnitude of SIV-specific T cells	+	+	[11,29,30]
• Temporal association between CTL appearance and decline of peak viremia	Yes	Yes	[11,29,30]
• Polyfunctional SIV-specific $CD8^+$ T	++	+	[11]
• Kinetics of SIV clearance from lymphoid tissue	Fast	Slow	[11,12]
IMMUNE RESPONSES IN CHRONIC SIV INFECTION			
• Magnitude of SIV-specific T cells	+	+	[26]
• Durable control of set-point viremia	Yes	No	[7,8]
• Loss of $CD4^+$ T cell responses to recall antigens	No	Yes	[20]

Continued

TABLE 10.2 Comparison of CD4+ and CD8+ T Lymphocyte-Mediated Immune Responses in Natural and Non-natural Hosts of SIV Infection—cont'd

	Natural Host	Non-natural Host	References
IN VIVO CD8+ DEPLETION			
• Increased peak viremia	No	Yes	[40]
• Delayed decline in post-peak viremia	Yes	Yes	[39–41]
• Increased chronic set-point viremia	Yes	Yes	[39,42]
• Disease progression to AIDS	No	Yes	[40]
• Increased lymph node associated SIV	Yes	Yes	[12]
• Disappearance of productive SIV infection in lymph node on CD8 recovery	Yes	No	[12]

T lymphocytes containing the cytolytic granule granzyme B compared to SIV-negative rhesus macaques [11]. A similar observation was reported in SIV-negative AGM [19]. In vitro studies have shown that sooty mangabey T lymphocytes display a greater degree of responsiveness, manifested by increased IL-2 production and resistance to anergy on T-cell receptor (TCR) stimulation [20,21]. Central memory CD4+ T cells were found to be resistant to anergy, a subset recently shown to be relatively resistant to SIV infection in SM [22,23]. A trend towards greater Th2 cytokine production has also been observed in mitogen-stimulated peripheral lymphocytes of SM as compared to rhesus macaques [17,24].

VIRUS-SPECIFIC CELLULAR IMMUNE RESPONSES

1. SIV-specific CD4+ and CD8+ T lymphocyte responses:

Detailed analysis of the SIV-specific cellular immune response has been performed during chronic SIV infection in naturally infected SM [25–27]. Using overlapping peptides spanning the entire SIV proteome in ELISPOT and intracellular cytokine staining assays, SIV-specific CD4+ and CD8+ T lymphocyte responses were readily detected in the peripheral blood of SIV-infected SM [26,27]. The magnitude and breadth of the SIV-specific cellular immune response in SM was comparable to experimentally-infected rhesus macaques with chronic SIV infection and was predominantly characterized by Th1-type effector memory SIV-specific CD8+ T lymphocytes secreting IFN-gamma, TNF-alpha and MIP1-beta and showing cytotoxic activity [25,26]. The similarities in cellular immunity between SIV-infected SM and SIV-infected rhesus macaques indicated that virus-specific CD8+ T lymphocytes do not contribute to AIDS-associated immunopathology. Neither is immune attenuation required for maintaining nonpathogenic lentiviral infection in its natural host. SIV-specific CD8+ T lymphocytes targeting Gag and Env are also detected in naturally infected AGM [19,28]; however, their magnitude tends to be lower compared to SIV-infected rhesus macaques [19]. Although longitudinal data are limited, it appears that natural hosts maintain their SIV-specific cellular immune response at steady levels, similar to maintenance of steady-state viremia levels in chronic infection (Figure 10.1).

Experimental SIV infection studies have further highlighted similarities in the cellular immune response to acute SIV infection in natural and non-natural hosts. An early report on experimental infection of SM with the macaque-passaged pathogenic SIVmac239 virus showed induction of a robust SIV-specific cytotoxic CD8+ T lymphocyte response targeting CTL epitopes in Gag, Env, and Nef [25,29]. However, SIVmac239 infection was atypical of natural infection because primary viremia was rapidly controlled to near undetectable levels [29]. More recently it was reported that in the setting of experimental infection with the primary SIVsmm isolate, SIVsmE041, which reproduced the set-point viremia levels seen in natural SIV infection, SM mounted an acute SIV-specific cellular immune response that was comparable in magnitude to rhesus macaques infected with SIVmac239 or SIVsmE041. Kinetic analysis of the cellular immune response in the first 6 months post-SIV infection showed comparable magnitude, breadth, and specificity of the SIV-specific IFN-gamma ELISPOT response in SM and rhesus macaques. However, SM showed a higher frequency of poly-functional Gag-specific CD8+ T lymphocytes in the first 10 weeks following SIV infection, a greater reduction in peak plasma viremia, and a more rapid disappearance of productively SIV-infected cells from the lymph node compared to SIVmac239-infected rhesus macaques [11]. Notably, even though the magnitude of virus-specific T cells was comparable in natural and non-natural hosts, levels of overall T cell activation were significantly higher in rhesus macaques compared to SM [30]. Rapid reduction of primary viremia and lymphoid tissue cell-associated SIV load was also observed in experimentally infected AGM [13]. These data suggest that qualitative differences in the early SIV-specific cellular immune response as well as a more focused T cell activation response dominated by activated anti-viral T cells might contribute to faster viral load reduction and absence of chronic immunopathology in lymphoid tissues of SIV-infected natural hosts (Table 10.2).

Even though cross-sectional analysis has not revealed a significant inverse relationship between viral load and the magnitude of SIV-specific cellular immune responses in naturally infected SM [26,27], this in itself does not exclude a role for cellular immune responses in maintaining effective, albeit partial, control of chronic viremia at steady-state levels. In this regard it is noteworthy that onset of CTL escape in a SIVmac239-infected sooty mangabey coincided with an increase in plasma viremia [31]. Similarities in the kinetics of control of primary viremia and the onset of the SIV-specific T lymphocyte response coinciding temporally with a decline in peak plasma viremia suggest that natural hosts are similar to non-natural hosts with regard to cellular immunity playing a causal role in reduction of post-peak viremia to steady-state set-point levels. Although target cell availability has also been postulated as a determinant of viremia control, it is noteworthy that studies thus far have implicated $CD8^+$ T-cell responses as the main factor responsible for control of primary viremia in SIV infection in both natural and non-natural host species [32,33].

2. Pathogen-specific cellular immune responses:

Loss of $CD4^+$ T lymphocyte proliferative responses against recall antigens is one of the earliest manifestations of immunodeficiency in pathogenic lentiviral infection [34,35]. Consistent with the lack of immunodeficiency in natural hosts, SIV-infected SM show normal primary and recall memory $CD4^+$ T proliferative cell responses to vaccination with tetanus antigen [20]. Intact CMV-specific $CD4^+$ T lymphocyte proliferative and ELISPOT responses are also seen in SIV-infected SM [36] (Table 10.2 and Kaur et al., unpublished observation).

IN VIVO $CD8^+$ DEPLETION STUDIES

The proof for a critical role of $CD8^+$ cells in viral containment was obtained when $CD8^+$ lymphocytes were depleted utilizing a cytolytic CD8-specific monoclonal antibody in vivo in SIV-infected rhesus macaques (Table 10.2). This finding is similar to data from several studies that have shown that SIV viremia was significantly increased when $CD8^+$ lymphocyte responses were temporarily depleted in SIV-infected rhesus macaques [37,38]. Moreover, $CD8^+$ lymphocyte depletion during primary SIV infection in macaques consistently resulted in a more rapid disease progression (Figure 10.2(A)).

$CD8^+$ lymphocyte depletion during acute SIV infection in AGM does not impact peak levels; instead there is a significant delay in decline of post-peak viremia to set-point levels [39–41]. In contrast, CD8-depleted macaques show higher levels of peak viremia, peak levels fail to plateau, and monkeys rapidly progress to AIDS (Figure 10.2(A)). These data suggest that even though CD8 responses have a viral inhibitory effect in natural hosts, short-duration CD8 depletion in acute infection neither affects

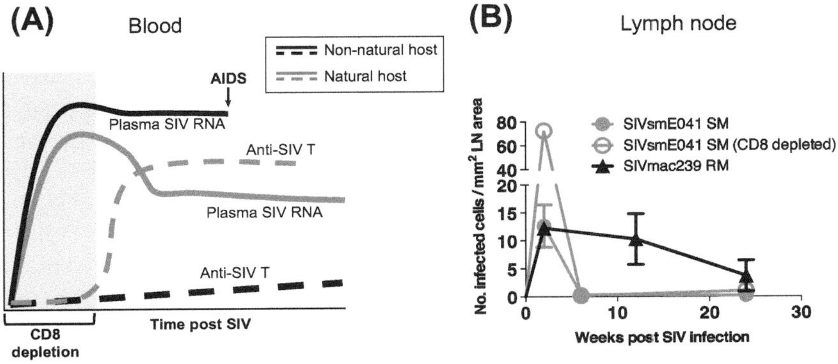

FIGURE 10.2 Impact of in vivo CD8⁺ lymphocyte depletion on course of primary SIV infection in natural and non-natural hosts. (A) Line diagram showing differential course of SIV viremia and anti-SIV T cell immunity following SIV infection in CD8-depleted natural and non-natural hosts. (B) Productive SIV infection in lymph nodes (LN) of sooty mangabeys (SM) and rhesus macaques (RM). Tissue SIV load determined by in situ hybridization for SIV RNA. Mean values and standard error of mean shown for SIVsmE041-infected SM ($n=4$), SIVmac239-infected RM ($n=6$) and CD8-depleted SIVsmE041-infected SM ($n=2$). LN figure adapted from previously published data [11,12].

subsequent chronic viremia levels, nor impacts outcome of disease progression (Figure 10.2(A)). Notably, the decline in post-peak viremia upon CD8 recovery in CD8-depleted AGM was associated with a significant increase in the magnitude of rebounding SIV-specific $CD8^+$ T lymphocytes compared to nondepleted animals suggesting that $CD8^+$ T lymphocytes were directly involved in partial containment of viremia [41] (Figure 10.2(A)). A transient effect of CD8 depletion on lymphoid cell-associated viral load is also seen in natural hosts (Figure 10.2(B)). CD8-depleted SM showed high levels of productive SIV infection in lymph nodes following acute infection; these levels were higher than in non-CD8-depleted rhesus macaques [12] (Figure 10.2(B)). However, there is a rapid disappearance of SIV load coincident with $CD8^+$ lymphocyte recovery indicating a central role of CD8-mediated immune control in containing cell-associated tissue SIV load (Figure 10.2(B)).

Published data on the effect of CD8 depletion in natural hosts during chronic infection are more limited. A modest increase following CD8 depletion has been observed in both SM and AGM [39,42]. A significant increase in plasma viremia as well as cell-associated viral load in lymphoid tissues was also observed in naturally infected SM following in vivo $CD8^+$ lymphocyte depletion (A. Kaur, unpublished data). Table 10.2 summarizes the differences in T lymphocyte-mediated cellular immunity and effect of in vivo $CD8^+$ lymphocyte depletion in natural and non-natural hosts.

ROLE OF INNATE T CELL-MEDIATED IMMUNITY IN NATURAL HOSTS

The innate immune system is the first line of defense against invading microorganisms and serves the dual function of restricting the pathogenic burden in the initial days of exposure as well as priming a specific adaptive immune response that may protect against reexposure (Figure 10.3). Support for an important role for innate immune mechanisms in the absence of disease progression in natural hosts of SIV comes from gene profiling studies in acute infection. Gene profiling studies in SM and AGM have shown that acute SIV infection is associated with robust immune activation and substantial upregulation of IFN-stimulated genes (ISGs) in both natural and non-natural hosts of SIV [43–45]. However, this immune activation and ISG upregulation is transient in natural hosts but persistent in SIV-infected non-natural hosts, suggesting that active immune regulatory mechanisms, rather than absence of innate immune responses, could underlie the low immune activation of chronically SIV-infected natural hosts.

Although T lymphocytes classically belong to the adaptive arm of the immune system, several unconventional T-cell subsets with innate properties have been discovered that lie in the gray zone where innate and adaptive immunity overlap. These innate T cells include regulatory T cells (Tregs), Th17cells, gamma/delta (γδ) T cells, natural killer T (NKT) cells, and mucosal-associated invariant T (MAIT) cells (Table 10.1). It is of interest to note that a number of these innate T-cell lineages recognize nonconventional antigens in the form of phosphoproteins, lipids, and vitamin metabolites that

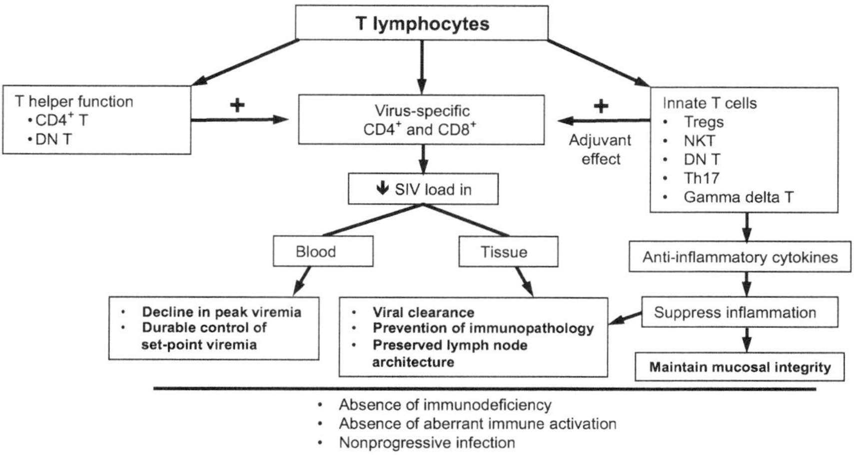

FIGURE 10.3 Schematic flow chart showing possible mechanisms whereby T cell-mediated immunity can influence outcome of SIV infection in natural hosts.

are constituents of microbial flora and predominantly present within the gut mucosa. These cell lineages could thus serve as sentinels and promote the regulation and/or the development of adaptive immune responses following sensing the need for the initiation of such responses. Innate immune activation could thus determine the quality of subsequent adaptive immune responses by enhancing antigen presentation, promoting dendritic cell recruitment and maturation, and providing the signals involved in T-cell polarization and priming. They could also play an important role in down-modulating nonspecific deleterious immune activation (Figure 3).

REGULATORY T CELLS

Tregs comprise about 2–5% of circulating $CD4^+$ T cells in humans and non-human primates. With their ability of regulating proliferative responses in antigen-specific T cells, Tregs play a key role in the maintenance of tolerance and in the control of immune activation during chronic infections. Considering the lack of chronic immune activation in natural hosts of SIV, it is possible that Tregs may play a major role. However, studies comparing circulating Tregs in SM and rhesus macaques showed no increase in frequency and function of Tregs in the natural host [46,47]. Tregs declined in SIV-infected rhesus macaques and an inverse relation between absolute numbers of circulating Tregs and viral loads was observed [47]. The low viremic rhesus macaques also had lower levels of T-cell activation compared to high viremic macaques, suggesting an association with maintained Treg numbers. Thus, although the study demonstrated a possible role of Tregs in controlling immune activation and viral loads in rhesus macaques, it concluded that there was no clear evidence of a role of Tregs in the absence of chronic immune activation in the natural host. It is likely that multifactorial mechanisms are involved in control of chronic immune activation in natural hosts and that Tregs may be one of the factors.

The maintained numbers and function of Tregs in SM could be a result of redistribution of responding Tregs to tissue sites of viral replication such as gut and lymph nodes, and increased homeostasis to maintain the circulating frequencies. It could also reflect a Treg response earlier than the 7-day time point first evaluated in this study. Indeed, a very early response within 1 day of SIV infection with increases in levels of $CD4^+CD25^+$ and $CD8^+CD25^+$ T cells and a strong induction of TGF-β1 and *FoxP3* expression has been reported in AGM [48]. Rhesus macaques but not SM show early evidence of lymph node TGF-β1-induced tissue fibrosis [49]. Another cross-sectional study in a large cohort of naturally SIV-infected SM showed a positive relationship between healthy $CD4^+$ T-cell counts and high levels of $CD4^+CD25^+$ "regulatory" T cells and low levels of $CD28^-CD95^+$ and/or $CD127^-$ effector T cells [50].

A study on $CD25^{+}FOXP3^{+}CD8^{+}$ Tregs ($CD8^{+}$ Tregs) demonstrated an expansion of $CD8^{+}$ Tregs in blood and colorectal mucosa following pathogenic SIV infection in rhesus macaques and HIV-infected humans but not in SIV-infected SM with high viremia [51]. These CD8 Tregs were shown to suppress proliferation of SIV-specific T cells in vitro and had limited killing potential owing to low expression of granzyme B and perforin. The expansion of $CD8^{+}$ Tregs was shown to correlate directly with acute phase viremia and inversely with the magnitude of antiviral T-cell response in this study, suggesting a detrimental role of $CD8^{+}$ Tregs in lentiviral infections in the non-natural hosts but not in the natural host. Our studies on peripheral Tregs in rhesus macaques and SM have shown that SM contain a subset of CD4/CD8 double-negative (DN) Tregs which are not present in rhesus macaques (Rout et al., unpublished data). The role of DN Tregs in reducing immune activation needs to be addressed.

Th17 CELLS

T helper 17 (Th17) cells, a more recently identified lineage of T cells, are characterized by expression of the IL-17 family cytokines, IL-17A, IL-17F, and IL-22. Th17 cells are enriched in mucosal tissues and are thought to play a central role in protecting the integrity of the mucosal barrier and maintaining immune homeostasis at mucosal sites [52]. Accompanied by severe depletion of $CD4^{+}$ T cells from mucosal sites, Th17 cells are lost very early during the course of pathogenic lentiviral infection, and this loss correlates with bacterial translocation [53,54]. Since SIV-infected natural hosts maintain mucosal immunity despite similar levels of CD4 T-cell depletion in the mucosal compartments as observed in humans and rhesus macaques, studies have compared Th17 cells in natural and non-natural hosts for HIV/SIV [55,56]. The depletion of mucosal Th17 cells is not observed during nonpathogenic SIV infection of SM and they maintain healthy frequencies of mucosal memory $CD4^{+}$ T cells that express IL-23R or produce IL-17 [55]. SIV-infected AGMs also maintain normal levels of functional Th17 cells in multiple anatomical sites including blood, spleen, lymph nodes and colon [56]. Therefore natural hosts of SIV appear to have evolved to preserve Th17 cells to avoid mucosal immune dysfunction and microbial translocation in the face of severe depletion of mucosal $CD4^{+}$ T cells [57]. Considering the finding that the severity of Th17 cell depletion correlates with the levels of systemic immune activation in pathogenic SIV infection [56], it is reasonable to assume that the preservation of Th17 cells in the mucosa of natural hosts may be central to the absence of chronic immune activation in nonpathogenic SIV infection.

γδ T CELLS

γδ T cells express the γδ T-cell receptor that recognizes unique sets of foreign antigens, such as bacterial phosphoantigens, nonclassical MHC-I molecules, and select unprocessed proteins [58]. These cells represent approximately 1–10% of human peripheral blood T cells. γδ T cells are important for mucosal immunity and response to bacterial antigens. Pathogenic lentiviral infection is associated with decreased effector γδ T cells and failure of these cells to respond effectively to opportunistic pathogens [59–61]. In contrast, γδ T cells in SIV-infected SM maintain their ability to respond to bacterial antigens [62]. Furthermore, γδ T cells in SIV-infected mangabeys were present at lymphoid and mucosal sites and retained the ability to produce the Th1 cytokines IFN-γ and TNF-α irrespective of the level of CD4 depletion [63]. γδ T cells may play a role in preventing opportunistic bacterial infections in the natural host, especially at the intestinal mucosa. Conversely a breach in gut mucosal integrity associated with hyper-immune activation in pathogenic SIV infection could result from dysfunction of γδ T cells.

NATURAL KILLER T CELLS

Natural Killer T (NKT) cells are an extremely rare subset of T cells, typically less than 1% in peripheral blood of humans and non-human primates. NKT cells are rapid responders of the innate immune system and mediate potent immunoregulatory and effector functions in a variety of disease settings [64]. NKT lymphocytes recognize antigen presented by the nonpolymorphic MHC class I-like CD1d molecules and are characterized by a restricted TCR repertoire due to the presence of an invariant TCR alpha chain paired with a limited number of TCR beta chains [64]. Human invariant NKT lymphocytes have a Vα24-Jα18 chain preferentially paired with Vβ11 [65]. Upon activation with glycolipids presented on CD1d molecules, NKT lymphocytes respond rapidly with the production of a diverse array of cytokines including IL-10, TGF-β, and several Th1 and Th2 cytokines. Several studies have shown the in vitro and in vivo impact of HIV/SIV infection on NKT lymphocytes. NKT lymphocytes are rapidly depleted in HIV-infected humans and SIV-infected pig-tailed macaques and this depletion appears to be due to both direct infection of CD4-expressing NKT cells and FAS-mediated apoptosis of CD4-negative NKT lymphocytes [66,67]. HIV proteins including Nef and Vpu downregulate CD1d expression on antigen presenting cells suggesting that lentiviruses have evolved strategies to inhibit NKT as well as conventional MHC class I-restricted T cells in vivo [68–70].

NKT cells in SM are unique in that they lack CD4 expression and display both effector and immunoregulatory functional capabilities by secreting IFN-γ, TNF-α, IL-2, IL-13, and IL-10 [71]. The frequencies of NKT cells in SM are also maintained during chronic SIV infection [71]. The absence of $CD4^+$ NKT cells protects SIV-infected SM from NKT cell depletion. Lack of $CD4^+$ NKT cells does not appear to have a functional impact because DN NKT cells in mangabeys are functionally similar to $CD4^+$ NKT cells [72]. The IL-10 cytokine-secreting ability of mangabey NKT cells along with their preservation following SIV infection suggests a role for NKT cells in down-modulating immune activation in SIV-infected SM. Comparison of NKT cells in SM and two macaque species showed that NKT cells from mangabeys have the capacity to synthesize greater amounts of IFN-γ, IL-13, and IL-17. Sooty mangabey NKT cells retained normal functionality following SIV infection and were more potent than macaque NKT cells in their ability to produce IFN-γ and secrete anti-inflammatory cytokines IL-10 and TGF-β [73]. In contrast, NKT cells of SIV-infected macaques were markedly hypofunctional with regard to secretion of anti-inflammatory and effector cytokines and showed an association between loss of $CD4^+$ NKT cells and increased immune activation [73]. These findings suggest that dysfunctional NKT cells may promote increased immune activation in AIDS-susceptible hosts while intact effector and anti-inflammatory NKT cells could help to prevent immunodeficiency and increased immune activation in natural hosts.

DOUBLE-NEGATIVE T CELLS

Within the T-cell compartment of mice and humans, a small population of cells (1–5%) express the αβ T-cell receptor (TCR), but lack the CD4 and CD8 coreceptors. These double negative (DN) T cells have been shown to display important immunological functions in human diseases [74]. Interestingly, lymphoid cells of natural hosts of SIV comprise high frequencies of $CD4^-CD8^-$DN T cells with CD4 T helper-like functionality [75–77]. DN T cells are capable of eliciting functions generally associated with $CD4^+$ T cells, including expression of FoxP3, IL-2, IL-17, and CD40L [77]. It is believed that the lack of clinical disease progression in natural hosts is associated with the presence of DN T cells that maintain helper T-cell function while remaining refractory to viral infection (Figure 10.3). Although the mechanism by which DN T cells provide T helper function in the absence of co-receptor expression is not known, comparative studies of DN T cells in SM and rhesus macaques have shown that the CD4-T helper-like function is unique to natural hosts of SIV (Rout et al., manuscript in preparation), strongly suggesting a role in the nonpathogenic outcome during SIV infection of natural hosts.

CONCLUSIONS

In summary, no single mechanism has yet been identified that can account for the maintenance of nonpathogenic infection in natural hosts. With regard to cellular immunity, natural hosts have several T lymphocyte subsets with varied functions that participate in both adaptive and innate immunity and could influence the course of SIV infection in natural hosts (Figure 10.3). On the one hand, natural hosts mount a virus-specific cellular immune response that partially contains peripheral blood viral replication and severely limits cell-associated viral load in lymphoid tissues. Natural hosts also have DN T cells that can substitute functionally for $CD4^+$ T cells and ensure that T helper function is not compromised even in the setting of $CD4^+$ T-cell depletion. As a result, it is likely that the antiviral T-cell response is more effective in natural hosts in terms of being more durable and being protected from dysfunction. Thus, unlike pathogenic infection, where persistent viremia leads to dysfunctional $CD8^+$ T cells and loss of immune control over time, natural hosts may be able to maintain a steady state of viral control without decompensating because of intact immune cells. Durable control, as evidenced by maintenance of steady-state set-point viral loads and minimal-to-absent levels of productive SIV infection in lymphoid tissue, may be important for nonpathogenicity. The preserved function of innate immunomodulatory cell functions in natural hosts of SIV, including those of NKT cells, Th17 cells, Tregs and $\gamma\delta$ T cells may all contribute to dampen nonspecific immune activation while maintaining normal immune responses as well as gut mucosal integrity.

References

[1] Sharp PM, Hahn BH. Origins of HIV and the AIDS pandemic. Cold Spring Harb Perspectives Medicine 2011;1(1): a006841.

[2] Pandrea I, Apetrei C. Where the wild things are: pathogenesis of SIV infection in African nonhuman primate hosts. Curr HIV/AIDS Rep 2010;7(1):28–36.

[3] Pandrea I, Sodora DL, Silvestri G, Apetrei C. Into the wild: simian immunodeficiency virus (SIV) infection in natural hosts. Trends Immunol 2008;29(9):419–28.

[4] Sodora DL, Allan JS, Apetrei C, Brenchley JM, Douek DC, Else JG, et al. Toward an AIDS vaccine: lessons from natural simian immunodeficiency virus infections of African nonhuman primate hosts. Nat Med 2009;15(8):861–5.

[5] Goulder PJ, Walker BD. HIV and HLA class I: an evolving relationship. Immunity 2012;37(3):426–40.

[6] Apetrei C, Gautam R, Sumpter B, Carter AC, Gaufin T, Staprans SI, et al. Virus subtype-specific features of natural simian immunodeficiency virus SIVsmm infection in sooty mangabeys. J Virol 2007;81(15):7913–23.

[7] Goldstein S, Ourmanov I, Brown CR, Beer BE, Elkins WR, Plishka R, et al. Wide range of viral load in healthy african green monkeys naturally infected with simian immunodeficiency virus. J Virol 2000;74(24):11744–53.

[8] Taaffe J, Chahroudi A, Engram J, Sumpter B, Meeker T, Ratcliffe S, et al. A five-year longitudinal analysis of sooty mangabeys naturally infected with simian immunodeficiency virus reveals a slow but progressive decline in CD4+ T-cell count whose magnitude is not predicted by viral load or immune activation. J Virol 2010;84(11):5476–84.
[9] Ling B, Apetrei C, Pandrea I, Veazey RS, Lackner AA, Gormus B, et al. Classic AIDS in a sooty mangabey after an 18-year natural infection. J Virol 2004;78(16):8902–8.
[10] Pandrea I, Onanga R, Rouquet P, Bourry O, Ngari P, Wickings EJ, et al. Chronic SIV infection ultimately causes immunodeficiency in African non-human primates. AIDS 2001;15(18):2461–2.
[11] Meythaler M, Wang Z, Martinot A, Pryputniewicz S, Kasheta M, McClure HM, et al. Early induction of polyfunctional simian immunodeficiency virus (SIV)-specific T lymphocytes and rapid disappearance of SIV from lymph nodes of sooty mangabeys during primary infection. J Immunol 2011;186(9):5151–61.
[12] Martinot AJ, Meythaler M, Pozzi LA, Dalecki Boisvert K, Knight H, Walsh D, et al. Acute SIV infection in sooty mangabey monkeys is characterized by rapid virus clearance from lymph nodes and absence of productive infection in germinal centers. PloS One 2013;8(3):e57785.
[13] Diop OM, Gueye A, Dias-Tavares M, Kornfeld C, Faye A, Ave P, et al. High levels of viral replication during primary simian immunodeficiency virus SIVagm infection are rapidly and strongly controlled in African green monkeys. J Virol 2000;74(16):7538–47.
[14] Brenchley JM, Vinton C, Tabb B, Hao XP, Connick E, Paiardini M, et al. Differential infection patterns of CD4+ T cells and lymphoid tissue viral burden distinguish progressive and nonprogressive lentiviral infections. Blood 2012;120(20):4172–81.
[15] Dyavar Shetty R, Velu V, Titanji K, Bosinger SE, Freeman GJ, Silvestri G, et al. PD-1 blockade during chronic SIV infection reduces hyperimmune activation and microbial translocation in rhesus macaques. J Clin Invest 2012;122(5):1712–6.
[16] Estes JD. Enhancing immune responses to limit chronic immune activation during SIV. J Clin Invest 2012;122(5):1611–4.
[17] Silvestri G, Sodora DL, Koup RA, Paiardini M, O'Neil SP, McClure HM, et al. Nonpathogenic SIV infection of sooty mangabeys is characterized by limited bystander immunopathology despite chronic high-level viremia. Immunity 2003;18(3):441–52.
[18] Kaur A, Di Mascio M, Barabasz A, Rosenzweig M, McClure HM, Perelson AS, et al. Dynamics of T- and B-lymphocyte turnover in a natural host of simian immunodeficiency virus. J Virol 2008;82(3):1084–93.
[19] Zahn RC, Rett MD, Korioth-Schmitz B, Sun Y, Buzby AP, Goldstein S, et al. Simian immunodeficiency virus (SIV)-specific CD8+ T-cell responses in vervet African green monkeys chronically infected with SIVagm. J Virol 2008;82(23):11577–88.
[20] Bostik P, Mayne AE, Villinger F, Greenberg KP, Powell JD, Ansari AA. Relative resistance in the development of T cell anergy in CD4+ T cells from simian immunodeficiency virus disease-resistant sooty mangabeys. J Immunol 2001;166(1):506–16.
[21] Bostik P, Noble ES, Stephenson ST, Villinger F, Ansari AA. CD4+ T cells from simian immunodeficiency virus disease-resistant sooty mangabeys produce more IL-2 than cells from disease-susceptible species: involvement of p300 and CREB at the proximal IL-2 promoter in IL-2 up-regulation. J Immunol 2007;178(12):7720–9.
[22] Bostik P, Noble ES, Mayne AE, Gargano L, Villinger F, Ansari AA. Central memory CD4 T cells are the predominant cell subset resistant to anergy in SIV disease resistant sooty mangabeys. AIDS 2006;20(2):181–8.
[23] Paiardini M, Cervasi B, Reyes-Aviles E, Micci L, Ortiz AM, Chahroudi A, et al. Low levels of SIV infection in sooty mangabey central memory CD(4)(+) T cells are associated with limited CCR5 expression. Nat Med 2011;17(7):830–6.
[24] Bostik P, Watkins M, Villinger F, Ansari AA. Genetic analysis of cytokine promoters in nonhuman primates: implications for Th1/Th2 profile characteristics and SIV disease pathogenesis. Clin Dev Immunol 2004;11(1):35–44.

[25] Kaur A, Yang J, Hempel D, Gritz L, Mazzara GP, McClure H, et al. Identification of multiple simian immunodeficiency virus (SIV)-specific CTL epitopes in sooty mangabeys with natural and experimentally acquired SIV infection. J Immunol 2000;164(2):934–43.
[26] Wang Z, Metcalf B, Ribeiro RM, McClure H, Kaur A. Th-1-type cytotoxic CD8+ T-lymphocyte responses to simian immunodeficiency virus (SIV) are a consistent feature of natural SIV infection in sooty mangabeys. J Virol 2006;80(6):2771–83.
[27] Dunham R, Pagliardini P, Gordon S, Sumpter B, Engram J, Moanna A, et al. The AIDS resistance of naturally SIV-infected sooty mangabeys is independent of cellular immunity to the virus. Blood 2006;108(1):209–17.
[28] Lozano Reina JM, Favre D, Kasakow Z, Mayau V, Nugeyre MT, Ka T, et al. Gag p27-specific B- and T-cell responses in Simian immunodeficiency virus SIVagm-infected African green monkeys. J Virol 2009;83(6):2770–7.
[29] Kaur A, Grant RM, Means RE, McClure H, Feinberg M, Johnson RP. Diverse host responses and outcomes following simian immunodeficiency virus SIVmac239 infection in sooty mangabeys and rhesus macaques. J Virol 1998;72(12):9597–611.
[30] Meythaler M, Martinot A, Wang Z, Pryputniewicz S, Kasheta M, Ling B, et al. Differential CD4+ T-lymphocyte apoptosis and bystander T-cell activation in rhesus macaques and sooty mangabeys during acute simian immunodeficiency virus infection. J Virol 2009;83(2):572–83.
[31] Kaur A, Alexander L, Staprans SI, Denekamp L, Hale CL, McClure HM, et al. Emergence of cytotoxic T lymphocyte escape mutations in nonpathogenic simian immunodeficiency virus infection. Eur J Immunol 2001;31(11):3207–17.
[32] Kouyos RD, Gordon SN, Staprans SI, Silvestri G, Regoes RR. Similar impact of CD8+ T cell responses on early virus dynamics during SIV infections of rhesus macaques and sooty mangabeys. PLoS Comput Biol 2010;6(8).
[33] Regoes RR, Antia R, Garber DA, Silvestri G, Feinberg MB, Staprans SI. Roles of target cells and virus-specific cellular immunity in primary simian immunodeficiency virus infection. J Virol 2004;78(9):4866–75.
[34] Komanduri KV, Feinberg J, Hutchins RK, Frame RD, Schmidt DK, Viswanathan MN, et al. Loss of cytomegalovirus-specific CD4+ T cell responses in human immunodeficiency virus type 1-infected patients with high CD4+ T cell counts and recurrent retinitis. J Infect Dis 2001;183(8):1285–9.
[35] McNeil AC, Shupert WL, Iyasere CA, Hallahan CW, Mican JA, Davey RT, Jr., et al. High-level HIV-1 viremia suppresses viral antigen-specific CD4(+) T cell proliferation. Proc Natl Acad Sci USA 2001;98(24):13878–83.
[36] Kaur A, Gauduin M-C, Johnson RP. Immune responses to nonhuman primate lentiviruses. In: PGaW BD, editor. Retroviral Immunology: Immune response and Restoration. Totowa, New Jersey: Humana Press; 2000. p. 249–79.
[37] Jin X, Bauer DE, Tuttleton SE, Lewin S, Gettie A, Blanchard J, et al. Dramatic rise in plasma viremia after CD8(+) T cell depletion in simian immunodeficiency virus-infected macaques. J Exp Med 1999;189(6):991–8.
[38] Schmitz JE, Kuroda MJ, Santra S, Sasseville VG, Simon MA, Lifton MA, et al. Control of viremia in simian immunodeficiency virus infection by CD8+ lymphocytes. Science 1999;283(5403):857–60.
[39] Gaufin T, Ribeiro RM, Gautam R, Dufour J, Mandell D, Apetrei C, et al. Experimental depletion of CD8+ cells in acutely SIVagm-infected African Green Monkeys results in increased viral replication. Retrovirology 2010;7:42.
[40] Schmitz JE, Zahn RC, Brown CR, Rett MD, Li M, Tang H, et al. Inhibition of adaptive immune responses leads to a fatal clinical outcome in SIV-infected pigtailed macaques but not vervet African green monkeys. PLoS Pathog 2009;5(12):e1000691.

[41] Zahn RC, Rett MD, Li M, Tang H, Korioth-Schmitz B, Balachandran H, et al. Suppression of adaptive immune responses during primary SIV infection of sabaeus African green monkeys delays partial containment of viremia but does not induce disease. Blood 2010;115(15):3070–8.

[42] Barry AP, Silvestri G, Safrit JT, Sumpter B, Kozyr N, McClure HM, et al. Depletion of CD8+ cells in sooty mangabey monkeys naturally infected with simian immunodeficiency virus reveals limited role for immune control of virus replication in a natural host species. J Immunol 2007;178(12):8002–12.

[43] Bosinger SE, Li Q, Gordon SN, Klatt NR, Duan L, Xu L, et al. Global genomic analysis reveals rapid control of a robust innate response in SIV-infected sooty mangabeys. J Clin Invest 2009;119(12):3556–72.

[44] Jacquelin B, Mayau V, Targat B, Liovat AS, Kunkel D, Petitjean G, et al. Nonpathogenic SIV infection of African green monkeys induces a strong but rapidly controlled type I IFN response. J Clin Invest 2009;119(12):3544–55.

[45] Lederer S, Favre D, Walters KA, Proll S, Kanwar B, Kasakow Z, et al. Transcriptional profiling in pathogenic and non-pathogenic SIV infections reveals significant distinctions in kinetics and tissue compartmentalization. PLoS Pathog 2009;5(2):e1000296.

[46] Ansari AA, Pereira LE, Mayne AE, Onlamoon N, Pattanapanyasat K, Mori K, et al. The role of disease stage, plasma viral load and regulatory T cells (Tregs) on autoantibody production in SIV-infected non-human primates. J Autoimmun 2007;28(2–3):152–9.

[47] Pereira LE, Villinger F, Onlamoon N, Bryan P, Cardona A, Pattanapanysat K, et al. Simian immunodeficiency virus (SIV) infection influences the level and function of regulatory T cells in SIV-infected rhesus macaques but not SIV-infected sooty mangabeys. J Virol 2007;81(9):4445–56.

[48] Kornfeld C, Ploquin MJ, Pandrea I, Faye A, Onanga R, Apetrei C, et al. Antiinflammatory profiles during primary SIV infection in African green monkeys are associated with protection against AIDS. J Clin Invest 2005;115(4):1082–91.

[49] Estes JD, Wietgrefe S, Schacker T, Southern P, Beilman G, Reilly C, et al. Simian immunodeficiency virus-induced lymphatic tissue fibrosis is mediated by transforming growth factor beta 1-positive regulatory T cells and begins in early infection. J Infect Dis 2007;195(4):551–61.

[50] Sumpter B, Dunham R, Gordon S, Engram J, Hennessy M, Kinter A, et al. Correlates of preserved CD4(+) T cell homeostasis during natural, nonpathogenic simian immunodeficiency virus infection of sooty mangabeys: implications for AIDS pathogenesis. J Immunol 2007;178(3):1680–91.

[51] Nigam P, Velu V, Kannanganat S, Chennareddi L, Kwa S, Siddiqui M, et al. Expansion of FOXP3+ CD8 T cells with suppressive potential in colorectal mucosa following a pathogenic simian immunodeficiency virus infection correlates with diminished antiviral T cell response and viral control. J Immunol 2010;184(4): 1690–701.

[52] Khader SA, Gaffen SL, Kolls JK. Th17 cells at the crossroads of innate and adaptive immunity against infectious diseases at the mucosa. Mucosal Immunol 2009;2(5):403–11.

[53] Cecchinato V, Trindade CJ, Laurence A, Heraud JM, Brenchley JM, Ferrari MG, et al. Altered balance between Th17 and Th1 cells at mucosal sites predicts AIDS progression in simian immunodeficiency virus-infected macaques. Mucosal Immunol 2008;1(4):279–88.

[54] Dandekar S. Pathogenesis of HIV in the gastrointestinal tract. Curr HIV/AIDS Rep 2007;4(1):10–5.

[55] Brenchley JM, Paiardini M, Knox KS, Asher AI, Cervasi B, Asher TE, et al. Differential Th17 CD4 T-cell depletion in pathogenic and nonpathogenic lentiviral infections. Blood 2008;112(7):2826–35.

[56] Favre D, Lederer S, Kanwar B, Ma ZM, Proll S, Kasakow Z, et al. Critical loss of the balance between Th17 and T regulatory cell populations in pathogenic SIV infection. PLoS Pathog 2009;5(2):e1000295.
[57] Paiardini M. Th17 cells in natural SIV hosts. Curr Opin HIV AIDS 2010;5(2):166–72.
[58] Ferreira LM. Gammadelta T cells: innately adaptive immune cells? Int Rev Immunol 2013;32(3):223–48.
[59] Martini F, Urso R, Gioia A, De Felici A, Narciso P, Amendola A, et al. gammadelta T-cell anergy in human immunodeficiency virus-infected persons with opportunistic infections and recovery after highly active antiretroviral therapy. Immunology 2000;100(4):481–6.
[60] Zhou D, Lai X, Shen Y, Sehgal P, Shen L, Simon M, et al. Inhibition of adaptive Vgamma2Vdelta2+ T-cell responses during active mycobacterial coinfection of simian immunodeficiency virus SIVmac-infected monkeys. J Virol 2003;77(5):2998–3006.
[61] Harris LD, Klatt NR, Vinton CA, Briant JA, Tabb B, Ladell K, et al. Mechanisms underlying gammadelta T-cell subset perturbations in SIV-infected Asian rhesus macaques. Blood 2010;116(20):4148–57.
[62] Milush JM, Reeves JD, Gordon SN, Zhou D, Muthukumar A, Kosub DA, et al. Virally induced CD4+ T cell depletion is not sufficient to induce AIDS in a natural host. J Immunol 2007;179(5):3047–56.
[63] Kosub DA, Lehrman G, Milush JM, Zhou D, Chacko E, Leone A, et al. Gamma/Delta T-cell functional responses differ after pathogenic human immunodeficiency virus and nonpathogenic simian immunodeficiency virus infections. J Virol 2008;82(3):1155–65.
[64] Bendelac A, Savage PB, Teyton L. The biology of NKT cells. Annu Rev Immunol 2007;25:297–336.
[65] Lantz O, Bendelac A. An invariant T cell receptor alpha chain is used by a unique subset of major histocompatibility complex class I-specific CD4+ and CD4-8- T cells in mice and humans. J Exp Med 1994;180(3):1097–106.
[66] Fernandez CS, Chan AC, Kyparissoudis K, De Rose R, Godfrey DI, Kent SJ. Peripheral NKT cells in simian immunodeficiency virus-infected macaques. J Virol 2009;83(4):1617–24.
[67] van der Vliet HJ, von Blomberg BM, Hazenberg MD, Nishi N, Otto SA, van Benthem BH, et al. Selective decrease in circulating V alpha 24+V beta 11+ NKT cells during HIV type 1 infection. J Immunol 2002;168(3):1490–5.
[68] Chen N, McCarthy C, Drakesmith H, Li D, Cerundolo V, McMichael AJ, et al. HIV-1 down-regulates the expression of CD1d via Nef. Eur J Immunol 2006;36(2):278–86.
[69] Cho S, Knox KS, Kohli LM, He JJ, Exley MA, Wilson SB, et al. Impaired cell surface expression of human CD1d by the formation of an HIV-1 Nef/CD1d complex. Virology 2005;337(2):242–52.
[70] Moll M, Andersson SK, Smed-Sorensen A, Sandberg JK. Inhibition of lipid antigen presentation in dendritic cells by HIV-1 Vpu interference with CD1d recycling from endosomal compartments. Blood 2010;116(11):1876–84.
[71] Rout N, Else JG, Yue S, Connole M, Exley MA, Kaur A. Paucity of CD4+ natural killer T (NKT) lymphocytes in sooty mangabeys is associated with lack of NKT cell depletion after SIV infection. PloS One 2010;5(3):e9787.
[72] Rout N, Else JG, Yue S, Connole M, Exley MA, Kaur A. Heterogeneity in phenotype and function of CD8+ and CD4/CD8 double-negative Natural Killer T cell subsets in sooty mangabeys. J Med Primatol 2010;39(4):224–34.
[73] Rout N, Greene J, Yue S, O'Connor D, Johnson RP, Else JG, et al. Loss of effector and anti-inflammatory natural killer T lymphocyte function in pathogenic simian immunodeficiency virus infection. PLoS Pathog 2012;8(9):e1002928.
[74] Thomson CW, Lee BP, Zhang L. Double-negative regulatory T cells: non-conventional regulators. Immunol Res 2006;35(1–2):163–78.

[75] Beaumier CM, Harris LD, Goldstein S, Klatt NR, Whitted S, McGinty J, et al. CD4 down-regulation by memory CD4+ T cells in vivo renders African green monkeys resistant to progressive SIVagm infection. Nat Med 2009;15(8):879–85.
[76] Sundaravaradan V, Saleem R, Micci L, Gasper MA, Ortiz AM, Else J, et al. Multifunctional double-negative T cells in sooty mangabeys mediate T-helper functions irrespective of SIV infection. PLoS Pathog 2013;9(6):e1003441.
[77] Vinton C, Klatt NR, Harris LD, Briant JA, Sanders-Beer BE, Herbert R, et al. CD4-like immunological function by CD4-T cells in multiple natural hosts of simian immunodeficiency virus. J Virol 2011;85(17):8702–8.

CHAPTER

11

Lack of Chronic Immune Activation During Natural Host SIV Infection

Vasudha Sundaravaradan[1], *Melanie A. Gasper*[1], *Kiran D. Mir*[2], *Donald L. Sodora*[1]

[1]Seattle Biomedical Research Institute, Seattle, WA, USA; [2]Emory Vaccine Center, Yerkes Primate Center, Atlanta, GA, USA

INTRODUCTION

The potential for chronic immune activation to impact acquired immune deficiency syndrome (AIDS) pathogenesis was first postulated by Ascher and Sheppard [1] and by Grossman and colleagues who arrived at similar conclusions during the late 1980s/early 1990s [2,3]. Subsequently, Giorgi and colleagues published a series of clinical studies supporting the concept that an excessive/aberrant activation of immune cells is a fundamental driving force for human immunodeficiency virus (HIV)-associated immune dysfunction [4]. Specifically, these studies identified the level of CD8 T-cell activation, as determined by CD38 and HLA-DR expression, as a better correlate of disease progression than viral load [5–7]. Since then, several studies by multiple groups have described increased levels of turnover and/or activation of multiple immune cell subtypes [8–11]. This body of work has expanded the paradigm of AIDS pathogenesis beyond the simple infection and depletion of $CD4^+$ T cells to include a prolonged hyperactivation of the immune system. The causes of this widespread immune cell activation are likely multifactorial and include both the direct and indirect effects of the virus and viral proteins, the depletion and dysfunction of regulatory immune cells, and the peripheral immune cell stimulation by bacterial products translocating from the intestinal lumen (reviewed in detail in [11]).

Natural Hosts of SIV.
http://dx.doi.org/10.1016/B978-0-12-404734-1.00011-5

Although a full understanding of HIV-associated chronic immune activation remains incomplete, a clearer picture of the consequences of elevated activation for different immune cell populations is emerging (Table 11.1). In T cells, immune activation can be observed through both the higher frequencies of activated and proliferating T cells [12–15] and as higher levels of activation-induced apoptosis of uninfected T cells [16–20] (most

TABLE 11.1

Cell Type	Characteristics of Cell Populations in Natural Hosts Following SIV Infection	Role in Reducing Immune Activation in SIV-Infected Natural Hosts	References
CD4 T cells	• Decrease in acute SIV infection • Generally maintained during chronic infection • Limited homeostatic proliferation • Preservation of Th17 subsets • Limited CCR5 expression on central memory (CM) subsets	• Limited proliferation and CCR5 expression • Treg function in SIV+ AGMs • Maintained Th17 function in SIV+ mangabeys	[9,14–20,33,37, 40,42,45–49]
NKT cells, γδT cells, and DN T cell subsets	• Limited expression of IL17 by NKT cells • Maintained Th1–Th2 function of γδT cells • Maintained helper function of CD8αdim cells • Limited apoptosis and maintained levels of DN T cells • Maintained Th1, Th2, Th17 function of DN T cells • Anti-inflammatory role of NKT and DN T cells	• Limit immune activation because of an anti-inflammatory cytokine production	[39,54–58,62–72]
pDCs	• Limited expression of β7 integrin • Low levels in the gut • Comparatively lower levels of IFNα + cells	• Reduced IFNα production in tissues during chronic phase of infection	[27,73–77]
Myeloid-derived cells	• Limited T cell stimulation • Limited TNFα production following LPS stimulation • Limited apoptosis	• Limited monocyte TNFα response to LPS stimulation in SIV+ mangabeys	[76,78,83]

studies focused on CD8 T cells as these cells were not impacted by direct HIV infection). However, the impact of an activated immune system can be observed in non-T immune cells as evidenced by the HIV-induced elevations in activation and proliferation of B cells, natural killer (NK) cells, monocytes, and macrophages [14,21,22]. Although these studies provide strong evidence that chronic immune activation plays a key role in disease progression observed during HIV infection, the studies are largely correlative. The cause-and-effect relationship between disease progression and immune activation is difficult to assess. It was the study of SIV infection in natural host monkey species that provided some of the most convincing evidence that immune activation is a contributing factor in the immune dysfunction and disease progression observed during pathogenic simian immunodeficiency virus (SIV)/HIV infections.

The first indication that sooty mangabeys' cells exhibited a reduced state of immune activation came from a 1996 study by Villinger and colleagues that identified reduced levels of apoptosis in mangabey T cells stimulated with an anti-CD3 antibody compared to macaques and humans [23]. Subsequently, Chakrabarti and colleagues reported that the high levels of T cell proliferation found during pathogenic SIV infection are not present in SIV-infected sooty mangabeys [8]. Early studies in African green monkeys (AGMs) demonstrated that these natural hosts exhibit comparable levels of viral replication and tropism as SIV-infected macaques, suggesting that the lack of disease progression in AGMs is mediated by immunologic rather than virologic factors [24]. Silvestri et al. furthered these findings by demonstrating that SIV-infected mangabeys displayed lower levels of immune activation compared to rhesus macaques (RMs) undergoing pathogenic SIV infection [25]. This disparity in immune activation between nonpathogenic and pathogenic SIV infections is most pronounced during the chronic phase of infection. Importantly, acute infection in both the pathogenic and nonpathogenic SIV infections is characterized by an increase in markers of immune activation. It is during the acute-to-chronic transition at 4–6 weeks postinfection that clear differences between the two disease outcomes arise [11,26–34]. Indeed, the ability of natural hosts of SIV to resolve the initial acute phase increase in immune activation is a signature characteristic and a key determinant of nonprogressive infection.

In this chapter, we will evaluate the immune cell subsets that have a potential to play key roles in lowering immune activation during the chronic phase of infection in SIV-infected natural hosts. Each section is designed to cover one or more of these immune subsets and elucidate how these cells might impact immune activation during pathogenic HIV/SIV infections as well as suppress immune activation during nonpathogenic SIV infection of natural host monkey species. The long-term goal of these analyses is to identify the precise phenotype and the function by which this subset of immune cells suppress the immune activation in the natural host SIV infections with the hope of developing a future immune therapeutic to modulate immune activation and inhibit disease progression during HIV infection.

CD4 T CELLS

CD4 T cells serve a key role in shaping both humoral and cellular immune responses and are likely to play an important role in HIV/SIV-induced immunologic activation. Their ability to produce inflammatory cytokines/chemokines is balanced by the capacity of specific CD4 T-cell subsets such as T regulatory cells (Tregs) and T helper (Th) 17 cells to control immune activation through the production of immunoregulatory cytokines.

CD4 T cells are the primary target cell of HIV/SIV infection during both pathogenic and nonpathogenic infections. The acute phase of nonpathogenic SIV infections is associated with an early peak in viral replication and decline in CD4 T cells that can be observed in both mangabeys and AGMs [9,35]. Still, these reduced CD4 T-cell levels are generally maintained within the "healthy" range (above 500 cells/μl of blood) throughout the disease course [12,25,33,36], with some striking exceptions observed in naturally infected captive mangabeys [37] and experimentally infected mangabeys [38,39]. Although peripheral CD4 T-cell levels decline during both pathogenic and nonpathogenic infections [40,41], the extent of homeostatic proliferation seen in pathogenic infections is much greater than in nonpathogenic infections (including mangabeys and mandrills) [42]. This increased proliferation of CD4 T cells observed in humans and macaques can act to "fuel the fire" of infection, whereas natural hosts are able to restrict the number of proliferating cells susceptible to infection.

During both pathogenic and nonpathogenic SIV infections, CD4 T-cell depletion is evident in multiple tissues. In natural hosts, CD4 T-cell depletion within the gut-associated lymphoid tissue (GALT) is rapid following SIV infection, similar to that observed during pathogenic infections [9,33]. Interestingly, GALT CD4 T cells do recover during the chronic phase of infection in AGMs [33] but generally remain relatively low in SIV-infected mangabeys [9]. Importantly, the experimental and virus-mediated depletion of CD4 T cells in the peripheral blood, lymph nodes (LNs), and GALT in mangabeys did not result in any evidence of immune activation or clinical signs of simian AIDS [28,38,39]. These findings suggest that the preservation of total CD4 T-cell levels is not essential to maintain low levels of immune activation in the SIV-infected natural host monkey species.

The contribution of specialized CD4 T-cell subsets, including Tregs and Th17 cells, to the resolution of immune activation in natural hosts is complex, and observations have varied among the different species. Evaluation of Tregs in sooty mangabeys uncovered no evidence that this immune subset (expressing either CD4 or CD8) controls immune activation during the chronic phase of infection [35,43]. In contrast, studies in AGMs identified maintained or increased levels of Tregs during acute SIV infection [36]. Kornfeld et al. identified increases in Foxp3 gene expression concurrent with increased plasma transforming growth factor beta (TGF-β) within 24h of SIV infection in AGMs. This early increase was followed by an increase in plasma interleukin-10 (IL-10) at day 6 [36]. Possibly, the presence of Tregs early during nonpathogenic SIV infection contributes to establishing an anti-inflammatory environment during acute infection, which suppresses the immune activation at later stages. However, there is no direct correlation between the levels of Tregs and resolution of immune activation in natural hosts [26]. Moreover, both SIV-infected pig tailed macaques (PTMs) and AGMs experience increases in lamina propria Tregs at the same time as colonic immune activation increases (45 days postinfection [dpi]), suggesting that these cells are not primarily responsible for controlling immune activation in the gut of either species [44]. Rather, evidence in the RM model indicates that a premature timing of Treg activation during pathogenic infection serves only to restrict cytotoxic T-cell responses and still does not control immune activation [44]. Similarly, in SIV+ PTMs, an increase in Foxp3+ Tregs occurs during the first 10 days of infection [45]. This increase was accompanied by a specific depletion of Th17 cells that led to a perturbed ratio of these subsets. By day 45, PTMs with lower Th17:Treg ratios had higher levels of T-cell activation [45]. In contrast, both Th17 and Treg cells were preserved in SIV-infected AGMs, and there was no correlation between the Th17:Treg ratio and T-cell activation. A similar preservation of Th17 cells was observed in the blood and gastrointestinal tract of SIV-infected mangabeys, whereas HIV patients had significant depletions in this subset [46]. These data suggest that the preferential preservation of Th17 T cells might be critical for maintaining mucosal health in SIV-infected natural hosts [45–47].

A key feature of SIV infection in natural hosts is a distinct infection profile in CD4 T-cell subsets. This feature likely plays a key role in the maintenance of effective helper CD4 T cells. One important observation in this regard is that there are low levels of CCR5 coreceptor expression on central memory CD4 T cells in natural SIV hosts [11,29,48]. Paiardini et al. demonstrated that central memory CD4 T cells in sooty mangabey resist CCR5 upregulation following activation and were relatively resistant to infection both in vivo and in vitro [49]. Thus, restricted and delayed expression of CCR5 on activated CD4 T cells may preserve the homeostasis of the pool of "resting" naïve and memory CD4 T cells while supporting high levels of virus production in activated effector CD4 cells that are likely to die as a consequence of activation-induced cell death [31].

It follows that the preserved pool of naïve and central memory T cells would be available to replenish a continuous source of these effector CD4 T cells capable of contributing to immune responses to both SIV and other pathogens [29,50].

In summary, CD4 T cells have the potential to impact the level of immune activation during the chronic phase in SIV-infected natural hosts through the use of regulatory and Th17 cell subsets; the production of immunoregulatory cytokines; and the alteration of the infectivity of central memory CD4 T cells, thereby preserving this key self-renewable helper subset.

NATURAL KILLER T CELLS, γδ T CELLS, AND DOUBLE-NEGATIVE T CELLS

T cells that do not express CD4 would be predicted to be resistant to HIV/SIV infection because of the absence of this receptor on their cell surface. Therefore, one might hypothesize that SIV-infected natural hosts might utilize these CD4-negative T-cell subsets to suppress the immune activation observed during chronic SIV infection. For example, natural killer T (NKT) cells are a relatively small subset of T cells that express both CD3 and the NK receptor CD16 and recognize nonpeptide antigens presented by nonpolymorphic major histocompatibility complex (MHC) class I-like molecules (CD1d) [51]. Many NKT cell subsets have been identified, with the best characterized being the invariant NKT lymphocytes that are defined as expressing only one T-cell receptor (TCR) α chain (Vα24 in humans) and a limited number of TCR β chains [52]. Upon activation of NKT cells by glycolipids presented by CD1d proteins, NKT cells produce a diverse array of cytokines, including IL-10 and TGF-β, as well as those influencing the Th1 and Th2 pathways [53]. During pathogenic SIV infection, an early increase in IL-17 production by NKT cells in the LNs was recently found to be associated with subsequent progression to simian AIDS in RMs [54]. IL-17 production peaked at 14 dpi and was correlated with increased IL-18 and TGF-β production. Moreover, these cytokines were continuously produced during chronic infection of noncontrolling RMs [54]. In contrast, SIV-infected AGMs did not exhibit any increase in IL-17 production in the LN NKT cells during acute infection [54]. In mangabeys, NKT cells are either $CD4^{-}CD8^{+}$ or $CD4^{-}CD8^{-}$; the lack of CD4 expression makes them inherently resistant to SIV/HIV infections [55–57]. NKT cells in mangabeys have been shown to produce many cytokines, including the anti-inflammatory mediator IL-10, implicating these cells in the low level of immune activation observed in the SIV+ mangabeys [58].

A second interesting T-cell subset are the gamma/delta (γδ) T cells that play key roles in innate immunity and initiation of the adaptive immune responses. In humans, γδ T cells comprise a relatively minor subset (1–5% on average) of circulating T cells but may represent as much as 50% of the T cells present within the mucosal-associated lymphoid tissue [59]. γδ T cells can produce both Th1 and Th2 cytokines, thereby influencing adaptive immune responses [60]. Two main γδ T cell subsets exist as defined by the expression of one of two δ variable regions, Vδ1 or Vδ2. The Vδ1+ γδ T cells are found predominantly at mucosal sites and can respond to nonclassical MHC molecules expressed on stressed cells, whereas Vδ2+ γδ T cells are predominantly in the peripheral circulation and respond to nonpeptide phosphoantigens [61]. γδ T cells are influenced by HIV infection as evidenced by a phenotypic switch from predominantly Vδ2 before infection to predominantly Vδ1 within the peripheral blood of HIV+ patients [62]. Following an oral SIV infection, decreased γδ T cell levels were observed at the oral and esophageal mucosa, whereas increased γδ T-cell levels were observed in the LNs of RMs [63]. In SIV-infected mangabeys, the γδ T cells maintain or increase their ability to produce cytokines (interferon gamma [IFN-γ] or tumor necrosis factor alpha [TNF-α) following stimulation with phosphoantigen, whereas human γδ T cells exhibit a decreased cytokine response following HIV infection [64].

A third subset of CD4-negative T cells are the $CD3^+CD4^-CD8^-$ or double-negative (DN) T cells. DN T cells in mangabeys and humans predominantly express the α/β T cell receptors and are not invariant NKT cells, as evidenced by their diverse usage of TCR loci [65]. Some studies in mice have suggested that these cells are a form of regulatory T cells; however, other studies have attributed effector type functions to DN T cells [66,67]. Interestingly, the peripheral blood from AGMs also contains a relatively high proportion of CD4-negative T cells that are CD8αdim [68]. Beaumier et al. found that these CD4-CD8αdim T cells display a predominantly memory phenotype and that they arise following downmodulation of the CD4 molecule on $CD4^+$ T cells [68]. Notably, these DN CD8αdim T cells retain Th cell function, including IL-2 and IL-17 production, are MHC class II restricted, and are resistant to SIV infection [68]. DN T cells in mangabeys do not express CD8α protein on their surface [39] and do not express CD4 protein either intra/extracellularly nor do they express CD4 mRNA [65]. Compared to humans, mangabeys have relatively high levels of DN T cells that are maintained following SIV infection [39,69]. The observation that some SIV-infected mangabeys could survive with very low $CD4^+$ T cell levels (CD4-low) and no evidence of clinical disease raised the speculation that these DN T cells have helper T-cell functions [39]. Recent studies have confirmed that DN T cells in both uninfected, SIV-infected, and in SIV-infected CD4-low mangabeys are multifunctional cells possessing Th1, Th2, and Th17 functions [65]. In mangabeys, DN T cells are also primarily effector memory cells that likely contain subpopulations that perform these different effector functions and can be restricted by either MHC class I or MHC class II [65]. DN T cells in SIV-infected mangabeys may also contribute to decreased systemic immune activation via the expression of IL-10 and yet would be predicted to have a reduced response to IL-10 because of the downmodulation of IL-10 receptor expression [65]. In contrast to natural hosts, SIV-infected RMs experience high rates of DN T cell apoptosis as early as 1 week postinfection [70]. In HIV-infected humans, recent evidence has shown a correlation among increased DN T cells early during infection, lower immune activation, and decreased disease progression later in infection [71,72]. The role of DN T cells in healthy humans, their role in HIV infection, and the mechanism of how they contribute to immune activation in populations with differential disease progression are still under investigation.

PLASMACYTOID DENDRITIC CELLS

Plasmacytoid dendritic cells (pDCs) are primed to be among the first responders to lentiviral infection, as they are tailored to detect viral pathogen-associated molecular patterns via a unique signature of pattern recognition receptors, including toll-like receptor (TLR)7, TLR8, and TLR9. Moreover, their capacity to produce more type I IFN on a per-cell basis than any other immune cell suggests that they have the potential to promote immune activation during pathogenic infection. It is clear that pathogenic HIV/SIV infection alters pDC dynamics very early during infection, as their levels in the blood decrease significantly within 2 weeks of infection [73,74]. Concurrent with the decline of pDCs in the blood, pDC numbers increase in LNs and plasma IFN-α levels reach their peak, suggesting that pDCs traffic to the LNs and encounter high levels of virus [73]. During pathogenic HIV/SIV infections, pDCs also traffic from the blood to the gut via the expression of α4β7 [50, 74]. In the gut, pDCs display a proinflammatory phenotype and likely contribute to the activation and proliferation of resident T cells [74]. In contrast, pDCs in the blood of SIV+ mangabeys do not increase their expression of β7 integrin and only low levels of pDCs are found in the gut [74]. Hence, regulating pDC trafficking may prevent one source of immune activation during nonpathogenic infection. In AGMs, the first weeks of SIV infection result in a transient decrease in the absolute numbers of blood pDCs, with a return to baseline by 42 dpi. The decreases in blood pDC levels during acute SIVagm infection were found to have a temporal association with pDC increases in LNs [75]. Just as in pathogenic infection, plasma IFN-α levels increase early in SIV-infected AGMs, although the magnitude of IFN-α-producing cells is lower [75,76].

Downmodulating aberrant IFN-α production by pDCs following acute infection of SIV natural host species may be critical for avoiding elevated levels of global immune activation during later stages of infection [27]. One report suggested that sooty mangabeys may achieve this goal via a reduced sensitivity to TLR7/TLR9 stimuli [77]. However, other studies have clearly demonstrated that pDCs from both SIV-infected mangabeys and AGMs display considerable sensing of viral TLR7/TLR8 and TLR9 ligands and produce levels of type I IFN comparable to RMs during acute infection [27,75,76]. Taken together, these data identify clear correlates with regard to pDCs and IFN-α in promoting immune activation during the acute phase of the infection and suggest that the resolution of immune activation during the chronic infection phase in SIV-infected natural hosts may involve a specific regulation of pDCs.

CELLS THAT PROMOTE INFLAMMATION: NEUTROPHILS, MYELOID DENDRITIC CELLS, MONOCYTES/MACROPHAGES

Myeloid cells, including neutrophils, monocytes, macrophages, myeloid dendritic cells (mDCs), and mast cells together compose a critical arm of the immune system, largely responsible for innate defense against an array of pathogens. Myeloid cells also play a key role in linking innate and adaptive immunity, primarily through antigen presentation and recruitment of adaptive immune cells. Comparative studies of pathogenic and nonpathogenic infections have revealed several key associations between myeloid cells and AIDS-related pathogenesis.

Neutrophils (or polymorphonuclear neutrophils [PMNs]) play a key role in promoting inflammation and therefore have the potential to influence immune activation during chronic SIV infections. Indeed, during acute infection SIV-infected RMs displayed significantly higher levels of neutrophil apoptosis, resulting in neutropenia that was highly correlated with viral load and disease progression [78]. This increased apoptosis during SIV infection of RMs was associated with higher expression of the SIV coreceptor BOB/GPR15 on the surface of neutrophils, suggesting that noninfectious binding of SIV to neutrophils may induce bystander apoptosis of the cells [79]. This increased neutrophil apoptosis continued throughout the chronic phase of infection and was not observed with an attenuated SIVΔnef strain [79]. Finally, this apoptosis was increased in macaques that progressed faster to AIDS compared to nonprogressors and correlated with the activation status of PMN reflected by increased CD11b expression and reactive oxygen species production [79]. In contrast, nonprogressing infection of AGMs (infected with SIVagm) did not exhibit any difference in neutrophil apoptosis during acute infection compared to preinfection levels [78]. Together, these findings suggest that neutrophil activation and apoptosis may also play a key role in promoting chronic immune activation and inflammation observed during pathogenic SIV infections.

mDCs play a key role in linking innate and adaptive immunity, primarily through antigen presentation. During pathogenic HIV and SIV infections, mDC numbers decrease in the peripheral blood, possibility resulting from increased homing to LNs [80,81]. However, Brown et al. demonstrated that both mDCs and pDCs were depleted in blood as well as LNs and spleens in monkeys with simian AIDS [82]. With regard to mDCs from natural host monkey species, in vitro analyses have determined that AGM mDCs are weak T-cell stimulators in contrast to mDCs cultured from RMs and humans [83]. Furthermore, in AGMs this decreased ability to stimulate T cells was associated with a higher ratio of anti-inflammatory (IL-10) vs. proinflammatory (IL-12, IFN-γ, TNF-α) cytokine production during mixed lymphocyte reaction [83]. Together, these data suggest that pathogenic HIV and SIV infections induce a depletion of mDCs and that these cells have a more stimulatory potential in humans and RMs, implicating a potential role for these cells in promoting AIDS-associated immune activation.

Although the AIDS-related dysfunction in monocytes has been known since 1984 [84], relatively little is known about these cells with regard to HIV/SIV-associated immune activation. Both precursor monocytes and mature macrophages have the potential to impact HIV-associated disease at several important tissue sites. It has been demonstrated that an increase in peripheral monocyte turnover, shown to be associated with tissue macrophage death, correlates with pathogenic SIV disease progression in RMs [10]. More recent studies have demonstrated that HIV virus may sensitize both monocyte-derived macrophages and dendritic cells to undergo apoptosis and that increased apoptosis of these cells may be a mechanism for early immune dysfunction in the context of pathogenic HIV/SIV infections [85]. It has also been shown that pathogenic HIV/SIV infections result in a redistribution of circulating monocyte subsets, including an increased proportion of proinflammatory $CD14^{+}CD16^{+}$ and $CD14^{+}CD16^{++}$ monocytes [86–88]. Interestingly, this same profile of monocyte subset redistribution has been observed in HIV-uninfected patients suffering from acute coronary syndrome, possibly providing a link between HIV-1 and increased cardiovascular disease [88]. Indeed, increased levels of the proinflammatory cytokines, such as IL-1 and TNF-α, have been found in monocytes from HIV-infected subjects [89], and IL-1, TNF-α, IL-6, and IL-8 production has been found to be increased in HIV-infected monocyte-derived macrophages [90]. In addition, exposure of monocytes to HIV-1 or HIV-1-derived single-stranded RNA sequences sensitized these cells for TLR4 stimulation, resulting in a significantly higher TNF-α production by monocytes in response to lipopolysaccharide

(LPS) [91]. Interestingly, ex vivo LPS stimulation of blood monocytes from SIV+ sooty mangabeys results in a decreased proportion of TNF-α-expressing monocytes compared to SIV-uninfected mangabeys [92]. Conversely, monocytes from HIV-infected patients and SIV-infected macaques responded similarly to their uninfected counterparts following LPS stimulation [92]. These results suggest that blood monocytes from SIV-infected mangabeys have a differential, controlled response to LPS stimulation that may have important implications in the context of translocation of microbial products from the gut lumen during acute infection and suppression of immune activation [93]. Finally, monocytes and macrophages may play a pivotal role in AIDS-related dementia, potentially by promoting the spread of infection to the brain via infected macrophages, as well as through the production of proinflammatory cytokines, including TNF-α, that weaken the blood-brain barrier [94]. Together, these data indicate that myeloid cells play key roles in the induction of inflammation associated with pathogenic HIV/SIV disease progression and that hosts of nonpathogenic infections may have evolved mechanisms to finely tune the effector functions of these highly inflammatory cells in order to maintain a state of limited immune activation.

IMPLICATIONS OF THESE STUDIES

Studies to date have identified a clear role for inflammation and immune activation in the disease outcome following an HIV infection [1,93]. Here, we have demonstrated that comparative studies of nonhuman primate models of AIDS pathogenesis are powerful tools for elucidating the contributions of specific immune cell subsets to distinct disease outcomes. Studies utilizing these models will likely identify novel targets for immunotherapies and vaccines, such as which immune cells/immune functions to bolster or suppress, and which cytokines/chemokines to augment or block. The current data support a model in which one or more immune cell subsets moderate immune activation and maintain/increase immune function during the critical acute-to-chronic transition in SIV-infected natural hosts but are unable to accomplish this feat during HIV/SIV pathogenic infections. Whether the reduced immune activation during the chronic phase of nonpathogenic SIV infection is primarily driven by one keystone immune cell subset or a combination of immune cells acting in concert remains to be determined. Clearly, future studies are warranted to determine the precise role of various immune cells in promoting or resolving immune activation in pathogenic and nonpathogenic models of AIDS. It is likely that such studies would enable the development of immune therapeutic approaches to limit immune activation in HIV-infected patients, thereby reducing disease progression and onset of AIDS.

References

[1] Ascher MS, Sheppard HW. AIDS as immune system activation: a model for pathogenesis. Clin Exp Immunol 1988;73(2):165–7.
[2] Grossman Z, Bentwich Z, Herberman RB. From HIV infection to AIDS: are the manifestations of effective immune resistance misinterpreted? Clin Immunol Immunopathol 1993;69(2):123–35.
[3] Grossman Z, Herberman RB. T-cell homeostasis in HIV infection is neither failing nor blind: modified cell counts reflect an adaptive response of the host. Nat Med 1997;3(5):486–90.
[4] Giorgi JV, Liu Z, Hultin LE, Cumberland WG, Hennessey K, Detels R. Elevated levels of CD38+ CD8+ T cells in HIV infection add to the prognostic value of low CD4+ T cell levels: results of 6 years of follow-up. The Los Angeles Center, Multicenter AIDS Cohort Study. J Acquir Immune Defic Syndr 1993;6(8):904–12.
[5] Giorgi JV, Hultin LE, McKeating JA, Johnson TD, Owens B, Jacobson LP, et al. Shorter survival in advanced human immunodeficiency virus type 1 infection is more closely associated with T lymphocyte activation than with plasma virus burden or virus chemokine coreceptor usage. J Infect Dis 1999;179(4):859–70.
[6] Giorgi JV, Lyles RH, Matud JL, Yamashita TE, Mellors JW, Hultin LE, et al. Predictive value of immunologic and virologic markers after long or short duration of HIV-1 infection. J Acquir Immune Defic Syndr 2002;29(4):346–55.
[7] Liu Z, Cumberland WG, Hultin LE, Prince HE, Detels R, Giorgi JV. Elevated CD38 antigen expression on CD8+ T cells is a stronger marker for the risk of chronic HIV disease progression to AIDS and death in the Multicenter AIDS Cohort Study than CD4+ cell count, soluble immune activation markers, or combinations of HLA-DR and CD38 expression. J Acquir Immune Defic Syndr Hum Retrovirol 1997;16(2):83–92.
[8] Chakrabarti LA, Lewin SR, Zhang L, Gettie A, Luckay A, Martin LN, et al. Normal T-cell turnover in sooty mangabeys harboring active simian immunodeficiency virus infection. J Virol 2000;74(3):1209–23.
[9] Gordon SN, Klatt NR, Bosinger SE, Brenchley JM, Milush JM, Engram JC, et al. Severe depletion of mucosal CD4+ T cells in AIDS-free simian immunodeficiency virus-infected sooty mangabeys. J Immunol 2007;179(5):3026–34.
[10] Hasegawa A, Liu H, Ling B, Borda JT, Alvarez X, Sugimoto C, et al. The level of monocyte turnover predicts disease progression in the macaque model of AIDS. Blood 2009;114(14):2917–25.
[11] Sodora DL, Silvestri G. Immune activation and AIDS pathogenesis. AIDS 2008; 22(4):439–46.
[12] Kovacs JA, Lempicki RA, Sidorov IA, Adelsberger JW, Herpin B, Metcalf JA, et al. Identification of dynamically distinct subpopulations of T lymphocytes that are differentially affected by HIV. J Exp Med 2001;194(12):1731–41.
[13] Orendi JM, Bloem AC, Borleffs JC, Wijnholds FJ, de Vos NM, Nottet HS, et al. Activation and cell cycle antigens in CD4+ and CD8+ T cells correlate with plasma human immunodeficiency virus (HIV-1) RNA level in HIV-1 infection. J Infect Dis 1998;178(5):1279–87.
[14] Rosenzweig M, DeMaria MA, Harper DM, Friedrich S, Jain RK, Johnson RP. Increased rates of CD4(+) and CD8(+) T lymphocyte turnover in simian immunodeficiency virus-infected macaques. Proc Natl Acad Sci USA 1998;95(11):6388–93.
[15] Sachsenberg N, Perelson AS, Yerly S, Schockmel GA, Leduc D, Hirschel B, et al. Turnover of CD4+ and CD8+ T lymphocytes in HIV-1 infection as measured by Ki-67 antigen. J Exp Med 1998;187(8):1295–303.
[16] Estaquier J, Idziorek T, de Bels F, Barre-Sinoussi F, Hurtrel B, Aubertin AM, et al. Programmed cell death and AIDS: significance of T-cell apoptosis in pathogenic and nonpathogenic primate lentiviral infections. Proc Natl Acad Sci USA 1994;91(20):9431–5.

[17] Finkel TH, Tudor-Williams G, Banda NK, Cotton MF, Curiel T, Monks C, et al. Apoptosis occurs predominantly in bystander cells and not in productively infected cells of HIV- and SIV-infected lymph nodes. Nat Med 1995;1(2):129–34.
[18] Katsikis PD, Wunderlich ES, Smith CA, Herzenberg LA, Herzenberg LA. Fas antigen stimulation induces marked apoptosis of T lymphocytes in human immunodeficiency virus-infected individuals. J Exp Med 1995;181(6):2029–36.
[19] Muro-Cacho CA, Pantaleo G, Fauci AS. Analysis of apoptosis in lymph nodes of HIV-infected persons. Intensity of apoptosis correlates with the general state of activation of the lymphoid tissue and not with stage of disease or viral burden. J Immunol 1995;154(10):5555–66.
[20] Gougeon ML, Lecoeur H, Dulioust A, Enouf MG, Crouvoiser M, Goujard C, et al. Programmed cell death in peripheral lymphocytes from HIV-infected persons: increased susceptibility to apoptosis of CD4 and CD8 T cells correlates with lymphocyte activation and with disease progression. J Immunol 1996;156(9):3509–20.
[21] Lane HC, Masur H, Edgar LC, Whalen G, Rook AH, Fauci AS. Abnormalities of B-cell activation and immunoregulation in patients with the acquired immunodeficiency syndrome. N Engl J Med 1983;309(8):453–8.
[22] Hellerstein MK, Hoh RA, Hanley MB, Cesar D, Lee D, Neese RA, et al. Subpopulations of long-lived and short-lived T cells in advanced HIV-1 infection. J Clin Invest 2003;112(6):956–66.
[23] Villinger F, Folks TM, Lauro S, Powell JD, Sundstrom JB, Mayne A, et al. Immunological and virological studies of natural SIV infection of disease-resistant nonhuman primates. Immunol Lett 1996;51(1–2):59–68.
[24] Broussard SR, Staprans SI, White R, Whitehead EM, Feinberg MB, Allan JS. Simian immunodeficiency virus replicates to high levels in naturally infected African green monkeys without inducing immunologic or neurologic disease. J Virol 2001;75(5):2262–75.
[25] Silvestri G, Sodora DL, Koup RA, Paiardini M, O'Neil SP, McClure HM, et al. Nonpathogenic SIV infection of sooty mangabeys is characterized by limited bystander immunopathology despite chronic high-level viremia. Immunity 2003;18(3):441–52.
[26] Estes JD, Gordon SN, Zeng M, Chahroudi AM, Dunham RM, Staprans SI, et al. Early resolution of acute immune activation and induction of PD-1 in SIV-infected sooty mangabeys distinguishes nonpathogenic from pathogenic infection in rhesus macaques. J Immunol 2008;180(10):6798–807.
[27] Harris LD, Tabb B, Sodora DL, Paiardini M, Klatt NR, Douek DC, et al. Downregulation of robust acute type I interferon responses distinguishes nonpathogenic simian immunodeficiency virus (SIV) infection of natural hosts from pathogenic SIV infection of rhesus macaques. J Virol 2010;84(15):7886–91.
[28] Klatt NR, Villinger F, Bostik P, Gordon SN, Pereira L, Engram JC, et al. Availability of activated CD4+ T cells dictates the level of viremia in naturally SIV-infected sooty mangabeys. J Clin Invest 2008;118(6):2039–49.
[29] Sodora DL, Allan JS, Apetrei C, Brenchley JM, Douek DC, Else JG, et al. Toward an AIDS vaccine: lessons from natural simian immunodeficiency virus infections of African nonhuman primate hosts. Nat Med 2009;15(8):861–5.
[30] Bosinger SE, Li Q, Gordon SN, Klatt NR, Duan L, Xu L, et al. Global genomic analysis reveals rapid control of a robust innate response in SIV-infected sooty mangabeys. J Clin Invest 2009;119(12):3556–72.
[31] Brenchley JM, Silvestri G, Douek DC. Nonprogressive and progressive primate immunodeficiency lentivirus infections. Immunity 2010;32(6):737–42.
[32] Jacquelin B, Mayau V, Targat B, Liovat AS, Kunkel D, Petitjean G, et al. Nonpathogenic SIV infection of African green monkeys induces a strong but rapidly controlled type I IFN response. J Clin Invest 2009;119(12):3544–55.
[33] Pandrea IV, Gautam R, Ribeiro RM, Brenchley JM, Butler IF, Pattison M, et al. Acute loss of intestinal CD4+ T cells is not predictive of simian immunodeficiency virus virulence. J Immunol 2007;179(5):3035–46.

[34] Mir KD, Gasper MA, Sundaravaradan V, Sodora DL. SIV infection in natural hosts: resolution of immune activation during the acute-to-chronic transition phase. Microbes Infect 2011;13(1):14–24.
[35] Nigam P, Velu V, Kannanganat S, Chennareddi L, Kwa S, Siddiqui M, et al. Expansion of FOXP3+ CD8 T cells with suppressive potential in colorectal mucosa following a pathogenic simian immunodeficiency virus infection correlates with diminished antiviral T cell response and viral control. J Immunol 2010;184(4):1690–701.
[36] Kornfeld C, Ploquin MJ, Pandrea I, Faye A, Onanga R, Apetrei C, et al. Antiinflammatory profiles during primary SIV infection in African green monkeys are associated with protection against AIDS. J Clin Invest 2005;115(4):1082–91.
[37] Sumpter B, Dunham R, Gordon S, Engram J, Hennessy M, Kinter A, et al. Correlates of preserved CD4(+) T cell homeostasis during natural, nonpathogenic simian immunodeficiency virus infection of sooty mangabeys: implications for AIDS pathogenesis. J Immunol 2007;178(3):1680–91.
[38] Milush JM, Mir KD, Sundaravaradan V, Gordon SN, Engram J, Cano CA, et al. Lack of clinical AIDS in SIV-infected sooty mangabeys with significant CD4+ T cell loss is associated with double-negative T cells. J Clin Invest 2011;121(3):1102–10.
[39] Milush JM, Reeves JD, Gordon SN, Zhou D, Muthukumar A, Kosub DA, et al. Virally induced CD4+ T cell depletion is not sufficient to induce AIDS in a natural host. J Immunol 2007;179(5):3047–56.
[40] Gordon SN, Dunham RM, Engram JC, Estes J, Wang Z, Klatt NR, et al. Short-lived infected cells support virus replication in sooty mangabeys naturally infected with simian immunodeficiency virus: implications for AIDS pathogenesis. J Virol 2008;82(7):3725–35.
[41] Taaffe J, Chahroudi A, Engram J, Sumpter B, Meeker T, Ratcliffe S, et al. A five-year longitudinal analysis of sooty mangabeys naturally infected with simian immunodeficiency virus reveals a slow but progressive decline in CD4+ T-cell count whose magnitude is not predicted by viral load or immune activation. J Virol 2010;84(11):5476–84.
[42] Chan ML, Petravic J, Ortiz AM, Engram J, Paiardini M, Cromer D, et al. Limited CD4+ T cell proliferation leads to preservation of CD4+ T cell counts in SIV-infected sooty mangabeys. Proc Biol Sci 2010;277(1701):3773–81.
[43] Pereira LE, Villinger F, Onlamoon N, Bryan P, Cardona A, Pattanapanysat K, et al. Simian immunodeficiency virus (SIV) infection influences the level and function of regulatory T cells in SIV-infected rhesus macaques but not SIV-infected sooty mangabeys. J Virol 2007;81(9):4445–56.
[44] Estes JD, Li Q, Reynolds MR, Wietgrefe S, Duan L, Schacker T, et al. Premature induction of an immunosuppressive regulatory T cell response during acute simian immunodeficiency virus infection. J Infect Dis 2006;193(5):703–12.
[45] Favre D, Lederer S, Kanwar B, Ma ZM, Proll S, Kasakow Z, et al. Critical loss of the balance between Th17 and T regulatory cell populations in pathogenic SIV infection. PLoS Pathog 2009;5(2):e1000295.
[46] Brenchley JM, Paiardini M, Knox KS, Asher AI, Cervasi B, Asher TE, et al. Differential Th17 CD4 T-cell depletion in pathogenic and nonpathogenic lentiviral infections. Blood 2008;112(7):2826–35.
[47] Cecchinato V, Trindade CJ, Laurence A, Heraud JM, Brenchley JM, Ferrari MG, et al. Altered balance between Th17 and Th1 cells at mucosal sites predicts AIDS progression in simian immunodeficiency virus-infected macaques. Mucosal Immunol 2008;1(4):279–88.
[48] Pandrea I, Apetrei C, Gordon S, Barbercheck J, Dufour J, Bohm R, et al. Paucity of CD4+CCR5+ T cells is a typical feature of natural SIV hosts. Blood 2007;109(3):1069–76.
[49] Paiardini M, Cervasi B, Reyes-Aviles E, Micci L, Ortiz AM, Chahroudi A, et al. Low levels of SIV infection in sooty mangabey central memory CD(4)(+) T cells are associated with limited CCR5 expression. Nat Med 2011;17(7):830–6.

[50] Reeves RK, Evans TI, Gillis J, Wong FE, Kang G, Li Q, et al. SIV infection induces accumulation of plasmacytoid dendritic cells in the gut mucosa. J Infect Dis 2012;206(9):1462–8.
[51] Rossjohn J, Pellicci DG, Patel O, Gapin L, Godfrey DI. Recognition of CD1d-restricted antigens by natural killer T cells. Nat Rev Immunol 2012;12(12):845–57.
[52] Godfrey DI, Kronenberg M. Going both ways: immune regulation via CD1d-dependent NKT cells. J Clin Invest 2004;114(10):1379–88.
[53] Coquet JM, Chakravarti S, Kyparissoudis K, McNab FW, Pitt LA, McKenzie BS, et al. Diverse cytokine production by NKT cell subsets and identification of an IL-17-producing CD4-NK1.1- NKT cell population. Proc Natl Acad Sci USA 2008;105(32):11287–92.
[54] Campillo-Gimenez L, Cumont MC, Fay M, Kared H, Monceaux V, Diop O, et al. AIDS progression is associated with the emergence of IL-17-producing cells early after simian immunodeficiency virus infection. J Immunol 2010;184(2):984–92.
[55] Motsinger A, Azimzadeh A, Stanic AK, Johnson RP, Van Kaer L, Joyce S, et al. Identification and simian immunodeficiency virus infection of CD1d-restricted macaque natural killer T cells. J Virol 2003;77(14):8153–8.
[56] Motsinger A, Haas DW, Stanic AK, Van Kaer L, Joyce S, Unutmaz D. CD1d-restricted human natural killer T cells are highly susceptible to human immunodeficiency virus 1 infection. J Exp Med 2002;195(7):869–79.
[57] Rout N, Else JG, Yue S, Connole M, Exley MA, Kaur A. Heterogeneity in phenotype and function of CD8+ and CD4/CD8 double-negative Natural Killer T cell subsets in sooty mangabeys. J Med Primatol 2010;39(4):224–34.
[58] Rout N, Else JG, Yue S, Connole M, Exley MA, Kaur A. Paucity of CD4+ natural killer T (NKT) lymphocytes in sooty mangabeys is associated with lack of NKT cell depletion after SIV infection. PLoS One 2010;5(3):e9787.
[59] Brenchley JM, Schacker TW, Ruff LE, Price DA, Taylor JH, Beilman GJ, et al. CD4+ T cell depletion during all stages of HIV disease occurs predominantly in the gastrointestinal tract. J Exp Med 2004;200(6):749–59.
[60] Dunham R, Pagliardini P, Gordon S, Sumpter B, Engram J, Moanna A, et al. The AIDS resistance of naturally SIV-infected sooty mangabeys is independent of cellular immunity to the virus. Blood 2006;108(1):209–17.
[61] Groh V, Steinle A, Bauer S, Spies T. Recognition of stress-induced MHC molecules by intestinal epithelial gammadelta T cells. Science 1998;279(5357):1737–40.
[62] Autran B, Triebel F, Katlama C, Rozenbaum W, Hercend T, Debre P. T cell receptor gamma/delta+ lymphocyte subsets during HIV infection. Clin Exp Immunol 1989;75(2):206–10.
[63] Kosub DA, Durudas A, Lehrman G, Milush JM, Cano CA, Jain MK, et al. Gamma/Delta T cell mRNA levels decrease at mucosal sites and increase at lymphoid sites following an oral SIV infection of macaques. Curr HIV Res 2008;6(6):520–30.
[64] Kosub DA, Lehrman G, Milush JM, Zhou D, Chacko E, Leone A, et al. Gamma/Delta T-cell functional responses differ after pathogenic human immunodeficiency virus and nonpathogenic simian immunodeficiency virus infections. J Virol 2008;82(3):1155–65.
[65] Sundaravaradan V, Saleem R, Micci L, Gasper MA, Ortiz AM, Else J, et al. Multifunctional double-negative T cells in sooty mangabeys Mediate T-helper functions Irrespective of SIV infection. PLoS Pathog 2013;9(6):e1003441.
[66] Antonelli LR, Dutra WO, Oliveira RR, Torres KC, Guimaraes LH, Bacellar O, et al. Disparate immunoregulatory potentials for double-negative (CD4- CD8-) alpha beta and gamma delta T cells from human patients with cutaneous leishmaniasis. Infect Immun 2006;74(11):6317–23.
[67] Ma Y, He KM, Garcia B, Min W, Jevnikar A, Zhang ZX. Adoptive transfer of double negative T regulatory cells induces B-cell death in vivo and alters rejection pattern of rat-to-mouse heart transplantation. Xenotransplantation 2008;15(1):56–63.

[68] Beaumier CM, Harris LD, Goldstein S, Klatt NR, Whitted S, McGinty J, et al. CD4 down-regulation by memory CD4+ T cells in vivo renders African green monkeys resistant to progressive SIVagm infection. Nat Med 2009;15(8):879–85.

[69] Vinton C, Klatt NR, Harris LD, Briant JA, Sanders-Beer BE, Herbert R, et al. CD4-like immunological function by CD4- T cells in multiple natural hosts of simian immunodeficiency virus. J Virol 2011;85(17):8702–8.

[70] Meythaler M, Martinot A, Wang Z, Pryputniewicz S, Kasheta M, Ling B, et al. Differential CD4+ T-lymphocyte apoptosis and bystander T-cell activation in rhesus macaques and sooty mangabeys during acute simian immunodeficiency virus infection. J Virol 2009;83(2):572–83.

[71] Liang Q, Jiao Y, Zhang T, Wang R, Li W, Zhang H, et al. Double Negative (DN) [CD3(+) CD4(-)CD8(-)] T cells correlate with disease progression during HIV infection. Immunol Invest 2013;42(5):431–7.

[72] Petitjean G, Chevalier MF, Tibaoui F, Didier C, Manea ME, Liovat AS, et al. Level of double negative T cells, which produce TGF-beta and IL-10, predicts CD8 T-cell activation in primary HIV-1 infection. AIDS 2012;26(2):139–48.

[73] Malleret B, Maneglier B, Karlsson I, Lebon P, Nascimbeni M, Perie L, et al. Primary infection with simian immunodeficiency virus: plasmacytoid dendritic cell homing to lymph nodes, type I interferon, and immune suppression. Blood 2008;112(12):4598–608.

[74] Kwa S, Kannanganat S, Nigam P, Siddiqui M, Shetty RD, Armstrong W, et al. Plasmacytoid dendritic cells are recruited to the colorectum and contribute to immune activation during pathogenic SIV infection in rhesus macaques. Blood 2011;118(10):2763–73.

[75] Diop OM, Ploquin MJ, Mortara L, Faye A, Jacquelin B, Kunkel D, et al. Plasmacytoid dendritic cell dynamics and alpha interferon production during Simian immunodeficiency virus infection with a nonpathogenic outcome. J Virol 2008;82(11):5145–52.

[76] Campillo-Gimenez L, Laforge M, Fay M, Brussel A, Cumont MC, Monceaux V, et al. Nonpathogenesis of simian immunodeficiency virus infection is associated with reduced inflammation and recruitment of plasmacytoid dendritic cells to lymph nodes, not to lack of an interferon type I response, during the acute phase. J Virol 2010;84(4):1838–46.

[77] Mandl JN, Barry AP, Vanderford TH, Kozyr N, Chavan R, Klucking S, et al. Divergent TLR7 and TLR9 signaling and type I interferon production distinguish pathogenic and nonpathogenic AIDS virus infections. Nat Med 2008;14(10):1077–87.

[78] Elbim C, Monceaux V, Mueller YM, Lewis MG, Francois S, Diop O, et al. Early divergence in neutrophil apoptosis between pathogenic and nonpathogenic simian immunodeficiency virus infections of nonhuman primates. J Immunol 2008;181(12):8613–23.

[79] Elbim C, Monceaux V, Francois S, Hurtrel B, Gougerot-Pocidalo MA, Estaquier J. Increased neutrophil apoptosis in chronically SIV-infected macaques. Retrovirology 2009;6:29.

[80] Grassi F, Hosmalin A, McIlroy D, Calvez V, Debre P, Autran B. Depletion in blood CD11c-positive dendritic cells from HIV-infected patients. AIDS 1999;13(7):759–66.

[81] Pacanowski J, Kahi S, Baillet M, Lebon P, Deveau C, Goujard C, et al. Reduced blood CD123+ (lymphoid) and CD11c+ (myeloid) dendritic cell numbers in primary HIV-1 infection. Blood 2001;98(10):3016–21.

[82] Brown KN, Trichel A, Barratt-Boyes SM. Parallel loss of myeloid and plasmacytoid dendritic cells from blood and lymphoid tissue in simian AIDS. J Immunol 2007;178(11):6958–67.

[83] Mortara L, Ploquin MJ, Faye A, Scott-Algara D, Vaslin B, Butor C, et al. Phenotype and function of myeloid dendritic cells derived from African green monkey blood monocytes. J Immunol Methods 2006;308(1–2):138–55.

[84] Smith PD, Ohura K, Masur H, Lane HC, Fauci AS, Wahl SM. Monocyte function in the acquired immune deficiency syndrome. Defective chemotaxis. J Clin Invest 1984;74(6):2121–8.

[85] Laforge M, Campillo-Gimenez L, Monceaux V, Cumont MC, Hurtrel B, Corbeil J, et al. HIV/SIV infection primes monocytes and dendritic cells for apoptosis. PLoS Pathog 2011;7(6):e1002087.

[86] Han J, Wang B, Han N, Zhao Y, Song C, Feng X, et al. CD14(high)CD16(+) rather than CD14(low)CD16(+) monocytes correlate with disease progression in chronic HIV-infected patients. J Acquir Immune Defic Syndr 2009;52(5):553–9.

[87] Ansari AW, Meyer-Olson D, Schmidt RE. Selective expansion of pro-inflammatory chemokine CCL2-loaded CD14+CD16+ monocytes subset in HIV-infected therapy naive individuals. J Clin Immunol 2013;33(1):302–6.

[88] Funderburg NT, Mayne E, Sieg SF, Asaad R, Jiang W, Kalinowska M, et al. Increased tissue factor expression on circulating monocytes in chronic HIV infection: relationship to in vivo coagulation and immune activation. Blood 2010;115(2):161–7.

[89] Estcourt C, Rousseau Y, Sadeghi HM, Thieblemont N, Carreno MP, Weiss L, et al. Flow-cytometric assessment of in vivo cytokine-producing monocytes in HIV-infected patients. Clin Immunol Immunopathol 1997;83(1):60–7.

[90] Esser R, Glienke W, von Briesen H, Rubsamen-Waigmann H, Andreesen R. Differential regulation of proinflammatory and hematopoietic cytokines in human macrophages after infection with human immunodeficiency virus. Blood 1996;88(9):3474–81.

[91] Mureith MW, Chang JJ, Lifson JD, Ndung'u T, Altfeld M. Exposure to HIV-1-encoded Toll-like receptor 8 ligands enhances monocyte response to microbial encoded Toll-like receptor 2/4 ligands. AIDS 2010;24(12):1841–8.

[92] Mir KD, Bosinger SE, Gasper M, Ho O, Else JG, Brenchley JM, et al. Simian immunodeficiency virus-induced alterations in monocyte production of tumor necrosis factor alpha contribute to reduced immune activation in sooty mangabeys. J Virol 2012;86(14):7605–15.

[93] Brenchley JM, Price DA, Schacker TW, Asher TE, Silvestri G, Rao S, et al. Microbial translocation is a cause of systemic immune activation in chronic HIV infection. Nat Med 2006;12(12):1365–71.

[94] Brabers NA, Nottet HS. Role of the pro-inflammatory cytokines TNF-alpha and IL-1beta in HIV-associated dementia. Eur J Clin Invest 2006;36(7):447–58.

CHAPTER

12

Distinct Cellular Targets of SIV Infection in Natural and Non-natural Hosts of SIV

Thomas H. Vanderford[1, 2], *Mirko Paiardini*[1, 2]

[1]Department of Pathology and Laboratory Medicine, Emory University, School of Medicine, Atlanta, GA, USA; [2]Division of Microbiology and Immunology, Yerkes National Primate Research Center, Emory University, Atlanta, GA, USA

INTRODUCTION

Non-human primate (NHP) models of AIDS have contributed significantly to our understanding not only of how HIV/Simian Immunodeficiency Virus (SIV)-specific immunity could be established but also of the pathogenic mechanisms responsible for lentivirus-induced progression to AIDS. While the study of pathogenic SIV infection of the prototype non-natural hosts of SIV that include Asian macaques (rhesus, pig-tail, cynomolgus, etc.), both vaccinated and unvaccinated, has provided invaluable insight into the immunopathological effects of lentiviral infection that have confounded our efforts to develop an effective vaccine [1], investigation of the natural hosts of SIV has likewise been indispensable for our understanding of precisely why lentivirus infection causes AIDS in non-natural hosts and what immunological pathways should be targeted for future therapies [2].

Currently, over 40 species of African non-human primate have been shown to harbor simian immunodeficiency viruses that, despite their similarity to HIV, do not typically lead to immunodeficiency in their natural host species [3–5]. This non-pathogenic nature is not a fundamental property of the virus that infects these NHPs, but is likely the result of an evolutionary arms race between virus and host fought over millennia

Natural Hosts of SIV.
http://dx.doi.org/10.1016/B978-0-12-404734-1.00012-7

that has resulted in a host that avoids immunodeficiency despite a lifelong infection characterized by moderate to high viral loads and maintenance of normal immune function [5–10]. However, upon transmission of these viruses to an Asian NHP species (which presumably has never come into contact with these viruses), SIV typically results in a progressive immunodeficiency similar to what is observed in HIV-infected humans [11,12]. In fact, the virus isolates (SIVmac and its derivatives) most often used as a challenge virus to test the efficacy of candidate pre-clinical lentiviral vaccines and to study the pathogenic mechanisms responsible for AIDS were derived directly from the virus that naturally infects the sooty mangabey (SIVsmm), a natural NHP host that exhibits a non-pathogenic relationship with its strain of SIV [3,11,13].

Sooty mangabeys (SM), along with another natural host species, the African green monkey (AGM), are among the best studied non-pathogenic models of SIV infection [2]. Early observations of both natural and experimental SIV infection of these animals showed that they exhibit much lower levels of immune activation than that observed in pathogenic SIV infection of NHP and HIV infection of humans as measured by the proliferation marker, Ki-67, as well as more classical activation markers such as Human Leukocyte Antigen (HLA)-DR [6,8,10,12,14,15]. It is thought that the low levels of generalized immune activation in these animals helps prevent the premature decline of immune memory [10,16,17], the destruction of lymphoid tissue architecture [10], and damage to non-lymphoid, bystander tissues [18–22]. Some evidence suggests that the natural hosts have evolved mechanisms that directly "turn down" immune responses against the virus [23,24]. The mechanisms behind this low level of immune activation are still being explored.

While an immunoregulatory mechanism likely contributes to avoiding the progressive decline of immune function associated with lentiviral infection, emerging evidence suggests that the natural host species may also avoid long-term permanent immune damage by limiting virus replication to largely expendable subsets of CD4 T cells, thus allowing the preservation of more important, long-lived immune cell populations [25]. This target cell restriction model of HIV/SIV pathogenesis is the focus of some of the most promising research in the natural host species and will be discussed at length in this chapter.

CD4 T-CELL SUBSETS

T-Cell Memory Differentiation

The mammalian immune system has not only evolved an ability to generate enormous repertoires of receptor specificity with the potential to recognize a multitude of antigenic patterns encoded by pathogens, but it has also acquired the capability of storing this information for quick and easy recall in the case of a second encounter with the same pathogen. Both T and B lymphocytes maintain pools of long-lived memory cells that recognize many of the invading pathogens already encountered during the lifetime of an organism [26,27]. Herein, we will focus on the development of CD4+ T lymphocytes as the CD4+ T lymphocyte subset are the primary cell infected by HIV/SIV.

Upon egress from the thymus, appropriately licensed naïve CD4+ T cells (T_N) circulate throughout the periphery and move through secondary lymphoid tissues awaiting contact with their cognate antigens in the context of Major Histocompatibility Complex (MHC)-II on an antigen-presenting cell [28] (Figure 12.1(A)). Recognition of this peptide-MHC-II complex combined with signaling through costimulatory and cytokine receptors on its surface causes the CD4+ T-cell to begin dividing, produce cytokines (and in some cases cytotoxic granules) that mediate the primary function of that particular cell type, and upregulate Fas-ligand (CD95) [29]. The precise nature of the costimulatory and cytokine signals determines the functional fate of the now antigen-experienced $CD4^+$ T-cell [30]. Regardless of this fate, the dividing lineage of these cells differentiates into effector (T_{EFF}) and memory cell subsets [27].

During the early phase of an immune response, the pathogen-specific CD4+ T-cell pool is dominated by a subset of effector cells that are short-lived producers of cytokines and other effector molecules that activate mononuclear phagocytes and provide help to both B cells and CD8 T cells among other functions [31]. As the infection is cleared, most of these short-lived effector cells die off, leaving behind a heterogeneous T-cell memory pool capable of responding to a second infection by the same pathogen [32]. These memory cell populations are primarily split into three lineages based on functional and phenotypic markers and tissue distribution: effector memory (T_{EM}), central memory (T_{CM}), and stem-cell memory (T_{SCM}) [27,33,34].

Effector memory T cells primarily reside in peripheral tissues (Figure 12.1(B)) and are defined by their lack of expression of CCR7 and CD62L (lymphoid tissue homing markers) [35]. At steady state these cells express high levels of chemokine receptors (most notably CCR5) and other homing molecules that facilitate migration to peripheral, non-lymphoid mucosal sites where they represent the first line of defense against reinfection [34,35]. Upon restimulation at the site of an infection or inflammation, these cells proliferate and produce high levels of effector molecules such as IL-2, Interferon (IFN)-g, IL-4, and IL-5 [36]. However, they are not multipotent, as antigen stimulation results only in the production of more effector and effector memory T cells [33]. Compared to other memory populations, they are short-lived and must constantly be replenished in

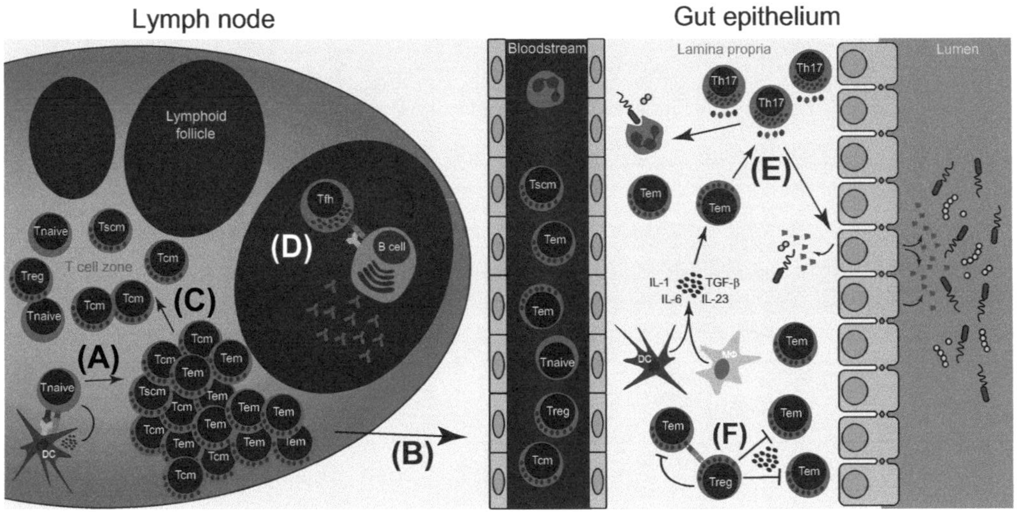

FIGURE 12.1 **CD4+ T-cell development and differentiation in lymphoid and peripheral tissues.** (A) Naïve CD4+ T cells become activated by encountering cognate antigen presented on class II MHC by professional antigen-presenting cells. (B) The antigen-experienced cell divides and differentiates mostly into short-lived effector and effector memory CD4+ T cells, which migrate to sites of inflammation. (C) A small subset of these cells differentiate into central memory and stem cell memory cells providing long-term immunological memory. (D) T follicular helper cells facilitate the maturation and survival of antigen-specific B cells. (E) Some CD4+ T cells in the gut mucosa differentiate into Th17 cells, which provide signals that help maintain gut mucosal integrity and recruit phagocytic cells to fight infection. (F) T regulatory cells help keep inflammation in check by providing anti-inflammatory signals. (See color plate at the back of the book.)

peripheral tissues by homeostatic proliferation of effector memory cells in the periphery and central memory cells in lymphoid tissues [37].

Central memory T cells are a long-lived subset of memory cells that reside in secondary lymphoid tissues (Figure 12.1(C)) and are defined by the expression of CCR7 and/or CD62L in conjunction with CD95 [34–36]. These memory cells express high levels of the IL-7 receptor alpha chain (CD127) and are the largest pool of self-renewing antigen-specific T-cell memory [34]. Upon restimulation with cognate antigen, these cells divide very rapidly and have the potential to differentiate into effector memory T cells, which then migrate to sites of inflammation [33]. The maintenance of a functional central memory cell population is critical for long-term immunity.

The third memory subset, stem-cell memory T cells, is extremely long-lived, is self-renewing, and provides a continual reservoir for the maintenance of T-cell memory [33]. Despite serving a similar function to central memory T cells, these cells exhibit a more naïve-like cell surface marker phenotype with high levels of CD45RA, CD27, CD28, CD122, and CD127 in addition to the central memory markers CCR7 and CD95 [38]. Like naïve T cells, they have high proliferative capacity and are extremely multipotent, as they are the only memory T-cell capable of producing all memory cell subsets following T-cell receptor stimulation [33]. However, unlike naïve T cells, they express high levels of CD95, marking them as antigen-experienced [33]. In non-human primates these cells reside primarily in lymph nodes and other secondary lymphoid tissues, and are absent almost entirely from mucosal surfaces [33]. Thus these cells perform a similar function to central memory cells but are less differentiated and less numerous (only 2–3% of circulating CD4 or CD8 T cells are stem-cell memory cells) [33].

T-Cell Functional Differentiation

The primary function of antigen-experienced $CD4^+$ T cells is to provide help in the form of cytokines and chemokines for other immune cell populations. Traditionally, only two types of CD4+ T cells were identified, were labeled T helper type 1 (Th1) and T helper type 2 (Th2), and were defined primarily by the spectrum of cytokine production [39,40]. Th1 cells produce mostly IFN-g and IL-2, and predominantly induced cell-mediated immunity by providing help to macrophages and CD8+ T cells [39,40]. Th2 cells, on the other hand, are primarily responsible for inducing antibody-mediated immunity by providing help to B cells through the production of IL-4, IL-5, IL-10, and a number of other cytokines [31]. In more recent years, however, a range of highly specialized and highly differentiated new helper CD4+ T-cell subsets have been identified.

The first CD4+ T-cell functional subset with a potential impact on HIV/SIV infection is the T follicular helper cell (Tfh; Figure 12.1(D)). Tfh cells migrate to the germinal centers of secondary lymphoid tissues (SLT) via the expression of CXCR5 coupled with a loss of CCR7 expression. Tfh cells within the SLT facilitate the maturation and survival of antigen-specific B cells during the germinal center reaction, likely through the expression of CD40L and the secretion of IL-21 and IL-10 [41]. These signals provided by Tfh are also required to promote isotype switching and somatic hypermutation in the antigen-binding pockets of antibodies encoded by these B cells [41]. High expression of the transcription factor BCL-6 and c-Maf [42] and stimulation via IL-6 are also required to induce and maintain their function [43]. A recent study that characterized Tfh cells in lymph nodes of NHP defined them immunophenotypically as $CCR7^{lo}$, $CXCR5^{hi}$, $PD\text{-}1^{hi}$, and $ICOS^{hi}$ [44].

Th17 cells not only are important components of antibacterial immunity but are also implicated in the maintenance of mucosal integrity, especially in the gut (Figure 12.1(E)). The differentiation of Th17 cells requires a complex combination of cytokines, with IL-6, transforming growth factor-β (TGF-β), IL-1β and IL-21 being identified as the more critical ones, and with IL-23 subsequently serving as an amplifying and survival factor for committed Th17 cells [45,46]. These cells are characterized by the production of IL-17 and IL-22, cytokines that are responsible for promoting epithelial regeneration by stimulating the production of tight junction proteins, inducing the production of antibacterial defensins, and in the recruitment of neutrophils to sites of bacterial infection [47,48]. Although dysregulation of these cells has been implicated in a number of autoimmune diseases, they are primarily beneficial components of a healthy intestinal mucosal barrier [46].

Finally, CD4+ T regulatory cells (Tregs) are a T-cell subset that is critical for suppressing autoimmunity by down-modulating the effector activity of multiple other immune cell types [49] (Figure 12.1(F)). The best studied and most abundant Treg population is identified by the expression of CD4 and CD25, and the transcription factor Forkhead Box P3 (FoxP3). Tregs also exhibit intermediate to low levels of CD127 [50]. It is thought that they mediate their anti-inflammatory function through the secretion of IL-10 and TGF-β, or by their ability to ligate CTLA-4 [49,51,52]. The function and importance of these cells have mostly been demonstrated through observations of autoimmune pathology associated with their absence, and through in vitro experiments that highlight their ability to suppress effector T-cell responses [53].

CELLULAR TARGETS OF SIV INFECTION

Recent work in NHP models of AIDS has demonstrated several differences in how SIV infection impacts the balance of the different CD4+ T-cell subsets in natural versus non-natural hosts. While the main cell type

TABLE 12.1 CD4+ T-Cell Memory and Functional Subset

Cell Type	Phenotypic Markers	Primary Anatomic Location	Infection Status	
			Natural Hosts	Non-natural Hosts
MEMORY SUBSETS				
Central memory T cells (T_{CM})	CCR7+ CD62L+ CD95+	Secondary lymphoid tissues	+	+++
Effector memory T cells (T_{EM})	CCR7− CD62L− CD95+	Mucosal and peripheral tissues	++	++
Stem cell memory cells (T_{SCM})	CD45RA+ CD95+ CCR7+ CD27+ CD28+ CD127+	Secondary lymphoid tissues	No data	No data
FUNCTIONAL SUBSETS				
T follicular helper cells (Tfh)	$CXCR5^{hi}$, $PD\text{-}1^{hi}$, $ICOS^{hi}$, $CCR7^{lo}$	Germinal center follicles	+	++
T helper 17 cells (Th17)	IL-17 and IL-22 production, no IFN-γ production	Gut mucosa	No data	++
T regulatory cells (Treg)	$CD25^{hi}$, $CD127^{lo}$, FoxP3+	Mucosal, peripheral, and secondary lymphoid tissues	No data	+[a]

[a] *Data from humans only.*

infected by SIV in both pathogenic and non-pathogenic infection is a population of short-lived, activated CD4+ T cells, natural hosts have developed several mechanisms to protect other, more critical CD4+ T-cell populations (see Table 12.1) that promote the long-term survival of the host [25].

SIV Receptor Modulation

SIV targets and infects CD4+ T cells and macrophages of NHP similar to the targeting of human CD4+ T cells by HIV-1. Thus, SIV binds to CD4 and the primary coreceptor CCR5 [54], which interestingly exhibits higher levels of expression on more differentiated subsets of these cell populations [17]. Despite similar high viral loads [12], and similar distribution of memory subsets (i.e., $T_{EFF}/T_{EM} > T_{CM} > T_{SCM} > T_N$) [15,17], CD4+ T cells in natural hosts exhibit much lower frequencies of CCR5 expression when compared to non-natural hosts [55]. In this setting, it is possible that the relative low level of CCR5 expression may still be sufficient to support viral entry and even replication in many CD4+ T-cell subsets, but major viral replication is likely supported by the CD4+ T-cell population with the highest CCR5 expression. It is also possible that SIVs can use other than CCR5 as alternative coreceptors to infect cells. This latter possibility appears to be quite likely in SMs where a rare (~8%) subpopulation of animals homozygous for a CCR5-null mutant has been observed to support levels of viral replication that are only 0.5 log lower than in CCR5-wt SMs [56]. In these animals, the CCR5 gene contains a nonsense deletion that causes improper folding and subsequent exclusion of the CCR5 molecule from the cell surface. Follow-up studies have demonstrated that the SIVs that infect these animals are capable of using several alternative coreceptors (most notably CXCR6, GPR15, GPR1, and APJ), at least in vitro [57]. Further characterization of the cell-type distribution of expression of these receptors will help to better elucidate the nature of the CD4+ T cells supporting the replication of SIV in SMs.

Interestingly, in addition to exhibiting low frequencies of CCR5 expression, AGMs have developed a novel mechanism of active CD4 down-modulation to prevent SIV infection [58]. Upon T-cell receptor stimulation, CD4+ T cells entering the memory and effector pools down-regulate the expression of CD4 on their cell surface. Although these cells lack CD4, they still exhibit MHC class II restriction and maintain typical CD4+ T-cell memory function, such as IL-2 production, CD40L expression, IL-17 production (Th17 cells), and FoxP3 expression (Tregs). Most important, though, the lack of CD4 on their cell surface protects these cells from SIVagm infection in vivo. Thus, AGMs have evolved a mechanism that protects a portion of the memory CD4+ T-cell population from infection that not only preserves the recall capabilities of the CD4+ T-cell memory pool but also prevents the depletion by infection-induced cell death of several very important functional CD4+ T-cell subsets (i.e., Th17 and Tregs). A similar phenomenon has also been observed in a few SMs that experience severe loss of CD4+ T-cell during SIV infection [59]. In these animals, an expanded CD4(−)CD8(−) lymphocyte population performs the typical functions of CD4+ T cells and partially compensates for their loss.

Preservation of Central Memory CD4+ T-Cell Compartment

The importance of preserving CD4+ T_{CM} cell homeostasis is clearly highlighted by in vivo studies in SIV-infected natural and non-natural hosts. Indeed, two recent studies have shown that the infection frequencies of CD4+ T_{CM} cells of SMs are low as compared to CD4+ T_{CM} of RMs in both blood and lymph nodes [17,60,61]. Consistent with this relative preservation from infection, CD4+ T_{CM} cells from SIV-infected SMs are significantly more stable and proliferate to a lower extent than in SIV-infected RMs (Paiardini M, unpublished). Furthermore, in SIV-infected RMs, progressive depletion of CD4+ T_{CM} is the key factor dictating the tempo of progression to AIDS, both in the natural history of infection and in the context of vaccination [62–68]. Of note, as in SIV-infected RMs, in the majority of HIV-infected humans, a high frequency of the circulating CD4+ T_{CM} cells are infected with HIV-1 both in vivo and in vitro and represent the largest reservoir of infected CD4+ T cells [69].

Altogether, these findings highlight the important contribution of CD4+ T_{CM} loss to AIDS progression. Several mechanisms have been proposed to explain why infection and depletion of CD4+ T_{CM} cells may contribute to the establishment of chronic immune activation and disease progression more significantly than does infection of the more expendable CD4+ T_{EM} cells [2,25]. First, CD4+ T_{CM} cells are critical for the maintenance of CD4+ T-cell homeostasis [63]. Their depletion results in defective CD4+ T-cell homeostasis coupled with increased T-cell homeostatic proliferation. Second, CD4+ T_{CM} cells are concentrated in secondary lymphoid tissues like the lymph node. Their higher levels of infection translate into higher viral replication in the main anatomic sites in which immune responses are initiated, thus chronically sustaining virus-induced immune activation in pathogenic HIV/SIV infections [61]. It is likely that high levels of infection of CD4+ T_{CM} cells and increased levels of immune activation support each other in a vicious cycle whose establishment is central for the immune deficiency that follows HIV/SIV infections of humans and RMs. On the one hand, high levels of infected CD4+ T_{CM} may promote immune activation by increasing virus replication. On the other hand, high levels of immune activation may favor the loss of CD4+ T_{CM} cells by increasing direct virus infection, bystander cell death of uninfected CD4+ T_{CM} cells, and subsequent collagen deposition and fibrosis of lymphoid tissue, effectively destroying the anatomic niche of these cells [70].

Based on these findings, it is likely that reduced infection of CD4+ T_{CM} resulted from a mutually beneficial coadaptation between SIV and their African natural hosts that took place over the many thousands of years during which these viruses have infected these species of monkeys. Remarkably, recent data suggest that protection of the CD4+ T_{CM} compartment similar to that seen in SMs is also present in a few rare subsets of HIV-infected humans. Specifically, low levels of CD4+ T_{CM} infection have been described in (i) long-term nonprogressors with protective HLA alleles [71]; (ii) early-treated patients that show a prolonged control of viremia and preserve CD4+ T-cell after antiretroviral treatment (ART)-interruption [72]; and (iii) viremic nonprogressors (VNPs), a rare subset of HIV-infected patients that preserve CD4+ T-cell counts despite high levels of viral load, thus resembling the SM phenotype (Klatt NR, personal communication). In these patient sub-populations, preservation of the CD4+ T_{CM} pool is associated with preserved immune function, relatively low levels of chronic immune activation, and long-term, AIDS-free survival.

T Follicular Helper Cells

Several recent studies have investigated the levels, function, and susceptibility to infection of Tfh cells in HIV-infected humans and SIV-infected macaques [44,61,73–75]. In all of these studies, the frequency of Tfh cells within secondary lymphoid tissues is preserved or even increased in viremic HIV/SIV-infected subjects. This expansion corresponds with an increase in the number of activated germinal center B cells and plasma cells [44,73]. Interestingly, Tfh cells were infected at frequencies similar to or higher than that of non-Tfh CD4+ T cells, suggesting that an expansion of Tfh cells does not result from reduced susceptibility to HIV/SIV infection but is likely mediated by activation-induced proliferation and trafficking [44,61,75]. The impact of HIV/SIV infection on Tfh cell function is less well understood. In one study, SIV infection altered the gene expression profile of Tfh cells, and decreased their survival, cycling, and trafficking, while another study showed that Tfh cells remain capable of stimulating antibody production from B cells during SIV infection [44,75].

Characterization of Tfh cells in the natural hosts is still limited; however, one study has shown a differential level of infection of Tfh cells in humans, RMs, and SMs [61]. In this study, HIV/SIV-infected cells were observed at high levels in the B cell follicles of humans and RMs whereas SIV-infected cells were only noted in the extrafollicular T-cell zone in lymph nodes of SMs. Furthermore, when CD4+ naïve, central memory, effector memory, and follicular helper T cells were sorted from LNs of SIV-infected RMs and SMs, Tfh cells from RMs exhibited significantly higher frequencies of SIV DNA as both a percentage of the individual T-cell subset populations and as a percentage of the total number of cells sorted in these experiments. Despite these differential levels of infection of Tfh cells between pathogenic and non-pathogenic infections, it is still unclear what role the low level of Tfh cell infection might play in promoting AIDS avoidance in the natural hosts as SMs produce low levels, if any, of anti-SIV neutralizing antibodies during infection [76].

Preservation of Th17 Cells and Mucosal Barrier Integrity

In both humans and RMs, the HIV/SIV-induced depletion of intestinal CD4+ T cells preferentially involves Th17 cells, a recently described lineage of helper CD4+ T cells that are characterized by the production of IL-17 and IL-22 [18,77–85]. IL-17 and IL-22 function in vivo to stimulate epithelial cells to express cytokines (i.e., IL-6 and granulocyte-macrophage colony-stimulating factor [GM-CSF]), chemokines (i.e., IL-8, CXCL1, CXCL10, and CCL20), and metallo-proteinases involved in the recruitment, activation, and migration of neutrophils to areas of bacterial infection [86]. Indeed, when not properly regulated, the proinflammatory activity of Th17 cells may result in tissue damage, and a pathogenic role of Th17 cells has been suggested in several conditions of autoimmunity [46]. Moreover, IL-17 and IL-22 stimulate epithelial regeneration, production of tight junction proteins, and anti-bacterial defensins [47,48]. Therefore, Th17 cells are thought to be a crucial CD4+ T-cell subset for mucosal barrier integrity and anti-bacterial immunity, and their specific loss may render chronically HIV/SIV-infected subjects less capable of maintaining the physical and immunological integrity of the mucosal barrier.

While the mechanisms underlying the preferential loss of Th17 cells during pathogenic HIV/SIV infection remain unclear, the importance of this loss is highlighted by several pieces of evidence. First, and in contrast to the severe and generalized depletion found in SIV-infected RMs and PTMs, in SIV-infected SMs and AGMs the frequencies of functional Th17 cells were maintained at normal levels in all of the tissue sites studied to date [18,20,83]. Thus, SIV-infected natural hosts are characterized by their maintenance of intestinal Th17 associated with preservation of mucosal integrity, limited microbial translocation, and absence of chronic immune activation. HIV elite controllers and long-term nonprogressors also appear to maintain a significant pool of Th17 cells in the gut [87–89]. Second, the severity of Th17 cell depletion has been shown to correlate with microbial translocation, chronic immune activation, and disease progression in HIV/SIV-infected subjects [18,83,85,90,91]. Preservation of Th17 cells is also indicative of effective CD4+ T-cell restoration in gut-associated lymphoid tissue of HIV-infected patients on ART [92]. Third, a recent study showed that the size of the Th17 cell compartment prior to infection limited viral replication in SIV-infected RMs [93]. Finally, in a recent study we found that the in vivo administration of interleukin-21 in SIV-infected RMs resulted in the maintenance of intestinal Th17 cells during the acute/early chronic phase of infection. Remarkably, these higher levels of intestinal Th17 cells were associated with reduced levels of intestinal T-cell activation, microbial translocation, and systemic inflammation during chronic infection [94], providing direct in vivo confirmation of a molecular link between loss of Th17 cells, mucosal integrity, and chronic activation/inflammation in pathogenic SIV infection.

Similar to what has been described for Th17 cells, the frequencies of CD8+ T cells and innate lymphoid cells (ILC) that produce IL-17 and/or IL-22 are significantly lower within the gastrointestinal (GI) tracts of chronically SIV-infected macaques [47,48]. The fact that neither CD8+ T cells nor ILCs express CD4 on their cell surface suggests that the general loss of IL-17 and/or IL-22 producing cells within the GI tracts of SIV-infected macaques is likely due to complex mechanisms and not simply related to preferential infection by HIV/SIV [95].

Given all these findings, it is clear that the preservation of Th17 cells in the gut mucosa is an important factor contributing to the non-pathogenic outcome of SIV infection in the natural hosts. Although the fact that Th17 cells are maintained at a healthy frequency in natural hosts but preferentially depleted in non-natural hosts for lentiviral infections is well established [18], the mechanisms responsible for this differential regulation are not completely understood. Potential mechanisms include (i) Th17 cells being more resistant to HIV/SIV infection and death in natural than non-natural hosts; (ii) natural hosts being able to preserve Th17 differentiating cytokines at levels superior to those found in non-natural hosts, thus resulting in a more effective renewal and/or survival of Th17 cells [91]; and (iii) a more efficient homing of Th17 cell to mucosal tissues in natural hosts of SIV when compared to non-natural hosts.

Treg Controversy

The role of regulatory CD4+ T cells during pathogenic HIV/SIV-infection is still unclear. A number of factors including the use of different phenotypic definitions (i.e., different markers used for their quantification) and different measurements (frequencies versus absolute numbers) are responsible for this confusion. Additionally, some studies have shown the maintenance or accumulation of Tregs in the gut and lymph nodes, while others have not [96–98]. However, a consensus is emerging that the relative frequency of Tregs is increased during progressive HIV-1 infection, while their absolute counts are reduced as a consequence of lower total CD4+ T-cell counts [99]. Furthermore, the ratio of Th17 to Treg cells is decreased in both peripheral blood and GI tracts of HIV-infected individuals, and this ratio is correlated with increased plasma markers of microbial translocation and higher levels of activated CD8+ T cells [20].

Nevertheless, it is still unclear what functional role Tregs play in HIV/SIV infection, and how they might affect antiviral immune responses and chronic inflammation. On the one hand, Tregs may suppress HIV-specific immune responses [100–102]. In this scenario, their function would contribute to the failure of immune control. Interventions aimed at decreasing Treg numbers or their function might help bolster antiviral immune responses, resulting in lower levels of viral replication. On the other hand, Tregs could be beneficial by helping reduce the proinflammatory environment typical of pathogenic HIV/SIV infection and thus, interventions aimed at increasing the numbers and function of Tregs could consequently limit the deleterious effects of chronic immune activation and slow the pace of AIDS progression [103–106]. This latest hypothesis is supported by studies showing that Treg depletion is associated with both CD4+ and CD8+ T-cell activation in HIV-infected individuals, and that the in vitro suppressive activity of blood and lymph node derived Tregs correlates with high CD4+ T-cell counts and low levels of T-cell activation in SIV-infected cynomolgus macaques [104,105]. Interestingly, Tregs have been shown to be infected in vivo and in vitro in humans; however, their infection status in NHP has yet to be determined [107].

Consistent with a model in which Tregs function to positively impact HIV/SIV disease progression, a study comparing Tregs between pathogenic SIV infection of macaques and non-pathogenic SIV infection of AGMs suggests that these cells are critical to the establishment of an anti-inflammatory environment in AGMs shortly after infection, thus contributing to the low levels of immune activation and lack of disease progression observed in this species [108]. A similar study comparing Tregs in SMs and rhesus macaques (RMs) was unable to discern a positive impact of Tregs on limiting AIDS progression in SIV-infected SMs, although this study demonstrated an enhanced ability of SM Tregs to suppress allogeneic cell proliferation by thymidine uptake in a mixed leukocyte reaction compared to similar studies using RM Tregs [109].

MACROPHAGES AS TARGETS OF SIV INFECTION

Although CD4+ T cells, particularly those with a memory-activated phenotype, provide the most support for HIV and SIV replication in humans and NHP [17,110–112], macrophages have also been identified as a significant target for lentiviral infection [113–115]. Tissue macrophages represent a distinct subset of infected cells during infection due to their ability to survive for long periods of time postinfection, at least in vitro [116]. They are considered a major component of the stable, long-term viral reservoir due to this longevity and relative resistance in vivo to cytopathic effects of the virus and are considered the primary contributors to the maintenance of HIV viral burden in end-stage AIDS patients when the number of CD4+ T cells is reduced to very low levels [117–120].

Two studies demonstrate the importance of macrophages in supporting HIV/SIV infection in the context of low levels of CD4+ T cells. In the first study [119], RMs were infected with the highly pathogenic SHIV-DH12. This exclusively CD4+ T-cell tropic virus induced a rapid, severe, and irreversible depletion of CD4+ T cells within 3–5 weeks of infection. However, the animals survived for an additional 4–6 months with high viral loads. Examination of multiple tissues (lymph nodes, lung, brain, and liver) revealed that the sustained viral load was being primarily supported by SIV-infected macrophages. In the second study [115], RMs were experimentally depleted of CD4+ T cells via administration of an in vivo depleting anti-CD4 monoclonal antibody prior to infection. Contrary to expectation, the CD4-depleted animals exhibited higher viral loads than non-depleted animals during the post-acute phase. Interestingly, CD4 depletion induced the emergence of CD4-independent envelopes and faster disease progression. Examination of gut tissue by immunohistochemistry and in situ hybridization revealed that macrophages were likely the primary source of the high viral burden in the absence of CD4+ T cells.

Although macrophages are clear contributors to viral burden and persistence in pathogenic HIV/SIV infection, their impact on significant virus replication in natural host models of SIV infection is less well understood. Unlike in pathogenic infection of non-natural hosts, experimental depletion of CD4+ T cells in SMs resulted in a decrease in viral load [16]. Indeed, through the course of this study, plasma viral load most closely correlated with the levels of activated Ki67+ CD4+ T cells and not with the levels of blood monocytes, which remained stable during CD4 depletion. However, viruses derived from the few SMs that experienced profound CD4 loss during SIV infection exhibit CXCR4 and dual-tropic phenotypes suggesting they may be capable of replicating in macrophages as well as resting or naïve CD4+ T cells [59]. Of note, SIVs of the "smm" lineage contain an open reading frame encoding the accessory protein Vpx, which has recently been shown to antagonize SAMHD1, a viral restriction factor expressed in myeloid lineage cells [121–123]. Binding of Vpx (and the Vpr protein encoded by SIVagm [124]) to this restriction factor blocks SAMHD1's ability to deplete cellular stores of dNTPs, allowing viruses that enter that cell to complete reverse transcription, integration into the host cell genome, and the subsequent generation of progeny viruses [125]. Thus, it is theoretically possible for SIVsmm and SIVagm infecting their natural hosts to replicate in macrophages, although further study will be required to define their role in SIV persistence and transmission, and their contribution to the mechanisms of disease resistance in these natural host species.

CONCLUSION

A large body of evidence generated in the last few years strongly supports a working model of HIV pathogenesis in which changes in the balance of the different CD4+ T-cell subsets is one of the most critical factors in causing progression to AIDS. According to this model, the quality of the cells that get infected and killed by the virus is more important for inducing immunodeficiency than quantity of infected cells or plasma virus. In other words, the primate immune system can afford to lose a significant portion of its CD4+ T cells to lentiviral infection provided that those cells are not critical for maintaining long-term immune memory and immunocompetence. Furthermore, when infection is limited to these short-lived CD4+ T-cell populations, it is much more difficult for the virus to establish a stable, long-term reservoir. Thus, it is conceivable that HIV/SIV could persist longer during anti-retroviral therapy when integrated in long-lived and/or self-renewing CD4+ T cells, such as T_{SCM} and T_{CM}, as compared to short-lived, non-self-renewing cells. In conclusion, these findings suggest that natural hosts have evolved mechanisms to protect essential, long-lived memory CD4+ T-cell populations from SIV-induced depletion and/or loss of function. Better preservation of these cells may be critical for the maintenance of the overall CD4+ T-cell homeostasis, the lack of chronic immune activation, and the maintenance of normal immune competence.

References

[1] Lifson JD, Haigwood NL. Lessons in nonhuman primate models for AIDS vaccine research: from minefields to milestones. Cold Spring Harbor Perspect Med 2012;2(6): a007310.

[2] Chahroudi A, Bosinger SE, Vanderford TH, Paiardini M, Silvestri G. Natural SIV hosts: showing AIDS the door. Science 2012;335(6073):1188–93.

[3] Hahn BH, Shaw GM, De Cock KM, Sharp PM. AIDS as a zoonosis: scientific and public health implications. Science 2000;287(5453):607–14.

[4] VandeWoude S, Apetrei C. Going wild: lessons from naturally occurring T-lymphotropic lentiviruses. Clin Microbiol Rev 2006;19(4):728–62.

[5] Worobey M, Telfer P, Souquiere S, Hunter M, Coleman CA, Metzger MJ, et al. Island biogeography reveals the deep history of SIV. Science 2010;329(5998):1487.

[6] Apetrei C, Sumpter B, Souquiere S, Chahroudi A, Makuwa M, Reed P, et al. Immunovirological analyses of chronically simian immunodeficiency virus SIVmnd-1- and SIVmnd-2-infected mandrills (*Mandrillus sphinx*). J Virol 2011;85(24):13077–87.

[7] Goldstein S, Brown CR, Ourmanov I, Pandrea I, Buckler-White A, Erb C, et al. Comparison of simian immunodeficiency virus SIVagmVer replication and CD4+ T-cell dynamics in vervet and sabaeus African green monkeys. J Virol 2006;80(10): 4868–77.

[8] Pandrea I, Silvestri G, Onanga R, Veazey RS, Marx PA, Hirsch V, et al. Simian immunodeficiency viruses replication dynamics in African non-human primate hosts: common patterns and species-specific differences. J Med Primatol 2006;35(4–5):194–201.

[9] Rey-Cuille MA, Berthier JL, Bomsel-Demontoy MC, Chaduc Y, Montagnier L, Hovanessian AG, et al. Simian immunodeficiency virus replicates to high levels in sooty mangabeys without inducing disease. J Virol 1998;72(5):3872–86.

[10] Silvestri G, Sodora DL, Koup RA, Paiardini M, O'Neil SP, McClure HM, et al. Nonpathogenic SIV infection of sooty mangabeys is characterized by limited bystander immunopathology despite chronic high-level viremia. Immunity 2003;18(3):441–52.

[11] Letvin NL, Daniel MD, Sehgal PK, Desrosiers RC, Hunt RD, Waldron LM, et al. Induction of AIDS-like disease in macaque monkeys with T-cell tropic retrovirus STLV-III. Science 1985;230(4721):71–3.

[12] Silvestri G, Fedanov A, Germon S, Kozyr N, Kaiser WJ, Garber DA, et al. Divergent host responses during primary simian immunodeficiency virus SIVsm infection of natural sooty mangabey and nonnatural rhesus macaque hosts. J Virol 2005;79(7): 4043–54.

[13] Apetrei C, Lerche NW, Pandrea I, Gormus B, Silvestri G, Kaur A, et al. Kuru experiments triggered the emergence of pathogenic SIVmac. AIDS 2006;20(3):317–21.

[14] Chakrabarti LA, Lewin SR, Zhang L, Gettie A, Luckay A, Martin LN, et al. Normal T-cell turnover in sooty mangabeys harboring active simian immunodeficiency virus infection. J Virol 2000;74(3):1209–23.

[15] Sumpter B, Dunham R, Gordon S, Engram J, Hennessy M, Kinter A, et al. Correlates of preserved CD4(+) T cell homeostasis during natural, nonpathogenic simian immunodeficiency virus infection of sooty mangabeys: implications for AIDS pathogenesis. J Immunol 2007;178(3):1680–91.

[16] Klatt NR, Villinger F, Bostik P, Gordon SN, Pereira L, Engram JC, et al. Availability of activated CD4+ T cells dictates the level of viremia in naturally SIV-infected sooty mangabeys. J Clin Invest 2008;118(6):2039–49.
[17] Paiardini M, Cervasi B, Reyes-Aviles E, Micci L, Ortiz AM, Chahroudi A, et al. Low levels of SIV infection in sooty mangabey central memory CD(4)(+) T cells are associated with limited CCR5 expression. Nat Med 2011;17(7):830–6.
[18] Brenchley JM, Paiardini M, Knox KS, Asher AI, Cervasi B, Asher TE, et al. Differential Th17 CD4 T-cell depletion in pathogenic and nonpathogenic lentiviral infections. Blood 2008;112(7):2826–35.
[19] Brenchley JM, Price DA, Schacker TW, Asher TE, Silvestri G, Rao S, et al. Microbial translocation is a cause of systemic immune activation in chronic HIV infection. Nat Med 2006;12(12):1365–71.
[20] Favre D, Lederer S, Kanwar B, Ma ZM, Proll S, Kasakow Z, et al. Critical loss of the balance between Th17 and T regulatory cell populations in pathogenic SIV infection. PLoS Pathog 2009;5(2):e1000295.
[21] Gordon SN, Klatt NR, Bosinger SE, Brenchley JM, Milush JM, Engram JC, et al. Severe depletion of mucosal CD4+ T cells in AIDS-free simian immunodeficiency virus-infected sooty mangabeys. J Immunol 2007;179(5):3026–34.
[22] Pandrea IV, Gautam R, Ribeiro RM, Brenchley JM, Butler IF, Pattison M, et al. Acute loss of intestinal CD4+ T cells is not predictive of simian immunodeficiency virus virulence. J Immunol 2007;179(5):3035–46.
[23] Bosinger SE, Li Q, Gordon SN, Klatt NR, Duan L, Xu L, et al. Global genomic analysis reveals rapid control of a robust innate response in SIV-infected sooty mangabeys. J Clin Invest 2009;119(12):3556–72.
[24] Harris LD, Tabb B, Sodora DL, Paiardini M, Klatt NR, Douek DC, et al. Downregulation of robust acute type I interferon responses distinguishes nonpathogenic simian immunodeficiency virus (SIV) infection of natural hosts from pathogenic SIV infection of rhesus macaques. J Virol 2010;84(15):7886–91.
[25] Brenchley JM, Silvestri G, Douek DC. Nonprogressive and progressive primate immunodeficiency lentivirus infections. Immunity 2010;32(6):737–42.
[26] Dorner T, Radbruch A. Antibodies and B cell memory in viral immunity. Immunity 2007;27(3):384–92.
[27] Kaech SM, Wherry EJ. Heterogeneity and cell-fate decisions in effector and memory CD8+ T cell differentiation during viral infection. Immunity 2007;27(3):393–405.
[28] Gowans JL, Knight EJ. The route of re-circulation of lymphocytes in the rat. Proc R Soc London B 1964;159:257–82.
[29] Zhu J, Yamane H, Paul WE. Differentiation of effector CD4 T cell populations (*). Annu Rev Immunol 2010;28:445–89.
[30] Yamane H, Paul WE. Cytokines of the gamma(c) family control CD4+ T cell differentiation and function. Nat Immunol 2012;13(11):1037–44.
[31] Luckheeram RV, Zhou R, Verma AD, Xia B. CD4(+)T cells: differentiation and functions. Clin Dev Immunol 2012;2012:925135.
[32] Marshall HD, Chandele A, Jung YW, Meng H, Poholek AC, Parish IA, et al. Differential expression of Ly6C and T-bet distinguish effector and memory Th1 CD4(+) cell properties during viral infection. Immunity 2011;35(4):633–46.
[33] Lugli E, Dominguez MH, Gattinoni L, Chattopadhyay PK, Bolton DL, Song K, et al. Superior T memory stem cell persistence supports long-lived T cell memory. J Clin Invest 2013;123(2):594–9.
[34] Sallusto F, Geginat J, Lanzavecchia A. Central memory and effector memory T cell subsets: function, generation, and maintenance. Annu Rev Immunol 2004;22:745–63.
[35] Reinhardt RL, Khoruts A, Merica R, Zell T, Jenkins MK. Visualizing the generation of memory CD4 T cells in the whole body. Nature 2001;410(6824):101–5.

[36] Pepper M, Jenkins MK. Origins of CD4(+) effector and central memory T cells. Nat Immunol 2011;12(6):467–71.
[37] Seder RA, Ahmed R. Similarities and differences in CD4+ and CD8+ effector and memory T cell generation. Nat Immunol 2003;4(9):835–42.
[38] Lugli E, Gattinoni L, Roberto A, Mavilio D, Price DA, Restifo NP, et al. Identification, isolation and in vitro expansion of human and nonhuman primate T stem cell memory cells. Nat Protoc 2013;8(1):33–42.
[39] Killar L, MacDonald G, West J, Woods A, Bottomly K. Cloned, Ia-restricted T cells that do not produce interleukin 4(IL 4)/B cell stimulatory factor 1(BSF-1) fail to help antigen-specific B cells. J Immunol 1987;138(6):1674–9.
[40] Mosmann TR, Cherwinski H, Bond MW, Giedlin MA, Coffman RL. Two types of murine helper T cell clone. I. Definition according to profiles of lymphokine activities and secreted proteins. J Immunol 1986;136(7):2348–57.
[41] Ma CS, Deenick EK, Batten M, Tangye SG. The origins, function, and regulation of T follicular helper cells. J Exp Med 2012;209(7):1241–53.
[42] Kroenke MA, Eto D, Locci M, Cho M, Davidson T, Haddad EK, et al. Bcl6 and Maf cooperate to instruct human follicular helper CD4 T cell differentiation. J Immunol 2012;188(8):3734–44.
[43] Harker JA, Lewis GM, Mack L, Zuniga EI. Late interleukin-6 escalates T follicular helper cell responses and controls a chronic viral infection. Science 2011;334(6057):825–9.
[44] Petrovas C, Yamamoto T, Gerner MY, Boswell KL, Wloka K, Smith EC, et al. CD4 T follicular helper cell dynamics during SIV infection. J Clin Invest 2012;122(9):3281–94.
[45] Korn T, Bettelli E, Oukka M, Kuchroo VK. IL-17 and Th17 cells. Annu Rev Immunol 2009;27:485–517.
[46] Miossec P. IL-17 and Th17 cells in human inflammatory diseases. Microbes Infect 2009;11(5):625–30.
[47] Chen Y, Thai P, Zhao YH, Ho YS, DeSouza MM, Wu R. Stimulation of airway mucin gene expression by interleukin (IL)-17 through IL-6 paracrine/autocrine loop. J Biol Chem 2003;278(19):17036–43.
[48] Sugimoto K, Ogawa A, Mizoguchi E, Shimomura Y, Andoh A, Bhan AK, et al. IL-22 ameliorates intestinal inflammation in a mouse model of ulcerative colitis. J Clin Invest 2008;118(2):534–44.
[49] Campbell DJ, Koch MA. Phenotypical and functional specialization of FOXP3+ regulatory T cells. Nat Rev Immunol 2011;11(2):119–30.
[50] Dunham RM, Cervasi B, Brenchley JM, Albrecht H, Weintrob A, Sumpter B, et al. CD127 and CD25 expression defines CD4+ T cell subsets that are differentially depleted during HIV infection. J Immunol 2008;180(8):5582–92.
[51] Rubtsov YP, Rasmussen JP, Chi EY, Fontenot J, Castelli L, Ye X, et al. Regulatory T cell-derived interleukin-10 limits inflammation at environmental interfaces. Immunity 2008;28(4):546–58.
[52] Wing K, Onishi Y, Prieto-Martin P, Yamaguchi T, Miyara M, Fehervari Z, et al. CTLA-4 control over Foxp3+ regulatory T cell function. Science 2008;322(5899):271–5.
[53] Wing K, Sakaguchi S. Regulatory T cells exert checks and balances on self tolerance and autoimmunity. Nat Immunol 2010;11(1):7–13.
[54] Edinger AL, Amedee A, Miller K, Doranz BJ, Endres M, Sharron M, et al. Differential utilization of CCR5 by macrophage and T cell tropic simian immunodeficiency virus strains. Proc Natl Acad Sci USA 1997;94(8):4005–10.
[55] Pandrea I, Apetrei C, Gordon S, Barbercheck J, Dufour J, Bohm R, et al. Paucity of CD4+CCR5+ T cells is a typical feature of natural SIV hosts. Blood 2007;109(3):1069–76.
[56] Riddick NE, Hermann EA, Loftin LM, Elliott ST, Wey WC, Cervasi B, et al. A novel CCR5 mutation common in sooty mangabeys reveals SIVsmm infection of CCR5-null natural hosts and efficient alternative coreceptor use in vivo. PLoS Pathog 2010;6(8):e1001064.

[57] Elliott ST, Riddick NE, Francella N, Paiardini M, Vanderford TH, Li B, et al. Cloning and analysis of sooty mangabey alternative coreceptors that support simian immunodeficiency virus SIVsmm entry independently of CCR5. J Virol 2012;86(2):898–908.
[58] Beaumier CM, Harris LD, Goldstein S, Klatt NR, Whitted S, McGinty J, et al. CD4 downregulation by memory CD4+ T cells in vivo renders African green monkeys resistant to progressive SIVagm infection. Nat Med 2009;15(8):879–85.
[59] Milush JM, Mir KD, Sundaravaradan V, Gordon SN, Engram J, Cano CA, et al. Lack of clinical AIDS in SIV-infected sooty mangabeys with significant CD4+ T cell loss is associated with double-negative T cells. J Clin Invest 2011;121(3):1102–10.
[60] Stein BS, Gowda SD, Lifson JD, Penhallow RC, Bensch KG, Engleman EG. pH-independent HIV entry into CD4-positive T cells via virus envelope fusion to the plasma membrane. Cell 1987;49(5):659–68.
[61] Brenchley JM, Vinton C, Tabb B, Hao XP, Connick E, Paiardini M, et al. Differential infection patterns of CD4+ T cells and lymphoid tissue viral burden distinguish progressive and nonprogressive lentiviral infections. Blood 2012;120(20):4172–81.
[62] Letvin NL, Mascola JR, Sun Y, Gorgone DA, Buzby AP, Xu L, et al. Preserved CD4+ central memory T cells and survival in vaccinated SIV-challenged monkeys. Science 2006;312(5779):1530–3.
[63] Okoye A, Meier-Schellersheim M, Brenchley JM, Hagen SI, Walker JM, Rohankhedkar M, et al. Progressive CD4+ central memory T cell decline results in CD4+ effector memory insufficiency and overt disease in chronic SIV infection. J Exp Med 2007;204(9):2171–85.
[64] Picker LJ, Hagen SI, Lum R, Reed-Inderbitzin EF, Daly LM, Sylwester AW, et al. Insufficient production and tissue delivery of CD4+ memory T cells in rapidly progressive simian immunodeficiency virus infection. J Exp Med 2004;200(10):1299–314.
[65] Lifson JD, Rossio JL, Arnaout R, Li L, Parks TL, Schneider DK, et al. Containment of simian immunodeficiency virus infection: cellular immune responses and protection from rechallenge following transient postinoculation antiretroviral treatment. J Virol 2000;74(6):2584–93.
[66] Esser MT, Mori T, Mondor I, Sattentau QJ, Dey B, Berger EA, et al. Cyanovirin-N binds to gp120 to interfere with CD4-dependent human immunodeficiency virus type 1 virion binding, fusion, and infectivity but does not affect the CD4 binding site on gp120 or soluble CD4-induced conformational changes in gp120. J Virol 1999;73(5):4360–71.
[67] Mattapallil JJ, Douek DC, Buckler-White A, Montefiori D, Letvin NL, Nabel GJ, et al. Vaccination preserves CD4 memory T cells during acute simian immunodeficiency virus challenge. J Exp Med 2006;203(6):1533–41.
[68] Potter SJ, Lacabaratz C, Lambotte O, Perez-Patrigeon S, Vingert B, Sinet M, et al. Preserved central memory and activated effector memory CD4+ T-cell subsets in human immunodeficiency virus controllers: an ANRS EP36 study. J Virol 2007;81(24):13904–15.
[69] Chomont N, El-Far M, Ancuta P, Trautmann L, Procopio FA, Yassine-Diab B, et al. HIV reservoir size and persistence are driven by T cell survival and homeostatic proliferation. Nat Med 2009;15(8):893–900.
[70] Zeng M, Smith AJ, Wietgrefe SW, Southern PJ, Schacker TW, Reilly CS, et al. Cumulative mechanisms of lymphoid tissue fibrosis and T cell depletion in HIV-1 and SIV infections. J Clin Invest 2011;121(3):998–1008.
[71] Descours B, Avettand-Fenoel V, Blanc C, Samri A, Melard A, Supervie V, et al. Immune responses driven by protective human leukocyte antigen alleles from long-term nonprogressors are associated with low HIV reservoir in central memory CD4 T cells. Clin Infect Dis 2012;54(10):1495–503.
[72] Saez-Cirion A, Bacchus C, Hocqueloux L, Avettand-Fenoel V, Girault I, Lecuroux C, et al. Post-treatment HIV-1 controllers with a long-term virological remission after the interruption of early initiated antiretroviral therapy ANRS VISCONTI Study. PLoS Pathog 2013;9(3):e1003211.

[73] Hong JJ, Amancha PK, Rogers K, Ansari AA, Villinger F. Spatial alterations between CD4(+) T follicular helper, B, and CD8(+) T cells during simian immunodeficiency virus infection: T/B cell homeostasis, activation, and potential mechanism for viral escape. J Immunol 2012;188(7):3247–56.
[74] Lindqvist M, van Lunzen J, Soghoian DZ, Kuhl BD, Ranasinghe S, Kranias G, et al. Expansion of HIV-specific T follicular helper cells in chronic HIV infection. J Clin Invest 2012;122(9):3271–80.
[75] Perreau M, Savoye AL, De Crignis E, Corpataux JM, Cubas R, Haddad EK, et al. Follicular helper T cells serve as the major CD4 T cell compartment for HIV-1 infection, replication, and production. J Exp Med 2013;210(1):143–56.
[76] Li B, Stefano-Cole K, Kuhrt DM, Gordon SN, Else JG, Mulenga J, et al. Nonpathogenic simian immunodeficiency virus infection of sooty mangabeys is not associated with high levels of autologous neutralizing antibodies. J Virol 2010;84(12):6248–53.
[77] Lifson JD, Piatak Jr M, Rossio JL, Bess Jr J, Chertova E, Schneider D, et al. Whole inactivated SIV virion vaccines with functional envelope glycoproteins: safety, immunogenicity, and activity against intrarectal challenge. J Med Primatol 2002;31(4–5):205–16.
[78] Ling B, Veazey RS, Penedo C, Xu K, Lifson JD, Marx PA. Longitudinal follow up of SIVmac pathogenesis in rhesus macaques of Chinese origin: emergence of B cell lymphoma. J Med Primatol 2002;31(4–5):154–63.
[79] Allen TM, Jing P, Calore B, Horton H, O'Connor DH, Hanke T, et al. Effects of cytotoxic T lymphocytes (CTL) directed against a single simian immunodeficiency virus (SIV) Gag CTL epitope on the course of SIVmac239 infection. J Virol 2002;76(20):10507–11.
[80] Clements JE, Babas T, Mankowski JL, Suryanarayana K, Piatak Jr M, Tarwater PM, et al. The central nervous system as a reservoir for simian immunodeficiency virus (SIV): steady-state levels of SIV DNA in brain from acute through asymptomatic infection. J Infect Dis 2002;186(7):905–13.
[81] Larsson M, Fonteneau JF, Lirvall M, Haslett P, Lifson JD, Bhardwaj N. Activation of HIV-1 specific CD4 and CD8 T cells by human dendritic cells: roles for cross-presentation and non-infectious HIV-1 virus. AIDS 2002;16(10):1319–29.
[82] Campillo-Gimenez L, Cumont MC, Fay M, Kared H, Monceaux V, Diop O, et al. AIDS progression is associated with the emergence of IL-17-producing cells early after simian immunodeficiency virus infection. J Immunol 2010;184(2):984–92.
[83] Cecchinato V, Trindade CJ, Laurence A, Heraud JM, Brenchley JM, Ferrari MG, et al. Altered balance between Th17 and Th1 cells at mucosal sites predicts AIDS progression in simian immunodeficiency virus-infected macaques. Mucosal Immunol 2008;1(4):279–88.
[84] Klatt NR, Estes JD, Sun X, Ortiz AM, Barber JS, Harris LD, et al. Loss of mucosal CD103+ DCs and IL-17+ and IL-22+ lymphocytes is associated with mucosal damage in SIV infection. Mucosal Immunol 2012;5(6):646–57.
[85] Raffatellu M, Santos RL, Verhoeven DE, George MD, Wilson RP, Winter SE, et al. Simian immunodeficiency virus-induced mucosal interleukin-17 deficiency promotes Salmonella dissemination from the gut. Nat Med 2008;14(4):421–8.
[86] Ouyang W, Kolls JK, Zheng Y. The biological functions of T helper 17 cell effector cytokines in inflammation. Immunity 2008;28(4):454–67.
[87] Brandt L, Benfield T, Mens H, Clausen LN, Katzenstein TL, Fomsgaard A, et al. Low level of regulatory T cells and maintenance of balance between regulatory T cells and TH17 cells in HIV-1-infected elite controllers. J Acquired Immune Defic Syndr 2011;57(2):101–8.
[88] Ciccone EJ, Greenwald JH, Lee PI, Biancotto A, Read SW, Yao MA, et al. CD4+ T cells, including Th17 and cycling subsets, are intact in the gut mucosa of HIV-1-infected long-term nonprogressors. J Virol 2011;85(12):5880–8.
[89] Salgado M, Rallon NI, Rodes B, Lopez M, Soriano V, Benito JM. Long-term non-progressors display a greater number of Th17 cells than HIV-infected typical progressors. Clin Immunol 2011;139(2):110–4.

[90] Gordon SN, Cervasi B, Odorizzi P, Silverman R, Aberra F, Ginsberg G, et al. Disruption of intestinal CD4+ T cell homeostasis is a key marker of systemic CD4+ T cell activation in HIV-infected individuals. J Immunol 2010;185(9):5169–79.
[91] Micci L, Cervasi B, Ende ZS, Iriele RI, Reyes-Aviles E, Vinton C, et al. Paucity of IL-21-producing CD4(+) T cells is associated with Th17 cell depletion in SIV infection of rhesus macaques. Blood 2012;120(19):3925–35.
[92] Macal M, Sankaran S, Chun TW, Reay E, Flamm J, Prindiville TJ, et al. Effective CD4+ T-cell restoration in gut-associated lymphoid tissue of HIV-infected patients is associated with enhanced Th17 cells and polyfunctional HIV-specific T-cell responses. Mucosal Immunol 2008;1(6):475–88.
[93] Hartigan-O'Connor DJ, Abel K, Van Rompay KK, Kanwar B, McCune JM. SIV replication in the infected rhesus macaque is limited by the size of the preexisting TH17 cell compartment. Sci Transl Med 2012;4(136): 136ra69.
[94] Pallikkuth S, Micci L, Ende ZS, Iriele RI, Cervasi B, Lawson B, et al. Maintenance of intestinal Th17 cells and reduced microbial translocation in SIV-infected rhesus macaques treated with interleukin (IL)-21. PLoS Pathog 2013;9(7):e1003471.
[95] Kader M, Wang X, Piatak M, Lifson J, Roederer M, Veazey R, et al. Alpha4(+)beta7(hi) CD4(+) memory T cells harbor most Th-17 cells and are preferentially infected during acute SIV infection. Mucosal Immunol 2009;2(5):439–49.
[96] Chase AJ, Sedaghat AR, German JR, Gama L, Zink MC, Clements JE, et al. Severe depletion of CD4+ CD25+ regulatory T cells from the intestinal lamina propria but not peripheral blood or lymph nodes during acute simian immunodeficiency virus infection. J Virol 2007;81(23):12748–57.
[97] Epple HJ, Loddenkemper C, Kunkel D, Troger H, Maul J, Moos V, et al. Mucosal but not peripheral FOXP3+ regulatory T cells are highly increased in untreated HIV infection and normalize after suppressive HAART. Blood 2006;108(9):3072–8.
[98] Estes JD, Li Q, Reynolds MR, Wietgrefe S, Duan L, Schacker T, et al. Premature induction of an immunosuppressive regulatory T cell response during acute simian immunodeficiency virus infection. J Infect Dis 2006;193(5):703–12.
[99] Schulze Zur Wiesch J, Thomssen A, Hartjen P, Toth I, Lehmann C, Meyer-Olson D, et al. Comprehensive analysis of frequency and phenotype of T regulatory cells in HIV infection: CD39 expression of FoxP3+ T regulatory cells correlates with progressive disease. J Virol 2011;85(3):1287–97.
[100] Aandahl EM, Michaelsson J, Moretto WJ, Hecht FM, Nixon DF. Human CD4+ CD25+ regulatory T cells control T-cell responses to human immunodeficiency virus and cytomegalovirus antigens. J Virol 2004;78(5):2454–9.
[101] Kinter AL, Hennessey M, Bell A, Kern S, Lin Y, Daucher M, et al. CD25(+)CD4(+) regulatory T cells from the peripheral blood of asymptomatic HIV-infected individuals regulate CD4(+) and CD8(+) HIV-specific T cell immune responses in vitro and are associated with favorable clinical markers of disease status. J Exp Med 2004;200(3):331–43.
[102] Weiss L, Donkova-Petrini V, Caccavelli L, Balbo M, Carbonneil C, Levy Y. Human immunodeficiency virus-driven expansion of CD4+CD25+ regulatory T cells, which suppress HIV-specific CD4 T-cell responses in HIV-infected patients. Blood 2004;104(10):3249–56.
[103] Petitjean G, Chevalier MF, Tibaoui F, Didier C, Manea ME, Liovat AS, et al. Level of double negative T cells, which produce TGF-beta and IL-10, predicts CD8 T-cell activation in primary HIV-1 infection. AIDS 2012;26(2):139–48.
[104] Karlsson I, Malleret B, Brochard P, Delache B, Calvo J, Le Grand R, et al. Suppressive activity of regulatory T cells correlates with high CD4(+) T-cell counts and low T-cell activation during chronic simian immunodeficiency virus infection. AIDS 2011;25(5):585–93.

[105] Eggena MP, Barugahare B, Jones N, Okello M, Mutalya S, Kityo C, et al. Depletion of regulatory T cells in HIV infection is associated with immune activation. J Immunol 2005;174(7):4407–14.
[106] Chase AJ, Yang HC, Zhang H, Blankson JN, Siliciano RF. Preservation of FoxP3+ regulatory T cells in the peripheral blood of human immunodeficiency virus type 1-infected elite suppressors correlates with low CD4+ T-cell activation. J Virol 2008;82(17):8307–15.
[107] Chevalier MF, Weiss L. The split personality of regulatory T cells in HIV infection. Blood 2013;121(1):29–37.
[108] Kornfeld C, Ploquin MJ, Pandrea I, Faye A, Onanga R, Apetrei C, et al. Antiinflammatory profiles during primary SIV infection in African green monkeys are associated with protection against AIDS. J Clin Invest 2005;115(4):1082–91.
[109] Pereira LE, Villinger F, Onlamoon N, Bryan P, Cardona A, Pattanapanysat K, et al. Simian immunodeficiency virus (SIV) infection influences the level and function of regulatory T cells in SIV-infected rhesus macaques but not SIV-infected sooty mangabeys. J Virol 2007;81(9):4445–56.
[110] Brenchley JM, Hill BJ, Ambrozak DR, Price DA, Guenaga FJ, Casazza JP, et al. T-cell subsets that harbor human immunodeficiency virus (HIV) in vivo: implications for HIV pathogenesis. J Virol 2004;78(3):1160–8.
[111] Douek DC, Brenchley JM, Betts MR, Ambrozak DR, Hill BJ, Okamoto Y, et al. HIV preferentially infects HIV-specific CD4+ T cells. Nature 2002;417(6884):95–8.
[112] Schmitz JE, Ma ZM, Hagan EA, Wilks AB, Furr KL, Linde CH, et al. Memory CD4(+) T lymphocytes in the gastrointestinal tract are a major source of cell-associated simian immunodeficiency virus in chronic nonpathogenic infection of African green monkeys. J Virol 2012;86(20):11380–5.
[113] Gartner S, Markovits P, Markovitz DM, Kaplan MH, Gallo RC, Popovic M. The role of mononuclear phagocytes in HTLV-III/LAV infection. Science 1986;233(4760):215–9.
[114] Ho DD, Rota TR, Hirsch MS. Infection of monocyte/macrophages by human T lymphotropic virus type III. J Clin Invest 1986;77(5):1712–5.
[115] Ortiz AM, Klatt NR, Li B, Yi Y, Tabb B, Hao XP, et al. Depletion of CD4(+) T cells abrogates post-peak decline of viremia in SIV-infected rhesus macaques. J Clin Invest 2011;121(11):4433–45.
[116] Nicholson JK, Cross GD, Callaway CS, McDougal JS. In vitro infection of human monocytes with human T lymphotropic virus type III/lymphadenopathy-associated virus (HTLV-III/LAV). J Immunol 1986;137(1):323–9.
[117] Pantaleo G, Graziosi C, Butini L, Pizzo PA, Schnittman SM, Kotler DP, et al. Lymphoid organs function as major reservoirs for human immunodeficiency virus. Proc Natl Acad Sci USA 1991;88(21):9838–42.
[118] Orenstein JM, Fox C, Wahl SM. Macrophages as a source of HIV during opportunistic infections. Science 1997;276(5320):1857–61.
[119] Igarashi T, Imamichi H, Brown CR, Hirsch VM, Martin MA. The emergence and characterization of macrophage-tropic SIV/HIV chimeric viruses (SHIVs) present in CD4+ T cell-depleted rhesus monkeys. J Leukocyte Biol 2003;74(5):772–80.
[120] Hockett RD, Kilby JM, Derdeyn CA, Saag MS, Sillers M, Squires K, et al. Constant mean viral copy number per infected cell in tissues regardless of high, low, or undetectable plasma HIV RNA. J Exp Med 1999;189(10):1545–54.
[121] Sharova N, Wu Y, Zhu X, Stranska R, Kaushik R, Sharkey M, et al. Primate lentiviral Vpx commandeers DDB1 to counteract a macrophage restriction. PLoS Pathog 2008;4(5):e1000057.
[122] Laguette N, Sobhian B, Casartelli N, Ringeard M, Chable-Bessia C, Segeral E, et al. SAMHD1 is the dendritic- and myeloid-cell-specific HIV-1 restriction factor counteracted by Vpx. Nature 2011;474(7353):654–7.

[123] Goujon C, Arfi V, Pertel T, Luban J, Lienard J, Rigal D, et al. Characterization of simian immunodeficiency virus SIVSM/human immunodeficiency virus type 2 Vpx function in human myeloid cells. J Virol 2008;82(24):12335–45.
[124] Lim ES, Fregoso OI, McCoy CO, Matsen FA, Malik HS, Emerman M. The ability of primate lentiviruses to degrade the monocyte restriction factor SAMHD1 preceded the birth of the viral accessory protein Vpx. Cell Host Microbe 2012;11(2):194–204.
[125] Lahouassa H, Daddacha W, Hofmann H, Ayinde D, Logue EC, Dragin L, et al. SAMHD1 restricts the replication of human immunodeficiency virus type 1 by depleting the intracellular pool of deoxynucleoside triphosphates. Nat Immunol 2012;13(3):223–8.

CHAPTER

13

SIV Transmission in Natural Hosts

Ann Chahroudi[1,2], *Sallie Permar*[3], *Ivona Pandrea*[4,5]

[1]Department of Pediatrics, Emory University School of Medicine, Atlanta, GA, USA; [2]The Yerkes National Primate Research Center, Emory Vaccine Center, Atlanta, GA, USA; [3]Human Vaccine Institute, Duke University Medical Center, Durham, NC, USA; [4]Department of Pathology, School of Medicine, University of Pittsburgh, Pittsburgh, PA, USA; [5]Center for Vaccine Research, University of Pittsburgh, Pittsburgh, PA, USA

INTRODUCTION

It is general knowledge that natural hosts of simian immunodeficiency virus (SIV) generally do not progress to acquired immune deficiency syndrome (AIDS) after being infected with SIV despite the fact that they replicate this virus at higher levels than those reported in humans infected with human immunodeficiency virus type 1 (HIV-1) [1,2]. Identifying the causes behind the lack of pathogenicity of SIV infection in the natural hosts is one of the high-priority areas of HIV research and can be highly informative for the potential development of immune therapies for HIV-infected patients. It is largely acknowledged that the adaptations responsible for the lack of disease progression in natural hosts occurred during millennia of coevolution between these non-human primates (NHPs) and their species-specific SIVs [1,2].

Few may be aware, however, that the natural hosts developed additional mechanisms through which they adapted to become infected with SIV only at the age of sexual maturity [3]. These mechanisms provide protection of the offspring from infection both in utero and early in life either at the delivery or through breastfeeding. Although largely ignored to date, identifying the mechanisms through which the natural hosts block SIV

Natural Hosts of SIV.
http://dx.doi.org/10.1016/B978-0-12-404734-1.00013-9

transmission to their offspring may be invaluable for the design of strategies to impede HIV mother-to-infant transmission (MTIT) in humans.

This chapter will therefore review the evidence supporting adaptation of the natural hosts to block SIV transmission to their offspring and will give insight into the possible mechanisms through which vertical transmission is prevented in the natural hosts.

EVIDENCE SUPPORTING PAUCITY OF MTIT IN THE NATURAL HOSTS OF SIV

Studies in the Wild

Several studies performed in the mid-1990s were aimed at assessing both the prevalence and the routes of transmission of the SIV infection in the natural hosts [4]. An initial survey conducted in the Awash National Park in Ethiopia performed elegant correlations between the SIV serological status, age, and sex in grivet monkeys (*Chlorocebus ethiops ethiops*) and showed that both sexes remain almost entirely free of SIVagm until they become sexually active. Seropositivity always paralleled the age of sexual maturity and was found to be nearly universal in females of reproductive age [4]. These findings were reinforced by the case of a female that seroconverted at an age consistent with the first breeding. Similarly, the males were found to become infected later in life, as adults, which correlated with the onset of mating behavior, which occurs later in males, only when they reach full physical development and a high social status [4].

A follow-up study included samples collected over two decades from a different grivet population and confirmed the initial set of data by establishing that SIV is nearly exclusively transmitted in natural hosts through sexual contact; occasionally by trauma; and rarely, if ever, maternally [5].

Finally, an independent study [6] assessed the prevalence of anti-SIV antibodies in grivet offspring born from SIV-infected dams and established that although all the infants were seropositive at birth (consistent with passive antibody transfer from their mothers), their antibody titers continuously decreased during the first year of life, resulting in weak responses at 6 months and then complete negativity by 1 year of age [6]. As such, this study confirmed that MTIT is indeed negligible in natural hosts.

Another argument supporting the lack of MTIT in the African green monkeys (AGMs) is that the Caribbean AGMs, which were extensively imported from Africa as pets in the seventeenth and eighteenth centuries [7], are not infected with SIV [8,9]. This lack of SIV exposure has been attributed to the capture and movement of these animals as uninfected juveniles.

Several other studies also produced compelling evidence that SIV MTIT is minimal in other natural host species in the wild. As such, age-related differences in SIV prevalence have been reported for SIVmnd-1-infected mandrills (*Mandrillus sphinx*) [10] and for SIVsmm-infected sooty mangabeys (*Cercocebus atys*) [11–14]. All of these studies show an age-related increase in SIV seroprevalence in natural hosts, supporting sexual SIV transmission as the major route of transmission in these species [15].

More recently, phylogenetic studies suggested that MTIT may occur in sooty mangabeys in the wild. Thus, one study performed on samples collected from sooty mangabeys from the Tai Forest in the Ivory Coast [12] identified near-identical viruses in two mother/daughter pairs. However, in these studies the offspring were tested only when they were older than 4 years, so the precise pathway by which they acquired SIV from their mothers could not be established. Very relevant within this context are the observations made in a semifree colony of SIV-infected mandrills in Gabon, in which MTIT was not identified, but transmission of SIV from mothers to daughters was shown to occur later in life, probably through aggressive grooming or injuries [16–18].

MTIT was also reported as likely to have occurred in an SIV-infected male subjuvenile sooty mangabey (age was approximated based on weight) [19]. This study was carried out on bush meat samples from rural Sierra Leone and could not pair the juvenile sequence with a specific dam sequence. As such, this case could not be confirmed as a result of vertical transmission.

The most recent studies performed on 225 wild-trapped vervets from nine different parks from South Africa [20] and 121 sabaeus from the Gambia [21] brought compelling evidence for the occurrence of rare cases of MTIT in the natural hosts in the wild. These studies confirmed that SIVagm prevalence dramatically increases with the onset of sexual maturity in both AGM species, as supported not only by the significant increase in SIV prevalence with age, but also by the fact that the majority of the acute infections (defined as viral loads $>10^6$ copies in the absence of anti-SIV antibodies, Fiebig II) [20,21] occurred in juvenile females most likely at the time of their first sexual encounters. In contrast to previously published data on wild animals, these studies also established that 4–7% of infant AGMs are infected with SIV [20,21]. MTIT was further suggested by the observation that, in the phylogenetic trees, the viruses identified in a fraction of the SIVagm-infected infants and juveniles clustered together with those identified in adult females [20]. The MTIT rates identified in these studies are significantly lower than the 35–40% MTIT rates reported in HIV infection [22] or 40–70% transmission rates in the non-natural host species such as rhesus macaques (RMs) [23]. The rates of MTIT in wild AGMs are remarkably low considering the finding that the prevalence of SIV infection in the lactating females was more than 72% [20], pointing to a massive exposure through breastfeeding, and that 44% of the infants had passively transferred anti-SIV antibodies at the time of testing, supporting a significant in utero exposure [20] Figure 13.1. Altogether, these recent studies reinforce the view that there is resistance to a large extent to SIV infection in natural host offspring.

Studies in Captive Monkeys from Primate Centers

Surveys performed in the Primate Centers with large colonies of naturally infected natural hosts also reported a very low rate of vertical transmission in these species [6,14,18,24,25]. Only 6.8% (11/161) of the sooty mangabey infants from the Yerkes National Primate Research Center were reported to have acquired SIV by presumptive MTIT, based on the presence of anti-SIV antibodies without seroreversion and a viral load of >500 copies/ml of serum in the first year of life [25]. This study corroborates the data obtained in the wild AGMs and expands upon previous studies performed in captive NHPs [4,14,18,20,21,24]. Importantly, the viral loads of the 11 infants presumed to be infected by MTIT were 1–2 log lower than

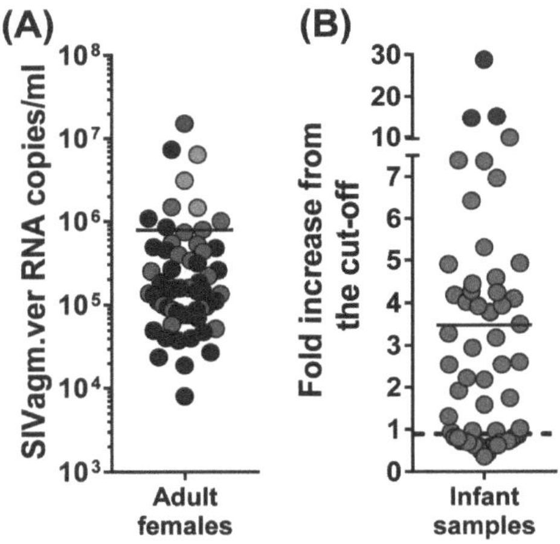

FIGURE 13.1 **Massive offspring exposure to the SIV infecting the dams.** (A) Massive exposure of the infants in the wild to the virus in milk was documented to occur by showing that the majority of the SIV-infected females were infected at the time of sample collection in a cross-sectional analysis. Plasma viral loads in both chronically and acutely infected females were high. Lactating dams are depicted as light circles. The samples identified as originating from acutely infected dams are illustrated as red (dark and light) circles. Detection limit, 100 copies/ml. (B) Massive in utero exposure of the offspring was indirectly documented by showing that a large proportion of the infants are seropositive for SIV, whereas the virus could only be amplified from a small fraction of infants (depicted as violet circles). This is highly suggestive for passively transmitted antibodies from their infected mothers, in agreement with the high levels of prevalence in the wild. (See color plate at the back of the book.)

that of sooty mangabeys infected as adults, suggesting not only a block to SIV transmission, but also restriction of SIV replication in infant natural hosts [25].

Altogether, studies on captive and wild African NHPs document a low rate of vertical transmission (4–7%) in the natural host compared to 40–70% in humans or macaques.

Experimental Testing of the Rate of SIV Transmission to Offspring in the Natural Hosts

Only one study directly investigated the rates of SIV transmission to offspring through breastfeeding in a natural host species [18]. In this study, mandrill dams were infected on the day of the delivery, and they were allowed to breastfeed their offspring throughout the lactation period. As such, the infants were naturally exposed during the lactation period and intensively tested to determine whether or not they became infected. Exposing females to the virus after delivery was performed to facilitate MTIT during acute infection of the dams, when viral loads are extremely high, and to avoid interference with potentially protective anti-SIV antibodies passively transferred from the mothers. Such a design also had the advantage to focus exclusively on the transmission of SIV via breastfeeding and to accurately establish the timing of offspring infection. Despite high viral loads detected in the milk and plasma of the dams, none of the offspring became infected by the end of the lactation period, in strong contrast with similar experiments performed in RMs, in which the post-delivery SIV infection of dams resulted in virus transmission to 62–71% of their offspring [23,26]. Therefore, this mandrill experiment [18] provided very strong support to the epidemiological studies that have reported the rare occurrence of MTIT in natural hosts.

Possible Mechanisms Involved in the Lack of MTIT in Natural Hosts

Understanding the mechanisms underlying the rarity of MTIT transmission in natural hosts of SIV could guide the development of interventions to impede peri- and postnatal HIV-1 transmission. However, only a few studies have been conducted in this area, the majority of which have focused only on the mechanisms of transmission of SIV via breastfeeding.

Breast Milk Factors: Viral and Immunologic Characteristics

Comparative studies of the breast milk between natural hosts and non-natural hosts focused on features of milk that may be unique to the natural hosts to better understand the contribution of milk on the different levels of MTIT between the two models. Human milk contains many factors that can block HIV infection in vitro or that have been associated epidemiologically with lower rates of vertical transmission (reviewed in [27]). The key missing piece of information is whether natural hosts possess additional and/or "super factor(s)" that may be responsible for protecting infants from the acquisition of SIV via breastfeeding.

Lactation was pharmacologically induced in SIV-infected AGMs to allow for comparisons of the systemic and breast milk virus populations with the aim to determine whether natural SIV hosts harbor distinct immune responses or virus populations in breast milk compared to nonnatural hosts [28]. The breast milk cell-associated virus load was found to be approximately 0.5 log lower in AGMs compared to RMs, possibly due to the paucity of $CD4^+$ SIV target cells in this compartment. Therefore, if cell-associated virus is primarily responsible for postnatal infant transmission, the lower levels of infected cells could potentially contribute to the lower rate of breast milk transmission in natural SIV hosts [28]. However, milk cell-free viral loads were similar in AGMs and RMs, and both cell-free and cell-associated milk viral loads were similar in sooty mangabeys and RMs [29], suggesting that the low rate of postnatal transmission of SIV in natural hosts is not a result of low milk virus content.

The viral load in breast milk from humans and RMs is typically 1–2 log lower than that of plasma [30–32], but in AGMs milk and plasma viral loads are similar [28]. The same studies showed that in non-natural hosts of HIV/SIV, milk virus variants exhibit limited compartmentalization from plasma virus variants [30,31], whereas the virus populations in breast milk in AGMs revealed restricted diversity and transient compartmentalization from the blood virus variants [28,33]. These results suggest that the origin and mechanisms of replication of virus maybe distinct in breast milk of natural and non-natural SIV hosts, although the implications of this finding for MTIT remain unclear.

The immunologic composition of breast milk has also been compared between non-natural SIV hosts (RMs) and natural SIV hosts (AGMs) [28]. The milk immunoglobulin (Ig)G concentration is 1–2 log lower than plasma in both AGMs and macaques. However, the amount of IgA is similar in plasma and milk of AGMs, and this pattern is distinct from that of macaques and humans where the levels of milk IgA are typically 0.5–1 log lower than that in plasma [34–36]. The milk T lymphocyte populations are also very different in AGMs and RMs, with milk of both infected and uninfected AGMs containing a small pool of $CD4^+CCR5^+$ T cells and a large population of $CD4^{neg}CD8^{neg}$ T cells, which may represent $CD4^+$ T cells that have downregulated the CD4 molecule [37,38]. In contrast, milk $CD8^+$ T-cell populations are highly comparable between both infected and uninfected AGMs and RMs. Moreover, the frequencies of SIV-specific $CD8^+$ T-cell responses in milk of infected AGMs and RMs are also very similar in terms of magnitude and profile [28], indicating that the quantitative nature of the anti-SIV cellular response may not be the source of the natural protection against infant SIV transmission in natural hosts. Importantly, a strong autologous neutralizing IgG response has been observed in breast milk of AGMs [28,39] that has not been detected in breast milk of RMs [35] or humans [40]. Therefore, this robust SIV-specific antibody response is a possible mechanism for the prevention of postnatal virus transmission in the natural hosts of SIV (Figure 13.2).

Infant Factors: Target Cell Restriction

Further studies in NHPs have focused on the role of infant factors in driving the efficacy of SIV breastfeeding transmission, a less explored area of research in humans for obvious ethical reasons. In the experimental breastfeeding exposure study in mandrills referenced above, the lack of SIV transmission to offspring in the context of high virus loads in milk and absence of putative protective maternal antibodies suggested resistance to lentiviral infection in the mandrill infants [18]. The hypothesis that has been advanced is that low levels of $CD4^+$ T cells expressing the chemokine receptor CCR5 (the main targets of SIV/HIV) observed in a number of natural host species (mandrill, AGM, and sooty mangabey) may protect them against vertical SIV acquisition. Target cell restriction has also been proposed as a mechanism limiting viral replication in the rare sooty mangabey infants who do acquire SIV from their mothers and show a 2 log lower viremia than adult mangabeys who acquire SIV horizontally [25].

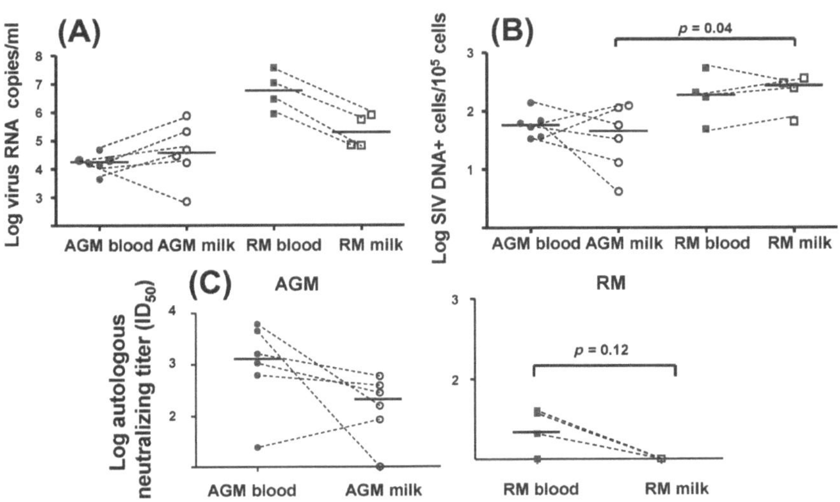

FIGURE 13.2 SIV RNA (A) and DNA (B) load and autologous neutralizing antibody responses (C) in breast milk (open symbols) and plasma (closed symbols) of chronically SIV-infected natural SIV hosts AGMs and non-natural hosts RMs.

A subsequent study performed on a large cohort of AGMs of different ages confirmed that the expression CCR5 on CD4+ T cells is indeed very low in AGM infants not only in blood but also at mucosal sites, such as jejunum and esophagus [21], the latter being a site that may be involved in SIV breastfeeding transmission [3]. A strong correlation between increases in the levels of CCR5 expressing CD4+ T cells (Figure 13.3) and the efficacy of SIV mucosal transmission could be established. Furthermore, experimental oral exposure of infant AGMs and rectal exposure of juvenile and adult AGMs with the same dose of virus resulted in the lack of SIV transmission in the infants, partial transmission (50%) in juveniles, and infection of all the adults. The juveniles that did not become infected had lower levels of CCR5-expressing CD4+ T cells in the blood and mucosal sites compared with those that tested positive after SIV exposure [3]. The number of transmitted viral variants was higher in adults than in juveniles, correlating with the number of target cells at mucosal sites and demonstrating that older individuals are relatively more susceptible to SIV infection compared to juveniles [3]. Altogether, these data show that the strongest correlate to explain the difference in susceptibility to mucosal SIV transmission among AGMs of different ages is the relative abundance of CD4+ T cells expressing CCR5 at mucosal sites [3].

A comparative study performed to directly analyze both breast milk factors and infant factors in the natural host sooty mangabey and the

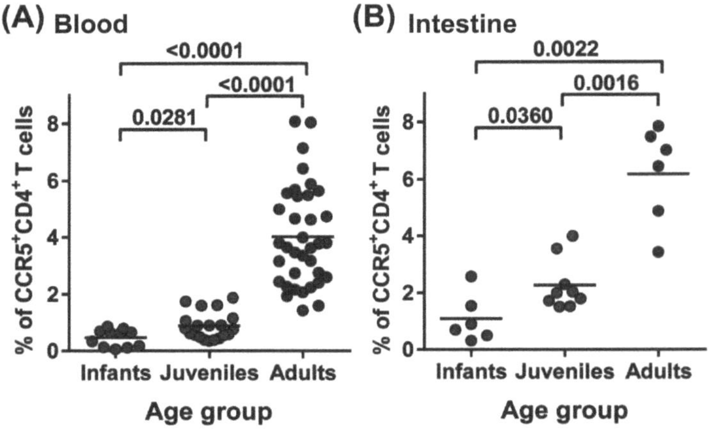

FIGURE 13.3 **Significantly higher levels of CCR5 expression on CD4+ T cells are observed in adult AGMs.** Cross-sectional assessment of CD4+CCR5+ T target cells both in peripheral blood (A) and at mucosal sites (B) in a cohort of uninfected AGMs demonstrate that in both peripheral blood (A) and jejunum (B), the levels of CD4+ T cells expressing CCR5 are very low in infants and juveniles and dramatically increase only after 4 years of age, corresponding to sexual maturity. This increase is highly significant and parallels increases in SIV prevalence in the wild. The *x* axis represents the age group of the animals; the *y* axis represents the fraction of CD4+ T cells expressing CCR5, as determined by flow cytometry.

non-natural host RM further highlights the importance of target cell restriction in determining susceptibility to mother-to-infant transmission of SIV [29]. In this work, no differences were seen in the level of virus as well as a number of potentially protective cytokines and chemokines in the milk of the two species. Furthermore, breast milk from both mangabeys and macaques was equally able to inhibit SIV infection of TZM-bl cells in vitro, suggesting that mangabey milk does not contain a particular factor or factors that have a superior ability to prevent SIV transmission to infants. The major distinction between the natural and non-natural hosts was the high level of $CD4^+CCR5^+$ target cells present within memory T cell subsets (central and effector) in blood as well as lymphoid and gastrointestinal tissues of infant macaques but that were almost absent in mangabey infants [29]. Altogether, these data strongly support a role of the infant mucosal milieu in blocking MTIT in natural hosts and provide a foundation for future investigations in this subject area.

The role of target cell availability determining susceptibility to infection is also supported by several other observations. In RMs, a species in which MTIT rates are very high, the numbers of $CD4^+CCR5^+$ T cells in circulation and at the mucosal sites are also correlated with the rates of transmission. Thus, the percentage of $CD4^+CCR5^+$ T cells is very low at birth in macaques [41] but increases rapidly with age, reaching levels comparable to those observed in adult macaques at only 6 months after birth. This observation may explain the results of breastfeeding transmission studies in macaques that reported that SIV transmission via breastfeeding occurs late during the lactation period [23]. This maturation of CCR5 expression by $CD4^+$ T cells may also be the reason behind the observation that the SIV doses required to infect newborn RMs are 10 times higher than those used for older infants [42]. Target cell availability at mucosal sites also appears to be important for SIV acquisition in adults. A comparison of the infectious dose required for successful experimental rectal transmission in AGMs and pigtailed macaques (PTMs) showed that the doses needed to infect adult AGMs are 10 times higher than those that successfully infected PTMs by the same route [3]. This susceptibility to infection was again correlated with the availability of $CD4^+CCR5^+$ target cells, whose numbers are 10 times lower in the rectum in the AGMs than in the PTMs [3].

Genetic Factors in SIV Transmission in Natural Hosts

Genetic factors have also been proposed to be involved in SIV transmission in natural hosts [24]. In the mandrill colony from Centre International de Recherches Medicales, Gabon, it was reported that transmission more frequently occurs in the same families. Some of these cases were transmissions from mothers to female offspring, but the majority were not vertical transmissions and were possibly the result of aggressive grooming (leading to small bites) or wound care or through premastication of food.

Although involvement of genetic factors in MTIT in natural hosts cannot be discarded, it appears that this is probably more critical for SIV sexual transmission. Indeed, it was recently documented that a fraction of multiparous AGM females remain negative in the wild despite numerous sexual encounters [20,21]. These AGMs may represent an interesting population of exposed, uninfected individuals, and their study may bring new insights into the mechanisms developed by these animals that allow them to remain negative despite numerous SIV exposures.

Lack of disease progression in natural hosts causes these species to be frequently considered nonrepresentative as models for HIV infection. However as shown by the past 20 years of AIDS research, natural host of SIVs were at the foundation of the new paradigms of SIV/HIV pathogenesis. The new focus on the determinants of SIV transmission in the natural hosts is a fascinating avenue of research that can help us to develop successful strategies to block HIV MTIT as well as horizontal HIV transmission.

References

[1] Pandrea I, Apetrei C. Where the wild things are: pathogenesis of SIV infection in African nonhuman primate hosts. Curr HIV/AIDS Rep 2010;7(1):28–36.

[2] Chahroudi A, Bosinger SE, Vanderford TH, Paiardini M, Silvestri G. Natural SIV hosts: showing AIDS the door. Science 2012;335(6073):1188–93. [Epub 2012/03/10].

[3] Pandrea I, Parrish NF, Raehtz K, Gaufin T, Barbian HJ, Ma D, et al. Mucosal simian immunodeficiency virus transmission in African green monkeys: susceptibility to infection is proportional to target cell availability at mucosal sites. J Virol 2012;86(8):4158–68. [Epub 2012/02/10].

[4] Phillips-Conroy JE, Jolly CJ, Petros B, Allan JS, Desrosiers RC. Sexual transmission of SIVagm in wild grivet monkeys. J Med Primatol 1994;23(1):1–7.

[5] Jolly CJ, Phillips-Conroy JE, Turner TR, Broussard S, Allan JS. SIVagm incidence over two decades in a natural population of Ethiopian grivet monkeys (*Cercopithecus aethiops aethiops*). J Med Primatol 1996;25(2):78–83.

[6] Otsyula MG, Gettie A, Suleman M, Tarara R, Mohamed I, Marx P. Apparent lack of vertical transmission of simian immunodeficiency virus (SIV) in naturally infected African green monkeys, *Cercopithecus aethiops*. Ann Trop Med Parasitol 1995;89(5):573–6.

[7] Denham WW. History of green monkeys in West Indies: migration from Africa. J BHMS 1981;36:210–29.

[8] Daniel MD, Letvin NL, Sehgal PK, Schmidt DK, Silva DP, Solomon KR, et al. Prevalence of antibodies to 3 retroviruses in a captive colony of macaque monkeys. Int J Cancer 1988;41(4):601–8.

[9] Hendry RM, Wells MA, Phelan MA, Schneider AL, Epstein JS, Quinnan GV. Antibodies to simian immunodeficiency virus in African green monkeys in Africa in 1957–62. Lancet 1986;2(8504):455.

[10] Souquiere S, Bibollet-Ruche F, Robertson DL, Makuwa M, Apetrei C, Onanga R, et al. Wild *Mandrillus sphinx* are carriers of two types of lentivirus. J Virol 2001;75(15):7086–96.

[11] Apetrei C, Kaur A, Lerche NW, Metzger M, Pandrea I, Hardcastle J, et al. Molecular epidemiology of SIVsm in US primate centers unravels the origin of SIVmac and SIVstm. J Virol 2005;79(14):8991–9005.
[12] Santiago ML, Range F, Keele BF, Li Y, Bailes E, Bibollet-Ruche F, et al. Simian immunodeficiency virus infection in free-ranging sooty mangabeys (*Cercocebus atys atys*) from the Tai Forest, Cote d'Ivoire: implications for the origin of epidemic human immunodeficiency virus type 2. J Virol 2005;79(19):12515–27.
[13] Chen Z, Telfier P, Gettie A, Reed P, Zhang L, Ho DD, et al. Genetic characterization of new West African simian immunodeficiency virus SIVsm: geographic clustering of household-derived SIV strains with human immunodeficiency virus type 2 subtypes and genetically diverse viruses from a single feral sooty mangabey troop. J Virol 1996;70(6):3617–27.
[14] Fultz PN, Gordon TP, Anderson DC, McClure HM. Prevalence of natural infection with simian immunodeficiency virus and simian T-cell leukemia virus type I in a breeding colony of sooty mangabey monkeys. AIDS 1990;4(7):619–25.
[15] VandeWoude S, Apetrei C. Going wild: lessons from T-lymphotropic naturally occurring lentiviruses. Clin Microbiol Rev 2006;19:728–62.
[16] Estaquier J, Peeters M, Bedjabaga L, Honore C, Bussi P, Dixson A, et al. Prevalence and transmission of simian immunodeficiency virus and simian T-cell leukemia virus in a semi-free-range breeding colony of mandrills in Gabon. AIDS 1991;5(11):1385–6.
[17] Georges-Courbot MC, Moisson P, Leroy E, Pingard AM, Nerrienet E, Dubreuil G, et al. Occurrence and frequency of transmission of naturally occurring simian retroviral infections (SIV, STLV, and SRV) at the CIRMF primate center, Gabon. J Med Primatol 1996;25(5):313–26.
[18] Pandrea I, Onanga R, Souquiere S, Mouinga-Ondéme A, Bourry O, Makuwa M, et al. Paucity of CD4+CCR5+ T-cells may prevent breast-feeding transmission of SIV in natural non-human primate hosts. J Virol 2008;82(11):5501–9.
[19] Apetrei C, Metzger MJ, Robinson D, Ling B, Telfer PT, Reed P, et al. Detection and partial characterization of new simian immunodeficiency virus (SIVsm) strains from bush meat samples from rural Sierra Leone. J Virol 2005;79(4):2631–6.
[20] Ma D, Jasinska A, Kristoff J, Grobler JP, Turner T, Jung Y, et al. SIVagm infection in wild African green monkeys from South Africa: epidemiology, natural history, and evolutionary considerations. PLoS Pathog 2013;9(1):e1003011. [Epub 2013/01/26].
[21] Ma D, Jasinska A, Feyertag F, Wijewardana V, Kristoff J, He T, et al. Factors associated with SIV transmission in a natural African nonhuman primate host in the wild. J Virol 88:5687–705.
[22] Aldrovandi GM, Kuhn L. What infants and breasts can teach us about natural protection from HIV infection. J Infect Dis 2010;202(Suppl. 3):S366–70. [Epub 2010/10/05].
[23] Amedee AM, Rychert J, Lacour N, Fresh L, Ratterree M. Viral and immunological factors associated with breast milk transmission of SIV in rhesus macaques. Retrovirology 2004;1(1):17.
[24] Fouchet D, Verrier D, Ngoubangoye B, Souquiere S, Makuwa M, Kazanji M, et al. Natural simian immunodeficiency virus transmission in mandrills: a family affair? Proc Biol Sci 2012;279(1742):3426–35. [Epub 2012/06/08].
[25] Chahroudi A, Meeker T, Lawson B, Ratcliffe S, Else J, Silvestri G. Mother-to-infant transmission of simian immunodeficiency virus is rare in sooty mangabeys and is associated with low viremia. J Virol 2011;85(12):5757–63. [Epub 2011/04/01].
[26] Amedee AM, Lacour N, Ratterree M. Mother-to-infant transmission of SIV via breast-feeding in rhesus macaques. J Med Primatol 2003;32(4–5):187–93.
[27] Van de Perre P, Rubbo PA, Viljoen J, Nagot N, Tylleskar T, Lepage P, et al. HIV-1 reservoirs in breast milk and challenges to elimination of breast-feeding transmission of HIV-1. Sci Transl Med 2012;4(143):143sr3. [Epub 2012/07/21].

[28] Wilks AB, Perry JR, Ehlinger EP, Zahn RC, White R, Gauduin MC, et al. High cell-free virus load and robust autologous humoral immune responses in breast milk of simian immunodeficiency virus-infected African green monkeys. J Virol 2011;85(18):9517–26. [Epub 2011/07/08].
[29] Chahroudi A, Cartwright E, Carnathan DG, Lee ST, Lawson B, Carnathan PM, et al. Target cell availability, rather than breast milk specific factors, dictates mother-to-infant transmission of SIV. Plos Pathog March 6, 2014;10(3):e1003958.
[30] Permar SR, Kang HH, Wilks AB, Mach LV, Carville A, Mansfield KG, et al. Local replication of simian immunodeficiency virus in the breast milk compartment of chronically-infected, lactating rhesus monkeys. Retrovirology 2010;7:7. [Epub 2010/02/04].
[31] Salazar-Gonzalez JF, Salazar MG, Learn GH, Fouda GG, Kang HH, Mahlokozera T, et al. Origin and evolution of HIV-1 in breast milk determined by single-genome amplification and sequencing. J Virol 2011;85(6):2751–63. [Epub 2010/12/31].
[32] Pillay K, Coutsoudis A, York D, Kuhn L, Coovadia HM. Cell-free virus in breast milk of HIV-1-seropositive women. J Acquired Immune Defic Syndr 2000;24(4):330–6. [Epub 2000/10/03].
[33] Ho C, Wu S, Amos JD, Colvin L, Smith SD, Wilks AB, et al. Transient compartmentalization of simian immunodeficiency virus variants in the breast milk of African green monkeys. J Virol 2013;87(20):11292–9. [Epub 2013/08/09].
[34] Rychert J, Amedee AM. The antibody response to SIV in lactating rhesus macaques. J Acquired Immune Defic Syndr 2005;38(2):135–41. [Epub 2005/01/27].
[35] Permar SR, Wilks AB, Ehlinger EP, Kang HH, Mahlokozera T, Coffey RT, et al. Limited contribution of mucosal IgA to Simian immunodeficiency virus (SIV)-specific neutralizing antibody response and virus envelope evolution in breast milk of SIV-infected, lactating rhesus monkeys. J Virol 2010;84(16):8209–18. [Epub 2010/06/04].
[36] Mabuka J, Nduati R, Odem-Davis K, Peterson D, Overbaugh J. HIV-specific antibodies capable of ADCC are common in breast milk and are associated with reduced risk of transmission in women with high viral loads. PLoS Pathog 2012;8(6):e1002739. [Epub 2012/06/22].
[37] Beaumier CM, Harris LD, Goldstein S, Klatt NR, Whitted S, McGinty J, et al. CD4 downregulation by memory CD4+ T cells in vivo renders African green monkeys resistant to progressive SIVagm infection. Nat Med 2009;15(8):879–85.
[38] Vinton C, Klatt NR, Harris LD, Briant JA, Sanders-Beer BE, Herbert R, et al. CD4-like immunological function by CD4- T cells in multiple natural hosts of simian immunodeficiency virus. J Virol 2011;85(17):8702–8. [Epub 2011/07/01].
[39] Amos JD, Wilks AB, Fouda GG, Smith SD, Colvin L, Mahlokozera T, et al. Lack of B cell dysfunction is associated with functional, gp120-dominant antibody responses in breast milk of simian immunodeficiency virus-infected African green monkeys. J Virol 2013;87(20):11121–34. [Epub 2013/08/09].
[40] Fouda GG, Yates NL, Pollara J, Shen X, Overman GR, Mahlokozera T, et al. HIV-specific functional antibody responses in breast milk mirror those in plasma and are primarily mediated by IgG antibodies. J Virol 2011;85(18):9555–67. [Epub 2011/07/08].
[41] Veazey RS, Lifson JD, Pandrea I, Purcell J, Piatak Jr M, Lackner AA. Simian immunodeficiency virus infection in neonatal macaques. J Virol 2003;77(16):8783–92.
[42] Chenine AL, Siddappa NB, Kramer VG, Sciaranghella G, Rasmussen RA, Lee SJ, et al. Relative transmissibility of an R5 clade C simian-human immunodeficiency virus across different mucosae in macaques parallels the relative risks of sexual HIV-1 transmission in humans via different routes. J Infect Dis 2010;201(8):1155–63. [Epub 2010/03/11].

CHAPTER

14

Immunogenetics of NK Cell Receptors and MHC Class I Ligands in Non-human Primates

Lutz Walter

Primate Genetics Laboratory, German Primate Center, Leibniz Institute for Primate Research, Kellnerweg, Göttingen, Germany

GENERAL ASPECTS OF THE MAJOR HISTOCOMPATIBILITY COMPLEX

In general, the term "Immunogenetics" encompasses all genes that contribute to immune responses and immunity. Certain immune genes, however, have a major impact on the immune response and control immune responsiveness and hence were formerly designated "IR genes" (immune response genes). Further hallmarks of IR genes are the extraordinary allelic variability, rapid evolution, and a certain tendency to form gene families (polygeny). More than 25 years ago, researchers realized that the IR genes are actually major histocompatibility complex (MHC) class II genes. The MHC is a region of about 4 Mb in mammals and contains more than 200 genes [1,2]. The genes in this complex are distinguished as class I, class II, and class III genes [3,4] that have different functions in the immune response (Figure 14.1). The number of alleles of certain genes of the human MHC (HLA system) is enormous: for example, more than 3000 alleles are known for the class I gene HLA-B (for actual numbers see Immuno Polymorphism Database IPD at http://www.ebi.ac.uk/ipd/imgt/hla/stats.html).

The most distinguished features of the MHC are its enormous genetic variability and its rapid evolution. To discriminate between different types of variability, there is agreement to define allelic variability as *polymorphism* and copy-number variation (CNV) of genes as *diversity*. On a population

Natural Hosts of SIV.
http://dx.doi.org/10.1016/B978-0-12-404734-1.00014-0

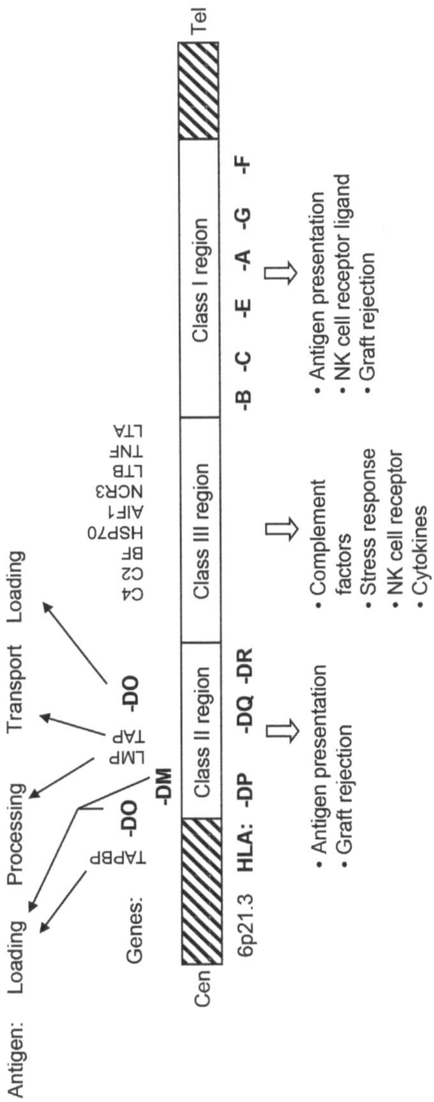

FIGURE 14.1 Scheme of the MHC and associated functions. The MHC encompasses roughly 4 Mb on human chromosome 6 in 6p21.3 and 5.3 Mb on rhesus macaque chromosome 6 in 6q24. A selection of nonclass I, nonclass II genes is shown that perform essential functions in adaptive and innate immune responses. Class I and class II genes are in bold. The centromeric and telomeric orientation of the chromosome is indicated.

level, forms of balancing selection such as heterozygote advantage or frequency-dependent selection are thought to maintain both polymorphism and diversity and thereby protect the population from pathogenic threat. On the level of individuals, certain alleles or combination of maternally or paternally inherited alleles (haplotypes) may result in susceptibility to infectious diseases. Indeed, data acquired on population genetics indicate that certain alleles or haplotypes are more frequently found in patients than in controls in not only a variety of infectious diseases, but also in a number of autoimmune diseases, allergies, and hematological malignancies [5]. One of the most prominent examples of the role of the MHC in infectious disease is its role in human immunodeficiency virus (HIV)-1 infection, where full heterozygosity at HLA-A, HLA-B, and HLA-C genes is significantly associated with prolonged time to acquire acquired immunodeficiency syndrome (AIDS) and presence of HLA-B*35 and HLA-Cw4 with rapid progression [6]. This study was also an important confirmation of the "heterozygous advantage" of the MHC, a hypothesis already put forward by Nobel laureates Doherty and Zinkernagel in 1975 [7].

THE MHC OF MACAQUE SPECIES

The MHC loci of the rhesus macaque (*Macaca mulatta*) is designated by the symbol Mamu. A single Mamu haplotype has been sequenced completely and spans 5.3 Mb [8] on chromosome 6q24 [9]. Discrepant data have been published in the past on the chromosomal assignment of Mamu: chromosome 2 or chromosome 5 [10], or chromosome 4 in public databases such as University of California, Santa Cruz (UCSC) genome browser (http://genome.ucsc.edu/cgi-bin/hgGateway) or in the GenBank genome database. However, fluorescence in situ hybridization-mapping with MHC gene-containing cosmid clones combined with classical RBA and distamycin A/4′,6-diamidino-2-phenylindole-like banding patterns clearly identified the localization of the Mamu gene complex to the long arm of chromosome six at 6q24 [9].

The Mamu system exhibits a similar genomic organization as the HLA complex (Figure 14.2) [8,9,11]. However, there were also important differences seen that could significantly contribute to differences in the immune response to certain antigens. Such differences were encountered both in the MHC class I and class II regions. Due to frequent gene duplications, the Mamu class II region is characterized by a larger number of MHC-DRB genes contributing to extensive Mamu class II region configurations (diversity). In contrast, the number of MHC class II alleles (polymorphism) is rather low in rhesus macaques [12,13]. Thus, variability of Mamu class II genes may not be focused on polymorphism as in humans, but rather on

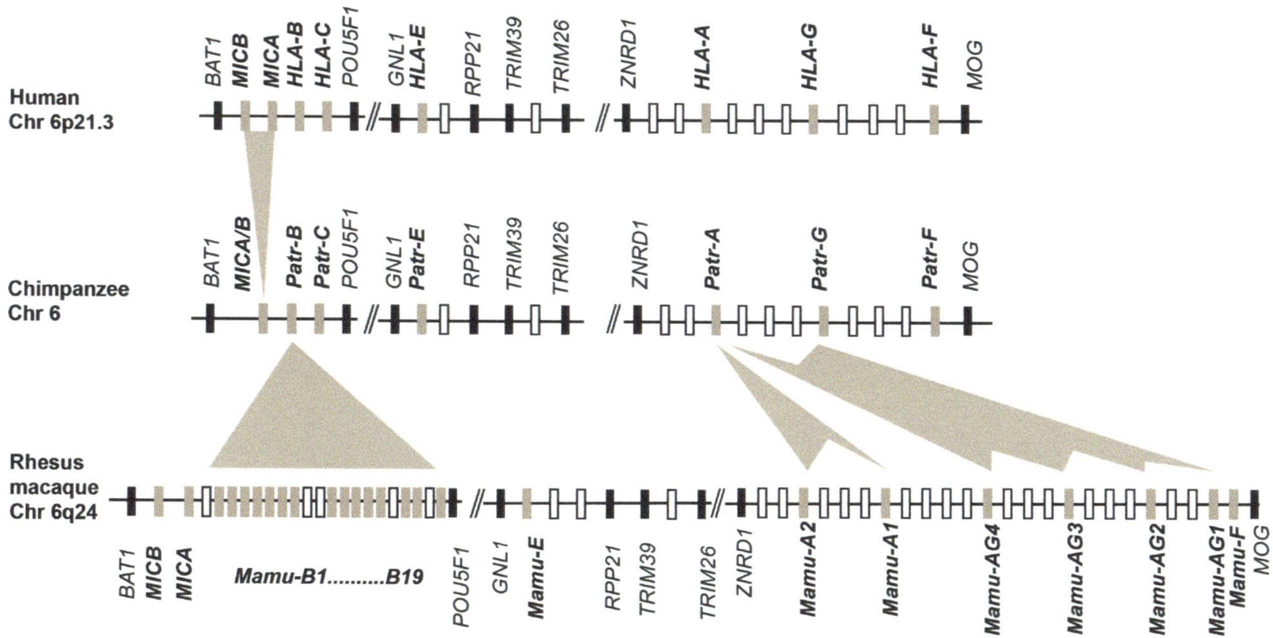

FIGURE 14.2 Comparison of human, chimpanzee, and rhesus macaque MHC class I regions. Nonclass I genes flanking the MHC class I genes are shown as black boxes, and class I genes and MHC class I-like MIC genes are shown as gray boxes. Open symbols represent MHC class I pseudogenes. Gene fusions (MICA/B in chimpanzee) and expansions (Mamu-B, Mamu-A, Mamu-AG in rhesus macaques) are indicated by gray triangles and polygons. Genomic organization of human, chimpanzee, and rhesus macaque MHC is extracted from references [1,8,106].

haplotypic diversity [13]. The Mamu-DQ and Mamu-DP regions are quite similar to HLA-DQ and HLA-DP, respectively, except that only one copy of Mamu-DQA and Mamu-DQB is present [8].

The Mamu MHC class I region shows the same overall structure as in humans (Figure 14.2) [8,14] and follows the general pattern of the MHC class I region in mammals, with clusters of class I genes being embedded between clusters of highly conserved nonclass I genes [3]. The only true orthologous MHC class I genes are Mamu-E and HLA-E as well as Mamu-F and HLA-F. The Mamu-A, Mamu-B, and Mamu-G genes show characteristics of the single copy HLA-A, HLA-B, and HLA-G genes but have expanded considerably (Figure 14.2) and are therefore paralogous to these human class I genes. The Mamu-G genes represent pseudogenes [15], and their function was replaced by Mamu-AG genes that share sequence characteristics of Mamu-A genes and show a restricted expression pattern like HLA-G [16].

The Mamu-A and Mamu-B regions of rhesus and other macaque species are characterized by substantial diversity (Figure 14.2). Four and seven Mamu-A genes were described for Indian and Chinese rhesus macaques, respectively [17,18]. Notably, the Mamu-A genes show both polymorphism and diversity: the Mamu-A1 gene is present on almost every haplotype (low diversity) and exhibits a considerable number of alleles, whereas the other six Mamu-A genes (Mamu-A2 to Mamu-A7) are oligomorphic but show substantial diversity (CNVs) [17,18]. Whereas Mamu-A1, Mamu-A2, and Mamu-A3-encoded proteins are densely expressed upon transfection, Mamu-A4 shows only a very low level of cell-surface expression [19]. The Mamu-B region is distinguished by considerable diversity (Figure 14.2). Mamu-B haplotypes show substantial CNVs of Mamu-B genes [17,18,20]. The complexity of the Mamu-B subregion is enormous, as exemplified by the sequenced haplotype by Daza-Vamenta and colleagues [8] who found 19 Mamu-B genes, of which 14 appear to be functional based on their sequence criteria or on identity to known cDNA, and by cDNA cloning and family studies [17,18]. Assignment of individual Mamu-B sequences to Mamu-B genes has turned out much more difficult compared to Mamu-A genes where assignment of novel sequences to individual genes is straightforward. Therefore, Mamu-B sequences still have an allele-like nomenclature. The Mamu-B region is rich in repetitive elements and is, therefore, highly prone to segmental duplications, giving rise to CNVs and differential Mamu-B gene content [8,14,21]. The Mamu-B region also contains the so-called Mamu-I genes that are a subtype of Mamu-B and were previously designated Mamu-B*09 [22].

The plethora of Mamu-A and Mamu-B genes in single rhesus macaque individuals challenges a current dogma in T-cell immunology: mammals should come up with only a few MHC class I genes/proteins to avoid severe limitations of their $CD8^+$ T-cell repertoire. So, how is this issue resolved in macaques? The answer to this problem may lie in the transcriptional control and the subcelluar localization of many Mamu-B genes/proteins. Certain Mamu-B genes are either not transcribed at all (or only under certain conditions), and if transcribed only to a low degree [17,18], or only in a special cell type [23]. In addition, certain Mamu-B molecules such as B*012, B*028, B*030, B*57, or I*010 are almost exclusively found intracellularly and show only a very low level of expression at the cell surface [19]. Thus, macaque MHC class I genes are functionally diversified, with some showing regular expression patterns and performing classical functions and others being strictly regulated and likely performing special functions. Thus, the macaque CD8+ T-cell repertoire is obviously shaped by only few (classical) MHC class I molecules as is found in other mammals; thereby, establishment of a broad T-cell receptor repertoire may be secured.

Although significant detailed structural knowledge of the rhesus macaque (Mamu) MHC has already been acquired, similar knowledge of the MHC is currently not available for other macaque species and, unfortunately for any of the natural hosts of simian immunodeficiency virus (SIV). However, considerable progress was achieved recently in the analysis of MHC polymorphism and diversity of long-tailed (also called cynomolgus) [24–28] and pig-tailed macaques [29–33], which both are important NHP models of human AIDS The MHC of long-tailed macaques (*Macaca fascicularis*) is called Mafa and that of pig-tailed macaques (*Macaca nemestrina*) is designated Mane. Both Mafa and Mane are similar to Mamu with duplications of genes in the Mafa-DRB, Mafa-A, and Mafa-B regions [27]. Only very limited knowledge of the MHC is available for stump-tailed macaques (*Macaca arctoides*) [34] and lion-tailed macaques (*Macaca silenus*) [35].

THE LEUKOCYTE RECEPTOR COMPLEX

Natural killer (NK) cells represent an important and essential part of the innate immune system. In contrast to B and T cells, these lymphocytes express germline-encoded receptors on their surface that interact with a variety of different ligands. The functions of NK cells include the killing of infected or transformed cells and the regulation of adaptive immune reactions via release of cytokines and chemokines. They scan other cells for presence of appropriate ligands and usually receive signals via their inhibitory and activating receptors. Inhibitory signaling actively suppresses signals from activating receptors [36,37], whereby healthy cells are normally protected from lysis by NK cells. However, viruses have developed sophisticated mechanisms to evade the recognition by T cells through downregulation of most host MHC class I molecules on the surface of infected cells [38,39]. MHC class I molecules are important ligands for inhibitory NK cell receptors, and such virally induced downregulation leads to loss of interaction between inhibitory NK receptors and cognate ligands, finally resulting in activation of the NK cell and killing of virus-infected target cells. However, viruses can counteract NK cell recognition by expression of virally encoded MHC class I-like proteins that enable interaction with inhibitory NK cell receptors [38,39]. Downregulation of MHC class I molecules is also seen for tumor cells as a result of selective pressure of the adaptive immune system against transformed cells.

NK cell receptors belong to two structurally distinct families: immunoglobulin (Ig)-like and C-type lectin-like receptors. Primates have developed a considerable degree of polymorphism and diversity in their so-called killer cell immunoglobulin-like (Ig-like) receptors (KIR) with variable numbers of Ig-like domains [40,41]. This is also reflected in the nomenclature of these proteins: KIR1D, KIR2D, or KIR3D, where D stands for domain. The prevailing KIRs in humans and great apes are of the KIR2D type [42–45], whereas the dominating KIRs in macaques are KIR3D [46–50]. Inhibitory NK cell receptors have long cytoplasmic tails containing inhibitory motifs (ITIMs), and activating receptors have short tails and interact with adaptor molecules that carry activating motifs (ITAMs) and transmit activating signals. These characteristics are also found in the designation of KIR genes: KIR3DL and KIR3DS encode inhibitory (long) and activating (short) receptors, respectively. With the exception of prosimian primates [51], KIR genes build a family of duplicated and expanded genes in primates that form a single cluster [52] that is part of the leukocyte receptor complex (LRC).

Due to their tight functional linkage, there is strong coevolution of MHC class I and KIR genes. MHC-C molecules are important ligands for KIR2D receptors in humans and great apes [43,53–56]. There is accumulating evidence that the emergence of MHC-C, which is fixed in gorillas, chimpanzees, and humans and is present in 50% of orangutan individuals [57], and the expansion of KIR2D genes are linked [45,53,58]. Along the same lines, the expansion of KIR3D genes and the amplification of MHC-A and MHC-B genes are believed to be correlated in macaques [41]. Notably, results of studies of the gibbon species also support the concept of KIR/ligand coevolution: gibbon KIR haplotypes are characterized by their small size, lack of expansion of KIR2D and lack of MHC-C (the potential receptor for KIR2D), and lack of MHC-G and deletions of its potential receptor KIR2DL4 [59].

Macaque KIR haplotypes reveal substantial expansions of KIR3D genes [46–50,60,61]. The complexity of rhesus macaque KIR [49,61,62] and MHC haplotypes [8,18,63–65] is remarkable, making typing of these two highly variable genomic regions a difficult and technically challenging task. In general, it is difficult to discriminate KIR sequences as being derived from

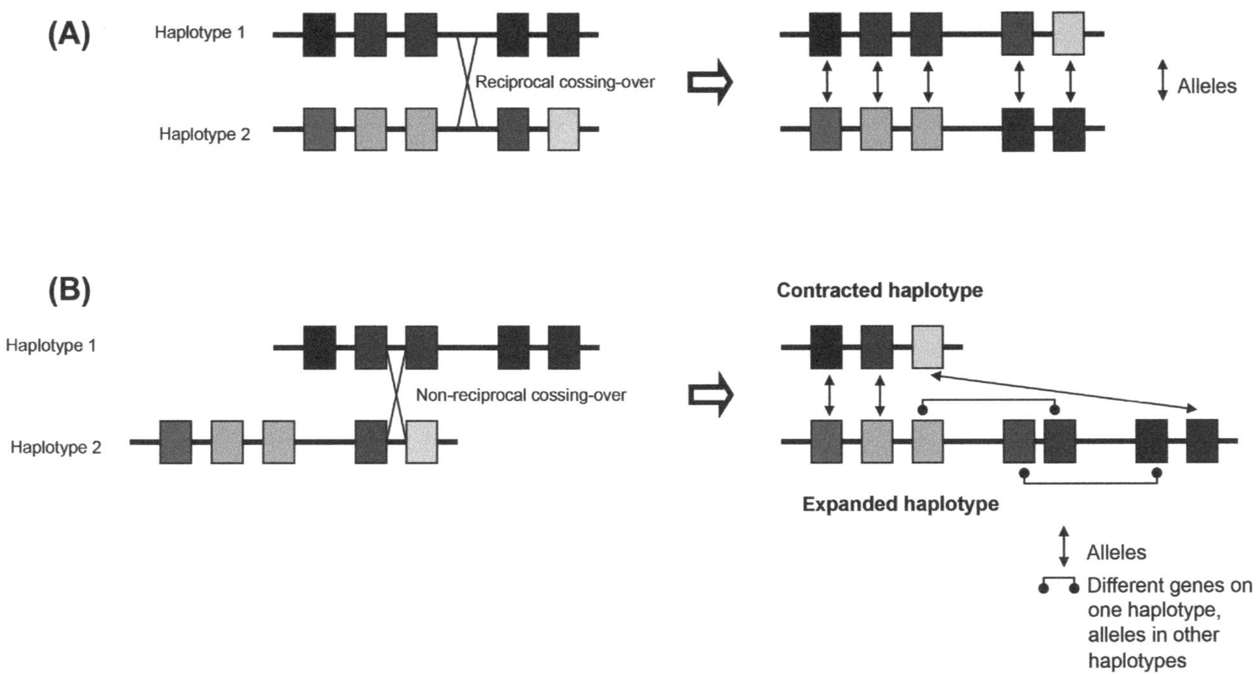

FIGURE 14.3 Scheme of reciprocal (A) and nonreciprocal (B) crossing-over events during meiosis, leading to stable and differential numbers of KIR genes in haplotypes. (See color plate at the back of the book.)

different KIR genes or from the same gene (allelic relationship) due to high sequence similarity. Because only a single rhesus macaque KIR haplotype was sequenced [47], knowledge of the structural organization of the KIR region in macaques is generally lacking but is essential for the establishment of locus-specific primers to perform polymerase chain reaction-based typing using genomic DNA. Currently, sophisticated typing of both KIR and MHC class I is performed by next-generation sequencing of gene transcripts, a method developed by David O'Connor's lab [32,61,66]. A further complexity in the discrimination of KIR genes and alleles arises from nonreciprocal crossing-over in meiosis [67–69]: sequences that are alleles on most haplotypes can become separate genes on the same haplotype (Figure 14.3). The genetic variability is additionally enhanced via shuffling of exons between distinct KIR genes and recurrent formation of activating KIRs from inhibitory KIRs [70–72]. In concert with their polymorphic MHC class I ligands, the highly variable KIR proteins form extremely complex genetic configurations that are associated with susceptibility to various diseases [73]. Recently, a few monoclonal antibodies against rhesus macaque KIR proteins were established that possess either broad reactivity (pan-rhesus macaque KIR3D) or are specific for only a single KIR3D protein [74].

THE NATURAL KILLER COMPLEX

The natural killer complex (NKC) encodes a variety of C-type lectin-like receptors. The best-characterized ones are CD94 and the natural killer group 2 (NKG2) family of receptors. The human CD94 molecule forms heterodimers with NKG2A, NKG2C, or NKG2E. Whereas CD94/NKG2A forms an inhibitory receptor due to the presence of immunoreceptor tyrosin-based inhibitory motifs (ITIM) in NKG2A, CD94/NKG2C and CD94/NKG2E are activating receptors with NKG2C and NKG2E, having short cytoplasmic tails and the ability to associate with activating adaptor molecules (see above for KIR). The CD94/NKG2 heterodimeric receptors are highly conserved in humans, apes, Old and New World monkeys [51,75,76], and in humans interact with the nonpolymorphic HLA-E ligand [77]. The function of this conserved and invariable NK cell receptor/ligand system is believed to detect viral interference with host MHC class I protein expression [58,77,78] and to function as a back-up system to ensure tolerance of NK cells toward self, if an interacting MHC class I ligand is not available for KIR-expressing NK cells. Recently, human NK cells expressing the activating CD94/NKG2C receptor were shown to be particularly frequent in cytomegalovirus (CMV) seropositive individuals [79], indicating that this receptor plays a crucial role in the control of CMV infection. It remains to be shown whether similar observations can be made in macaques naturally infected with CMV.

Gene transcripts of CD94, NKG2A, NKG2C, and NKG2D were described in rhesus macaques [80–82] and long-tailed macaques [82]. Monoclonal antibodies raised against human NKG2A and NKG2C proteins were tested for cross-reactivity with corresponding macaque proteins. Antibody clone P25 detects both CD94/NKG2A and CD94/NKG2C in humans but does not react with similar macaque receptors. In contrast, the monoclonal antibody clone Z199 not only specifically binds to human CD94/NKG2A, but also reacts with both macaque CD94/NKG2A and CD94/NKG2C [82].

Interestingly, lemurs (and possibly other strepsirrhine primates and tarsier) use the CD94/NKG2 system and not KIR as their polymorphic and diverse NK cell receptors [51]. In accord with strong coevolution between NK cell receptors and MHC class I, also the MHC class I genomic region is bizarre compared to other mammals: all functional MHC class I genes of the gray mouse lemur (*Microcebus murinus*) are found on chromosome 26, in contrast to the MHC class II region that maps to *M. murinus* chromosome 6. Intriguingly, all MHC class I genes in the "old" MHC class I region on chromosome six were inactivated a long time ago and represent pseudogenes [51].

The conserved NKG2D receptor is only distantly related to the other NKG2 proteins and forms homodimers. NKG2D is an activating receptor that interacts with a variety of different ligands that are all distantly related members of the MHC class I protein family. Among these ligands are the highly polymorphic MHC class I chain related MICA and MICB proteins and the UL16-binding proteins ULBP1 to ULBP6. These ligands are inducible by stress such as viral infection or DNA damage and thus play an important role in the detection and elimination of infected or transformed cells [83]. Human MICA and MICB genes are considerably polymorphic [84]. Although the biological significance of the majority of MICA or MICB polymorphisms remains unclear, a methionine-to-valine substitution at amino acid position 129 of human MICA is associated with different affinity for the NKG2D receptor [85]. Rhesus macaques also possess two functional MIC genes, MICA and MICB [86]. MICB was previously designated MIC2 and MICA is expressed as two divergent allelic lineages that were previously designated MIC1 and MIC3 [86]. Although MIC1 and MIC3 unambiguously segregate as alleles in offspring [86] and were never observed together on the same haplotype [86,87], others regard it as a separate gene that was created by fusion between a MICA and MICB gene [87].

ASSOCIATION OF MHC AND LRC WITH INFECTIOUS DISEASES

There are numerous rhesus macaque models of human infectious diseases that were reviewed in an excellent report by Gardner and Luciw [88] (article available via open access at http://ilarjournal.oxfordjournals.org/content/49/2/220.full.pdf). Disease association studies have been particularly performed for the MHC in rhesus macaques that were experimentally infected with SIV, and study suggested that the clinical course of SIV-mediated disease was correlated with the MHC class I genotype [89]. Four MHC class I alleles have been reported to be associated with control of SIV infection: Mamu-A1*001 [90–94], Mamu-A3*13:03 [92], Mamu-B*008 [94,95], and Mamu-B*017 [96]. Also MHC class I haplotypes have been identified that mediate either rapid, intermediate, or slow progression to AIDS in SIV-infected rhesus macaques of Indian origin [97]. SIV infection studies in long-tailed macaques of Mauritian origin experimentally infected with SHIVsbg [98] and SIVmac251 [99], identified MHC haplotypes M3 and M6 as playing a role in the control of infection, respectively.

Specific interactions between KIR3D and Mamu class I proteins have been described for rhesus and pig-tailed macaques. In rhesus macaques, robust interactions were identified for KIR3DLW03 with Mamu-A1 allotypes 001 and 011 [100] and for KIR3DL05 with Mamu-A1 allotypes 001 [100], 002 [100,101], and Mamu-A3*13 [100]. The KIR3DL protein encoded by the KIR049-4 allele interacts with both Mane-A1 allotypes *082 and *084 in pig-tailed macaques [102]. Thus, the A1-encoded MHC class I proteins obviously play an important role as interaction partners of KIR3D receptors in these two macaque species. In accord with such importance is the presence of the A1 MHC class I gene on almost all Mamu class I haplotypes [64]. Human KIR3D proteins show restricted specificity: KIR3DL1 interacts only with HLA-B molecules carrying the serological epitope Bw4 and KIR3DL2 binds to the HLA-A allotypes A3 and A11. Those rhesus and pig-tailed macaque KIR3D for which interaction data are available do not show such restrictions and interact with both Bw4- and Bw6-containing MHC class I ligands [100–102], indicating some level of degeneracy in the recognition of class I molecules by macaque KIR3D proteins.

Peptides presented by MHC class I proteins not only determine interaction with T-cell receptors, but also have an obvious influence on interaction of these ligands with KIR proteins. This peptide dependency was demonstrated in macaques by using MHC class I tetramers loaded with different peptides and testing the interaction with KIR proteins ectopically expressed by cells [101,102]. These findings suggest that small changes in peptide content of MHC class I molecules may regulate the activity of human NK cells [103]. Thus, polymorphisms that influence the peptide repertoire of MHC class I proteins and pathogen-induced changes in peptide content might not only be important for T-cell responses, but also for NK cell recognition. Such slight changes in presented peptides might indeed be functionally relevant as a recent study showed specific clustering of MHC class I molecules presenting the same peptide on the cell surface [104]. Interestingly, mutations in select HIV-1 sequences obtained from patients when presented by appropriate MHC class I molecules were shown to enhance interaction with inhibitory KIR, leading to reduced NK cell recognition of infected cells in vitro [105]. Whether a similar evasion of NK cell responses is also present in SIV infection models of macaques is currently not known.

Our knowledge of the immunogenetic variability both in terms of polymorphism and diversity in NHP model species is constantly increasing. However, we are only beginning to understand the specific interactions between MHC class I ligands and NK cell receptors and the role of genetic variability in this interaction. The enormous expansion of genes of both ligands and receptors and their considerable copy-number variations in macaques represent a further challenge. Beyond doubt, this knowledge further contributes to the constant refinement of NHP models of human infectious diseases. Because rodents do not possess corresponding KIR proteins, these NHP models are particularly important for studies of those human diseases where KIR proteins play an important role such as in HIV-1 infection.

References

[1] The MHC sequencing consortium. Complete sequence and gene map of a human major histocompatibility complex. Nature 1999;401:921–3.

[2] Hurt P, Walter L, Sudbrak R, Klages S, Müller I, Shiina T, et al. The genomic sequence and comparative analysis of the rat major histocompatibility complex. Genome Res 2004;14:631–9.

[3] Günther E, Walter L. Comparative genomic aspects of rat, mouse and human MHC class I gene regions. Cytogenet Cell Genet 2000;91:107–12.

[4] Kelley J, Walter L, Trowsdale J. Comparative genomics of major histocompatibility complexes. Immunogenetics 2005;56:683–95.

[5] Thorsby E. Invited anniversary review: HLA associated diseases. Hum Immunol 1997;53:1–11.

[6] Carrington M, Nelson GW, Martin MP, Kissner T, Vlahov D, Goedert JJ, et al. HLA and HIV-1: heterozygote advantage and B*35-Cw*04 disadvantage. Science 1999;283:1748–52.

[7] Doherty PC, Zinkernagel RM. Enhanced immunological surveillance in mice heterozygous at the H-2 gene complex. Nature 1975;256:50–2.

[8] Daza-Vamenta R, Glusman G, Rowen L, Guthrie B, Geraghty DE. Genetic divergence of the rhesus macaque major histocompatibility complex. Genome Res 2004;14:1501–15.

[9] Huber I, Walter L, Wimmer R, Pasantes JJ, Günther E, Schempp W. Cytogenetic mapping and orientation of the rhesus macaque MHC. Cytogenet Genome Res 2004;103:144–9.

[10] Hirai M, Takahashi E, Ishida T, Hori T. Chromosomal localization of the major histocompatibility complex (MHC) in the rhesus monkey and chimpanzee by fluorescence in situ hybridization. Cytogenet Cell Genet 1991;57:204–5.

[11] Sudbrak R, Reinhardt R, Hennig S, Lehrach H, Günther E, Walter L. Comparative and evolutionary analysis of the rhesus macaque extended MHC class II region. Immunogenetics 2003;54:699–704.

[12] Doxiadis GG, Otting N, de Groot NG, Noort R, Bontrop RE. Unprecedented polymorphism of Mhc-DRB region configurations in rhesus macaques. J Immunol 2000;164:3193–9.

[13] Doxiadis GG, Otting N, de Groot NG, Bontrop RE. Differential evolutionary MHC class II strategies in humans and rhesus macaques: relevance for biomedical studies. Immunol Rev 2001;183:76–85.

[14] Kulski JK, Anzai T, Shiina T, Inoko H. Rhesus macaque class I duplicon structures, organization, and evolution within the alpha block of the major histocompatibility complex. Mol Biol Evol 2004;21:2079–91.

[15] Boyson JE, Shufflebotham C, Cadavid LF, Urvater JA, Knapp LA, Hughes AL, et al. The MHC class I genes of the rhesus monkey. Different evolutionary histories of MHC class I and II genes in primates. J Immunol 1996;156:4656–65.

[16] Boyson JE, Iwanaga KK, Golos TG, Watkins DI. Identification of a novel MHC class I gene, Mamu-AG, expressed in the placenta of a primate with an inactivated G locus. J Immunol 1997;159:3311–21.

[17] Otting N, Heijmans CMC, Noort RC, de Groot NG, Doxiadis GGM, van Rood JJ, et al. Unparalleled complexity of the MHC class I region in rhesus macaques. Proc Natl Acad Sci USA 2005;102:1626–31.

[18] Otting N, Heijmans C, van der Wiel M, de Groot N, Doxiadis G, Bontrop R. A snapshot of the Mamu-B genes and their allelic repertoire in rhesus macaques of Chinese origin. Immunogenetics 2008;60(9):507–14. http://dx.doi.org/10.1007/s00251-008-0311-5.

[19] Rosner C, Kruse PH, Lübke T, Walter L. Rhesus macaque MHC class I molecules show differential subcellular localizations. Immunogenetics 2010;62:149–58.

[20] Karl JA, Wiseman RW, Campbell KJ, Blasky AJ, Hughes AL, Ferguson B, et al. Identification of MHC class I sequences in Chinese-origin rhesus macaques. Immunogenetics 2008;60:37–46.

[21] Bonhomme M, Doxiadis GG, Heijmans CM, Vervoort V, Otting N, Bontrop RE, et al. Genomic plasticity of the immune-related Mhc class I B region in macaque species. BMC Genomics 2008;9:514.

[22] Urvater JA, Otting N, Loehrke JH, Rudersdorf R, Slukvin II, Piekarczyk MS, et al. Mamu-I: a novel primate MHC class I B-related locus with unusually low variability. J Immunol 2000;164:1386–98.

[23] Greene JM, Wiseman RW, Lank SM, Bimber BN, Karl JA, Burwitz BJ, et al. Differential MHC class I expression in distinct leukocyte subsets. BMC Immunol 2011;12:39.

[24] Watanabe A, Shiina T, Shimizu S, Hosomichi K, Yanagiya K, Kita YF, et al. A BAC-based contig map of the cynomolgus macaque (Macaca fascicularis) major histocompatibility complex genomic region. Genomics 2007;89:402–12.

[25] Kita YF, Hosomichi K, Kohara S, Itoh Y, Ogasawara K, Tsuchiya H, et al. MHC class I A loci polymorphism and diversity in three Southeast Asian populations of cynomolgus macaque. Immunogenetics 2009;61:635–48.

[26] Mitchell JL, Mee ET, Almond NM, Cutler K, Rose NJ. Characterisation of MHC haplotypes in a breeding colony of Indonesian cynomolgus macaques reveals a high level of diversity. Immunogenetics 2012;64:123–9.

[27] Otting N, de Groot N, de Vos-Rouweler AJM, Louwerse A, Doxiadis GGM, Bontrop RE. Multilocus definition of MHC haplotypes in pedigreed cynomolgus macaques (*Macaca fascicularis*). Immunogenetics 2012;64(10):755–65.

[28] Budde ML, Wiseman RW, Karl JA, Hanczaruk B, Simen BB, O'Connor DH. Characterization of Mauritian cynomolgus macaque major histocompatibility complex class I haplotypes by high-resolution pyrosequencing. Immunogenetics 2010;62:773–80.

[29] Lafont BAP, Buckler-White A, Plishka R, Buckler C, Martin MA. Characterization of pig-tailed macaque classical MHC class I genes: implications for MHC evolution and antigen presentation in macaques. J Immunol 2003;171:875–85.

[30] Pratt BF, O'Connor DH, Lafont BAP, Mankowski JL, Fernandez CS, Triastuti R, et al. MHC class I allele frequencies in pigtail macaques of diverse origin. Immunogenetics 2006;58:995–1001.

[31] Wu J, Bassinger S, Montoya GD, Chavez L, Jones CE, Holder-Lockyer B, et al. Allelic diversity within the high frequency Mamu-A2*05/Mane-A2*05 (Mane-A*06)/Mafa-A2*05 family of macaque MHC-A loci. Tissue Antigens 2008;72:29–38.
[32] O'Leary CE, Wiseman RW, Karl JA, Bimber BN, Lank SM, Tuscher JJ, et al. Identification of novel MHC class I sequences in pig-tailed macaques by amplicon pyrosequencing and full-length cDNA cloning and sequencing. Immunogenetics 2009;61:689–701.
[33] Fernandez CS, Reece JC, Saepuloh U, De Rose R, Ishkandriati D, O'Connor DH, et al. Screening and confirmatory testing of MHC class I alleles in pig-tailed macaques. Immunogenetics 2011;63:511–21.
[34] Otting N, de Groot NG, Doxiadis GGM, Bontrop RE. Extensive Mhc-DQB variation in humans and non-human primate species. Immunogenetics 2002;54:230–9.
[35] Blankenburg A, Kaup FJ, Sauermann U. Mhc class II DRB sequences of lion-tailed macaques (*Macaca silenus*). Tissue Antigens 2003;62:267–9.
[36] Binstadt BA, Brumbaugh KM, Dick CJ, Scharenberg AM, Williams BL, Colonna M, et al. Sequential involvement of Lck and SHP-1 with MHC-recognizing receptors on NK cells inhibits FcR-initiated tyrosine kinase activation. Immunity 1996;5:629–38.
[37] Campbell KS, Dessing M, López-Botet M, Cella M, Colonna M. Tyrosine phosphorylation of a human killer inhibitory receptor recruits protein tyrosine phosphatase 1C. J Exp Med 1996;184:93–100.
[38] Jonjić S, Babić M, Polić B, Krmpotić A. Immune evasion of natural killer cells by viruses. Curr Opin Immunol 2008;20:30–8.
[39] Ploegh HL. Viral strategies of immune evasion. Science 1998;280:248–53.
[40] Parham P, Abi-Rached L, Matevosyan L, Moesta AK, Norman PJ, Older Aguilar AM, et al. Primate-specific regulation of natural killer cells. J Med Primatol 2010;39: 194–212.
[41] Walter L. Major histocompatibility complex class-I-interacting natural killer cell receptors of nonhuman primates. J Innate Immun 2011;3:236–41.
[42] Uhrberg M, Valiante NM, Shum BP, Shilling HG, Lienert-Weidenbach K, Corliss B, et al. Human diversity in killer cell inhibitory receptor genes. Immunity 1997;7:753–63.
[43] Khakoo SI, Rajalingam R, Shum BP, Weidenbach K, Flodin L, Muir DG, et al. Rapid evolution of NK cell receptor systems demonstrated by comparison of chimpanzees and humans. Immunity 2000;12:687–98.
[44] Rajalingam, Hong M, Adams EJ, Shum BP, Guethlein LA, Parham P. Short KIR haplotypes in pygmy chimpanzee (Bonobo) resemble the conserved framework of diverse human KIR haplotypes. J Exp Med 2001;193:135–46.
[45] Guethlein LA, Older Aguilar AM, Abi-Rached L, Parham P. Evolution of killer cell Ig-like receptor (KIR) genes: definition of an orangutan KIR haplotype reveals expansion of lineage III KIR associated with the emergence of MHC-C. J Immunol 2007;179:491–504.
[46] Hershberger KL, Shyam R, Miura A, Letvin NL. Diversity of the killer cell Ig-like receptors of rhesus monkeys. J Immunol 2001;166:4380–90.
[47] Sambrook JG, Bashirova A, Palmer S, Sims S, Trowsdale J, Abi-Rached L, et al. Single haplotype analysis demonstrates rapid evolution of the killer immunoglobulin-like receptor (KIR) loci in primates. Genome Res 2005;15:25–35.
[48] Bimber BN, Moreland AJ, Wiseman RW, Hughes AL, O'Connor DH. Complete characterization of killer Ig-like receptor (KIR) haplotypes in Mauritian cynomolgus macaques: novel insights into nonhuman primate KIR gene content and organization. J Immunol 2008;181:6301–8.
[49] Kruse PH, Rosner C, Walter L. Characterization of rhesus macaque KIR genotypes and haplotypes. Immunogenetics 2010;62:281–93.
[50] Blokhuis JH, van der Wiel MK, Doxiadis GGM, Bontrop RE. The mosaic of KIR haplotypes in rhesus macaques. Immunogenetics 2010;62:295–306.

[51] Averdam A, Petersen B, Rosner C, Neff J, Roos C, Eberle M, et al. A novel system of polymorphic and diverse NK cell receptors in primates. PLoS Genet 2009;5:e1000688.
[52] Kelley J, Walter L, Trowsdale J. Comparative genomics of natural killer cell receptor gene clusters. PLoS Genet 2005;1:129–39.
[53] Older Aguilar AM, Guethlein LA, Adams EJ, Abi-Rached L, Moesta AK, Parham P. Coevolution of killer cell Ig-like receptors with HLA-C to become the major variable regulators of human NK cells. J Immunol 2010;185:4238–51.
[54] Moesta AK, Parham P. Diverse functionality among human NK cell receptors for the C1 epitope of HLA-C: KIR2DS2, KIR2DL2, and KIR2DL3. Front Immunol 2012;3. http://dx.doi.org/10.3389/fimmu.2012.00336.
[55] Moesta AK, Abi-Rached L, Norman PJ, Parham P. Chimpanzees use more varied receptors and ligands than humans for inhibitory killer cell Ig-like receptor recognition of the MHC-C1 and MHC-C2 epitopes. J Immunol 2009;182:3628–37.
[56] Guethlein LA, Flodin LR, Adams EJ, Parham P. NK cell receptors of the orangutan (*Pongo pygmaeus*): a pivotal species for tracking the coevolution of killer cell Ig-like receptors with MHC-C. J Immunol 2002;169:220–9.
[57] Adams EJ, Thomson G, Parham P. Evidence for an HLA-C-like locus in the orangutan *Pongo pygmaeus*. Immunogenetics 1999;49:865–71.
[58] Parham P, Norman PJ, Abi-Rached L, Guethlein LA. Human-specific evolution of killer cell immunoglobulin-like receptor recognition of major histocompatibility complex class I molecules. Philos Trans R Soc Lond, B, Biol Sci 2012;367:800–11.
[59] Abi-Rached L, Kuhl H, Roos C, Hallers ten BFH, Zhu B, Carbone L, et al. A small, variable, and irregular killer cell Ig-like receptor locus accompanies the absence of MHC-C and MHC-G in gibbons. J Immunol 2010;184:1379–91.
[60] Grendell RL, Hughes AL, Golos TG. Cloning of rhesus monkey killer-cell Ig-like receptors (KIRs) from early pregnancy decidua. Tissue Antigens 2001;58:329–34.
[61] Moreland AJ, Guethlein LA, Reeves RK, Broman KW, Johnson RP, Parham P, et al. Characterization of killer immunoglobulin-like receptor genetics and comprehensive genotyping by pyrosequencing in rhesus macaques. BMC Genomics 2011;12:295.
[62] Blokhuis JH, van der Wiel MK, Doxiadis GGM, Bontrop RE. The extreme plasticity of killer cell Ig-like receptor (KIR) haplotypes differentiates rhesus macaques from humans. Eur J Immunol 2011;41:2719–28.
[63] Otting N, de Vos-Rouweler AJM, Heijmans CMC, de Groot NG, Doxiadis GGM, Bontrop RE. MHC class I A region diversity and polymorphism in macaque species. Immunogenetics 2007;59:367–75.
[64] Doxiadis GGM, de Groot N, Otting N, Blokhuis JH, Bontrop RE. Genomic plasticity of the MHC class I A region in rhesus macaques: extensive haplotype diversity at the population level as revealed by microsatellites. Immunogenetics 2011;63:73–83.
[65] Karl JA, Bohn PS, Wiseman RW, Nimityongskul FA, Lank SM, Starrett GJ, et al. Major histocompatibility complex class I haplotype diversity in chinese rhesus macaques. G3 (Bethesda) 2013;3(7):1195–201. http://dx.doi.org/10.1534/g3.113.006254.
[66] Wiseman RW, Karl JA, Bimber BN, O'Leary CE, Lank SM, Tuscher JJ, et al. Major histocompatibility complex genotyping with massively parallel pyrosequencing. Nat Med 2009;15(11):1322–6. http://dx.doi.org/10.1038/nm.2038.
[67] Norman PJ, Abi-Rached L, Gendzekhadze K, Hammond JA, Moesta AK, Sharma D, et al. Meiotic recombination generates rich diversity in NK cell receptor genes, alleles, and haplotypes. Genome Res 2009;19:757–69.
[68] Traherne JA, Martin M, Ward R, Ohashi M, Pellett F, Gladman D, et al. Mechanisms of copy number variation and hybrid gene formation in the KIR immune gene complex. Hum Mol Genet 2010;19:737–51.
[69] Uhrberg M. The KIR gene family: life in the fast lane of evolution. Eur J Immunol 2004;35:10–5.

[70] Rajalingam R, Parham P, Abi-Rached L. Domain shuffling has been the main mechanism forming new hominoid killer cell Ig-like receptors. J Immunol 2004;172:356–69.
[71] Abi-Rached L, Parham P. Natural selection drives recurrent formation of activating killer cell immunoglobulin-like receptor and Ly49 from inhibitory homologues. J Exp Med 2005;201:1319–32.
[72] Blokhuis JH, Wiel MK, Doxiadis GGM, Bontrop RE. Evidence for balancing selection acting on KIR2DL4 genotypes in rhesus macaques of Indian origin. Immunogenetics 2009;61:503–12.
[73] Parham P. MHC class I molecules and KIRs in human history, health and survival. Nat Rev Immunol 2005;5:201–14.
[74] Hermes M, Weil S, Groth A, Dressel R, Koch J, Walter L. Characterisation of mouse monoclonal antibodies against rhesus macaque killer immunoglobulin-like receptors KIR3D. Immunogenetics 2012;64:845–8.
[75] Shum BP, Flodin LR, Muir DG, Rajalingam R, Khakoo SI, Cleland S, et al. Conservation and variation in human and common chimpanzee CD94 and NKG2 genes. J Immunol 2001;168:240–52.
[76] Averdam A, Kuhl H, Sontag M, Becker T, Hughes AL, Reinhardt R, et al. Genomics and diversity of the common marmoset monkey NK complex. J Immunol 2007;178:7151–61.
[77] Braud VM, Allan DS, O'Callaghan CA, Söderström K, D'Andrea A, Ogg GS, et al. HLA-E binds to natural killer cell receptors CD94/NKG2A, B and C. Nature 1998;391:795–9.
[78] Tomasec P, Braud VM, Rickards C, Powell MB, McSharry BP, Gadola S, et al. Surface expression of HLA-E, an inhibitor of natural killer cells, enhanced by human cytomegalovirus gpUL40. Science 2000;287:1031.
[79] Béziat V, Dalgard O, Asselah T, Halfon P, Bedossa P, Boudifa A, et al. CMV drives clonal expansion of NKG2C(+) NK cells expressing self-specific KIRs in chronic hepatitis patients. Eur J Immunol 2012;42:447–57.
[80] LaBonte ML, Levy DB, Letvin NL. Characterization of rhesus monkey CD94/NKG2 family members and identification of novel transmembrane-deleted forms of NKG2-A, B, C, and D. Immunogenetics 2000;51:496–9.
[81] Kravitz RH, Grendell RL, Slukvin II , Golos TG. Selective expression of NKG2-A and NKG2-C mRNAs and novel alternative splicing of 5′ exons in rhesus monkey decidua. Immunogenetics 2001;53:69–73.
[82] Biassoni R, Fogli M, Cantoni C, Costa P, Conte R, Koopman G, et al. Molecular and functional characterization of NKG2D, NKp80, and NKG2C triggering NK cell receptors in rhesus and cynomolgus macaques: monitoring of NK cell function during simian HIV infection. J Immunol 2005;174:5695–705.
[83] Raulet DH, Gasser S, Gowen BG, Deng W, Jung H. Regulation of ligands for the NKG2D activating receptor. Annu Rev Immunol 2013;31:413–41.
[84] Bahram S. MIC genes: from genetics to biology. Adv Immunol 2000;76:1–60.
[85] Steinle A, Li P, Morris DL, Groh V, Lanier LL, Strong RK, et al. Interactions of human NKG2D with its ligands MICA, MICB, and homologs of the mouse RAE-1 protein family. Immunogenetics 2001;53:279–87.
[86] Averdam A, Seelke S, Grützner I, Rosner C, Roos C, Westphal N, et al. Genotyping and segregation analyses indicate the presence of only two functional MIC genes in rhesus macaques. Immunogenetics 2007;59:247–51.
[87] Doxiadis GGM, de Groot N, Claas FHJ, Doxiadis IIN, van Rood JJ, Bontrop RE. A highly divergent microsatellite facilitating fast and accurate DRB haplotyping in humans and rhesus macaques. Proc Natl Acad Sci USA 2007;104:8907–12.
[88] Gardner MB, Luciw PA. Macaque models of human infectious disease. ILAR J 2008;49:220–55.
[89] Goulder PJR, Watkins DI. Impact of MHC class I diversity on immune control of immunodeficiency virus replication. Nat Rev Immunol 2008;8:619–30.

[90] Pal R, Venzon D, Letvin NL, Santra S, Montefiori DC, Miller NR, et al. ALVAC-SIV-gag-pol-env-based vaccination and macaque major histocompatibility complex class I (A*01) delay simian immunodeficiency virus SIVmac-induced immunodeficiency. J Virol 2002;76:292–302.
[91] Zhang Z-Q, Fu T-M, Casimiro DR, Davies M-E, Liang X, Schleif WA, et al. Mamu-A*01 allele-mediated attenuation of disease progression in simian-human immunodeficiency virus infection. J Virol 2002;76:12845–54.
[92] Mühl T, Krawczak M, Haaft ten P, Hunsmann G, Sauermann U. MHC class I alleles influence set-point viral load and survival time in simian immunodeficiency virus-infected rhesus monkeys. J Immunol 2002;169:3438–46.
[93] Mothé BR, Weinfurter J, Wang C, Rehrauer W, Wilson N, Allen TM, et al. Expression of the major histocompatibility complex class I molecule Mamu-A*01 is associated with control of simian immunodeficiency virus SIVmac239 replication. J Virol 2003;77:2736–40.
[94] O'Connor DH, Mothé BR, Weinfurter JT, Fuenger S, Rehrauer WM, Jing P, et al. Major histocompatibility complex class I alleles associated with slow simian immunodeficiency virus disease progression bind epitopes recognized by dominant acute-phase cytotoxic-T-lymphocyte responses. J Virol 2003;77:9029–40.
[95] Loffredo JT, Maxwell J, Qi Y, Glidden CE, Borchardt GJ, Soma T, et al. Mamu-B*08-positive macaques control simian immunodeficiency virus replication. J Virol 2007;81:8827–32.
[96] Yant LJ, Friedrich TC, Johnson RC, May GE, Maness NJ, Enz AM, et al. The high-frequency major histocompatibility complex class I allele Mamu-B*17 is associated with control of simian immunodeficiency virus SIVmac239 replication. J Virol 2006;80:5074–7.
[97] Sauermann U, Siddiqui R, Suh Y-S, Platzer M, Leuchte N, Meyer H, et al. Mhc class I haplotypes associated with survival time in simian immunodeficiency virus (SIV)-infected rhesus macaques. Genes Immun 2008;9:69–80.
[98] Mee ET, Berry N, Ham C, Aubertin A, Lines J, Hall J, et al. Mhc haplotype M3 is associated with early control of SHIVsbg infection in Mauritian cynomolgus macaques. Tissue Antigens 2010;76:223–9.
[99] Mee ET, Berry N, Ham C, Sauermann U, Maggiorella MT, Martinon F, et al. Mhc haplotype H6 is associated with sustained control of SIVmac251 infection in Mauritian cynomolgus macaques. Immunogenetics 2009;61:327–39.
[100] Rosner C, Kruse PH, Hermes M, Otto N, Walter L. Rhesus macaque inhibitory and activating KIR3D interact with Mamu-A-encoded ligands. J Immunol 2011;186:2156–63.
[101] Colantonio AD, Bimber BN, Neidermyer WJ, Reeves RK, Alter G, Altfeld M, et al. KIR polymorphisms Modulate peptide-dependent binding to an MHC class I ligand with a Bw6 motif. PLoS Pathog 2011;7:e1001316.
[102] Maloveste SM, Chen D, Gostick E, et al. Degenerate recognition of MHC class I molecules with Bw4 and Bw6 motifs by a killer cell Ig-like receptor 3DL expressed by macaque NK cells. J Immunol 2012;189(9):4338–48. http://dx.doi.org/10.4049/jimmunol.1201360.
[103] Fadda L, Borhis G, Ahmed P, et al. Peptide antagonism as a mechanism for NK cell activation. Proc Natl Acad Sci USA 2010;107(22):10160–5. http://dx.doi.org/10.1073/pnas.0913745107.
[104] Lu X, Gibbs JS, Hickman HD, et al. Endogenous viral antigen processing generates peptide-specific MHC class I cell-surface clusters. Proc Natl Acad Sci 2012;109(38):15407–12. http://dx.doi.org/10.1073/pnas.1208696109.
[105] Alter G, Heckerman D, Schneidewind A, et al. HIV-1 adaptation to NK-cell-mediated immune pressure. Nature 2011;476:96–100.
[106] Anzai T, Shiina T, Kimura N, et al. Comparative sequencing of human and chimpanzee MHC class I regions unveils insertions/deletions as the major path to genomic divergence. Proc Natl Acad Sci USA 2003;100:7708–13.

CHAPTER

15

The Different Modes of Resistance to AIDS: Lessons from HIV/SIV Controllers and SIV Natural Hosts

Lisa A. Chakrabarti

Unité de Pathogénie Virale, Institut Pasteur, Paris, France

INTRODUCTION: THE MANY WAYS TO RESIST AIDS

Human immunodeficiency virus (HIV) infection in humans does not always lead to AIDS, even if left untreated. Rare HIV-1-infected individuals do not show signs of disease progression in the long term, even after two or three decades of seropositivity. This protection from disease is associated with a spontaneous and persistent control of HIV replication, as indicated by an undetectable viral load in the plasma by standard HIV detection assays. These rare individuals have been called HIV controllers [1], elite controllers (ECs) [2], elite suppressors [3], or natural viral suppressors [4]. Large-scale efforts are being devoted to identify and study these rare patients, with the aim of elucidating mechanisms that prevent progression to AIDS and harnessing these findings for the development of immunotherapeutic and vaccine approaches against HIV.

Non-human primates (NHPs) also provide several distinct models of resistance to AIDS. Simian immunodeficiency virus (SIV) inoculation in macaques usually causes AIDS but sometimes results in a controlled viral infection with a low or undetectable viral load in plasma, in a situation reminiscent of that seen in HIV controllers. The SIV controller model has provided important clues on the host and viral determinants of control, which have been validated experimentally. Simian species that are naturally infected by SIV in the wild provide an apparent radically different

Natural Hosts of SIV.
http://dx.doi.org/10.1016/B978-0-12-404734-1.00015-2

model of resistance to AIDS, as the viral load remains persistently high in these species without obvious pathogenic consequences. As SIV appears to have cytopathic potential for $CD4^+$ T cells from its natural hosts, the persistence of this virus at high levels in the absence of major $CD4^+$ T-cell loss poses a paradox, which has been the subject of intense scrutiny.

The purpose of this review is to compare the mechanisms of AIDS resistance in the different human and simian models. Because natural SIV infections are analyzed in depth in the different chapters of this book, the present article will primarily focus on viral and host factors that may contribute to limit pathogenicity in HIV/SIV controllers. A point-by-point comparison with the salient features of natural SIV infection will then be carried out to identify common mechanisms involved in avoiding AIDS.

VIRAL PARAMETERS IN HIV/SIV CONTROL

Definition of HIV Control

It was recognized early on that about 5% of HIV-1-infected individuals do not show signs of $CD4^+$ T-cell decline and appeared to have a very slow progression to AIDS. These individuals, called long-term nonprogressors (LTNPs), were identified based on high and stable CD4 counts, generally corresponding to levels >500 or >600 $CD4^+$ T cells/mm^3 for a duration of longer than 5–10 years [5,6]. Further studies showed that the LTNP group was actually heterogeneous and that a significant percentage of LTNPs eventually progressed to disease. With the generation of advanced viral load testing techniques, it was recognized that virological criteria were more efficient at predicting lack of progression and that the subgroup of patients characterized by an undetectable viral load in plasma had a very low risk of progressing to AIDS. HIV controllers are thought to represent about 0.5% of seropositive individuals, with estimates ranging from 0.2% to 1% depending on the chosen definition of "viral control" [2,7,8]. The case of the second AIDS virus, HIV-2, is of interest, as it is associated clearly with a higher level of viral control, ranging from 9% in a French cohort up to 37% in a community cohort in Guinea-Bissau [9,10].

The viral load threshold and the duration of control used to define the HIV controller status vary depending on the cohort studied and should be considered carefully, as more stringent criteria are associated with better clinical outcome [5]. The HIV controller cohort organized by the National Agency for AIDS Research (ANRS) in France currently includes patients infected for >5 years with <400 copies of HIV-1 RNA/ml in the absence of antiretroviral therapy (ART) [1]. Of note, >75% of these patients have a median viral load of <50 copies/ml, which corresponds to the current threshold of detection in routine assays. The International HIV

Controllers Study coordinated by the Ragon Institute in Boston defines ECs as untreated individuals with an HIV-1 viral load of <50 copies/ml for >1 year [11]. Viremic controllers (VCs), who have viral loads between 50 and 2000 copies/ml for >1 year, are also studied as a distinct group. The cohort of ECs being studied by the University of California, San Francisco (UCSF) are defined as untreated patients with viral loads <75 copies/ml for >5 years [12]. In the case of the Baltimore cohort, natural viral suppressors are defined as untreated patients with viral loads <400 copies for >2 years [4]. In the U.S. Department of Defense HIV Natural History Study, ECs are defined as untreated patients with viral loads <50 copies/ml for >1 year [5]. Some groups have kept the LTNP nomenclature for patients with viral loads <50 copies/ml for several years [13], although, in general, the term "LTNP" refers to patients with stable $CD4^+$ T-cell counts.

Of note, the majority of HIV controllers maintain high and stable $CD4^+$ T-cell counts and thus also qualify as LTNPs [14]. However, the controller group is not completely included within the LTNP group, as about one quarter of the controllers have $CD4^+$ T-cell counts below the 500 cells/mm^3 threshold [7]. Intriguingly, very rare controllers can show persistent $CD4^+$ T-cell decline and occurrence of opportunistic infections in the absence of detectable HIV-1 RNA in the periphery, suggesting that immune exhaustion may occur in the absence of massive viral replication [15]. Conversely, very rare LTNPs maintain a high viral load without progressing to disease. Such patients are defined as viremic nonprogressors (VNPs) if they maintain a median viral load >100,000 copies/ml for >3 years without showing signs of $CD4^+$ T-cell decline [16]. This unusual pattern of infection, which may share similarities with nonpathogenic SIV infection in its natural hosts, is examined in depth in another chapter of this book.

Residual Viral Replication in HIV Controllers

Importantly, viral replication is not absent in HIV controllers but rather is maintained at very low levels. High-sensitivity viral load assays show a median viral concentration of 2 copies of HIV-1 RNA/ml, with <10% of controllers having a viral load persistently below the detection threshold of 0.2 copy/ml [17]. These levels are comparable to those found in patients receiving efficient ART [18]. A majority of controllers also have detectable cell-associated HIV viral RNA and DNA [12] and show a degree of viral sequence evolution in longitudinal studies, suggestive of ongoing viral replication [19].

In most instances, HIV controllers are identified after several years of silent infection. However, a few cases have been documented of primary infection in individuals who later became controllers [5,20–22]. These reports suggest that controllers do show an early peak in viral load, although it appears to reach lower values than in the majority of patients.

Thus, HIV does spread systemically in controllers and does not remain contained locally, as may be the case in exposed seronegative individuals [23]. The recent Concerted Action on SeroConversion to AIDS and Death in Europe (CASCADE) study based on pooled data from seroconverter cohorts of Europe, Canada, and Australia shows HIV control is not immediate but rather established progressively after the acute infection stage, with a median delay of 10 months in patients who were monitored frequently [22]. The delay to HIV control appears quite variable, ranging from <1 month to up to 3 years of detectable viral replication [5,20,22]. A prospective study shows that short-term viral control after the acute infection stage is actually not a rare phenomenon, as 6.7% of seroconverters achieved a viral load <500 copies/ml and maintained these low levels of viral replication for a median duration of 11.9 months [24]. Once viral control has been sustained for a few years, it appears to become a more stable phenotype. For instance, patients from the CASCADE study who were controllers 5 years after seroconversion had a 0.74 chance of retaining their controller status at 20 years postseroconversion [22]. A retrospective analysis over 15 years indicates that controllers with viral "blips", or intermittent episodes of limited viral replication, are more likely to show a negative $CD4^+$ T-cell slope than are controllers with persistently undetectable viral loads [25]. With the advent of ultrasensitive viral load assays, it has become apparent that even extremely low levels of viral replication may have a deleterious effect in the long term, as controllers with viral loads over the 1 copy/ml threshold are more likely to have a measurable $CD4^+$ T-cell decrease per year than are those without detectable virus [17].

SIV Controllers

Macaques infected with pathogenic strains of SIVmac generally progress to simian AIDS between 6 months and 3 years postinoculation, although the rate of progression is quite variable and depends on the species infected. Pig-tailed macaques (*Macaca nemestrina*) are highly susceptible to simian AIDS, while rhesus macaques (*Macaca mulatta*) show intermediate susceptibility, and cynomolgus macaques (*Macaca fascicularis*) are more resistant [26,27]. Among rhesus macaques, animals of Indian origin progress more rapidly than do those of Chinese origin, indicating that subtle genetic differences may have a significant impact on SIV disease progression [28]. Even among macaques sharing the same origin, inoculation of a single SIVmac viral clone or strain may have very different outcomes, ranging from rapid progression in <6 months to long-term SIV control with viral loads <30–100 copies SIV RNA/ml [29,30]. It has to be kept in mind that the SIV being used has been adapted to rhesus macaques of Indian origin that could influence these findings. In addition, while studies in the SIV/macaque model have revealed that host genetics is not the only parameter influencing the probability of viral control, the inoculum dose and the route of inoculation may also contribute to this variability in the outcome. In particular, mucosal inoculation of low doses of SIVmac via the vaginal or rectal routes results in a high frequency of controlled infections [31,32]. A similar phenomenon is observed after low-dose inoculation of SHIVs, which are chimeric SIVmac clones carrying an HIV-1 envelope gene [33]. Interestingly, seroconversion is not observed in the most controlled infections, and virus is detected intermittently or not at all using highly sensitive polymerase chain reaction (PCR)-based assays. However, cellular responses to the virus can still be detected, especially when they are assayed in cells from lymphoid organs draining the site of inoculation [31,33]. Such "occult" infections suggest that in certain cases, SIV or SHIV infections can be contained early and locally, rather than after a stage of systemic viral spread. Occult infections are reminiscent of the situations observed in HIV-exposed seronegative individuals (ESN), who have little or no detectable virus, do not seroconvert, but often have low but detectable HIV-specific cellular responses [34]. Thus, there may be a continuum in the degree of viral dissemination between ESN and typical HIV controllers, a notion that is supported by the finding that controllers with the lowest viral loads have weaker HIV-specific antibody responses and do not fully seroconvert [17].

Viral Replication in Natural SIV Infection

SIV infection in its natural hosts starkly contrasts with situations of SIV or HIV control in terms of frequency and viremia levels. Resistance to AIDS is the rule rather than the exception in African monkey species that are infected in the wild with SIV (see review [35]). More than 40 species of monkeys and apes have been found to harbor SIV strains, with a seroprevalence that can be close to 50% in some species, such as the African green monkeys (AGM; *Chlorocebus* genus) infected with SIVagm, the mandrills (*Mandrillus sphinx*) infected with SIVmnd, and the sooty mangabeys (SM; *Cercocebus atys*) infected with SIVsm [36]. Long-term follow-up of captive animals from these species has shown that SIVagm, SIVmnd, and SIVsm do not appear to shorten the life span of infected animals. However, isolated cases of pathogenesis similar to AIDS have been reported for a few animals with an advanced age [37,38]. SM are of particular interest as they are infected by SIV strains that are the origin of HIV-2, the second human AIDS virus that emerged in West Africa [39–41]. SIVsm is also the direct ancestor of SIVmac, which appeared in U.S. primate centers in the 1970s when SM tissues were inoculated to macaques for experimental purposes [42,43]. The ape viruses SIVcpz found in chimpanzees and SIVgor found in gorillas are, for their part, the origin of HIV-1 [44–46]. The seroprevalence of SIVcpz is variable, ranging from 0% to close to 50% depending on the group studied. Of note, field studies in the Gombe reservation suggest that SIVcpz significantly reduces the life expectancy of infected chimpanzees [47]. Phylogenetic evidence suggests that SIVcpz is actually a recombinant virus derived from guenon and mangabey viruses, and it has been speculated that chimpanzees may have acquired SIVcpz relatively recently through predation of monkeys and may not have yet evolved to become fully resistant to the pathogenic effects of SIV [48].

A key point is that SIV replicates at high levels in its diverse natural hosts, with viral loads generally between 10^3 and 10^5 SIV RNA copies/ml [49–51]. In addition, SIV mutation rates are high in natural hosts, indicative of active viral replication. Viremia remains orders of magnitude higher in natural hosts than in SIV or HIV controllers, while pathogenesis is minimal in all these situations. Considering that the lifespan of productively infected $CD4^+$ T cells in SM appears to be short in vivo [52,53], a key issue that remains undefined concerns the mechanisms by which the naturally infected monkeys avoid progressing to AIDS despite low levels of ongoing $CD4^+$ T-cell depletion.

Viral Reservoirs

Viral Reservoirs in HIV/SIV Controllers

Latently infected $CD4^+$ T cells containing silent proviruses represent the major HIV reservoir in efficiently treated patients [54]. Because latently infected cells do not express viral antigens, they are not targeted by the immune system and can thus persist in the long term. The persistence of latently infected cells with the potential to reactivate HIV expression represents the major obstacle to HIV eradication with current treatment regimens. A useful surrogate marker of the viral reservoir is provided by the measurement of total HIV DNA in peripheral blood mononucleated cells (PBMCs), with the limitations that this measurement also takes into account productively infected cells and that the presence of HIV DNA does not necessarily indicates that of infectious HIV, as a significant fraction of proviruses may be defective. A key finding is that HIV controllers have a particularly low viral reservoir, as indicated by a median HIV DNA content $<10^2$ copies/10^6 PBMCs, which is lower than the HIV DNA content seen in LTNPs and in efficiently treated patients [1,55–57]. This notion is supported by the observation that HIV isolation in vitro is less frequent and requires a longer propagation time in culture when assaying cells from controllers compared with cells from ART-treated patients [1,58]. A low viral reservoir thus represents a correlate of HIV control, a relevant finding in light of current research efforts aimed at decreasing the viral reservoir in treated patients [59]. A recent study suggests that some patients treated early during primary infection and for a sufficiently long time may control HIV after cessation of therapy [60]. Importantly, the HIV reservoir is particularly low in these "post-treatment controllers" and, in a few cases, keeps decreasing after therapy interruption. Thus, the human immune system seems to have the capacity to control HIV when the number of target cells able to reactivate the virus is kept at low levels.

The gastrointestinal tract is a major site for HIV replication and CD4$^+$ T-cell depletion in progressor patients, due to the physiologically high levels of activated memory T cells in this compartment [61]. There is evidence that this compartment is spared in controllers and LTNPs, as indicated by a low HIV DNA content in rectal biopsy samples [62]. Still, in a model of SIV control in Chinese rhesus macaques, the mucosal tissues of the colon remain a significant viral reservoir even in SIV controllers, where a median concentration of 10^3 SIV DNA copies/10^6 cells can be detected, which is about 1 log higher than the SIV DNA content measured in peripheral blood [63]. These and other findings (see review [64]) indicate that viral DNA measurements in the periphery do not provide a full picture and that studies of tissues are needed to fully evaluate the HIV/SIV reservoir. It was recognized early that lymph nodes also constitute a major HIV reservoir in progressor patients, due to the high concentration of CD4$^+$ T target cells within these organs and due to the trapping of viral particles as immune complexes within germinal centers. HIV DNA content is about 1 log higher in lymph node tissue than in peripheral blood in patients with active HIV replication [65]. There have been limited studies of the viral reservoir in lymph nodes of controllers primarily due to the scarcity of tissue samples. The limited evidence available suggests that while the tissue reservoir is generally lower in LTNPs than in progressors, the virus can still be isolated from the Tfh fraction of the lymph node CD4$^+$ T-cell population [66]. Evidence from the Chinese macaque model nevertheless suggests that the viral DNA reservoir remains low in lymph nodes of SIV controllers [63].

Viral Reservoirs in Natural SIV Hosts

The viral reservoir is relatively high in the blood and intestinal mucosa of natural SIV hosts, consistent with the high viremia observed in these species [50,51,67,68]. Interestingly, the SIV reservoir is comparatively decreased in lymph node tissue from naturally infected AGM and SM. In particular, the SIV DNA content does not appear proportionately higher in lymph nodes than in the blood of SIVagm-infected AGM [50,69], while lymph nodes have an approximately 1 log higher SIV/HIV DNA concentration in pathogenic infections. Experimental SIVsm infection in SM shows that productively infected cells are abundant in SM lymph nodes during the acute stage but that productive infection is rapidly controlled in lymphoid tissue as the chronic stage is established [70]. In contrast, productively infected cells persist in lymph nodes of

SIVmac-infected RM after the acute stage and tend to accumulate within germinal centers [70,71]. Analysis of SIV DNA content in sorted $CD4^+$ T-cell subsets confirms that the lymph node viral reservoir is substantially lower in SM than in RM and shows that the difference is most marked in the Tfh subset, which corresponds to the resident $CD4^+$ T-cell population within germinal centers [72]. Thus, restricted viral replication and low viral reservoir in lymph nodes represent distinctive features of natural SIV infection that may help explain the lack of disease progression in the AGM and SM species. Of note, lymph nodes also appear to be spared in SIV controllers, suggesting that the preservation of lymphoid tissue architecture and function may be a requisite to avoid AIDS progression.

Viral Fitness

A Degree of Viral Attenuation in HIV Controllers

Whether viral strains of reduced virulence are needed for the establishment of HIV control remains a debated issue. In one notorious series of attenuated HIV transmission cases, one patient infected with an HIV-1 strain carrying a large deletion in the *nef* gene donated blood that infected eight recipients, who became known as the Sydney Blood Bank Cohort [73]. All six recipients who could be followed in the long term showed slow disease progression and low viral replication levels, implicating the inactivation of *nef* as a determinant of HIV control. Three patients showed low to medium levels of viremia and eventually progressed to disease, while the three others maintained undetectable viral loads and remained asymptomatic for >25 years. However, cases with such obvious HIV gene-inactivating mutations occur only sporadically [74–76]. The viral strains that are isolated from HIV controllers appear to be replication competent in most cases and do not carry large deletions [58,77,78]. On the other hand, evidence for more subtle impairments in viral fitness have been obtained in multiple instances. The *gag* and *pol* sequences amplified from controller viral strains confer a reduced fitness when inserted into the backbone of a fully replication competent HIV-1 molecular clone [79,80]. The reduced viral replication capacity has been mainly ascribed to the presence of cytotoxic T lymphocyte (CTL) escape mutations in Gag, which target conserved regions of the capsid that appear intolerant to structural change. Env sequences from controller viral strains confer reduced infectivity and slower entry kinetics than those of progressors [81]. Nef functions such as CD4 and MHC I downregulation are also suboptimal in controller-derived HIV clones, with some of the Nef polymorphisms corresponding to CTL escape mutations [82]. A likely interpretation of these findings is that efficient

CTL-based and possibly antibody-based immune pressure constrains HIV to undergo mutations that carry a significant fitness cost. Thus, the low fitness of controller viral strains could be a consequence rather than a cause of efficient antiviral responses. The picture is more complex, however, as signs of viral attenuation could be found at high frequency in very early HIV sequences obtained from controllers who could be studied during the primary infection stage [21]. Importantly, some of the attenuating mutations corresponded to drug escape mutations that were unlikely to have been selected in the untreated individuals who became controllers. Other mutations corresponded to HLA-dependent CTL escape mutations that were either selected very early or were directly transmitted from the infected partner. Taken together, these findings suggest that slight attenuation of the transmitted virus may be a factor favoring the establishment of HIV control, which can be subsequently further strengthened by host immune pressure.

Viral Fitness in NHP Models

In NHP models where experimental SIV inoculation is feasible, there is clear evidence that viral attenuation can induce control. RM inoculated with the SIVmac293 clone deleted in the *nef* gene show slow disease progression with a pattern comparable to that observed in the Sydney Blood Bank Cohort, namely a very low viral load, preserved lymphoid architecture, and immune functions, but in some cases a gradually progressive $CD4^+$ T-cell depletion that can lead to disease in the very long term [83,84]. In contrast, there is no conclusive evidence for generalized viral attenuation in natural SIV hosts, as primary SIVsm and SIVagm isolates replicate efficiently in vitro in PBMCs from the original host species. It has been proposed that SIV isolated from natural hosts limit pathogenesis due to their capacity to downregulate CD3 via the Nef protein, resulting in a decreased susceptibility of infected target $CD4^+$ T cells to undergo hyperactivation and apoptosis [85]. However, Nef from pathogenic SIVmac strains also causes CD3 downregulation in susceptible RM, indicating that this property does not strictly correlate with lack of pathogenicity. Of note, cross-species transmission may dramatically alter SIV fitness in vivo. For instance, an SIVagm strain which replicates to high levels in its natural host species, the sabaeus monkey (*Chlorocebus sabaeus*), becomes controlled when inoculated in RM, with a viremia that remains below detectable levels after the acute infection stage [86]. Conversely, an SIVagm strain from vervets (*Chlorocebus pygerythrus*) still replicates to high levels when inoculated in pig-tailed macaques, but in addition becomes highly pathogenic [87]. Thus, SIV cross-species transmission can generate models of SIV control or models of AIDS, depending on finely tuned viral/host interactions.

CELLULAR RESPONSES IN HIV/SIV CONTROL

Protective MHC Alleles

Influence of HLA-B and HLA-C in HIV Control

Genome-wide analyses of single nucleotide polymorphisms (SNP) have been launched on large cohorts to identify genetic factors that influence the level of HIV replication. Previous studies had shown that the viral load achieved after the primary infection stage (viral setpoint) is a relatively stable parameter during the asymptomatic phase of HIV infection and is a good predictor of the risk of progression to AIDS [88]. The genetic studies showed that the only significant association with the HIV setpoint were SNPs localized to the MHC region on chromosome 6, with a clear contribution from MHC class I genes [89], and a possible contribution from MHC class III genes [90]. The other polymorphisms known to influence HIV replication such as genes that controlled the level of expression of the HIV co-receptor CCR5 [91], had a smaller effect and did not reach genome-wide significance. These findings were consistent with the repeated observation that certain MHC I alleles are highly enriched in controllers cohorts, with HLA-B*57 and -B*27 having the clearest protective effects [92–94]. Indeed, the frequency of protective B*57 alleles (B*5701, B*5702, B*5703, and B*5801) is often close to or above 50% in large controller cohorts [8]. These findings, coupled with the observation that CD8 responses restricted by HLA-B*57 and -B*27 appear immunodominant and particularly efficient in controllers (see later), have led to the notion that $CD8^+$ T cells likely play a major role in HIV control.

While the dominant genetic association with the HIV setpoint has been localized to HLA-B, another signal that has been identified is located 35 kilobases upstream of HLA-C [95]. This upstream region was recently found to be the target of a miRNA that regulates the level of HLA-C expression, thus providing a possible mechanistic explanation for the effect of this region on antiviral immune responses [96]. Of note, antigenic peptide bearing MHC I molecules serve as ligands for TCR expressed by $CD8^+$ T cells, but also for receptors such as the killer cell immunoglobulin-like receptors (KIR) on NK cells and leukocyte immunogobulin-like receptors (LILR) on dendritic cells (DC). Thus, MHC I influence of HIV control may also result from a more efficient activity of NK cells or DC. The combination of MHC class I molecules belonging to the Bw4 public allotype (which include the HLA-B*27 and -B*57 alleles) with the KIR3DL1 and KIR3DS1 receptors is more protective against HIV disease progression than the MHC class I molecules alone [97]. These epistatic interactions between MHC I and KIR receptors suggest that NK cells could indeed play a role in HIV control. On the other hand, a recent genetic association study that included a large number of controllers showed that the determinants of HIV control mapped precisely to the peptide binding groove of MHC I molecules rather than to the more conserved regions of these molecules, suggesting that MHC I presentation of HIV peptides may play a more important role in control than MHC I recognition by NK cells [98]. Intriguingly, while most protective MHC I alleles, such as HLA-B*57, -B*27, -B*13, and -C*14, are associated with an early containment of HIV replication, the haplotype B*14-C*08 appears to promote long-term control without decreasing the early viral load [99]. Thus, MHC I molecules may influence HIV control through multiple mechanisms that remain to be fully elucidated.

Influence of MHC Class I in NHP Models

The models of SIV control in macaques provide interesting comparisons, as a clear protective role of certain MHC I alleles has also been reported in these animals. For instance, 50% of RM of Indian origin positive for the MHC I allele Mamu B*08 show a degree of control after infection with the highly pathogenic SIVmac293 clone [100]. The allele Mamu B*17 is protective to a lower extent, with about 20% of B*17$^+$ RM able to control SIVmac293 replication [101]. Mamu A*01 only slightly decreases the viral load but is associated with a slower CD4$^+$ T-cell depletion after SIVmac infection and confers a significant degree of protection in SIV vaccination trials [102,103]. In Burmese rhesus macaques, slow or rapid progression after SIVmac239 infection is also associated with particular MHC I haplotypes, the combination of MHC I A1*043:01 and B*061:03 being protective [104]. Remarkably, certain structural characteristics are shared between human and simian MHC I alleles associated with HIV/SIV control, in spite of the extensive genetic diversity characteristic of this gene family. Specifically, the Mamu B*08 and HLA-B*27 alleles bind similar peptides: they both require an arginine as the first anchor residue in position P2 of the bound peptide and have a preference for basic residues at positions P1 and P9 [100]. Peptides with two N-terminal basic residues appear more resistant to cellular peptidases [105], which may lead to more efficient presentation and activation of cognate CD8$^+$ T cells and may explain in part the protective effect of HLA-B*27-related alleles.

Studies on the genetic factors that influence SIV infection in natural hosts remain few, but it is interesting to note that chimpanzee MHC I alleles appear to preferentially target the same conserved Gag peptides as the protective human alleles B*27 and B*57 [106]. As chimpanzees have a relatively restricted MHC I repertoire, a "selective sweep" hypothesis has been proposed, based on the notion that widespread SIVcpz infection may have selected protective MHC I alleles at the population level [107]. Whether SIV infection was the actual selective force that narrowed the chimpanzee MHC I repertoire remains to be ascertained, however. SIV infection still decreases the lifespan and reproductive success of contemporary chimpanzee populations, suggesting that SIVcpz and its host have not yet reached a stable evolutionary equilibrium [47].

A Possible Role for MHC Class II in HIV/SIV Control

A role for MHC class II alleles in HIV-1 control has been proposed, based on a protective effect that appears significant but smaller than that observed for MHC I. The frequency of HLA-DRB1*13 is increased in HIV controllers [108,109], with the DRB1*1303 allele in particular being associated with lower viral loads [110]. A recent analysis of >1000 infected individuals suggests that HLA-DRB1*1502 is associated with a low viral load, and that conversely HLA-DRB1*0301 predicts a high viral load [111]. The allele DRB1*1303 previously implicated in control also showed a protective effect, though not significant after correction for multiple comparisons. The protective alleles were able to present a larger breadth of HIV peptides, predominantly from Gag and Nef, suggesting a possible mechanism for the role of DRB1 in HIV control [111]. However, no associations were found between HLA class II alleles and the in vitro replication capacity of recombinant viruses carrying the *gag* and *protease* genes from chronically infected individuals, suggesting that HLA class II may not exert its protective effect by inducing the selection of deleterious escape mutations [112]. In the later study, carried out in South African patients infected with HIV-1 clade C, HLA-DRB1*1303 was again the MHC II allele associated with the lowest viral replication capacity, but it proved to be in linkage disequilibrium with the strongly protective MHC I allele HLA-B*5703, highlighting the difficulty of teasing out the effects of highly polymorphic genes situated at neighboring loci. Animal models also provide hints that MHC II may influence SIV replication. In particular, study of a cohort of 196 Indian RM infected with SIVmac239 showed an association between alleles Mamu DRB1*1003 and DRB1*0306 with SIV control [113]. Taken together, these findings suggest that certain MHC II alleles promote HIV/SIV control, possibly through their capacity to present a wider diversity of viral peptides, with the caveat that similar studies performed by other labs are needed to entirely rule out effects of linkage disequilibrium with MHC I.

$CD8^+$ T-Cell Response

Potent Cytotoxic Responses in HIV Controllers

Multiple lines of evidence suggest that $CD8^+$ T cells play a key role in HIV control. First and foremost, controller $CD8^+$ T cells are remarkably efficient at killing autologous $CD4^+$ T cells infected with HIV in vitro [114]. This killing activity, also called HIV suppressive capacity, appears to rely on classical cytotoxic mechanisms, as it is contact dependent and MHC I restricted. Interestingly, the cytotoxic effect is readily detectable in vitro without the need to pre-activate $CD8^+$ T cells prior to the co-culture with infected CD4+ T cells, suggesting that primed CD8+ T cells with rapidly inducible effector functions continuously recirculate in the peripheral blood of HIV controllers. Sorting of CD8+ T-cell subsets shows that effector and effector memory (EM) cells have better HIV suppressive capacity than central memory (CM) cells [115]. However, after specific peptide stimulation and amplification in culture, CM CD8+ T cells also become able to efficiently kill infected autologous CD4+ T cells [116]. Importantly, CD8+ T cells from controllers can eliminate the infected CD4+ T cells that reactivate latent HIV proviruses after in vitro activation, while CD8+ T cells from treated patients do not [117]. This capacity to rapidly eliminate latently infected cells that sporadically reactivate may help explain how controllers avoid or limit HIV replication episodes.

Efficient loading of cytotoxic granules with effector molecules such as granzyme B are thought to contribute to the efficient cytotoxic activity of controller CD8+ T cells [13]. Perforin is also rapidly upregulated after stimulation of controller CD8+ T cells with HIV peptides [118]. These properties suggest that HIV-specific CD8+ T cells from controllers are capable of advanced effector differentiation, a finding compatible with their higher expression of the transcription factor T-bet [119]. Controller CD8+ T cells also maintain a high expression of the co-stimulatory molecule NKG2D [120] and express low levels of the negative regulatory molecules such as PD-1 [121]. In contrast, CD8+ T cells from viremic patients show clear signs of exhaustion, with high PD-1, CD160, and 2B4 expression [122]. Expression of these negative regulatory markers is not entirely ablated in treated patients with undetectable viral loads, suggesting that viral antigenemia is not the only factor driving immune exhaustion. Remarkably, HIV-specific CD8+ T cell from controllers also possess properties associated with long-term CM cells, such as a high proliferative capacity [123]. They also have longer tetomeres, suggesting a high potential for self-renewal [124]. In addition, controller CD8+ T cells were found to be highly poly-functional, meaning they had the capacity to secrete multiple cytokines, including IL-2, upon stimulation with HIV antigens [125,126]. This set of properties may help explain why HIV-specific CD8+ T cells can accumulate at high levels in certain HIV controllers, with a median of 6% of CD8+ T cells responding to Gag peptides in the intestinal mucosa [127]. Thus, controllers seem to be characterized by the co-existence of CM CD8+ T cells that maintain long-term immunological memory against HIV antigens and cytotoxic effector CD8+ T cells that directly target HIV.

Importance of CTL Escape Mutations and Select TCR Clonotypes

Controller CD8+ T cells appear to predominantly target HIV-1 Gag antigens. Indeed, an inverse correlation between Gag CD8+ T-cell responses measured by IFN-γ production and viral load was reported in a large South African cohort, while Env CD8+ T-cell responses were not protective [128]. Of note, the correlation was even more marked in HIV-2 infected individuals, whose Gag-specific CD8$^+$ T-cell responses were polyfunctional and of high magnitude in most patients with controlled infection [129]. The suppressive activity of CD8$^+$ T cells against HIV-1 infected targets strongly correlate with Gag-specific but not Nef-specific responses [130,131]. One reason may be the more rapid action of Gag-specific CTL as compared with CTL targeting less abundant HIV proteins. Kinetic measurements suggest that CTL can detect Gag antigens originating from incoming virions, and do not have to wait for neo-synthesized Gag expressed from integrated proviruses [132]. Another explanation may be that the protective MHC I alleles HLA-B*57 and -B*27 preferentially restrict immunodominant epitopes located within conserved regions of Gag. Protective HLA class I alleles disproportionately contribute to the initial CD8$^+$ T-cell response during acute HIV-1 infection, supporting a role for the responding CD8$^+$ T cells in rapidly establishing viral control [133]. The presence of HLA-B*57 is associated with potent CD8$^+$ T-cell responses directed to a well characterized and limited set of Gag specific immunodominant epitopes [134], and less frequently in Nef [135]. CD8$^+$ T cells exert an efficient selection pressure, as these epitopes rapidly undergo escape mutations during the early stages of infection. For instance the TW10 epitope located at Gag position 240–249 in the viral capsid shows a high frequency of the T242N mutation. Fitness analyses have shown that this escape mutation significantly reduces HIV replication capacity, likely explaining why B*57 retains a protective effect in spite of the occurrence of escape mutations [79]. Viral fitness can be recovered to a certain extent by the occurrence of compensatory mutations within the epitope (G248A) and in an upstream region called the cyclophilin binding loop [136]. The acquisition of the T242N escape mutation and of the associated compensatory mutations can occur in controllers, indicating that HIV replication is not entirely blocked and that the virus may still evolve to avoid being targeted by CTL [137,138]. However, the T242N mutation tends to occur less frequently in B*57$^+$ controllers than in B*57$^+$ patients that progress to disease, with frequencies of 65% versus 83%, reported in one study [139]. The frequency of compensatory mutations is also lower in B*57 controllers, possibly explaining why HIV still replicates poorly after acquiring the T242N escape mutation. The general picture that emerges is that the low levels of HIV replication in controllers limits the probability of the virus acquiring all the required escape mutations to recover for optimal fitness. In addition, controllers often develop de novo CD8$^+$ T-cell responses that target the escaped viruses [137]. Last, some B*57 controllers show rare CTL escape mutations such as G248D that lead to particularly poor viral fitness, suggesting that the CD8-dependent selection pressure drove the virus into an evolutionary impasse [139,140].

In the case of the protective allele HLA-B*27, the pattern of CTL escape is different in the sense that several concomitant capsid mutations are needed for escape. The B*2705 allele is associated with the induction of a highly immunodominant CTL response directed against the KK10 epitope located at Gag 263–272 [133]. The process of CTL escape is generally initiated by the occurrence of the L268M mutation, but full escape with loss of MHC I binding requires the R264K mutation [134,141]. However, the later change impairs fitness so drastically that mutated HIV clones will not replicate in vivo in the absence of an additional compensatory mutation, usually at S173A. The probability of acquiring these combined mutations is low in patients with controlled HIV replication, with the consequence that CTL activity targeting the KK10 epitope is usually maintained in controllers. Importantly, escape in the KK10 epitope has been associated with loss of viral control [142,143]. Recent evidence suggests that the number of required mutations is not the only factor limiting escape in KK10, and that the particular nature of the TCR recognizing this epitope also plays a role. When compared with $CD8^+$ T cells of other specificities, KK10-specific cells were found to have superior function, as measured by the level of IFN-γ production, the degree of poly-functionality, and the capacity to respond to very low amounts of antigenic peptide, a property called high functional avidity [143]. This sensitive detection of antigen was counterbalanced by a high turnover rate, presumably because low amounts of virus were sufficient to trigger the activation and senescence of KK10-specific cells. These properties were due to the expression of TCR with a high avidity for its cognate peptide-MHC complex. Importantly, $CD8^+$ T-cell clones expressing these high-avidity TCR were more efficient at inhibiting HIV infection in autologous $CD4^+$ T cells, thus recapitulating an essential property of controller $CD8^+$ T cells [144]. The same KK10-specific TCR can be identified in different patients, thus meeting the definition of public clonotypes [145]. Importantly, certain of these public clonotypes have the capacity to restrict both the wild type KK10 and the mutant epitope carrying the early L268M mutation, and are able to inhibit the growth of both wild type and mutant viruses in vitro [146–148]. These particularly efficient TCR are enriched in HIV controllers, suggesting that they play a key role in preventing the emergence of viral variants engaged in the KK10-specific escape pathway. There are also indications that clonotypes targeting the B*57 restricted epitope TW10 are also more cytotoxic and cross-reactive in controllers than in progressor patients, suggesting that the selection of particularly efficient TCR clonotypes may be a characteristic of HIV control [147]. Of note, artificially increasing the avidity of a cloned TCR directed to the HIV-1 Gag p17 epitope SL9 (amino acid 77–85) made that TCR cross-reactive against common SL9 escape mutants and conferred a stronger cytotoxicity against $CD4^+$ targets infected by both wild-type HIV-1 and SL9 mutant strains [149]. HIV-2 controllers also harbor high-avidity $CD8^+$

T-cell clones and show little CTL escape mutations [150], supporting the notion that a highly avidity $CD8^+$ T-cell response may prevent viral escape and thus maintain an efficient viral control.

$CD8^+$ T-Cell Responses in SIV Controller Macaques

Findings in models of SIV control mostly support observations made in human controllers, and provide further mechanistic insights. A key finding is that experimental depletion of $CD8^+$ T cells by injection of an in vivo depleting anti-CD8 monoclonal antibody leads to a transient increase in SIV viral load, even in animals with previously well controlled infections [151,152]. These results support an essential role for $CD8^+$ T cells in SIV control, though bystander effects of the CD8 depletion renders interpretation of the findings more complex than initially thought. Due to the almost complete obliteration of $CD8^+$ T cells, homeostatic cytokines such as IL-7, IL-2, and IL-15 are less consumed and become available to fuel the activation and proliferation of $CD4^+$ T cells, which become more prone to SIV infection and may contribute to the observed peak in viral load. Limiting IL-15 availability by adding an anti-IL-15 blocking antibody does reduce $CD4^+$ T-cell proliferation without decreasing the peak in viral load, favoring a role for $CD8^+$ T cells, but not entirely ruling out a role for activated but non-proliferating $CD4^+$ T cells [153]. CD8 responses in macaques carrying protective MHC I alleles are highly focused on a few immunodominant epitopes, as in humans. However, the immunodominant SIV epitopes are not predominantly found in SIV Gag, but are frequent in SIV Nef, and are also present in SIV Vif, Tat and Env [27,154,155]. The reasons for the different localization of immunodominant epitopes in the SIV and HIV proteome are not clear, but one may speculate that SIV Nef is a more constrained protein than HIV Nef, due to the overlap between SIV Nef and Env, which may limit the sequence space available for CTL escape and thus favor the persistence and efficiency of anti-Nef CTL responses. Importantly, a vaccine targeting solely the three major immunodominant Nef and Vif epitopes recognized in Mamu B*08 animals can protect them from disease progression after SIVmac239 challenge, reinforcing the notion that $CD8^+$ T-cell responses to a few key epitopes can make a major contribution to SIV control [156]. In addition, experiments where immunodominant epitopes are purposefully mutated in an infectious SIV molecular clone suggest that MHC-matched macaques inoculated with the mutant SIV less frequently control infection [157,158]. CTL escape mutations are more frequently detected in B^*08^+ RM that progress to disease than in B^*08^+ SIV controllers, reminiscent of the situation observed in B^*27^+ human controllers [154]. SIV control also correlates with a better capacity of $CD8^+$ T cells to kill infected autologous $CD4^+$ T cells, further highlighting the importance of $CD8^+$ T-cell cytotoxic activity [159]. Lastly, some cross-reactive public clonotypes may be associated with particularly efficient CD8 anti-Gag responses in Mamu A*01 macaques [160]. Thus, convergent evidence in monkeys and humans point to the importance of focused, highly efficient CD8$^+$ T-cell responses in establishing and maintaining HIV/SIV control.

Whether CD8$^+$ T cells control HIV and SIV by actually killing infected target cells, or by non-cytotoxic mechanisms that dampen viral expression, is still debated [161–163]. The mechanisms associated with CD8$^+$ T-cell priming are not entirely elucidated either. For instance, vaccination of monkeys with a recombinant rhCMV vector expressing SIV Gag generates an unusually broad CD8 response mediated predominantly by unconventional CD8$^+$ T cells restricted by MHC class II [164]. Interestingly, this recombinant CMV vaccine confers protection from disease to close to 50% RM challenged with SIVmac 293, with the protected animals achieving profound SIV control [165]. Thus, unconventional CD8 responses that do not target the classical SIV immunodominant epitopes may also induce control.

A Subgroup of Controllers with Low CD8$^+$ T-Cell Responses

Another point that needs clarification is whether efficient CD8$^+$ T-cell responses are absolutely required for HIV and SIV control. A subgroup of HIV controllers show weak CD8$^+$ T-cell responses when measured by IFN-γ production by ELISPOT assay and by the capacity to suppress HIV replication in autologous CD4$^+$ T cells [130,166]. These "weak responders" show particularly good control, as indicated by a very low occurrence of viral load blips, but nevertheless harbor replication competent HIV that can be isolated in culture [130]. Transcriptional profiling of CD4$^+$ T cells from controllers with weak and strong CD8 responses show that they segregate in two distinct groups, with the profile of the weak responders being indistinguishable from that of uninfected individuals [167]. In contrast, the two groups of controllers are not distinguished by the frequency of protective HLA alleles. One interpretation is that controllers with weak CD8 responses have achieved such a profound degree of control that the amount of viral antigens left is insufficient to boost memory CD8$^+$ T cells, resulting in a progressive waning of the CD8 response (Figure 15.1(C)). This memory response may still be present and recoverable, as suggested by experiments where peptide stimulated CD8$^+$ T cells from weak responders gain the capacity to produce IFN-γ and suppress HIV replication in autologous CD4$^+$ T cells after amplification in culture [116]. On the other hand, CD8-independent mechanisms may also contribute to HIV/SIV control. The possibility that CD4$^+$ T cells from controllers are slightly less susceptible to infection has been raised (see below), and the contribution of other arms of the immune response has also not been ruled out.

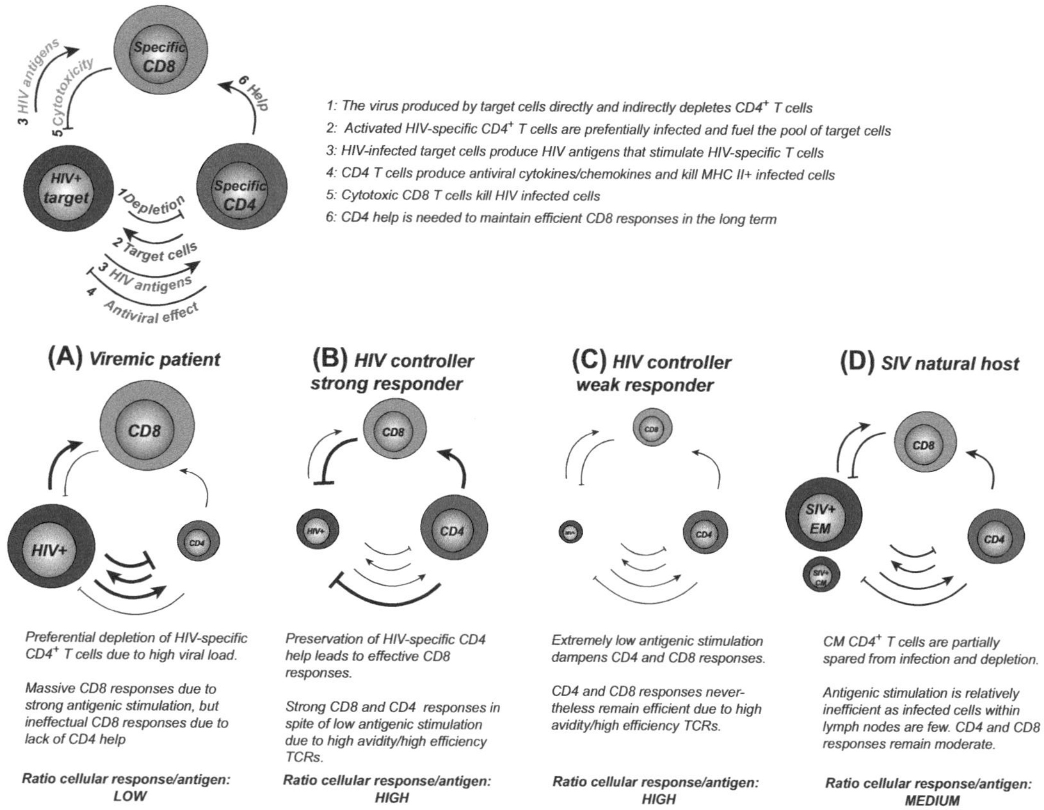

FIGURE 15.1 Interplay between T-cell responses and target cells in pathogenic and non-pathogenic HIV/SIV infections. The 4 schemes correspond to the cases of a progressive HIV infection (A), a controlled HIV infection with a strong CD8 response (B), a controlled HIV infection with a weak CD8 response (C), and an SIV infection in its natural host (D). The strength of interactions 1 to 6 are indicated by the thickness of corresponding arrows, while the size of the specific CD4+, specific CD8+, and target cells populations are indicated by circle areas.

CD8⁺ T-Cell Responses in Natural SIV Hosts

CD8⁺ T-cell responses in natural SIV hosts do not appear as strikingly efficient as those detected in the majority of HIV and SIV controllers. CD8⁺ T cells of SM do produce cytokines in response to a variety of SIV peptides, but the number of responding cells appears in general lower than that observed in SIV infected macaques [168,169]. The cytokine response in SM is poly-functional and is induced during the early stages of primary SIVsm infection, concomitant with a decrease of productively infected cells in lymph nodes [70]. Thus, the CD8 response may contribute to contain SIVsm replication, even though it does not drive the viral load to undetectable levels. CD8⁺ T cells from infected SM also have the capacity to inhibit SIV replication in vitro to a certain extent [170]. The SIV-specific CD8 response in AGM is also detectable but quantitatively lower than that observed in macaques [171]. Depletion of CD8⁺ T cells in naturally infected SM induces a modest increase in viremia but does not trigger disease [172]. Depletion of CD8⁺ T cells during the acute phase of experimental SIVagm infection does cause a more prolonged viremia increase but still does not lead to disease progression [173]. Combined depletion of CD8⁺ T cells and B cells is fatal in pig-tailed macaques infected with SIVagm, but does not cause disease in AGM infected with the same virus [87]. These findings suggest a limited role for adaptive immune responses in the non-pathogenic nature of SIV infection in its natural hosts.

CD4⁺ T-Cell Response

HIV-specific CD4⁺ T cells are needed to orchestrate the antiviral response, but they may also fuel the infection [174], as they represent preferential targets for HIV replication [175]. Their role in HIV control has thus been debated [176,177], and one of the key challenges in HIV vaccinology is to understand how to strike the right balance between providing long-term CD4⁺ T-cell help and avoiding the facilitation of HIV spread. The study of CD4⁺ T-cell responses in HIV controllers should yield clues, as these rare individuals spontaneously achieved such a balance.

Multiple CD4⁺ T-Cell Functions in Mouse and NHP Models

Mouse models of chronic viral infection show that CD4⁺ T cells are not always needed to initiate the antiviral response, but that they are required to sustain viral control and to avoid the exhaustion of CD8⁺ T cells [178]. Virus specific CD4⁺ T cells appear to play a dominant role in the termination of Hepatitis A virus infection in chimpanzees [179] and their persistence correlates with viral control in human Hepatitis C virus infection [180]. Virus-specific CD4⁺ T cells are also abundant producers of IFN-γ and chemokines, which are needed to activate macrophages, and have a recently recognized role in attracting effector CD8⁺ T cells within tissues [181].

The specialized T follicular helper (Tfh) subset is required for the affinity maturation of the antibody response and the generation of long-lived plasma cells [182,183]. Experimental depletion or modification of $CD4^+$ T cells in NHP models of AIDS highlights their importance in SIV and SHIV control. In particular, $CD4^+$ T-cell depletion in RM six weeks prior to SIVmac251 inoculation abrogates the post-peak viral load decline and results in more rapid disease progression [184]. Thus, weakening the CD4 response without abrogating the pool of $CD4^+$ target cells does appear to have deleterious effects. In contrast, depletion of $CD4^+$ T cells during the chronic phase of SIV infection does not cause viral rebound in Mamu B*08 controller RM, suggesting that other immune effectors can temporarily maintain control once it is established [185]. In a gene therapy experiment where a fraction of $CD4^+$ T cells were protected by expression of a fusion peptide inhibitor, SHIV infection resulted in amplification of transgenic $CD4^+$ T cells post-challenge and in a decreased viral load at setpoint [186]. Interestingly, non-transgenic $CD4^+$ T cells also recovered after the acute phase, suggesting that the transgenic $CD4^+$ T-cell subset promoted the establishment of an efficient antiviral response that, in turn, protected other $CD4^+$ T cells. Indeed, CD4 responses were spared, CD8 responses were stronger and broader, and B cell responses were more efficient at SHIV neutralization in the transgenic animals. Thus, an early difference in the CD4 response may tip the balance toward viral control.

Persistence of both CM and Effector CD4 Responses in HIV Controllers

HIV infection is characterized by an early loss in HIV-specific $CD4^+$ T-cell number and function, this loss being detectable as early as the primary infection stage [187–189]. In contrast, controllers maintain abundant and highly functional $CD4^+$ T cells that preferentially recognize the Gag antigen [190], in the blood and in the gastrointestinal mucosa [108]. The amplitude of the CD4 response was found to correlate positively with the amount of HIV-1 DNA present in PBMC, raising the possibility that HIV-specific $CD4^+$ T cells play a deleterious role by providing a constant supply of activated $CD4^+$ T cells that would be preferentially infected by HIV [191]. On the other hand, the correlation may result from a stronger induction of the CD4 response when HIV-1 antigens are more abundant. The proliferative capacity of $CD4^+$ T cells from controllers is remarkable, and is associated with the preservation of the CM $CD4^+$ T-cell population [192]. Proliferation can be sustained due to a high IL-2 secretion capacity [193,194], protection from apoptosis by inactivation of the FOXO3a pathway [195], and a low expression of negative regulatory receptors such as CTLA-4 and PD-1 [196,197]. It has been argued that proliferative capacity may simply be a consequence of a low viral load, as the balance between CM and EM $CD4^+$ T cells naturally shifts toward the former subset in situations of low antigenemia [198,199]. Treated patients have generally lower $CD4^+$ T-cell proliferative responses than controllers, but can recover these responses to a large extent after 10 years of efficient viral suppression [200]. On the other hand, $CD4^+$ T-cell responses of controllers remain qualitatively and quantitatively different from those of patients treated for a long time when other parameters are considered. Controller $CD4^+$ T cells specific for HIV are more poly-functional than those of treated patients, with a better capacity to produce IL-2, IFN-γ, TNF-α, MIP-1β and to degranulate (CD107a expression) simultaneously, in the blood and in the intestinal mucosa [108,126]. Comparison of the frequency of MHC II tetramer binding cells specific for the immunodominant Gag293-312 epitope and that of IFN-γ secreting cells responding to the same epitope shows that a high proportion of specific $CD4^+$ T cells remains functional in controllers, while specific $CD4^+$ T cells appear to lose their IFN-γ secretion capacity in patients treated in the long-term, and show signs of negative immunoregulation as indicated by increased IL-10 secretion [201]. Thus, a key parameter distinguishing controllers from patients treated in the long term is the persistence of Th1 effector cells with IFN-γ secretion capacity.

Presence of $CD4^+$ T Cells with Cytotoxic Potential in HIV/SIV Controllers

Recent evidence suggest that HIV-specific $CD4^+$ T cells can have direct cytotoxic activity, and that a high frequency of $CD4^+$ T cells expressing the cytotoxic granule protein granzyme A during the acute infection stage is associated with slower disease progression [188]. Cytotoxic $CD4^+$ T-cell activity is focused on cells expressing high levels of MHC II, mainly antigen presenting cells (APC) such as macrophages and dendritic cells. It is conceivable that the destruction of infected APC limits HIV spread, as these cells have been shown to be particularly efficient at transmitting the virus to new $CD4^+$ T cells through a structure called a virological synapse [202]. Highly differentiated cytotoxic $CD4^+$ T cells with a high content in lytic granules were reported in a few controllers [203,204], but the generality of this finding remains to be ascertained. Mutational escape from the CD4 response, which could help the virus escape direct cytotoxic activity, has been reported in a few instances [205]. $CD4^+$ T-cell clones from controller macaques were shown to kill SIV infected macrophages, but not SIV infected $CD4^+$ T-cells, perhaps explaining why escape from the CD4 response is not a generalized phenomenon [206]. Cytotoxic $CD4^+$ T cells have also been reported in macaques infected with the attenuated strain SIVmac239Δnef [207]. Thus, evidence for cytotoxic $CD4^+$ T cells has been obtained in several models of HIV and SIV control. Mechanistically, cytotoxic $CD4^+$ T cells are thought to emerge through a reprogramming of mature $CD4^+$ T cells, which downregulate expression of the transcription factor ThPOK, and consequently upregulate a subset of CD8 lineage genes [208]. Findings in models of HIV and SIV control thus suggest that $CD4^+$ T cells can achieve an advanced maturation stage compatible with re-programming to possess cytotoxic potential.

High Avidity $CD4^+$ T-Cell Responses in HIV Controllers

It is noteworthy that controllers show signs of optimal $CD4^+$ T-cell responses in both the CM and the EM compartments. CM cells endowed with a high proliferative capacity and a long half-life ensure long-term immunological memory in controllers. In contrast, CM $CD4^+$ T cells are progressively depleted in pathogenic SIV and HIV infection, a process thought to play a critical role in disease progression [209,210]. As CM cells fail to replenish the EM pool, the residual EM cells become progressively dysfunctional and exhausted through repeated stimulation by abundant antigen. In contrast, EM $CD4^+$ T cells appear fully functional in controlled HIV infection, with broad cytokine secretion capacity and cytotoxic potential. It is intriguing that specific EM cells persist at all in controlled HIV infection, as the amount of viral antigen available to boost these responses is minimal. The explanation may reside in the fact that controllers harbor $CD4^+$ T cells with a high avidity for HIV antigens, as such cells can receive chronic antigenic stimulation in spite of the low viral load, and can thus persist in an EM/effector state. Indeed, HIV controllers were found to harbor $CD4^+$ T cells with high-avidity TCR directed against immunodominant Gag peptides, while such cells were absent in viremic and efficiently treated patients [211]. Models of chronic viral infection show that T cells with high TCR avidity are endowed with a set of properties that result in high quality immune responses: (1) high avidity T cells proliferate more efficiently than low avidity T cells, resulting in a progressive enrichment and immunodominance of the high avidity population (2) high avidity T cells are more poly-functional, meaning that at a given level of antigen stimulation, they will produce a greater variety of cytokines (3) In the case of the $CD8^+$ T-cell subset, high avidity T cells show more efficient cytotoxicity and have been associated with viral control [144,212–215]. It is striking that this set of properties parallels that observed in HIV controllers, suggesting that many of the immune characteristics of controllers may be explained by the particular nature of their HIV-specific TCR.

$CD4^+$ T-Cell Responses in Natural SIV Hosts

SIVsm-specific $CD4^+$ T cells directed toward multiple viral antigens are present in naturally infected SM, though the intensity of the response tends to remain lower than that seen in SIVmac infected RM [168,169]. Similarly, SIV-specific $CD4^+$ T cells are detectable but few in naturally infected AGM [216]. Experimental SIVsm inoculation results in a somewhat different picture, as SM mount a $CD4^+$ T-cell response that is early and of comparable intensity as that seen in RM inoculated with SIVsm or

SIVmac [70]. Thus, the early CD4 response may be efficient in SM, and may then be downregulated as the animals enter the chronic phase of infection. Taken together, these studies did not reveal a particularly broad or potent CD4 response that can account for the lack of disease progression in SIV natural hosts. On the other hand, the studies showed that both SM and AGM did mount a CD4 response to the virus, excluding the possibility that immune tolerance could have protected $CD4^+$ T cells by avoiding their activation. The general picture is that of a balanced cellular response that may play a role in viral containment during acute infection, but that avoids over activation and immune exhaustion during the chronic stage of infection (Figure 15.1(D)).

Innate Response

NK Cells in HIV Controllers

Natural killer (NK) cells from HIV controllers show a degree of immune activation as compared with those of healthy individuals, as indicated by a higher expression of CD69 and a more efficient production of IFN-γ after cytokine stimulation [217]. From a functional point of view, NK cells from controllers maintain an efficient cytotoxicity against MHC I negative K562 target cells, while this activity is decreased in progressor patients [217]. NK cells from controllers do not show the signs of dysfunction seen in progressors, such as the loss of the cytokine producing CD56hi $CD16^-$ subset and the accumulation of an exhausted cytotoxic $CD56^-$ $CD16^+$ $Siglec7^-$ subset in the periphery [218]. However, the capacity of NK cells to control HIV replication in autologous $CD4^+$ T cells infected in vitro remains generally limited, is not observed for all controllers, and is clearly lower than the suppressive activity mediated by $CD8^+$ T cells [219]. Recently, the notion that NK cells may contribute to adaptive responses has emerged, based on the finding that certain NK cell subsets can be primed and amplified by viral antigens [220]. Mechanistically, it is conceivable that the recognition of MHC molecules by NK cells may be influenced by MHC bound peptides. A cytokine response can be detected in the CD3-negative lymphocyte subset after stimulation of patient blood with certain HIV peptide pools, and this NK cell specific response appears to correlate with HIV control [221]. However, it is difficult to rule out a contribution of specific T-cell responses in these whole blood experiments, as γc cytokines produced by specific T cells may be sufficient to activate bystander NK cells. Thus, the extent to which NK cell contribute to HIV control remains to be determined.

The genetic epistasis observed between the KIR3DL1/KIR3DS1 and HLA Bw4-80Ile genes in decreasing the viral load argues for an actual role of NK cells in HIV control [97], though the mechanisms remain to be fully elucidated. The combined expression of the activating allele KIR3DS1 on NK cells and of its ligand HLA Bw4-80Ile on target cells results in a more efficient inhibition of HIV replication in vitro, which could account for the protective effect of this combination [222]. Other combinations of certain inhibitory alleles of KIR3DL1 with HLA Bw4-Ile80 are also protective, one hypothesis being that an efficient inhibition mediated by KIR3DL1 during NK cell development, at the "licensing stage", would result in a more efficient release of NK cell activity once these cells are triggered at the mature stage [97].

NK Cells in NHP Models

A genetic study also points to a role of NK cells in the control of pathogenic SIVmac infection, as an association between the number of KIR3DH alleles and the early containment of SIVmac replication was detected in RM [223]. NK cell phenotype and function are perturbed in SIVmac infection in RM, with a decrease of the $CD56^+$ cytokine secreting subset, and increase of the $CD16^+$ cytolytic subset, and a trafficking away from secondary lymphoid organs [224]. In contrast, the activation of NK cells appears lower in natural SIV infection in SM, as indicated by a lower proportion of proliferating NK cells in SM than in RM during the acute stage [70]. However, the activation of the $CD16^+$ cytolytic NK cell subset occurs earlier in SM than in RM that progress to disease, suggesting a possible role for NK cell in the rapid resolution of acute infection in SM [225].

Other Innate Effectors

Several minor lymphocyte populations share T-cell characteristics, such as the expression of a TCR, but recognize conserved ligands and carry "first line of defense" functions that places them at the border between innate and adaptive immunity. These include $\gamma\delta$ T cells, which express TCR of restricted diversity that target phosphoantigens produced by bacteria and cancerous cells. The well-characterized Vδ2 subset, which recognizes a small microbial compound in the isopentenyl pyrophosphate biosynthesis pathway, is decreased and loses Th1 cytokine secretion capacity in pathogenic HIV infection [226]. In contrast, the Vδ2 subset is increased and maintains good cytokine secretion in HIV controllers compared with ethnically-matched healthy donors [227,228]. Interestingly, naturally infected SM also maintain good $\gamma\delta$ T-cell function [226]. Natural killer T cells (NK T cells), another T-cell population characterized by a semi-invariant TCR, share characteristics with NK cells, such as CD16, CD56, and CD161 expression. Most of these cells recognize lipids and glycolipids presented by the CD1d molecule, and as such contribute to the early response to mycobacterial infections. The $CD161^+$ NK T-cell population is decreased in pathogenic HIV infection, especially in the $CD4^+$ $CD161^+$ subset, which can be directly infected with HIV [229,230]. NK T cells from patients also show reduced IFN-γ secretion when stimulated with α-galactosyl ceramide combined to recombinant CD1d [230].

Similarly, macaques infected with pathogenic SIV show a decrease in the $CD4^+$ NK T-cell subset, and decreased secretion of IFN-γ, IL-2, IL-10, and TGF-β by stimulated NK T cells [231]. In contrast, NK T cells from SM do not express the CD4 receptor and show no depletion nor dysfunction upon SIV infection. The persistence of anti-inflammatory NK T-cell functions, such as IL-10 and TGF-β secretion, may contribute to lessen SIV pathogenesis in these natural hosts [231]. Mucosal associated invariant T cells (MAIT cells) are a recently identified $CD161^+$ T-cell subset that recognizes vitamin B metabolites found in bacteria and yeasts [232]. MAIT cells show signs of exhaustion during pathogenic HIV infection [233,234], but their fate in non-pathogenic HIV/SIV infections remains to be explored.

HUMORAL RESPONSE IN HIV/SIV CONTROL

Low Intensity but High-Quality Antibody Response in HIV Controllers

HIV controllers mount an antibody response directed against HIV, as indicated by seroconversion and recognition of multiple HIV proteins in western blot assays. However, the titers of antibodies able to bind and neutralize HIV remain generally low as compared with those of progressor patients [235]. Among controllers, the titers and the breadth of antibodies correlate positively with HIV viremia measured by highly sensitive assays, suggesting that the availability of viral antigens is the limiting factor in antibody induction [17]. HIV controllers very rarely develop the broadly neutralizing antibodies (bNAb) able to inhibit the growth of highly divergent HIV strains, while viremic patients may do so after 2–3 years of infection [236]. Thus, the consensus so far has been that antibodies do not play a major role in the spontaneous control of HIV replication. Several elements should however mitigate this view, one being that bNAb are of little benefit to patients as they generally do not neutralize the autologous HIV strain present at the time of antibody isolation. Rather, the selection of bNAbs appears to be driven by continuous viral escape, the humoral response having to cope with successive waves of viral escape mutations but rarely succeeding in catching up and blocking the contemporaneous HIV strain. The HIV research field has rightly focused on bNAbs because of their strong potential as immunotherapeutic and vaccine tools [237], but from a natural history perspective autologous neutralizing antibodies (aNAb) appear more relevant. Thus, it is interesting that viremic controllers (with a viral load comprised between 50 and 2000 HIV RNA copies/ml) show higher titers of aNAbs than highly viremic patients [238,239], pointing to a better efficiency of the humoral response in situations of relative viral control. While aNAbs titers remain low in controllers with a viral load of <50 copies/ml, the possibility that these few aNAb contribute to control should not be ruled out considering that the amount of circulating virus that needs to be neutralized is exceedingly low. Also, immunoglobulins from controllers appear more efficient at mediating cell-dependent functions, as suggested by the higher antibody-dependent cytotoxicity induced by plasma from controllers than from viremic patients [240], and the higher capacity of controller antibodies to mediate immune complex phagocytosis through the activating receptor FcγR2a [241]. A skewed glycosylation pattern of the antibody Fc region may play a role in these functional differences [242]. Last, controllers produce more gp41-specific antibodies of the IgG2 subtype, pointing to possible differences in isotype-dependent antibody functions [243]. Of note, HIV-2 infected patients produce particularly broad and efficient aNAbs [244,245], reinforcing the notion that antibodies may well play a role in HIV control.

Role of Tfh Cells

One reason why HIV-1 controllers may harbor a low intensity but high quality antibody response is the persistent availability of help from follicular helper T cells (Tfh). The essential role of this $CD4^+$ T-cell subset in sustaining long-term B cell responses against viral infections continues to be increasingly recognized [182,183]. Intriguingly, Tfh cells are not depleted but rather amplified within germinal centers in the course of progressive HIV infection [66,246]. However, recent evidence indicates that Tfh cells isolated from the lymph nodes of viremic patients are dysfunctional and induce lower levels of IgG when co-cultivated with autologous B cells, due to the inhibitory effect of high PD-L1 expression on B cells [247]. Another reason for better Tfh function in controllers may be a higher level of IL-21 production [248,249], as this $CD4^+$ T cell-derived cytokine reinforces Tfh differentiation in addition to helping B cells, Th17 cells, and $CD8^+$ T cells [182]. More efficient Tfh/B cell interactions in HIV controllers may favor efficient somatic hyper-mutation and affinity maturation of anti-HIV antibodies within germinal centers. In this respect, it is interesting that the molecular analysis of IgG genes amplified from Env-specific B cells from a few HIV controllers has revealed a high degree of somatic hyper-mutation, with a median of 25–30 mutations per VH region [250], consistent with the notion of a highly matured antibody response.

Sequential analysis of lymph nodes from SIV inoculated macaques shows an initial stage of SIV replication in the T-cell area, followed by an accumulation of productively infected cells within germinal centers as soon as secondary follicles develop [71]. Infection of Tfh cells occurs rapidly and massively at this stage, probably sustained by the accumulation of infectious viral particles trapped as immune complexes within germinal centers [251–253]. The relative exclusion of $CD8^+$ T cells from germinal centers may also facilitate Tfh infection by limiting the cytotoxic destruction of productively infected cells [253]. Attenuated SIVmacΔnef infection shows a contrasting situation, where productive infection is reduced but germinal centers develop more rapidly and anti-SIV antibodies are detected earlier as compared with pathogenic SIVmac infection [254]. The finding of a rapid humoral response in attenuated SIV infection is unexpected given the lower amounts of SIV antigens available to trigger this response. By contrast, this finding suggests that the humoral response is delayed in pathogenic SIV infection, possibly due to the massive infection of Tfh cells within developing germinal centers. Further studies will be required to determine whether the kinetics of the antibody response is indeed faster in the different models of SIV control.

Humoral Response in Natural SIV Hosts

Natural SIV infection is characterized by a detectable antiviral antibody response of relatively low intensity, which at first sight may seem comparable to the response detected in HIV controllers. It was recognized early on that the level of SIV binding antibodies appeared lower in infected SM than infected RM [255]. It is now established that the aNAbs titers in naturally infected SM are significantly lower than those measured in pathogenic HIV-1 infection in humans [256]. The titers of SIV binding antibodies also appear relatively low in naturally infected AGM, one peculiarity being the particularly weak or absent antibody response directed to the Gag protein, while T-cell responses to the same protein can be detected [216]. On the other hand, the levels of binding antibodies to Env appear robust in the plasma and milk of lactating female AGM, which may contribute to limit vertical SIV transmission in this species [257]. One shared features of natural SIV infection and HIV control are the lack of signs for non-specific B cell activation, the normal structure of lymphoid tissues, and the limited accumulation of viral particles trapped as immune complexes in germinal centers [49,72,258,259]. Naturally infected SM also show a clearly decreased frequency of Tfh infection as compared with SIVmac infected RM [72]. This contrasts with pathogenic HIV/SIV infections, which are characterized by non-specific hyper-gammaglobulinemia, major lymphoid hyperplasia, accumulation of virions within germinal centers, and massive Tfh infection [66,246,260].

Though SIV natural hosts and HIV controllers share a lack of abnormal B cell activation, they differ in other aspects of the B cell response. Comparison of SIV and HIV genetic variability suggest that the selective pressure exerted by antibodies is weaker in natural SIV infection. Phylogenetic analyses reveal a lower degree of positive selection in SIVsm lineages as compared with HIV-1 lineages, including in the env gene where selection is mainly driven by escape from antibodies [261]. Signs of positive selection in env are detected in HIV controllers, suggesting the occurrence of antibody-driven viral evolution even in situations of low HIV replication [238]. Another element that points to the moderate effect of antibodies in natural SIV infection is the limited effect of B cell depletion through the administration of cytolytic anti-CD20 antibody injections. AGM treated with anti-CD20 during the primary infection stage to up to 6 month post-SIVagm inoculation showed a delayed seroconversion and lower neutralizing antibody titers but no major changes in viral load or disease progression [262]. The combination of anti-CD20 and anti-CD8 antibody treatment did not cause disease in AGM either, while it induced a rapid immunodeficiency with massive CMV reactivation in SIVagm-infected pig-tailed macaques [87]. Anti-CD20 treatment alone is sufficient to increase the viral load and accelerate disease progression in SIVmac293 infected RM [263]. In one case report, anti-CD20 antibody treatment for a low grade lymphoma in an HIV-1 infected patient led to a 1.7 log rise in viral load, pointing to a role of antibodies in HIV-1 control in humans [264]. Thus, the contrasting effects of anti-CD20 treatment in the different species argues for a limited efficiency of the humoral response in natural SIV infection.

IMMUNE ACTIVATION IN HIV/SIV CONTROL

Chronic immune activation is thought to play a central role in AIDS pathogenesis. Pathogenic HIV/SIV infections are characterized by a persistent activation of T cells and other leukocyte subsets, including B cells, NK cells, and DC, which in concert likely results in a progressive exhaustion of the immune system. In addition, chronic activation of $CD4^+$ T cells maintains a pool of highly susceptible target cells that can fuel HIV replication. The degree of T-cell activation in the early asymptomatic stage of HIV infection is a better predictor of the risk of disease progression than the viral load, emphasizing the importance of immune activation in the pathogenic process [265]. Moreover, the degree of residual immune activation in efficiently treated patients with undetectable viremia inversely correlates with the recovery of the $CD4^+$ T-cell pool and is predictive of the risk of death from non-AIDS related causes [266]. Residual immune activation leads to deleterious effects associated with chronic inflammation, including an increased risk of cardiovascular disease, bone frailty, cognitive impairment, and cancer occurrence. Thus, persisting immune activation is of major concern in HIV-infected patients, and is currently the focus of utilizing experimental anti-inflammatory approaches in SIV models and select human clinical trials [267]. Importantly, both HIV controllers and natural SIV hosts show generally low levels of chronic immune activation, which may be a requirement to avoid progression to AIDS.

T-Cell Activation and Proliferation Markers

Moderate Immune Activation in HIV Controllers

HIV controllers show lower levels of activated T cells, as indicated by expression of the CD38 and/or HLA-DR markers, and lower levels of proliferating Ki-67^+ T cells than viremic patients, both in the $CD4^+$ and $CD8^+$ T-cell compartments [15,193,194]. $CD4^+$ T cells from controllers also express lower levels of annexin-V, indicating limited apoptotic cell death [268]. However, signs of a moderate immune activation are present in most controllers, as indicated by a higher expression of activation markers than in uninfected blood donors and most groups of efficiently treated patients [15,193,194]. The plasma levels of the cytokine IL-1α and of the chemokine IP-10 are also increased in controllers as compared with uninfected donors and treated patients, even though levels of these inflammatory mediators remain lower than those observed in viremic patients (our unpublished observation). These findings indicate that HIV infection is not immunologically silent in controllers, and rather suggest that the efficient antiviral response in these patients depends on a certain degree of ongoing immune activation. The presence of high-avidity T cells directed against HIV may contribute to fuel this chronic activation, considering the capacity of these cells to respond to persistently low viral antigen amounts. Thus, a high avidity T-cell response may contribute to control but may also have an immunological cost in the long term.

Rare Cases of $CD4^+$ T-Cell Loss with High Immune Activation in HIV Controllers

Controllers usually show signs of a moderate $CD4^+$ T-cell decline, with median CD4 counts in the 600–800 cells/mm^3 range [4,5,11,15,25]. Intriguingly, a small proportion of controllers show a protracted decrease of $CD4^+$ T cells below the 350 cells/mm^3 threshold, which correlates with increased levels of immune activation [15]. In two reported cases, $CD4^+$ T-cell loss was shown to lead to actual immunodeficiency and occurrence of opportunistic infections (one tuberculosis and one Kaposi sarcoma), resulting in AIDS without detectable viremia [5,15]. The deleterious effects of chronic immune activation may be predominant in these paradoxical cases of AIDS. Immunological analyses point to a decrease in circulating $CD34^+$ hematopoietic stem cells in controllers with low CD4 counts, suggesting that exhaustion of the regenerative capacity of the immune system may occur at the progenitor level [269]. Reduced intra-thymic proliferation has also been reported in controllers with low CD4 counts, while thymic activity appears particularly efficient in the remainder of the controller population [270]. Still, a direct role of HIV should not be ruled out, as ART did seem to reduce immune activation in a controller with $CD4^+$ T-cell depletion [14]. Treatment of controllers with low CD4 counts has variable effects, with no improvement or a degree of CD4$^+$ T-cell recovery that tends to remain lower than that seen the majority of patients [14,271], pointing to a profound alteration of T-cell regeneration capacity in these extremely rare patients.

Cases of progressive CD4$^+$ T-cell loss in the absence of viremia have also been reported for HIV-2 infection, where this scenario appears relatively frequent. For instance, half of the controllers in a French cohort of HIV-2-infected individuals did not maintain their CD4 counts above the 500 cells/mm^3 threshold [9]. Of note, the viral DNA reservoir is generally comparable in HIV-2 and HIV-1 infections, while circulating viral RNA levels are lower in HIV-2-infected patients compared with HIV-1 infected patients at a comparable CD4$^+$ T-cell level [9]. One hypothesis is that HIV-2 can initially spread efficiently within its host but that released virions are kept out of the circulation at a later stage, possibly due to the highly efficient autologous neutralizing antibody response present in these patients [244,245]. Further studies are needed to determine whether all cases of CD4$^+$ T-cell loss in the absence of viremia can be linked to deleterious immune activation, or whether some cases may be explained by efficient antibody-mediated virus trapping. It is interesting to note that progression without viremia can also be observed in some NHP models, such as the infection of pig-tailed macaques with SIVlhoest [272], raising the possibility of addressing these issues experimentally.

Lack of Chronic Immune Activation in Natural SIV Hosts

Naturally infected SM do not show an increased turnover of either CD4$^+$ or CD8$^+$ T cells in the peripheral blood, while both subsets show persistent proliferation in SIVmac-infected RM during chronic infection [273–276]. Apoptosis remains limited in nonpathogenic SIV infection, while it is consistently increased in CD4$^+$ and CD8$^+$ T cells [274], but also in monocytes and DC [277], of RM infected with pathogenic SIVmac [277]. SIV-infected SM and AGM do show a transient increase in immune activation during acute SIV infection, but this increase is resolved as the chronic stage is established, while immune activation persists in the long term in SIV-infected RM [50,275,278]. AGM show a very early induction of anti-inflammatory markers transforming growth factor (TGF)-β and Foxp3 during the acute infection stage, which may help to rapidly dampen immune activation [279]. Resolution of early immune activation in SM is accompanied by a marked induction of the negative regulator PD-1 on T cells within lymph nodes, suggesting that T-cell autonomous mechanisms also contribute to contain chronic immune activation [280]. Trying to artificially induce immune activation by hyperimmunization with a variety of antigens did not trigger disease in SIV-infected SM but did accelerate progression in SIV-infected RM [281]. Using lipopolysaccharides (LPS) or an interleukin (IL)-2/diphtheria toxin fusion to induce immune activation in

SIV-infected AGM resulted in an increase in viremia and caused transient $CD4^+$ T-cell loss but did not lead to overt disease [282]. Thus, the mechanisms that control immune activation in natural SIV hosts are induced early and are not easily perturbed.

Role of Regulatory T Cells

The possibility that regulatory T cells (Tregs) could contribute to HIV control, by dampening deleterious immune activation or, conversely, by avoiding suppression of antiviral responses, has been explored by several groups. Results varied depending on the study, with Treg frequencies in HIV controllers that were higher [283], equivalent [126,284], or even lower [191] than those found in uninfected blood donors. These differences, which may be due to the variable criteria used to define Tregs, preclude definitive conclusions. Still, several of these studies concur in reporting a positive association between Treg frequency and T-cell activation [191,283], suggesting that Tregs are induced as a homeostatic response to activation, in controllers as in other patients. Interestingly, HIV-specific $CD8^+$ T cells restricted by HLA-$B57^+$ or -$B27^+$ in controllers appear less sensitive to Treg-mediated suppression than $CD8^+$ T cells restricted by nonprotective HLA alleles [285]. Resistance to Treg suppression is associated with a lower expression of the negative regulator Tim3 on the $CD8^+$ T cells and with the capacity of these $CD8^+$ T cells to kill the Tregs in a granzyme B-dependent manner. Thus, the intrinsic efficacy of HIV-specific $CD8^+$ T cells, rather than particular Treg properties, appears to sustain the antiviral response in HIV control. Of note, recent findings in an influenza model suggest that $CD4^+$ T helper cells can act by rendering $CD8^+$ T cells resistant to Treg suppression [286], raising the possibility that $CD4^+$ and $CD8^+$ T cells cooperate in establishing a Treg-resistant response in HIV controllers.

Tregs are induced earlier in AGM than in macaques during the course of primary SIV infection, suggesting that these cells may help dampen immune activation in the former species [287]. However, Treg frequency remains stable and Treg suppressive capacity appears low in experimentally infected SM, which does not support the notion of Treg suppression playing a key protective role in SIV natural hosts [288].

Microbial Translocation and Role of Th17 Cells

Microbial Translocation May Drive Chronic Immune Activation

Increased levels of bacterial products, and in particular of the bacterial cell wall component LPS, are detected in the blood of HIV-1-infected patients [289]. The passage of components of the gut microbiota into the general circulation, called microbial translocation, points to a compromised integrity of the intestinal mucosa in HIV-infected patients, a notion that is supported by histologic analyses in patients and SIV-infected macaques [61,290]. The potent inflammatory properties of LPS and other microbial products may fuel chronic immune activation in patients and infected macaques, leading to the proposal that microbial translocation may contribute significantly to AIDS pathogenesis [289]. The Th17 $CD4^+$ T-cell subset plays a key role in maintaining antimicrobial immunity at mucosal and epithelial barriers, by producing IL-17, which helps recruit neutrophils in case of bacterial and fungal invasion, and IL-22, which promotes the integrity of epithelial cells. Importantly, Th17 are preferentially depleted in the intestinal mucosa during progressive HIV and SIVmac infections, which may help explain the increase in microbial translocation [287,291,292]. Of note, other cells such as NK T cells produce IL-17 during primary SIVmac infection, perhaps explaining why mucosal translocation is not detected at the primary stage, while mucosal $CD4^+$ T cells are already severely depleted [293,294]. In contrast, the multiple lymphocyte subsets that produce IL-17 and IL-22 are depleted from the intestinal mucosa at the chronic stage of SIVmac infection, likely contributing to the loss of mucosal integrity in advanced SIV disease [295].

Limited Microbial Translocation in HIV Controllers

Intriguingly, HIV controllers show a degree of microbial translocation, as indicated by increased concentration of LPS in peripheral blood compared with healthy donors, even though LPS levels remain lower than those seen in viremic patients [15,289,296]. Histological analyses are few, but the cytofluorometric analyses of cells recovered from rectal biopsies show either a moderate decrease [127] or a preservation of mucosal $CD4^+$ T-cell frequencies in controllers [296,297]. The Th17 population, measured by the frequency of IL-17-producing cells after nonspecific stimulation, is generally spared in the blood and rectal mucosa of LTNPs and controllers [283,296]. It is possible that a moderate decrease of mucosal $CD4^+$ T-cell numbers in controllers is sufficient to account for a degree of microbial translocation, even though the frequency of Th17 among $CD4^+$ T cells does not appear decreased. On the other hand, other parameters, such as the activation and functional status of macrophages [298] or the composition of the microbiota, may determine the extent of microbial translocation, which remains an incompletely understood process.

Absence of Microbial Translocation in Natural SIV Hosts

The circulating concentration of LPS is not increased in the blood of chronically infected SM and AGM, consistent with a low level of systemic immune activation and a generally preserved Th17 population in these animals [67,291]. Still, experimental SIV inoculation in SM and AGM does cause a marked loss of $CD4^+$ T cells in the intestinal mucosa during the acute infection stage, which can be accompanied by transient increases in microbial translocation [68,289]. The recovery of mucosal $CD4^+$ T cells is often incomplete, but this does not seem sufficient to trigger chronic microbial translocation and deleterious immune activation. It is possible that other cell populations that produce IL-17 and/or IL-22, such as innate lymphoid cells, NK T cells, and $\gamma\delta$ T cells, help maintain the integrity of the intestinal epithelium in SIV natural hosts. The cytokine IL-21 may also play a protective role, as it reinforces Th17 differentiation and shows preserved levels in SIV-infected SM but not RM [299]. Injection of IL-21 could limit the depletion of Th17 cells and decrease microbial translocation in SIVmac-infected RM, emphasizing the relevance of this cytokine for mucosal integrity [300]. Of note, levels of circulating IL-21 and frequencies of IL-21-producing $CD4^+$ T cells appear intact in HIV controllers and thus do not account for the triggering of microbial translocation in this group of patients [248,249]. Taken together, these findings suggest that natural SIV hosts are protected from potentially deleterious microbial translocation, while this process is not entirely absent in HIV controllers, which may contribute to the increased immune activation levels detected in this group.

Interferon Response

A Link between the Interferon Response and Immune Activation

HIV infection triggers the synthesis of interferon (IFN), which in turn activate a variety of interferon stimulated genes (ISG) with diverse antiviral activities. Type I IFN also promotes the development of antiviral immune responses, as shown by increased antibody titers in patients who were treated with IFN-α2 during primary HIV infection [301]. Direct IFN-α2 injection does cause a reduction of about 1 log in viral load in both HIV-infected patients [302] and natural SIV hosts [303], clearly demonstrating the antiviral effect of IFN-α. However, the IFN response is only partially efficient, as indicated by the persistence of HIV/SIV replication after the acute stage and by the refractory or exhausted state of plasmacytoid dendritic cells (pDC), which are the main IFN-α producers [304–306]. Recent evidence suggests that a prolonged IFN response may even have deleterious effects in the chronic stage of viral infections, by contributing to drive chronic immune activation and exhaustion of the immune system. For instance, in the LCMV model in mice, persistent-type I IFN signaling contributes to the induction of negative immunoregulatory molecules, such as PD-L1 and IL-10, that dampen T-cell responses. Conversely, type I IFN receptor blockade during the chronic stage of LCMV infection leads to viral clearance and to restoration of the lymphoid tissue architecture, in a process dependent on $CD4^+$ T cells [307,308]. In HIV infection, IFN-α activity in plasma correlates positively with the expression of the immune activation marker CD38 and negatively with the CD4 count, indicating that a strong IFN response is associated with disease progression [309].

IFN-α injection increases CD38 expression on $CD8^+$ T cells in humans [309] and CCR5 expression on thymocytes in humanized mice [310], thus demonstrating a direct link between IFN-α and immune activation.

Moderate IFN Response in HIV Controllers

A defining feature of progressive HIV and SIV infection is the broad and persistent activation of ISG, while IFN-α per se is often less easily detectable. Importantly, the induction of ISGs is minimal in HIV controllers, indicating that the IFN response abates when viremia is controlled to sufficiently low levels [167,311–313]. The IFN-α secretion capacity of pDC is also preserved in controllers, indicating a lack of pDC exhaustion [314,315]. Thus, HIV controllers do not appear subjected to the deleterious effects of a persistent IFN response but rather maintain the capacity to mount a rapid IFN response if needed. In vitro experiments suggest that HIV-2 may be more sensitive to IFN-α than HIV-1, which may contribute to the high frequency of control seen in HIV-2 infection [316]. It is also intriguing that pDC from women produce more IFN-α in response to TLR7 stimulation than pDC derived from men, and as a consequence induces stronger secondary activation of $CD8^+$ T cells [317]. This difference may help explain why chronically infected women show stronger immune activation and more rapid disease progression than men for a given level of viremia. On the other hand, there is no evidence for a particularly high frequency of men in the HIV controller population. Rather, there are indications for a reverse trend, with a higher percentage of women in controller cohorts compared with reference cohorts [5,11,25]. These observations may be reconciled by considering that a high IFN-α secretion capacity could be an asset at the earliest stage of HIV infection, when control is being established, but could become a liability at later stages, by promoting immune activation. Whether individual differences in IFN response may contribute to HIV control remains to be fully explored.

Rapidly Contained IFN Response in SIV Natural Hosts

Natural SIV hosts mount an IFN response, as indicated by a robust induction of ISG during the acute infection stage, but show a rapid downregulation of these genes early in the chronic stage, as shown by transcriptomics studies of experimentally infected AGM [318] and SM [319]. In contrast, SIVmac-infected RM show a persistent induction of ISG and of negative regulatory molecules such as indoleamine-2,3-dioxygenase and lymphocyte activation gene-3 (LAG-3) [306,318,319]. It is significant that natural SIV hosts downregulate the ISG response while viremia persists, suggesting the presence of potent immunoregulatory mechanisms that avoid immune activation and exhaustion in these species. Injection of recombinant IFN-α2 results in a decrease of the SIV load by approximately 1 log but does not increase T-cell activation or turnover in SM, indicating that the mechanisms that maintain a low immune activation baseline are resistant to perturbation [303]. Taken together, studies in natural SIV hosts and HIV controllers concur in showing that a self-contained IFN response is a correlate of nonpathogenic HIV/SIV infection.

TARGET CELLS IN HIV/SIV CONTROL

Preservation of Critical CD4$^+$ T-Cell Subsets in Nonpathogenic SIV/HIV Infections

Target Cells in HIV Controllers

HIV controllers show a low frequency of infected cells in the circulation, as indicated by a low HIV DNA reservoir [1,55–57]. Monocytes are rarely infected in controllers [320], suggesting that most of the target cells belong to the CD4$^+$ T-cell population. Analysis of the distribution of HIV DNA in the different resting CD4$^+$ T-cell subsets has shed light on the dynamics of the viral reservoir. In efficiently treated patients, HIV DNA concentration is higher in the CM subset (CD45RA$^-$ CCR7$^+$ CD27$^+$) and then progressively lower in the transitional memory subset (TM; CD45RA$^-$ CCR7$^-$ CD27$^+$), the EM subset (CD45RA$^-$ CCR7$^-$ CD27$^-$), and the terminally differentiated subset (TD; CD45RA$^+$ CCR7$^-$ CD27$^-$) [321]. This distribution may be explained by the respective life span of the resting CD4$^+$ T-cell subsets, which decreases from CM to TM, EM, and TD. As viral replication is blocked by therapy, latently infected cells with a short life span are more rapidly lost, and the HIV DNA reservoir becomes progressively enriched in long-lived CM CD4$^+$ T cells. Interestingly, the distribution of the reservoir appears skewed in HLA-B57$^+$/B27$^+$ HIV controllers, with a proportionally higher concentration of HIV-1 DNA in the TM than in the CM subset [322]. Why the CM subset would be spared in controllers remains to be elucidated, as this subset did not show a particularly low expression of the HIV-1 co-receptor CCR5, compared with CM cells of other patients. A possibility is that CM cells are intrinsically much less susceptible to infection than EM cells in situations of low immune activation and that this difference is attenuated in progressor patients where CM cells, and even naive cells, reach an activation status that makes them susceptible to HIV replication [323]. Regardless of the mechanism, the finding that CM CD4$^+$ T cells may be protected from infection in controllers is important, considering the key role of this subset in ensuring long-term immunological memory, and the fact that a loss of the CM subset is a hallmark of HIV and SIV pathogenesis [209,324]. Of note, CM CD4$^+$ T cells also contain less HIV DNA than TM cells in post-treatment controllers, suggesting that protection of the CM population is a consistent feature in HIV control [60]. Why early treatment preferentially spared the CM population in these patients is not yet known, but one may speculate that most HIV target cells have a highly activated phenotype, and hence a short life span, during the acute infection stage due to the ongoing cytokine storm [325], and that blocking infection at this early stage through therapy may avoid the infection of long-lived cells. Of note, a T-cell population with an even longer life span and a stronger repopulation capacity than CM cells has been recently described [326]. These cells, called stem cell memory T cells, are phenotypically close to naive T cells but express a subset of activation markers (CD95, CD122, CXCR3, LFA-1) and are endowed with long-term immunological memory in the SIV/macaque model [327]. Analyzing the HIV DNA content of this subset in controllers should help determine whether the protection of long-lived $CD4^+$ T cells from infection is essential to HIV control.

Target Cells in Natural SIV Hosts

Experimentally inducing $CD4^+$ T-cell activation in SM and AGM causes a transient rise in SIV load, suggesting that SIV preferentially replicates in activated $CD4^+$ T cells in its natural hosts [282,328]. This notion is consistent with the fact that productively infected cells are short-lived in SM, according to measurements of viral RNA decay on ART [52]. Importantly, the CM $CD4^+$ T-cell population is proportionally less infected than the EM $CD4^+$ T-cell population in SIV-infected SM, likely due to a naturally low CCR5 expression in the CM subset and to a low inducibility of this receptor on CM activation [329]. This finding helps understand how SM can retain immunocompetence in the face of a high viral load, as the spared CM population may have the capacity to maintain immunological memory and to continuously replenish an EM population depleted by active SIV replication (Figure 15.1(D)). Thus, a common feature of nonpathogenic SIV and HIV infections appears to be the protection of the CM $CD4^+$ T-cell population from infection.

The relative protection of CM $CD4^+$ T cells and the preferential targeting of EM $CD4^+$ T cells have implications for SIV intrahost dissemination. CM cells preferentially recirculate to secondary lymphoid organs, while EM cells home to peripheral tissues and are abundant in the intestinal mucosa. Consistent with this pattern, lymph nodes are relatively protected from infection in SIV natural hosts, while the intestinal mucosa can show a significant degree of $CD4^+$ T-cell depletion [67–70]. Recent findings indicate that all lymph node $CD4^+$ T-cell types are less infected in SM than in RM, with a particularly marked difference in the Tfh subset [72]. The protection of the Tfh subset in SM may not be solely explained by a low CCR5 expression, but may also result from the low concentration of virions trapped within germinal centers [72,259]. The limitation of SIV spread within lymph nodes may well have major consequences on the outcome of infection. Lymph nodes are specialized lymphoid organs aimed at facilitating APC/T-cell encounters and at promoting the induction of adaptive immune responses and thus can be viewed as the cradle of immune activation. Accordingly, decreased infection within lymph nodes may explain why systemic immune activation remains limited in natural SIV hosts. It is also relevant that lymph nodes constitute the epicenter of the IFN response, as pDC migrate to these organs en masse early in the course of pathogenic SIV infection [306], and IFN-α mRNA is mainly detected within lymphoid tissue in the course of HIV infection [309]. Restricted lymph node infection may help contain the IFN response and thus further limit immune activation in natural SIV hosts. Studies of lymphoid tissue remain few in HIV controllers due to the limited availability of samples, but such studies are warranted to determine whether the protection of lymph nodes from infection is a correlate of HIV control.

Determinants of Target Cell Susceptibility

A Reduced Susceptibility to Infection in HIV Controllers?

The possibility that $CD4^+$ T cells from controllers are intrinsically less susceptible to HIV infection has been raised. One study found that both anti-CD3-stimulated $CD4^+$ T cells and macrophages from controllers were less susceptible to in vitro challenge by HIV, due to an early impairment of HIV replication at the reverse transcription stage [330]. A low susceptibility to HIV infection in vitro was associated with a low HIV DNA reservoir in vivo, emphasizing the potential physiological relevance of this finding. Another group reported that the low susceptibility of $CD4^+$ T cells from controllers to HIV infection was due to a high expression of the cyclin-dependent kinase inhibitor p21 [331], though the role of p21 could not be confirmed in a follow-up study [332]. Other groups found that $CD4^+$ T cells from controllers could be robustly infected in vitro [56] and that they could even be more susceptible to infection than $CD4^+$ T cells from progressor patients, even though the $CD4^+$ T cells from the controllers ended up producing less virions because of a lower level of immune activation [268,333]. Although no definitive conclusions can be drawn at this stage, it is interesting to note that one study of the HIV DNA reservoir in controllers reported a proportionally higher content in two LTR DNA circles in this group of patients, pointing to a possible impairment in HIV integration [334]. Of note, even a slightly decreased infectability of $CD4^+$ T cells may have a major influence on the outcome of HIV infection, considering that sparing a few $CD4^+$ T cells from infection early on may not only reduce the pool of target cells but also allow an earlier emergence of the CD4 antiviral response, leading in turn to an increased efficacy and a more rapid recruitment of multiple other antiviral immune effectors, including $CD8^+$ T cells, B cells, and macrophages.

Reasons for a possibly lower susceptibility of controller target cells to HIV infection remain to be elucidated. Genetic polymorphisms that decrease expression of the HIV co-receptor CCR5 are associated with a slower rate of disease progression [91], raising the possibility that CCR5 expression levels could play a role in HIV control. Although the protective polymorphisms CCR5Δ32 and CCR2-64I are more frequent in LTNPs [335,336], they do not appear enriched in a large controller cohort [11]. The regulation of CCR5 expression is complex, as it depends on the genetic background but also on the differentiation status of the cell subset (naive $<<$ CM $<$ TM $<$ EM $>$ TD), on its activation status, its capacity to secrete CCR5 ligands, and its exposure to CCR5 inducers such as IFN-α [310]. Highly polarized Th1 cells can be protected from HIV infection through the autocrine production of β-chemokines that internalize CCR5 [337,338]. It is possible that such a mechanism is at work in controllers considering the persistence of differentiated Th1 effectors in these patients [188,201], but further studies are needed to document possible activation-dependent resistance mechanisms.

Decreased Expression of Viral Receptors in SIV Natural Hosts

Multiple lines of evidence suggest that CD4 and/or CCR5 expression can limit SIV spread in its natural hosts. $CD4^+$ T-cell counts naturally decline with age in AGM and can reach very low values in adult monkeys. $CD4^+$ T cells from AGM can downregulate the CD4 receptor on activation, perhaps explaining the progressive loss of CD4 expression in this species [339]. SIV-infected AGM with little or no $CD4^+$ T cells nevertheless remain healthy and maintain T-cell helper function, which can be provided by double negative or $CD8\alpha\alpha^+$ T cells [339–341]. SM show a moderate age-dependent $CD4^+$ T-cell decline that is usually slightly amplified on natural SIV infection [342–344]. However, in rare cases, SIV infected SM show a drastic and persistent $CD4^+$ T-cell loss, that was associated with the emergence of an SIVsm strain of wider cellular tropism [345]. These monkeys remained healthy over several years, which could be explained by the presence of double negative T cells expressing a pattern of cytokines compatible with helper function [345]. A recent study shows that helper T-cell subsets that do not express the canonical CD4 receptor are commonly found in SIV natural host species of the *Cercocebus, Chlorocebus,* and *Erythrocebus* genera but are not detected in disease susceptible macaque species [341]. Importantly, the noncanonical helper T cells show a lower frequency of infection than $CD4^+$ T cells, which provides a strong clue to the lack of disease in SIV natural hosts. These findings also raise intriguing issues in fundamental immunology, such as understanding how TCR/peptide–MHC II interactions can remain efficient in the absence of CD4/MHC II interactions.

Paucity of CCR5 expression is also a notable characteristic of $CD4^+$ T cells in SIV natural hosts [346,347]. CCR5 expression is clearly lower in memory $CD4^+$ T cells from AGM and SM than in comparable subsets in macaques and humans. The difference is particularly marked in neonates and juvenile animals, which may explain the scarcity of vertical transmission in natural SIV hosts [348–350]. However, there is a relative flexibility in co-receptor use in natural SIV strains, and lack of CCR5 does not always prevent infection. For instance, up to 8% of SM in two major U.S. colonies do not express CCR5 at the cell surface due to small genetic deletions, but these animals can still be infected by SIVsm strains that use alternate co-receptors (CXCR6, GPR15, and GPR1) and show only slightly decreased viral loads [351]. The in-depth analysis of natural SIV infection thus suggests an unexpected degree of evolutionary adaptation on both host and virus sides. Natural host species may have adapted to widespread SIV infection by downregulating CD4 and CCR5 expression in critical cellular subsets, while SIV may have responded by the use of alternate co-receptors, with all these evolutionary changes leading to a balanced interaction characterized by low pathogenicity but relatively efficient viral spread. Of note, drugs blocking CCR5 and CD4 use by HIV have been developed for human use [352], and there is a major ongoing research effort aimed at emulating the protective effect of inactivating CCR5 mutations through gene therapy [59]. These approaches may also be tested in macaques, to determine whether lowering CCR5 and/or CD4 use could lead to a non-pathogenic SIVmac infection. In this respect, it is interesting that in a gene therapy experiment in humanized mice, where zinc finger nucleases were used to inactivate CCR5 expression, treated mice were protected from CD4 depletion but not from high viral loads [353], a situation reminiscent of natural SIV infection.

Viral Restriction Factors in HIV/SIV Control

Viral restriction factors are proteins that exhibit direct antiviral activity against HIV [354]. Most of these proteins, which include APOBEC3, TRIM5-α, Tetherin, and SAMHD1, are upregulated by type I IFN and thus participate in the innate response to HIV infection. Genetic variants of TRIM5α [355], the Vif and APOBEC3-interacting protein Cullin5 [356], and proteins of the APOBEC3 family [357,358] have been associated with the risk of disease progression or that of $CD4^+$ T-cell loss, raising the possibility that some of these variants may be involved in HIV control. However, these associations still need to be replicated in large scale genetic studies, as HLA associations are the only ones to have reached genome-wide significance so far [97].

APOBEC3

The APOBEC3 proteins are cytidine deaminases that can target single-stranded viral DNA APOBEC3G/F, when incorporated into an HIV particle, is able to deaminate cytosines in the first HIV DNA strand synthesized during retrotranscription, resulting in guanosine (G)-to-adenosine (A) mutations in the provirus. As a countermeasure, the HIV Vif protein is able to target APOBEC3G/F for proteasomal degradation. Incomplete Vif activity results in an inhibition of reverse transcription and in an accumulation of hypermutated proviruses carrying multiple G-to-A transitions. HIV controllers can harbor a significant proportion of hypermutated proviral genomes, indicative of inefficient Vif function or particularly efficient APOBEC3G/F activity [359,360]. Whether the hypermutation frequency is higher than in efficiently treated patients remains, however, to be ascertained. It is interesting that Th1 cells express more APOBEC3G than Th2 cells and that this difference affects the susceptibility to HIV infection [361], considering the persistence of Th1 effectors in HIV controllers [188,201]. Taken together, there are hints but not yet conclusive evidence that APOBEC3 proteins may play a role in HIV control. Of note, SIV Vif proteins tend to target APOBEC3G proteins of their natural host but not those of closely related monkey species, suggesting a role for the APOBEC3 family in defining SIV species-specificity [362].

TRIM5α

TRIM5α is a restriction factor that is thought to bind to incoming retroviral capsids and cause their premature disassembly, resulting in impaired retrotranscription and innate antiviral signaling [354,363,364]. Studies of TRIM5α function have so far provided the clearest evidence for the implication of a restriction factor in limiting HIV and SIV disease progression. In rhesus macaques, different TRIM5α alleles control the extent of the viral load induced by sensitive SIVmac strains. For instance, a 1.3 log difference in viral setpoint is observed in SIVmac251-infected Indian rhesus macaques homozygous for potent and weak TRIM5α alleles, respectively [365]. The TRIM5α effect is even more marked on the replication of the SIVsmE543-3 viral strain, as the most potent TRIM5α allele combinations are associated with a viral setpoint of 10^3 SIV copies/ml, which approximates viral control [29]. In addition, these TRIM5α alleles appear responsible for the control of a primary SIV strain from SM when transferred to Indian rhesus macaques, suggesting an important role of TRIM5α in suppressing SIV cross-species transmission. Of note, TRIM5α appears to have little effect on the replication of the highly pathogenic strain SIVmac239, suggesting that the virus can adapt to escape TRIM5α-mediated selective pressure [29].

There is suggestive evidence that human TRIM5α contributes to restrict HIV-2 replication. In vitro, chimeric viruses expressing HIV-2 capsids replicate less efficiently in the presence than in the absence of functional human TRIM5α, with up to 9-fold differences in the extent of viral replication [366,367]. Analysis of HIV-2 capsid sequences in a West African community cohort showed that certain proline residues in capsid were associated with a lower viral load and that these residues altered the susceptibility of HIV-2 to TRIM5α [368], although the role of these capsid polymorphisms has been debated [366]. Evidence for an effect of human TRIM5α on HIV-1 replication is scarce, as most laboratory-adapted HIV-1 strains are not efficiently restricted by TRIM5α [364]. However, capsids from certain clinical HIV-1 strains can confer a 4- to 6-fold restriction in the presence of functional human TRIM5α [369]. Notably, the TRIM5α-sensitive capsids carry the typical CTL resistance mutations associated with the HLA-B*57 and -B*27 alleles, suggesting that CTL-mediated immune pressure may force the virus to mutate toward vulnerable capsid conformations. Recent findings indicate that capsids derived from HLA-B*57$^+$ patients are significantly more sensitive to TRIM5α than are those derived from B*57$^-$ patients [370]. The difference in TRIM5α sensitivity is accentuated for capsids derived from B*57$^+$ HIV controllers and is also significant for B*27$^+$ controllers, possibly due to the particularly strong immune pressure exerted by CTL in these patients. The T242N escape mutation frequently found in B*57$^+$ patients was consistently associated with increased TRIM5α sensitivity, while the compensatory G248A mutation played a role in attenuating this phenotype. Trying to avoid TRIM5α restriction also likely explains the occurrence of rare escape mutations, such as Q244R/G248D, that are found almost exclusively in controller viruses and that apparently confer deleterious properties, such as poor viral fitness and only transient CTL escape [139,370]. Capsid is a viral protein demonstrating extreme genetic frailty, meaning that it is unusually intolerant to mutations [371]. This protein is subjected to strong structural constraints to produce functional mature viral cores and is additionally targeted by both CTL and TRIM5α. Available evidence suggests that the convergence of these three selective pressures, combined with a low viral replication level that limits mutation frequency, results in suboptimal evolutionary choices for the virus. Thus, multiple selective pressures focused on the viral capsid may contribute to HIV control.

Conclusion: Common Threads in Nonpathogenic HIV and SIV Infections

Infections in HIV controllers and SIV natural hosts differ in several important points, such as the extent of the plasma viral load and the strength of the antiviral cellular response. However, close examination of these two models also reveals common threads, highlighting key elements involved in resisting AIDS: these include contained levels of chronic immune activation, a downregulation of the interferon response, a low viral load in lymph nodes, and a relative protection of certain CD4$^+$ T-cell subsets, including CM and possibly Tfh cells, from infection (summarized in Table 15.1). One can thus propose a general

TABLE 15.1 Comparison of Viral and Immunological Parameters in Nonpathogenic and Pathogenic HIV/SIV Infections

	Nonpathogenic HIV/SIV Infection		Pathogenic HIV/SIV Infection
	HIV/SIV Controllers	**SIV Natural Hosts**	**Humans or Macaques**
CD4⁺ T-CELL DEPLETION			
In blood	±	±	++
In lymphoid organs	±	±	++
In gut mucosa	±	± or +	++
VIRAL PARAMETERS			
Viral Load			
In blood	±	++	++
In lymphoid organs	±	+	++
Viral Reservoir	±	+	++
Viral Fitness	+ or ++	++	++
ANTIVIRAL RESPONSES			
Cellular Antiviral Responses			
CD4⁺ T-Cell Response			
Intensity	+	+	+
Quality	++	+	±
CD8⁺ T-Cell Response			
Intensity	+	+	++
Quality	++	+	±
NK Cell Response			
Intensity	+	+	++
Quality	+	+	±
Protective MHC-I Alleles	+	?	±
Antibody Responses			
Intensity	+	+	++
Quality	+	+	±

TABLE 15.1 Comparison of Viral and Immunological Parameters in Nonpathogenic and Pathogenic HIV/SIV Infections — cont'd

	Nonpathogenic HIV/SIV Infection		Pathogenic HIV/SIV Infection
	HIV/SIV Controllers	**SIV Natural Hosts**	**Humans or Macaques**
IMMUNE ACTIVATION			
Ki-67 expression in blood	±	±	++
Activation markers	± or +	±	++
Microbial translocation	+	±	++
Induction of interferon stimulated genes	±	±	++
TARGET CELL SUSCEPTIBILITY			
Target Cells			
CM $CD4^+$ T cells	+	+	++
EM $CD4^+$ T cells	+	++	++
Tfh cells	?	±	++
Macrophages	±	±	± or +
Viral Receptor Expression			
CD4 expression	++	+ or ++	++
CCR5 expression	++	+	++
Viral Restriction Factors	+?	±	±

±, Low; +, medium; ++, high; ?, unknown.

model where primate lentiviral infections remain nonpathogenic if they spare lymph nodes and other lymphoid tissues that function as immune inductive sites. In contrast, if HIV or SIV replicates abundantly in cells that traffic (CM) or reside (Tfh) within inductive sites, it will trigger a massive immune response that will fuel chronic immune activation, resulting in a continuously expanding pool of activated $CD4^+$ target cells and in the ultimate exhaustion of the immune system.

In the case of controlled HIV and SIV infections, in-depth analyses of cellular responses have shown that these responses are of a remarkable intensity in the context of the amount of viral antigen available (Figure 15.1(B) and (C)). However, the general amplitude of the responses remains moderate, due to the efficient elimination of viral antigens. The presence of an efficient ongoing immune response within inductive sites is likely reflected by the non-negligible signs of immune activation that can be detected in

HIV controllers, including significant rises of inflammatory molecules in plasma, the expression of activation markers on T cells, and possibly the occurrence of microbial translocation. The containment of viral replication appears primarily mediated by $CD8^+$ and $CD4^+$ T cells with a high avidity for viral antigens, although other immune effectors such as B cells and NK cells also appear highly functional and likely contribute to viral control. How controllers initially succeed in establishing an antiviral immune response that contains HIV replication and avoids spiraling into hyperactivation remains a major field of investigation. Current studies highlight a clear protective role for certain MHC I alleles, that may function in concert with the restriction factor TRIM5α in constraining the viral capsid toward unfit conformations. A slightly lower infectability of $CD4^+$ T cells from controllers may also play a role, by giving the host a critical edge during the acute infection stage and allowing the antiviral response to develop in the presence of sufficient help. A moderate attenuation of the infecting viral strain, which has been documented in some cases, would have the same effect. The recent description of post-treatment controllers suggests that containing HIV replication at an early stage can indeed be sufficient to allow the development of an immune response able to control HIV.

Natural SIV hosts present a more clear-cut picture, where viral replication is contained in lymph nodes by a limited capacity to infect target cells rather than by a particularly efficient antiviral response. The realization that both SIV receptor and major co-receptor can be downregulated helps explain the protection of CM cells, which become less likely to disseminate SIV as they traffic within lymphoid organs. The limited influx of infected cells may in turn account for the lower infection of resident cells such as Tfh. This would then limit the amplitude of the antibody response and the development of germinal centers, perhaps explaining why viral particle trapping remains barely detectable in natural SIV hosts, and why the virus persists in the circulation of these hosts. Overall, the limited availability of viral antigens in immune inductive sites is consistent with the low levels of systemic immune activation and with the relatively moderate amplitude of the SIV-specific immune response.

Studies of nonpathogenic HIV and SIV infections have been rich in lessons and will likely inform the development of novel vaccine and immunotherapeutic approaches against HIV. For instance, inducing high avidity T-cell responses that mimic those seen in controllers may well enhance the protective efficacy of candidate HIV vaccines. Interventions that would reduce CCR5 expression may be beneficial in treated patients by limiting residual viral replication in the CM subset. Very early ARTs may benefit patients by limiting the seeding of lymph node reservoirs. Further exploration of the nonpathogenic models is also warranted to uncover novel restriction factors that may then be used to confer broad HIV resistance to the $CD4^+$ T-cell population.

Acknowledgments

I thank Cecilia Cheng-Mayer for critical reading of the manuscript. I apologize to the authors of many interesting studies that could not be cited due to space limitations. The author is supported by grants from ANRS, Sidaction, and the Pasteur Institute.

References

[1] Lambotte O, Boufassa F, Madec Y, Nguyen A, Goujard C, Meyer L, et al. HIV controllers: a homogeneous group of HIV-1-infected patients with spontaneous control of viral replication. Clin Infect Dis 2005;41(7):1053–6.

[2] Deeks SG, Walker BD. Human immunodeficiency virus controllers: mechanisms of durable virus control in the absence of antiretroviral therapy. Immunity 2007;27(3):406–16.

[3] Buckheit 3rd RW, Salgado M, Martins KO, Blankson JN. The implications of viral reservoirs on the elite control of HIV-1 infection. Cell Mol Life Sci 2013;70(6):1009–19.

[4] Sajadi MM, Constantine NT, Mann DL, Charurat M, Dadzan E, Kadlecik P, et al. Epidemiologic characteristics and natural history of HIV-1 natural viral suppressors. J Acquir Immune Defic Syndr 2009;50(4):403–8.

[5] Okulicz JF, Marconi VC, Landrum ML, Wegner S, Weintrob A, Ganesan A, et al. Clinical outcomes of elite controllers, viremic controllers, and long-term nonprogressors in the US Department of Defense HIV natural history study. J Infect Dis 2009;200(11):1714–23.

[6] Candotti D, Costagliola D, Joberty C, Bonduelle O, Rouzioux C, Autran B, et al. Status of long-term asymptomatic HIV-1 infection correlates with viral load but not with virus replication properties and cell tropism. French ALT Study Group. J Med Virol 1999;58(3):256–63.

[7] Grabar S, Selinger-Leneman H, Abgrall S, Pialoux G, Weiss L, Costagliola D. Prevalence and comparative characteristics of long-term nonprogressors and HIV controller patients in the French Hospital Database on HIV. AIDS 2009;23(9):1163–9.

[8] Okulicz JF, Lambotte O. Epidemiology and clinical characteristics of elite controllers. Curr Opin HIV AIDS 2011;6(3):163–8.

[9] Thiebaut R, Matheron S, Taieb A, Brun-Vezinet F, Chene G, Autran B, et al. Long-term nonprogressors and elite controllers in the ANRS CO5 HIV-2 cohort. AIDS 2011;25(6):865–7.

[10] van der Loeff MF, Larke N, Kaye S, Berry N, Ariyoshi K, Alabi A, et al. Undetectable plasma viral load predicts normal survival in HIV-2-infected people in a West African village. Retrovirology 2010;7:46.

[11] Pereyra F, Addo MM, Kaufmann DE, Liu Y, Miura T, Rathod A, et al. Genetic and immunologic heterogeneity among persons who control HIV infection in the absence of therapy. J Infect Dis 2008;197(4):563–71.

[12] Hatano H, Delwart EL, Norris PJ, Lee TH, Dunn-Williams J, Hunt PW, et al. Evidence for persistent low-level viremia in individuals who control human immunodeficiency virus in the absence of antiretroviral therapy. J Virol 2009;83(1):329–35.

[13] Migueles SA, Osborne CM, Royce C, Compton AA, Joshi RP, Weeks KA, et al. Lytic granule loading of $CD8^+$ T cells is required for HIV-infected cell elimination associated with immune control. Immunity 2008;29(6):1009–21.

[14] Sedaghat AR, Rastegar DA, O'Connell KA, Dinoso JB, Wilke CO, Blankson JN. T cell dynamics and the response to HAART in a cohort of HIV-1-infected elite suppressors. Clin Infect Dis 2009;49(11):1763–6.

[15] Hunt PW, Brenchley J, Sinclair E, McCune JM, Roland M, Page-Shafer K, et al. Relationship between T cell activation and $CD4^+$ T cell count in HIV-seropositive individuals with undetectable plasma HIV RNA levels in the absence of therapy. J Infect Dis 2008;197(1):126–33.

[16] Rotger M, Dalmau J, Rauch A, McLaren P, Bosinger SE, Martinez R, et al. Comparative transcriptomics of extreme phenotypes of human HIV-1 infection and SIV infection in sooty mangabey and rhesus macaque. J Clin Invest 2011;121(6):2391–400.
[17] Pereyra F, Palmer S, Miura T, Block BL, Wiegand A, Rothchild AC, et al. Persistent low-level viremia in HIV-1 elite controllers and relationship to immunologic parameters. J Infect Dis 2009;200(6):984–90.
[18] Dinoso JB, Kim SY, Wiegand AM, Palmer SE, Gange SJ, Cranmer L, et al. Treatment intensification does not reduce residual HIV-1 viremia in patients on highly active antiretroviral therapy. Proc Natl Acad Sci USA 2009;106(23):9403–8.
[19] Mens H, Kearney M, Wiegand A, Shao W, Schonning K, Gerstoft J, et al. HIV-1 continues to replicate and evolve in patients with natural control of HIV infection. J Virol 2010;84(24):12971–81.
[20] Goujard C, Chaix ML, Lambotte O, Deveau C, Sinet M, Guergnon J, et al. Spontaneous control of viral replication during primary HIV infection: when is "HIV controller" status established? Clin Infect Dis 2009;49(6):982–6.
[21] Miura T, Brumme ZL, Brockman MA, Rosato P, Sela J, Brumme CJ, et al. Impaired replication capacity of acute/early viruses in persons who become HIV controllers. J Virol 2010;84(15):7581–91.
[22] Madec Y, Boufassa F, Porter K, Prins M, Sabin C, Monforte AD, et al. Natural history of HIV control since seroconversion. AIDS 2013;27(15):2451–60.
[23] Yao XD, Omange RW, Henrick BM, Lester RT, Kimani J, Ball TB, et al. Acting locally: innate mucosal immunity in resistance to HIV-1 infection in Kenyan commercial sex workers. Mucosal Immunol 2013;7(2):268–79.
[24] Madec Y, Boufassa F, Rouzioux C, Delfraissy JF, Meyer L, Group SS. Undetectable viremia without antiretroviral therapy in patients with HIV seroconversion: an uncommon phenomenon? Clin Infect Dis 2005;40(9):1350–4.
[25] Boufassa F, Saez-Cirion A, Lechenadec J, Zucman D, Avettand-Fenoel V, Venet A, et al. CD4 dynamics over a 15 year-period among HIV controllers enrolled in the ANRS French observatory. PLoS One 2011;6(4):e18726.
[26] Klatt NR, Canary LA, Vanderford TH, Vinton CL, Engram JC, Dunham RM, et al. Dynamics of simian immunodeficiency virus SIVmac239 infection in pigtail macaques. J Virol 2012;86(2):1203–13.
[27] Budde ML, Greene JM, Chin EN, Ericsen AJ, Scarlotta M, Cain BT, et al. Specific $CD8^+$ T cell responses correlate with control of simian immunodeficiency virus replication in Mauritian cynomolgus macaques. J Virol 2012;86(14):7596–604.
[28] Ling B, Veazey RS, Luckay A, Penedo C, Xu K, Lifson JD, et al. SIV(mac) pathogenesis in rhesus macaques of Chinese and Indian origin compared with primary HIV infections in humans. AIDS 2002;16(11):1489–96.
[29] Kirmaier A, Wu F, Newman RM, Hall LR, Morgan JS, O'Connor S, et al. TRIM5 suppresses cross-species transmission of a primate immunodeficiency virus and selects for emergence of resistant variants in the new species. PLoS Biol 2010;8(8).
[30] Hirsch VM, Santra S, Goldstein S, Plishka R, Buckler-White A, Seth A, et al. Immune failure in the absence of profound $CD4^+$ T-lymphocyte depletion in simian immunodeficiency virus-infected rapid progressor macaques. J Virol 2004;78(1):275–84.
[31] McChesney MB, Collins JR, Lu D, Lu X, Torten J, Ashley RL, et al. Occult systemic infection and persistent simian immunodeficiency virus (SIV)-specific CD4(+)-T-cell proliferative responses in rhesus macaques that were transiently viremic after intravaginal inoculation of SIV. J Virol 1998;72(12):10029–35.
[32] Trivedi P, Horejsh D, Hinds SB, Hinds II PW, Wu MS, Salvato MS, et al. Intrarectal transmission of simian immunodeficiency virus in rhesus macaques: selective amplification and host responses to transient or persistent viremia. J Virol 1996;70(10):6876–83.

[33] Tasca S, Tsai L, Trunova N, Gettie A, Saifuddin M, Bohm R, et al. Induction of potent local cellular immunity with low dose X4 SHIV(SF33A) vaginal exposure. Virology 2007;367(1):196–211.
[34] Horton RE, McLaren PJ, Fowke K, Kimani J, Ball TB. Cohorts for the study of HIV-1-exposed but uninfected individuals: benefits and limitations. J Infect Dis 2010;202(Suppl. 3):S377–81.
[35] Sodora DL, Allan JS, Apetrei C, Brenchley JM, Douek DC, Else JG, et al. Toward an AIDS vaccine: lessons from natural simian immunodeficiency virus infections of African nonhuman primate hosts. Nat Med 2009;15(8):861–5.
[36] Locatelli S, Peeters M. Cross-species transmission of simian retroviruses: how and why they could lead to the emergence of new diseases in the human population. AIDS 2012;26(6):659–73.
[37] Ling B, Apetrei C, Pandrea I, Veazey RS, Lackner AA, Gormus B, et al. Classic AIDS in a sooty mangabey after an 18-year natural infection. J Virol 2004;78(16):8902–8.
[38] Pandrea I, Onanga R, Rouquet P, Bourry O, Ngari P, Wickings EJ, et al. Chronic SIV infection ultimately causes immunodeficiency in African non-human primates. AIDS 2001;15(18):2461–2.
[39] Hirsch VM, Olmsted RA, Murphey-Corb M, Purcell RH, Johnson PR. An African primate lentivirus (SIVsm) closely related to HIV-2. Nature 1989;339(6223):389–92.
[40] Gao F, Yue L, White AT, Pappas PG, Barchue J, Hanson AP, et al. Human infection by genetically diverse SIVSM-related HIV-2 in west Africa. Nature 1992;358(6386):495–9.
[41] Chen Z, Telfier P, Gettie A, Reed P, Zhang L, Ho DD, et al. Genetic characterization of new West African simian immunodeficiency virus SIVsm: geographic clustering of household-derived SIV strains with human immunodeficiency virus type 2 subtypes and genetically diverse viruses from a single feral sooty mangabey troop. J Virol 1996;70(6):3617–27.
[42] Chakrabarti L, Guyader M, Alizon M, Daniel MD, Desrosiers RC, Tiollais P, et al. Sequence of simian immunodeficiency virus from macaque and its relationship to other human and simian retroviruses. Nature 1987;328(6130):543–7.
[43] Apetrei C, Kaur A, Lerche NW, Metzger M, Pandrea I, Hardcastle J, et al. Molecular epidemiology of simian immunodeficiency virus SIVsm in U.S. primate centers unravels the origin of SIVmac and SIVstm. J Virol 2005;79(14):8991–9005.
[44] Keele BF, Van Heuverswyn F, Li Y, Bailes E, Takehisa J, Santiago ML, et al. Chimpanzee reservoirs of pandemic and nonpandemic HIV-1. Science 2006;313(5786):523–6.
[45] Van Heuverswyn F, Li Y, Neel C, Bailes E, Keele BF, Liu W, et al. Human immunodeficiency viruses: SIV infection in wild gorillas. Nature 2006;444(7116):164.
[46] Plantier JC, Leoz M, Dickerson JE, De Oliveira F, Cordonnier F, Lemee V, et al. A new human immunodeficiency virus derived from gorillas. Nat Med 2009;15(8):871–2.
[47] Keele BF, Jones JH, Terio KA, Estes JD, Rudicell RS, Wilson ML, et al. Increased mortality and AIDS-like immunopathology in wild chimpanzees infected with SIVcpz. Nature 2009;460(7254):515–9.
[48] Sharp PM, Hahn BH. The evolution of HIV-1 and the origin of AIDS. Philos Trans R Soc Lond B Biol Sci 2010;365(1552):2487–94.
[49] Rey-Cuille MA, Berthier JL, Bomsel-Demontoy MC, Chaduc Y, Montagnier L, Hovanessian AG, et al. Simian immunodeficiency virus replicates to high levels in sooty mangabeys without inducing disease. J Virol 1998;72(5):3872–86.
[50] Diop OM, Gueye A, Dias-Tavares M, Kornfeld C, Faye A, Ave P, et al. High levels of viral replication during primary simian immunodeficiency virus SIVagm infection are rapidly and strongly controlled in African green monkeys. J Virol 2000;74(16):7538–47.
[51] Goldstein S, Ourmanov I, Brown CR, Beer BE, Elkins WR, Plishka R, et al. Wide range of viral load in healthy african green monkeys naturally infected with simian immunodeficiency virus. J Virol 2000;74(24):11744–53.

[52] Gordon SN, Dunham RM, Engram JC, Estes J, Wang Z, Klatt NR, et al. Short-lived infected cells support virus replication in sooty mangabeys naturally infected with simian immunodeficiency virus: implications for AIDS pathogenesis. J Virol 2008;82(7):3725–35.
[53] Pandrea I, Ribeiro RM, Gautam R, Gaufin T, Pattison M, Barnes M, et al. Simian immunodeficiency virus SIVagm dynamics in African green monkeys. J Virol 2008;82(7):3713–24.
[54] Eisele E, Siliciano RF. Redefining the viral reservoirs that prevent HIV-1 eradication. Immunity 2012;37(3):377–88.
[55] Lewin SR, Rouzioux C. HIV cure and eradication: how will we get from the laboratory to effective clinical trials? AIDS 2011;25(7):885–97.
[56] Julg B, Pereyra F, Buzon MJ, Piechocka-Trocha A, Clark MJ, Baker BM, et al. Infrequent recovery of HIV from but robust exogenous infection of activated CD4(+) T cells in HIV elite controllers. Clin Infect Dis 2010;51(2):233–8.
[57] Sajadi MM, Heredia A, Le N, Constantine NT, Redfield RR. HIV-1 natural viral suppressors: control of viral replication in the absence of therapy. AIDS 2007;21(4):517–9.
[58] Blankson JN, Bailey JR, Thayil S, Yang HC, Lassen K, Lai J, et al. Isolation and characterization of replication-competent human immunodeficiency virus type 1 from a subset of elite suppressors. J Virol 2007;81(5):2508–18.
[59] International ASSWGoHIVC, Deeks SG, Autran B, Berkhout B, Benkirane M, Cairns S, et al. Towards an HIV cure: a global scientific strategy. Nat Rev Immunol 2012;12(8):607–14.
[60] Saez-Cirion A, Bacchus C, Hocqueloux L, Avettand-Fenoel V, Girault I, Lecuroux C, et al. Post-treatment HIV-1 controllers with a long-term virological remission after the interruption of early initiated antiretroviral therapy ANRS VISCONTI Study. PLoS Pathog 2013;9(3):e1003211.
[61] Brenchley JM, Schacker TW, Ruff LE, Price DA, Taylor JH, Beilman GJ, et al. $CD4^+$ T cell depletion during all stages of HIV disease occurs predominantly in the gastrointestinal tract. J Exp Med 2004;200(6):749–59.
[62] Avettand-Fenoel V, Prazuck T, Hocqueloux L, Melard A, Michau C, Kerdraon R, et al. HIV-DNA in rectal cells is well correlated with HIV-DNA in blood in different groups of patients, including long-term non-progressors. AIDS 2008;22(14):1880–2.
[63] Ling B, Mohan M, Lackner AA, Green LC, Marx PA, Doyle LA, et al. The large intestine as a major reservoir for simian immunodeficiency virus in macaques with long-term, nonprogressing infection. J Infect Dis 2010;202(12):1846–54.
[64] Cory TJ, Schacker TW, Stevenson M, Fletcher CV. Overcoming pharmacologic sanctuaries. Curr Opin HIV AIDS 2013;8(3):190–5.
[65] Pantaleo G, Graziosi C, Demarest JF, Butini L, Montroni M, Fox CH, et al. HIV infection is active and progressive in lymphoid tissue during the clinically latent stage of disease. Nature 1993;362(6418):355–8.
[66] Perreau M, Savoye AL, De Crignis E, Corpataux JM, Cubas R, Haddad EK, et al. Follicular helper T cells serve as the major CD4 T cell compartment for HIV-1 infection, replication, and production. J Exp Med 2013;210(1):143–56.
[67] Pandrea IV, Gautam R, Ribeiro RM, Brenchley JM, Butler IF, Pattison M, et al. Acute loss of intestinal $CD4^+$ T cells is not predictive of simian immunodeficiency virus virulence. J Immunol 2007;179(5):3035–46.
[68] Gordon SN, Klatt NR, Bosinger SE, Brenchley JM, Milush JM, Engram JC, et al. Severe depletion of mucosal $CD4^+$ T cells in AIDS-free simian immunodeficiency virus-infected sooty mangabeys. J Immunol 2007;179(5):3026–34.
[69] Beer B, Scherer J, zur Megede J, Norley S, Baier M, Kurth R. Lack of dichotomy between virus load of peripheral blood and lymph nodes during long-term simian immunodeficiency virus infection of African green monkeys. Virology 1996;219(2):367–75.

[70] Meythaler M, Wang Z, Martinot A, Pryputniewicz S, Kasheta M, McClure HM, et al. Early induction of polyfunctional simian immunodeficiency virus (SIV)-specific T lymphocytes and rapid disappearance of SIV from lymph nodes of sooty mangabeys during primary infection. J Immunol 2011;186(9):5151–61.
[71] Chakrabarti L, Cumont MC, Montagnier L, Hurtrel B. Variable course of primary simian immunodeficiency virus infection in lymph nodes: relation to disease progression. J Virol 1994;68(10):6634–43.
[72] Brenchley JM, Vinton C, Tabb B, Hao XP, Connick E, Paiardini M, et al. Differential infection patterns of $CD4^+$ T cells and lymphoid tissue viral burden distinguish progressive and nonprogressive lentiviral infections. Blood 2012;120(20):4172–81.
[73] Zaunders J, Dyer WB, Churchill M. The Sydney Blood Bank Cohort: implications for viral fitness as a cause of elite control. Curr Opin HIV AIDS 2011;6(3):151–6.
[74] Kirchhoff F, Greenough TC, Brettler DB, Sullivan JL, Desrosiers RC. Brief report: absence of intact nef sequences in a long-term survivor with nonprogressive HIV-1 infection. N Engl J Med 1995;332(4):228–32.
[75] Huang Y, Zhang L, Ho DD. Characterization of gag and pol sequences from long-term survivors of human immunodeficiency virus type 1 infection. Virology 1998; 240(1):36–49.
[76] Alexander L, Aquino-DeJesus MJ, Chan M, Andiman WA. Inhibition of human immunodeficiency virus type 1 (HIV-1) replication by a two-amino-acid insertion in HIV-1 Vif from a nonprogressing mother and child. J Virol 2002;76(20):10533–9.
[77] Lamine A, Caumont-Sarcos A, Chaix ML, Saez-Cirion A, Rouzioux C, Delfraissy JF, et al. Replication-competent HIV strains infect HIV controllers despite undetectable viremia (ANRS EP36 study). AIDS 2007;21(8):1043–5.
[78] Miura T, Brockman MA, Brumme CJ, Brumme ZL, Carlson JM, Pereyra F, et al. Genetic characterization of human immunodeficiency virus type 1 in elite controllers: lack of gross genetic defects or common amino acid changes. J Virol 2008;82(17): 8422–30.
[79] Martinez-Picado J, Prado JG, Fry EE, Pfafferott K, Leslie A, Chetty S, et al. Fitness cost of escape mutations in p24 Gag in association with control of human immunodeficiency virus type 1. J Virol 2006;80(7):3617–23.
[80] Miura T, Brockman MA, Brumme ZL, Brumme CJ, Pereyra F, Trocha A, et al. HLA-associated alterations in replication capacity of chimeric NL4-3 viruses carrying gag-protease from elite controllers of human immunodeficiency virus type 1. J Virol 2009;83(1):140–9.
[81] Lassen KG, Lobritz MA, Bailey JR, Johnston S, Nguyen S, Lee B, et al. Elite suppressor-derived HIV-1 envelope glycoproteins exhibit reduced entry efficiency and kinetics. PLoS Pathog 2009;5(4):e1000377.
[82] Mwimanzi P, Markle TJ, Martin E, Ogata Y, Kuang XT, Tokunaga M, et al. Attenuation of multiple Nef functions in HIV-1 elite controllers. Retrovirology 2013;10:1.
[83] Alexander L, Illyinskii PO, Lang SM, Means RE, Lifson J, Mansfield K, et al. Determinants of increased replicative capacity of serially passaged simian immunodeficiency virus with nef deleted in rhesus monkeys. J Virol 2003;77(12):6823–35.
[84] Ho Tsong Fang R, Khatissian E, Monceaux V, Cumont MC, Beq S, Ameisen JC, et al. Disease progression in macaques with low SIV replication levels: on the relevance of TREC counts. AIDS 2005;19(7):663–73.
[85] Schindler M, Schmokel J, Specht A, Li H, Munch J, Khalid M, et al. Inefficient Nef-mediated downmodulation of CD3 and MHC-I correlates with loss of $CD4^+$ T cells in natural SIV infection. PLoS Pathog 2008;4(7):e1000107.
[86] Pandrea I, Gaufin T, Gautam R, Kristoff J, Mandell D, Montefiori D, et al. Functional cure of SIVagm infection in rhesus macaques results in complete recovery of $CD4^+$ T cells and is reverted by $CD8^+$ cell depletion. PLoS Pathog 2011;7(8):e1002170.

[87] Schmitz JE, Zahn RC, Brown CR, Rett MD, Li M, Tang H, et al. Inhibition of adaptive immune responses leads to a fatal clinical outcome in SIV-infected pigtailed macaques but not vervet African green monkeys. PLoS Pathog 2009;5(12):e1000691.
[88] Mellors JW, Rinaldo Jr CR, Gupta P, White RM, Todd JA, Kingsley LA. Prognosis in HIV-1 infection predicted by the quantity of virus in plasma. Science 1996;272(5265):1167–70.
[89] Fellay J, Shianna KV, Ge D, Colombo S, Ledergerber B, Weale M, et al. A whole-genome association study of major determinants for host control of HIV-1. Science 2007;317(5840):944–7.
[90] Guergnon J, Dalmasso C, Broet P, Meyer L, Westrop SJ, Imami N, et al. Single-nucleotide polymorphism-defined class I and class III major histocompatibility complex genetic subregions contribute to natural long-term nonprogression in HIV infection. J Infect Dis 2012;205(5):718–24.
[91] Mummidi S, Ahuja SS, Gonzalez E, Anderson SA, Santiago EN, Stephan KT, et al. Genealogy of the CCR5 locus and chemokine system gene variants associated with altered rates of HIV-1 disease progression. Nat Med 1998;4(7):786–93.
[92] Kaslow RA, Carrington M, Apple R, Park L, Munoz A, Saah AJ, et al. Influence of combinations of human major histocompatibility complex genes on the course of HIV-1 infection. Nat Med 1996;2(4):405–11.
[93] Carrington M, Nelson GW, Martin MP, Kissner T, Vlahov D, Goedert JJ, et al. HLA and HIV-1: heterozygote advantage and B*35-Cw*04 disadvantage. Science 1999;283(5408):1748–52.
[94] Migueles SA, Sabbaghian MS, Shupert WL, Bettinotti MP, Marincola FM, Martino L, et al. HLA B*5701 is highly associated with restriction of virus replication in a subgroup of HIV-infected long term nonprogressors. Proc Natl Acad Sci USA 2000;97(6):2709–14.
[95] Thomas R, Apps R, Qi Y, Gao X, Male V, O'HUigin C, et al. HLA-C cell surface expression and control of HIV/AIDS correlate with a variant upstream of HLA-C. Nat Genet 2009;41(12):1290–4.
[96] Apps R, Qi Y, Carlson JM, Chen H, Gao X, Thomas R, et al. Influence of HLA-C expression level on HIV control. Science 2013;340(6128):87–91.
[97] Carrington M, Walker BD. Immunogenetics of spontaneous control of HIV. Annu Rev Med 2012;63:131–45.
[98] International HIVCS, Pereyra F, Jia X, McLaren PJ, Telenti A, de Bakker PI, et al. The major genetic determinants of HIV-1 control affect HLA class I peptide presentation. Science 2010;330(6010):1551–7.
[99] Antoni G, Guergnon J, Meaudre C, Samri A, Boufassa F, Goujard C, et al. MHC-driven HIV-1 control on the long run is not systematically determined at early times post-HIV-1 infection. AIDS 2013;27(11):1707–16.
[100] Loffredo JT, Sidney J, Bean AT, Beal DR, Bardet W, Wahl A, et al. Two MHC class I molecules associated with elite control of immunodeficiency virus replication, Mamu-B*08 and HLA-B*2705, bind peptides with sequence similarity. J Immunol 2009;182(12):7763–75.
[101] Yant LJ, Friedrich TC, Johnson RC, May GE, Maness NJ, Enz AM, et al. The high-frequency major histocompatibility complex class I allele Mamu-B*17 is associated with control of simian immunodeficiency virus SIVmac239 replication. J Virol 2006;80(10):5074–7.
[102] Pal R, Venzon D, Letvin NL, Santra S, Montefiori DC, Miller NR, et al. ALVAC-SIV-gag-pol-env-based vaccination and macaque major histocompatibility complex class I (A*01) delay simian immunodeficiency virus SIVmac-induced immunodeficiency. J Virol 2002;76(1):292–302.
[103] Letvin NL, Rao SS, Montefiori DC, Seaman MS, Sun Y, Lim SY, et al. Immune and genetic correlates of vaccine Protection against mucosal infection by SIV in monkeys. Sci Trans Med 2011;3(81):81ra36.

[104] Nomura T, Yamamoto H, Shiino T, Takahashi N, Nakane T, Iwamoto N, et al. Association of major histocompatibility complex class I haplotypes with disease progression after simian immunodeficiency virus challenge in burmese rhesus macaques. J Virol 2012;86(12):6481–90.

[105] Herberts CA, Neijssen JJ, de Haan J, Janssen L, Drijfhout JW, Reits EA, et al. Cutting edge: HLA-B27 acquires many N-terminal dibasic peptides: coupling cytosolic peptide stability to antigen presentation. J Immunol 2006;176(5):2697–701.

[106] de Groot NG, Heijmans CM, Zoet YM, de Ru AH, Verreck FA, van Veelen PA, et al. AIDS-protective HLA-B*27/B*57 and chimpanzee MHC class I molecules target analogous conserved areas of HIV-1/SIVcpz. Proc Natl Acad Sci USA 2010;107(34):15175–80.

[107] de Groot NG, Bontrop RE. The HIV-1 pandemic: does the selective sweep in chimpanzees mirror humankind's future? Retrovirology 2013;10:53.

[108] Ferre AL, Hunt PW, McConnell DH, Morris MM, Garcia JC, Pollard RB, et al. HIV controllers with HLA-DRB1*13 and HLA-DQB1*06 alleles have strong, polyfunctional mucosal $CD4^+$ T-cell responses. J Virol 2010;84(21):11020–9.

[109] Malhotra U, Holte S, Dutta S, Berrey MM, Delpit E, Koelle DM, et al. Role for HLA class II molecules in HIV-1 suppression and cellular immunity following antiretroviral treatment. J Clin Invest 2001;107(4):505–17.

[110] Julg B, Moodley ES, Qi Y, Ramduth D, Reddy S, Mncube Z, et al. Possession of HLA class II DRB1*1303 associates with reduced viral loads in chronic HIV-1 clade C and B infection. J Infect Dis 2011;203(6):803–9.

[111] Ranasinghe S, Cutler S, Davis I, Lu R, Soghoian DZ, Qi Y, et al. Association of HLA-DRB1-restricted CD4 T cell responses with HIV immune control. Nat Medic 2013;19(7):930–3.

[112] Wright JK, Brumme ZL, Julg B, van der Stok M, Mncube Z, Gao X, et al. Lack of association between HLA class II alleles and in vitro replication capacities of recombinant viruses encoding HIV-1 subtype C Gag-protease from chronically infected individuals. J Virol 2012;86(2):1273–6.

[113] Giraldo-Vela JP, Rudersdorf R, Chung C, Qi Y, Wallace LT, Bimber B, et al. The major histocompatibility complex class II alleles Mamu-DRB1*1003 and -DRB1*0306 are enriched in a cohort of simian immunodeficiency virus-infected rhesus macaque elite controllers. J Virol 2008;82(2):859–70.

[114] Saez-Cirion A, Lacabaratz C, Lambotte O, Vermisse P, Urrutia A, Boufassa F, et al. HIV controllers exhibit potent CD8 T cell capacity to suppress HIV infection ex vivo and peculiar CTL activation phenotype. Proc Natl Acad Sci USA 2007;104(16):6776–81.

[115] Buckheit 3rd RW, Salgado M, Silciano RF, Blankson JN. Inhibitory potential of subpopulations of $CD8^+$ T cells in HIV-1-infected elite suppressors. J Virol 2012;86(24):13679–88.

[116] Ndhlovu ZM, Proudfoot J, Cesa K, Alvino DM, McMullen A, Vine S, et al. Elite controllers with low to absent effector $CD8^+$ T cell responses maintain highly functional, broadly directed central memory responses. J Virol 2012;86(12):6959–69.

[117] Shan L, Deng K, Shroff NS, Durand CM, Rabi SA, Yang HC, et al. Stimulation of HIV-1-specific cytolytic T lymphocytes facilitates elimination of latent viral reservoir after virus reactivation. Immunity 2012;36(3):491–501.

[118] Hersperger AR, Pereyra F, Nason M, Demers K, Sheth P, Shin LY, et al. Perforin expression directly ex vivo by HIV-specific CD8 T-cells is a correlate of HIV elite control. PLoS Pathog 2010;6(5):e1000917.

[119] Hersperger AR, Martin JN, Shin LY, Sheth PM, Kovacs CM, Cosma GL, et al. Increased HIV-specific $CD8^+$ T-cell cytotoxic potential in HIV elite controllers is associated with T-bet expression. Blood 2011;117(14):3799–808.

[120] Lecuroux C, Saez-Cirion A, Noel N, Ben-Lamine L, Girault I, Caillat-Zucman S, et al. NKG2D expression on HIV-specific $CD8^+$ T cells is reduced in viremic HIV-1-infected patients but maintained in HIV controllers. J Acquir Immune Defic Syndr 2013;62(1):17–20.

[121] Day CL, Kaufmann DE, Kiepiela P, Brown JA, Moodley ES, Reddy S, et al. PD-1 expression on HIV-specific T cells is associated with T-cell exhaustion and disease progression. Nature 2006;443(7109):350–4.
[122] Yamamoto T, Price DA, Casazza JP, Ferrari G, Nason M, Chattopadhyay PK, et al. Surface expression patterns of negative regulatory molecules identify determinants of virus-specific CD8+ T-cell exhaustion in HIV infection. Blood 2011;117(18):4805–15.
[123] Migueles SA, Laborico AC, Shupert WL, Sabbaghian MS, Rabin R, Hallahan CW, et al. HIV-specific CD8+ T cell proliferation is coupled to perforin expression and is maintained in nonprogressors. Nat Immunol 2002;3(11):1061–8.
[124] Lichterfeld M, Mou D, Cung TD, Williams KL, Waring MT, Huang J, et al. Telomerase activity of HIV-1-specific CD8+ T cells: constitutive up-regulation in controllers and selective increase by blockade of PD ligand 1 in progressors. Blood 2008;112(9):3679–87.
[125] Betts MR, Nason MC, West SM, De Rosa SC, Migueles SA, Abraham J, et al. HIV nonprogressors preferentially maintain highly functional HIV-specific CD8+ T cells. Blood 2006;107(12):4781–9.
[126] Owen RE, Heitman JW, Hirschkorn DF, Lanteri MC, Biswas HH, Martin JN, et al. HIV+ elite controllers have low HIV-specific T-cell activation yet maintain strong, polyfunctional T-cell responses. AIDS 2010;24(8):1095–105.
[127] Ferre AL, Hunt PW, Critchfield JW, Young DH, Morris MM, Garcia JC, et al. Mucosal immune responses to HIV-1 in elite controllers: a potential correlate of immune control. Blood 2009;113(17):3978–89.
[128] Kiepiela P, Ngumbela K, Thobakgale C, Ramduth D, Honeyborne I, Moodley E, et al. CD8+ T-cell responses to different HIV proteins have discordant associations with viral load. Nat Medic 2007;13(1):46–53.
[129] de Silva TI, Peng Y, Leligdowicz A, Zaidi I, Li L, Griffin H, et al. Correlates of T-cell-mediated viral control and phenotype of CD8(+) T cells in HIV-2, a naturally contained human retroviral infection. Blood 2013;121(21):4330–9.
[130] Saez-Cirion A, Sinet M, Shin SY, Urrutia A, Versmisse P, Lacabaratz C, et al. Heterogeneity in HIV suppression by CD8 T cells from HIV controllers: association with Gag-specific CD8 T cell responses. J Immunol 2009;182(12):7828–37.
[131] Julg B, Williams KL, Reddy S, Bishop K, Qi Y, Carrington M, et al. Enhanced anti-HIV functional activity associated with Gag-specific CD8 T-cell responses. J Virol 2010;84(11):5540–9.
[132] Sacha JB, Chung C, Rakasz EG, Spencer SP, Jonas AK, Bean AT, et al. Gag-specific CD8+ T lymphocytes recognize infected cells before AIDS-virus integration and viral protein expression. J Immunol 2007;178(5):2746–54.
[133] Altfeld M, Kalife ET, Qi Y, Streeck H, Lichterfeld M, Johnston MN, et al. HLA alleles associated with delayed progression to AIDS contribute strongly to the initial CD8(+) T cell response against HIV-1. PLoS Med 2006;3(10):e403.
[134] Goulder PJ, Walker BD. HIV and HLA class I: an evolving relationship. Immunity 2012;37(3):426–40.
[135] Bailey JR, Brennan TP, O'Connell KA, Siliciano RF, Blankson JN. Evidence of CD8+ T-cell-mediated selective pressure on human immunodeficiency virus type 1 nef in HLA-B*57+ elite suppressors. J Virol 2009;83(1):88–97.
[136] Brockman MA, Brumme ZL, Brumme CJ, Miura T, Sela J, Rosato PC, et al. Early selection in Gag by protective HLA alleles contributes to reduced HIV-1 replication capacity that may be largely compensated for in chronic infection. J Virol 2010;84(22):11937–49.
[137] Bailey JR, Williams TM, Siliciano RF, Blankson JN. Maintenance of viral suppression in HIV-1-infected HLA-B*57+ elite suppressors despite CTL escape mutations. J Exp Med 2006;203(5):1357–69.
[138] O'Connell KA, Brennan TP, Bailey JR, Ray SC, Siliciano RF, Blankson JN. Control of HIV-1 in elite suppressors despite ongoing replication and evolution in plasma virus. J Virol 2010;84(14):7018–28.

[139] Miura T, Brockman MA, Schneidewind A, Lobritz M, Pereyra F, Rathod A, et al. HLA-B57/B*5801 human immunodeficiency virus type 1 elite controllers select for rare gag variants associated with reduced viral replication capacity and strong cytotoxic T-lymphocyte [corrected] recognition. J Virol 2009;83(6):2743–55.
[140] O'Connell KA, Hegarty RW, Siliciano RF, Blankson JN. Viral suppression of multiple escape mutants by de novo CD8(+) T cell responses in a human immunodeficiency virus-1 infected elite suppressor. Retrovirology 2011;8:63.
[141] Kelleher AD, Long C, Holmes EC, Allen RL, Wilson J, Conlon C, et al. Clustered mutations in HIV-1 gag are consistently required for escape from HLA-B27-restricted cytotoxic T lymphocyte responses. J Exp Med 2001;193(3):375–86.
[142] Goulder PJ, Phillips RE, Colbert RA, McAdam S, Ogg G, Nowak MA, et al. Late escape from an immunodominant cytotoxic T-lymphocyte response associated with progression to AIDS. Nat Med 1997;3(2):212–7.
[143] Almeida JR, Price DA, Papagno L, Arkoub ZA, Sauce D, Bornstein E, et al. Superior control of HIV-1 replication by $CD8^+$ T cells is reflected by their avidity, polyfunctionality, and clonal turnover. J Exp Med 2007;204(10):2473–85.
[144] Almeida JR, Sauce D, Price DA, Papagno L, Shin SY, Moris A, et al. Antigen sensitivity is a major determinant of $CD8^+$ T-cell polyfunctionality and HIV-suppressive activity. Blood 2009;113(25):6351–60.
[145] Iglesias MC, Almeida JR, Fastenackels S, van Bockel DJ, Hashimoto M, Venturi V, et al. Escape from highly effective public $CD8^+$ T-cell clonotypes by HIV. Blood 2011;118(8):2138–49.
[146] Ladell K, Hashimoto M, Iglesias MC, Wilmann PG, McLaren JE, Gras S, et al. A molecular basis for the control of preimmune escape variants by HIV-specific $CD8^+$ T cells. Immunity 2013;38(3):425–36.
[147] Chen H, Ndhlovu ZM, Liu D, Porter LC, Fang JW, Darko S, et al. TCR clonotypes modulate the protective effect of HLA class I molecules in HIV-1 infection. Nat Immunol 2012;13(7):691–700.
[148] van Bockel DJ, Price DA, Munier ML, Venturi V, Asher TE, Ladell K, et al. Persistent survival of prevalent clonotypes within an immunodominant HIV gag-specific $CD8^+$ T cell response. J Immunol 2011;186(1):359–71.
[149] Varela-Rohena A, Molloy PE, Dunn SM, Li Y, Suhoski MM, Carroll RG, et al. Control of HIV-1 immune escape by CD8 T cells expressing enhanced T-cell receptor. Nat Med 2008;14(12):1390–5.
[150] Leligdowicz A, Onyango C, Yindom LM, Peng Y, Cotten M, Jaye A, et al. Highly avid, oligoclonal, early-differentiated antigen-specific $CD8^+$ T cells in chronic HIV-2 infection. Eur J Immunol 2010;40(7):1963–72.
[151] Schmitz JE, Kuroda MJ, Santra S, Sasseville VG, Simon MA, Lifton MA, et al. Control of viremia in simian immunodeficiency virus infection by $CD8^+$ lymphocytes. Science 1999;283(5403):857–60.
[152] Jin X, Bauer DE, Tuttleton SE, Lewin S, Gettie A, Blanchard J, et al. Dramatic rise in plasma viremia after CD8(+) T cell depletion in simian immunodeficiency virus-infected macaques. J Exp Med 1999;189(6):991–8.
[153] Okoye A, Park H, Rohankhedkar M, Coyne-Johnson L, Lum R, Walker JM, et al. Profound $CD4^+/CCR5^+$ T cell expansion is induced by $CD8^+$ lymphocyte depletion but does not account for accelerated SIV pathogenesis. J Exp Med 2009;206(7): 1575–88.
[154] Mudd PA, Ericsen AJ, Burwitz BJ, Wilson NA, O'Connor DH, Hughes AL, et al. Escape from CD8(+) T cell responses in Mamu-B*00801(+) macaques differentiates progressors from elite controllers. J Immunol 2012;188(7):3364–70.
[155] Weinfurter JT, May GE, Soma T, Hessell AJ, Leon EJ, Macnair CE, et al. Macaque long-term nonprogressors resist superinfection with multiple $CD8^+$ T cell escape variants of simian immunodeficiency virus. J Virol 2011;85(1):530–41.

[156] Mudd PA, Martins MA, Ericsen AJ, Tully DC, Power KA, Bean AT, et al. Vaccine-induced CD8+ T cells control AIDS virus replication. Nature 2012;491(7422):129–33.
[157] Valentine LE, Loffredo JT, Bean AT, Leon EJ, MacNair CE, Beal DR, et al. Infection with "escaped" virus variants impairs control of simian immunodeficiency virus SIVmac239 replication in Mamu-B*08-positive macaques. J Virol 2009;83(22):11514–27.
[158] Harris M, Burns CM, Becker EA, Braasch AT, Gostick E, Johnson RC, et al. Acute CD8 T cell responses that select for escape variants are needed to control live attenuated SIV. J Virol 2013;87(16):9353–64.
[159] Mendoza D, Migueles SA, Rood JE, Peterson B, Johnson S, Doria-Rose N, et al. Cytotoxic capacity of SIV-specific CD8(+) T cells against primary autologous targets correlates with immune control in SIV-infected rhesus macaques. PLoS Pathog 2013;9(2):e1003195.
[160] Price DA, Asher TE, Wilson NA, Nason MC, Brenchley JM, Metzler IS, et al. Public clonotype usage identifies protective Gag-specific CD8+ T cell responses in SIV infection. J Exp Med 2009;206(4):923–36.
[161] Klatt NR, Shudo E, Ortiz AM, Engram JC, Paiardini M, Lawson B, et al. CD8+ lymphocytes control viral replication in SIVmac239-infected rhesus macaques without decreasing the lifespan of productively infected cells. PLoS Pathog 2010;6(1):e1000747.
[162] Wong JK, Strain MC, Porrata R, Reay E, Sankaran-Walters S, Ignacio CC, et al. In vivo CD8+ T-cell suppression of siv viremia is not mediated by CTL clearance of productively infected cells. PLoS Pathog 2010;6(1):e1000748.
[163] Wick WD, Yang OO. Biologically-directed modeling reflects cytolytic clearance of SIV-infected cells in vivo in macaques. PLoS One 2012;7(9):e44778.
[164] Hansen SG, Sacha JB, Hughes CM, Ford JC, Burwitz BJ, Scholz I, et al. Cytomegalovirus vectors violate CD8+ T cell epitope recognition paradigms. Science 2013;340(6135):1237874.
[165] Hansen SG, Ford JC, Lewis MS, Ventura AB, Hughes CM, Coyne-Johnson L, et al. Profound early control of highly pathogenic SIV by an effector memory T-cell vaccine. Nature 2011;473(7348):523–7.
[166] Emu B, Sinclair E, Hatano H, Ferre A, Shacklett B, Martin JN, et al. HLA class I-restricted T-cell responses may contribute to the control of human immunodeficiency virus infection, but such responses are not always necessary for long-term virus control. J Virol 2008;82(11):5398–407.
[167] Vigneault F, Woods M, Buzon MJ, Li C, Pereyra F, Crosby SD, et al. Transcriptional profiling of CD4 T cells identifies distinct subgroups of HIV-1 elite controllers. J Virol 2011;85(6):3015–9.
[168] Wang Z, Metcalf B, Ribeiro RM, McClure H, Kaur A. Th-1-type cytotoxic CD8+ T-lymphocyte responses to simian immunodeficiency virus (SIV) are a consistent feature of natural SIV infection in sooty mangabeys. J Virol 2006;80(6):2771–83.
[169] Dunham R, Pagliardini P, Gordon S, Sumpter B, Engram J, Moanna A, et al. The AIDS resistance of naturally SIV-infected sooty mangabeys is independent of cellular immunity to the virus. Blood 2006;108(1):209–17.
[170] Powell JD, Yehuda-Cohen T, Villinger F, McClure HM, Sell KW, Ahmed-Ansari A. Inhibition of SIV/SMM replication in vitro by CD8+ cells from SIV/SMM infected seropositive clinically asymptomatic sooty mangabeys. J Med Primatol 1990;19(3–4):239–49.
[171] Zahn RC, Rett MD, Korioth-Schmitz B, Sun Y, Buzby AP, Goldstein S, et al. Simian immunodeficiency virus (SIV)-specific CD8+ T-cell responses in vervet African green monkeys chronically infected with SIVagm. J Virol 2008;82(23):11577–88.
[172] Barry AP, Silvestri G, Safrit JT, Sumpter B, Kozyr N, McClure HM, et al. Depletion of CD8+ cells in sooty mangabey monkeys naturally infected with simian immunodeficiency virus reveals limited role for immune control of virus replication in a natural host species. J Immunol 2007;178(12):8002–12.

[173] Gaufin T, Ribeiro RM, Gautam R, Dufour J, Mandell D, Apetrei C, et al. Experimental depletion of $CD8^+$ cells in acutely SIVagm-infected African green monkeys results in increased viral replication. Retrovirology 2010;7:42.
[174] Staprans SI, Barry AP, Silvestri G, Safrit JT, Kozyr N, Sumpter B, et al. Enhanced SIV replication and accelerated progression to AIDS in macaques primed to mount a CD4 T cell response to the SIV envelope protein. Proc Natl Acad Sci USA 2004;101(35):13026–31.
[175] Douek DC, Brenchley JM, Betts MR, Ambrozak DR, Hill BJ, Okamoto Y, et al. HIV preferentially infects HIV-specific $CD4^+$ T cells. Nature 2002;417(6884):95–8.
[176] Lichterfeld M, Pantaleo G, Altfeld M. Loss of HIV-1-specific T cell proliferation in chronic HIV-1 infection: cause or consequence of viral replication? AIDS 2005;19(11): 1225–7.
[177] Hunt PW, Hatano H, Sinclair E, Lee TH, Busch MP, Martin JN, et al. HIV-specific $CD4^+$ T cells may contribute to viral persistence in HIV controllers. Clin Infect Dis 2011;52(5):681–7.
[178] Yi JS, Cox MA, Zajac AJ. T-cell exhaustion: characteristics, causes and conversion. Immunology 2010;129(4):474–81.
[179] Zhou Y, Callendret B, Xu D, Brasky KM, Feng Z, Hensley LL, et al. Dominance of the CD4(+) T helper cell response during acute resolving hepatitis A virus infection. J Exp Med 2012;209(8):1481–92.
[180] Schulze Zur Wiesch J, Ciuffreda D, Lewis-Ximenez L, Kasprowicz V, Nolan BE, Streeck H, et al. Broadly directed virus-specific $CD4^+$ T cell responses are primed during acute hepatitis C infection, but rapidly disappear from human blood with viral persistence. J Exp Med 2012;209(1):61–75.
[181] Nakanishi Y, Lu B, Gerard C, Iwasaki A. CD8(+) T lymphocyte mobilization to virus-infected tissue requires CD4(+) T-cell help. Nature 2009;462(7272):510–3.
[182] Ma CS, Deenick EK, Batten M, Tangye SG. The origins, function, and regulation of T follicular helper cells. J Exp Med 2012;209(7):1241–53.
[183] Rasheed MA, Latner DR, Aubert RD, Gourley T, Spolski R, Davis CW, et al. Interleukin-21 is a critical cytokine for the generation of virus-specific long-lived plasma cells. J Virol 2013;87(13):7737–46.
[184] Ortiz AM, Klatt NR, Li B, Yi Y, Tabb B, Hao XP, et al. Depletion of CD4(+) T cells abrogates post-peak decline of viremia in SIV-infected rhesus macaques. J Clin Invest 2011;121(11):4433–45.
[185] Mudd PA, Ericsen AJ, Price AA, Wilson NA, Reimann KA, Watkins DI. Reduction of $CD4^+$ T cells in vivo does not affect virus load in macaque elite controllers. J Virol 2011;85(14):7454–9.
[186] Younan PM, Polacino P, Kowalski JP, Peterson CW, Maurice NJ, Williams NP, et al. Positive selection of mC46-expressing $CD4^+$ T cells and maintenance of virus specific immunity in a primate AIDS model. Blood 2013;122(2):179–87.
[187] Seth N, Kaufmann D, Lahey T, Rosenberg ES, Wucherpfennig KW. Expansion and contraction of HIV-specific CD4 T cells with short bursts of viremia, but physical loss of the majority of these cells with sustained viral replication. J Immunol 2005;175(10):6948–58.
[188] Soghoian DZ, Jessen H, Flanders M, Sierra-Davidson K, Cutler S, Pertel T, et al. HIV-specific cytolytic CD4 T cell responses during acute HIV infection predict disease outcome. Sci Transl Med 2012;4(123):123ra25.
[189] Lubong Sabado R, Kavanagh DG, Kaufmann DE, Fru K, Babcock E, Rosenberg E, et al. In vitro priming recapitulates in vivo HIV-1 specific T cell responses, revealing rapid loss of virus reactive CD4 T cells in acute HIV-1 infection. PLoS One 2009;4(1):e4256.
[190] Ranasinghe S, Flanders M, Cutler S, Soghoian DZ, Ghebremichael M, Davis I, et al. HIV-specific CD4 T cell responses to different viral proteins have discordant associations with viral load and clinical outcome. J Virol 2012;86(1):277–83.

[191] Hunt PW, Landay AL, Sinclair E, Martinson JA, Hatano H, Emu B, et al. A low T regulatory cell response may contribute to both viral control and generalized immune activation in HIV controllers. PLoS One 2011;6(1):e15924.
[192] Younes SA, Yassine-Diab B, Dumont AR, Boulassel MR, Grossman Z, Routy JP, et al. HIV-1 viremia prevents the establishment of interleukin 2-producing HIV-specific memory CD4+ T cells endowed with proliferative capacity. J Exp Med 2003;198(12):1909–22.
[193] Emu B, Sinclair E, Favre D, Moretto WJ, Hsue P, Hoh R, et al. Phenotypic, functional, and kinetic parameters associated with apparent T-cell control of human immunodeficiency virus replication in individuals with and without antiretroviral treatment. J Virol 2005;79(22):14169–78.
[194] Potter SJ, Lacabaratz C, Lambotte O, Perez-Patrigeon S, Vingert B, Sinet M, et al. Preserved central memory and activated effector memory CD4+ T-cell subsets in human immunodeficiency virus controllers: an ANRS EP36 study. J Virol 2007;81(24):13904–15.
[195] van Grevenynghe J, Procopio FA, He Z, Chomont N, Riou C, Zhang Y, et al. Transcription factor FOXO3a controls the persistence of memory CD4(+) T cells during HIV infection. Nat Med 2008;14(3):266–74.
[196] Kaufmann DE, Kavanagh DG, Pereyra F, Zaunders JJ, Mackey EW, Miura T, et al. Upregulation of CTLA-4 by HIV-specific CD4+ T cells correlates with disease progression and defines a reversible immune dysfunction. Nat Immunol 2007;8(11):1246–54.
[197] Porichis F, Kwon DS, Zupkosky J, Tighe DP, McMullen A, Brockman MA, et al. Responsiveness of HIV-specific CD4 T cells to PD-1 blockade. Blood 2011;118(4):965–74.
[198] Harari A, Petitpierre S, Vallelian F, Pantaleo G. Skewed representation of functionally distinct populations of virus-specific CD4 T cells in HIV-1-infected subjects with progressive disease: changes after antiretroviral therapy. Blood 2004;103(3):966–72.
[199] Tilton JC, Luskin MR, Johnson AJ, Manion M, Hallahan CW, Metcalf JA, et al. Changes in paracrine interleukin-2 requirement, CCR7 expression, frequency, and cytokine secretion of human immunodeficiency virus-specific CD4+ T cells are a consequence of antigen load. J Virol 2007;81(6):2713–25.
[200] Guihot A, Tubiana R, Breton G, Marcelin AG, Samri A, Assoumou L, et al. Immune and virological benefits of 10 years of permanent viral control with antiretroviral therapy. AIDS 2010;24(4):614–7.
[201] Vingert B, Benati D, Lambotte O, de Truchis P, Slama L, Jeannin P, et al. HIV controllers maintain a population of highly efficient Th1 effector cells in contrast to patients treated in the long term. J Virol 2012;86(19):10661–74.
[202] Sattentau Q. Avoiding the void: cell-to-cell spread of human viruses. Nat Rev Microbiol 2008;6(11):815–26.
[203] Norris PJ, Moffett HF, Yang OO, Kaufmann DE, Clark MJ, Addo MM, et al. Beyond help: direct effector functions of human immunodeficiency virus type 1-specific CD4(+) T cells. J Virol 2004;78(16):8844–51.
[204] Zaunders JJ, Dyer WB, Wang B, Munier ML, Miranda-Saksena M, Newton R, et al. Identification of circulating antigen-specific CD4+ T lymphocytes with a CCR5+, cytotoxic phenotype in an HIV-1 long-term nonprogressor and in CMV infection. Blood 2004;103(6):2238–47.
[205] Jones RB, Yue FY, Gu XX, Hunter DV, Mujib S, Gyenes G, et al. HIV-1 escapes from IL-2-producing CD4+ T cell responses without high-frequency fixation of mutations. J Virol 2009;83(17):8722–32.
[206] Sacha JB, Giraldo-Vela JP, Buechler MB, Martins MA, Maness NJ, Chung C, et al. Gag- and Nef-specific CD4+ T cells recognize and inhibit SIV replication in infected macrophages early after infection. Proc Natl Acad Sci USA 2009;106(24):9791–6.
[207] Gauduin MC, Yu Y, Barabasz A, Carville A, Piatak M, Lifson JD, et al. Induction of a virus-specific effector-memory CD4+ T cell response by attenuated SIV infection. J Exp Med 2006;203(12):2661–72.

[208] Mucida D, Husain MM, Muroi S, van Wijk F, Shinnakasu R, Naoe Y, et al. Transcriptional reprogramming of mature CD4(+) helper T cells generates distinct MHC class II-restricted cytotoxic T lymphocytes. Nat Immunol 2013;14(3):281–9.
[209] Okoye A, Meier-Schellersheim M, Brenchley JM, Hagen SI, Walker JM, Rohankhedkar M, et al. Progressive $CD4^+$ central memory T cell decline results in $CD4^+$ effector memory insufficiency and overt disease in chronic SIV infection. J Exp Med 2007;204(9): 2171–85.
[210] Douek DC, Roederer M, Koup RA. Emerging concepts in the immunopathogenesis of AIDS. Annu Rev Med 2009;60:471–84.
[211] Vingert B, Perez-Patrigeon S, Jeannin P, Lambotte O, Boufassa F, Lemaitre F, et al. HIV controller $CD4^+$ T cells respond to minimal amounts of Gag antigen due to high TCR avidity. PLoS Pathog 2010;6(2):e1000780.
[212] Williams MA, Ravkov EV, Bevan MJ. Rapid culling of the $CD4^+$ T cell repertoire in the transition from effector to memory. Immunity 2008;28(4):533–45.
[213] Zehn D, Lee SY, Bevan MJ. Complete but curtailed T-cell response to very low-affinity antigen. Nature 2009;458(7235):211–4.
[214] Alexander-Miller MA. High-avidity $CD8^+$ T cells: optimal soldiers in the war against viruses and tumors. Immunol Res 2005;31(1):13–24.
[215] Neveu B, Debeaupuis E, Echasserieau K, le Moullac-Vaidye B, Gassin M, Jegou L, et al. Selection of high-avidity CD8 T cells correlates with control of hepatitis C virus infection. Hepatology 2008;48(3):713–22.
[216] Lozano Reina JM, Favre D, Kasakow Z, Mayau V, Nugeyre MT, Ka T, et al. Gag p27-specific B- and T-cell responses in Simian immunodeficiency virus SIVagm-infected African green monkeys. J Virol 2009;83(6):2770–7.
[217] Vieillard V, Fausther-Bovendo H, Samri A, Debre P, French Asymptomatiques a Long Terme A-COSG. Specific phenotypic and functional features of natural killer cells from HIV-infected long-term nonprogressors and HIV controllers. J Acquir Immune Defic Syndr 2010;53(5):564–73.
[218] Berger CT, Alter G. Natural killer cells in spontaneous control of HIV infection. Curr Opin HIV AIDS 2011;6(3):208–13.
[219] O'Connell KA, Han Y, Williams TM, Siliciano RF, Blankson JN. Role of natural killer cells in a cohort of elite suppressors: low frequency of the protective KIR3DS1 allele and limited inhibition of human immunodeficiency virus type 1 replication in vitro. J Virol 2009;83(10):5028–34.
[220] Paust S, Gill HS, Wang BZ, Flynn MP, Moseman EA, Senman B, et al. Critical role for the chemokine receptor CXCR6 in NK cell-mediated antigen-specific memory of haptens and viruses. Nat Immunol 2010;11(12):1127–35.
[221] Tiemessen CT, Shalekoff S, Meddows-Taylor S, Schramm DB, Papathanasopoulos M, Gray G, et al. Natural killer cells that respond to human immunodeficiency virus type 1 (HIV-1) peptides are associated with control of HIV-1 infection. J Infect Dis 2010;202(9):1444–53.
[222] Alter G, Martin MP, Teigen N, Carr WH, Suscovich TJ, Schneidewind A, et al. Differential natural killer cell-mediated inhibition of HIV-1 replication based on distinct KIR/HLA subtypes. J Exp Med 2007;204(12):3027–36.
[223] Hellmann I, Lim SY, Gelman RS, Letvin NL. Association of activating KIR copy number variation of NK cells with containment of SIV replication in rhesus monkeys. PLoS Pathog 2011;7(12):e1002436.
[224] Reeves RK, Gillis J, Wong FE, Yu Y, Connole M, Johnson RP. CD16- natural killer cells: enrichment in mucosal and secondary lymphoid tissues and altered function during chronic SIV infection. Blood 2010;115(22):4439–46.
[225] Pereira LE, Johnson RP, Ansari AA. Sooty mangabeys and rhesus macaques exhibit significant divergent natural killer cell responses during both acute and chronic phases of SIV infection. Cell Immunol 2008;254(1):10–9.

[226] Kosub DA, Lehrman G, Milush JM, Zhou D, Chacko E, Leone A, et al. Gamma/Delta T-cell functional responses differ after pathogenic human immunodeficiency virus and nonpathogenic simian immunodeficiency virus infections. J Virol 2008;82(3):1155–65.

[227] Riedel DJ, Sajadi MM, Armstrong CL, Cummings JS, Cairo C, Redfield RR, et al. Natural viral suppressors of HIV-1 have a unique capacity to maintain gammadelta T cells. AIDS 2009;23(15):1955–64.

[228] Boudova S, Li H, Sajadi MM, Redfield RR, Pauza CD. Impact of persistent HIV replication on CD4 negative Vγ2Vδ2 T cells. J Infect Dis 2012;205(9):1448–55.

[229] Sandberg JK, Fast NM, Palacios EH, Fennelly G, Dobroszycki J, Palumbo P, et al. Selective loss of innate CD4(+) Vα24 natural killer T cells in human immunodeficiency virus infection. J Virol 2002;76(15):7528–34.

[230] Vasan S, Poles MA, Horowitz A, Siladji EE, Markowitz M, Tsuji M. Function of NKT cells, potential anti-HIV effector cells, are improved by beginning HAART during acute HIV-1 infection. Int Immunol 2007;19(8):943–51.

[231] Rout N, Greene J, Yue S, O'Connor D, Johnson RP, Else JG, et al. Loss of effector and anti-inflammatory natural killer T lymphocyte function in pathogenic simian immunodeficiency virus infection. PLoS Pathog 2012;8(9):e1002928.

[232] Le Bourhis L, Mburu YK, Lantz O. MAIT cells, surveyors of a new class of antigen: development and functions. Curr Opin Immunol 2013;25(2):174–80.

[233] Cosgrove C, Ussher JE, Rauch A, Gartner K, Kurioka A, Huhn MH, et al. Early and nonreversible decrease of CD161++/MAIT cells in HIV infection. Blood 2013;121(6):951–61.

[234] Leeansyah E, Ganesh A, Quigley MF, Sonnerborg A, Andersson J, Hunt PW, et al. Activation, exhaustion, and persistent decline of the antimicrobial MR1-restricted MAIT-cell population in chronic HIV-1 infection. Blood 2013;121(7):1124–35.

[235] Bailey JR, Lassen KG, Yang HC, Quinn TC, Ray SC, Blankson JN, et al. Neutralizing antibodies do not mediate suppression of human immunodeficiency virus type 1 in elite suppressors or selection of plasma virus variants in patients on highly active antiretroviral therapy. J Virol 2006;80(10):4758–70.

[236] Stamatatos L, Morris L, Burton DR, Mascola JR. Neutralizing antibodies generated during natural HIV-1 infection: good news for an HIV-1 vaccine? Nat Med 2009;15(8):866–70.

[237] Haynes BF, Kelsoe G, Harrison SC, Kepler TB. B-cell-lineage immunogen design in vaccine development with HIV-1 as a case study. Nat Biotechnol 2012;30(5):423–33.

[238] Mahalanabis M, Jayaraman P, Miura T, Pereyra F, Chester EM, Richardson B, et al. Continuous viral escape and selection by autologous neutralizing antibodies in drug-naive human immunodeficiency virus controllers. J Virol 2009;83(2):662–72.

[239] Deeks SG, Schweighardt B, Wrin T, Galovich J, Hoh R, Sinclair E, et al. Neutralizing antibody responses against autologous and heterologous viruses in acute versus chronic human immunodeficiency virus (HIV) infection: evidence for a constraint on the ability of HIV to completely evade neutralizing antibody responses. J Virol 2006;80(12):6155–64.

[240] Lambotte O, Ferrari G, Moog C, Yates NL, Liao HX, Parks RJ, et al. Heterogeneous neutralizing antibody and antibody-dependent cell cytotoxicity responses in HIV-1 elite controllers. AIDS 2009;23(8):897–906.

[241] Ackerman ME, Dugast AS, McAndrew EG, Tsoukas S, Licht AF, Irvine DJ, et al. Enhanced phagocytic activity of HIV-specific antibodies correlates with natural production of immunoglobulins with skewed affinity for FcγR2a and FcγR2b. J Virol 2013;87(10):5468–76.

[242] Ackerman ME, Crispin M, Yu X, Baruah K, Boesch AW, Harvey DJ, et al. Natural variation in Fc glycosylation of HIV-specific antibodies impacts antiviral activity. J Clin Invest 2013;123(5):2183–92.

[243] Martinez V, Costagliola D, Bonduelle O, N'go N, Schnuriger A, Theodorou I, et al. Combination of HIV-1-specific CD4 Th1 cell responses and IgG2 antibodies is the best predictor for persistence of long-term nonprogression. J Infect Dis 2005;191(12):2053–63.
[244] de Silva TI, Aasa-Chapman M, Cotten M, Hue S, Robinson J, Bibollet-Ruche F, et al. Potent autologous and heterologous neutralizing antibody responses occur in HIV-2 infection across a broad range of infection outcomes. J Virol 2012;86(2):930–46.
[245] Kong R, Li H, Georgiev I, Changela A, Bibollet-Ruche F, Decker JM, et al. Epitope mapping of broadly neutralizing HIV-2 human monoclonal antibodies. J Virol 2012; 86(22):12115–28.
[246] Lindqvist M, van Lunzen J, Soghoian DZ, Kuhl BD, Ranasinghe S, Kranias G, et al. Expansion of HIV-specific T follicular helper cells in chronic HIV infection. J Clin Invest 2012;122(9):3271–80.
[247] Cubas RA, Mudd JC, Savoye AL, Perreau M, van Grevenynghe J, Metcalf T, et al. Inadequate T follicular cell help impairs B cell immunity during HIV infection. Nat Med 2013;19(4):494–9.
[248] Iannello A, Boulassel MR, Samarani S, Debbeche O, Tremblay C, Toma E, et al. Dynamics and consequences of IL-21 production in HIV-infected individuals: a longitudinal and cross-sectional study. J Immunol 2010;184(1):114–26.
[249] Chevalier MF, Julg B, Pyo A, Flanders M, Ranasinghe S, Soghoian DZ, et al. HIV-1-specific interleukin-21+ $CD4^+$ T cell responses contribute to durable viral control through the modulation of HIV-specific $CD8^+$ T cell function. J Virol 2011;85(2):733–41.
[250] Scheid JF, Mouquet H, Feldhahn N, Seaman MS, Velinzon K, Pietzsch J, et al. Broad diversity of neutralizing antibodies isolated from memory B cells in HIV-infected individuals. Nature 2009;458(7238):636–40.
[251] Xu Y, Weatherall C, Bailey M, Alcantara S, De Rose R, Estaquier J, et al. Simian immunodeficiency virus infects follicular helper CD4 T cells in lymphoid tissues during pathogenic infection of pigtail macaques. J Virol 2013;87(7):3760–73.
[252] Petrovas C, Yamamoto T, Gerner MY, Boswell KL, Wloka K, Smith EC, et al. CD4 T follicular helper cell dynamics during SIV infection. J Clin Invest 2012;122(9):3281–94.
[253] Hong JJ, Amancha PK, Rogers K, Ansari AA, Villinger F. Spatial alterations between CD4(+) T follicular helper, B, and CD8(+) T cells during simian immunodeficiency virus infection: T/B cell homeostasis, activation, and potential mechanism for viral escape. J Immunol 2012;188(7):3247–56.
[254] Chakrabarti L, Baptiste V, Khatissian E, Cumont MC, Aubertin AM, Montagnier L, et al. Limited viral spread and rapid immune response in lymph nodes of macaques inoculated with attenuated simian immunodeficiency virus. Virology 1995;213(2):535–48.
[255] Chakrabarti LA. The paradox of simian immunodeficiency virus infection in sooty mangabeys: active viral replication without disease progression. Front Biosci 2004;9:521–39.
[256] Li B, Stefano-Cole K, Kuhrt DM, Gordon SN, Else JG, Mulenga J, et al. Nonpathogenic simian immunodeficiency virus infection of sooty mangabeys is not associated with high levels of autologous neutralizing antibodies. J Virol 2010;84(12):6248–53.
[257] Amos JD, Wilks AB, Fouda GG, Smith SD, Colvin L, Mahlokozera T, et al. Lack of B cell dysfunction is associated with functional, gp120-dominant antibody responses in breast milk of simian immunodeficiency virus-infected African green monkeys. J Virol 2013;87(20):11121–34.
[258] Pantaleo G, Menzo S, Vaccarezza M, Graziosi C, Cohen OJ, Demarest JF, et al. Studies in subjects with long-term nonprogressive human immunodeficiency virus infection. N Engl J Med 1995;332(4):209–16.
[259] Martinot AJ, Meythaler M, Pozzi LA, Dalecki Boisvert K, Knight H, Walsh D, et al. Acute SIV infection in sooty mangabey monkeys is characterized by rapid virus clearance from lymph nodes and absence of productive infection in germinal centers. PLoS One 2013;8(3):e57785.

[260] Moir S, Fauci AS. Insights into B cells and HIV-specific B-cell responses in HIV-infected individuals. Immunol Rev 2013;254(1):207–24.
[261] Fischer W, Apetrei C, Santiago ML, Li Y, Gautam R, Pandrea I, et al. Distinct evolutionary pressures underlie diversity in simian immunodeficiency virus and human immunodeficiency virus lineages. J Virol 2012;86(24):13217–31.
[262] Gaufin T, Pattison M, Gautam R, Stoulig C, Dufour J, MacFarland J, et al. Effect of B-cell depletion on viral replication and clinical outcome of simian immunodeficiency virus infection in a natural host. J Virol 2009;83(20):10347–57.
[263] Miller CJ, Genesca M, Abel K, Montefiori D, Forthal D, Bost K, et al. Antiviral antibodies are necessary for control of simian immunodeficiency virus replication. J Virol 2007;81(10):5024–35.
[264] Huang KH, Bonsall D, Katzourakis A, Thomson EC, Fidler SJ, Main J, et al. B-cell depletion reveals a role for antibodies in the control of chronic HIV-1 infection. Nat Commun 2010;1:102.
[265] Deeks SG, Kitchen CM, Liu L, Guo H, Gascon R, Narvaez AB, et al. Immune activation set point during early HIV infection predicts subsequent $CD4^+$ T-cell changes independent of viral load. Blood 2004;104(4):942–7.
[266] Deeks SG. Immune dysfunction, inflammation, and accelerated aging in patients on antiretroviral therapy. Top HIV Med 2009;17(4):118–23.
[267] Paiardini M, Muller-Trutwin M. HIV-associated chronic immune activation. Immunol Rev 2013;254(1):78–101.
[268] O'Connell KA, Rabi SA, Siliciano RF, Blankson JN. $CD4^+$ T cells from elite suppressors are more susceptible to HIV-1 but produce fewer virions than cells from chronic progressors. Proc Natl Acad Sci USA 2011;108(37):E689–98.
[269] Sauce D, Larsen M, Fastenackels S, Pauchard M, Ait-Mohand H, Schneider L, et al. HIV disease progression despite suppression of viral replication is associated with exhaustion of lymphopoiesis. Blood 2011;117(19):5142–51.
[270] Yang Y, Al-Mozaini M, Buzon MJ, Beamon J, Ferrando-Martinez S, Ruiz-Mateos E, et al. CD4 T-cell regeneration in HIV-1 elite controllers. AIDS 2012;26(6):701–6.
[271] Okulicz JF, Grandits GA, Weintrob AC, Landrum ML, Ganesan A, Crum-Cianflone NF, et al. CD4 T cell count reconstitution in HIV controllers after highly active antiretroviral therapy. Clin Infect Dis 2010;50(8):1187–91.
[272] Beer BE, Brown CR, Whitted S, Goldstein S, Goeken R, Plishka R, et al. Immunodeficiency in the absence of high viral load in pig-tailed macaques infected with Simian immunodeficiency virus SIVsun or SIVlhoest. J Virol 2005;79(22):14044–56.
[273] Chakrabarti LA, Lewin SR, Zhang L, Gettie A, Luckay A, Martin LN, et al. Normal T-cell turnover in sooty mangabeys harboring active simian immunodeficiency virus infection. J Virol 2000;74(3):1209–23.
[274] Silvestri G, Sodora DL, Koup RA, Paiardini M, O'Neil SP, McClure HM, et al. Nonpathogenic SIV infection of sooty mangabeys is characterized by limited bystander immunopathology despite chronic high-level viremia. Immunity 2003;18(3):441–52.
[275] Silvestri G, Fedanov A, Germon S, Kozyr N, Kaiser WJ, Garber DA, et al. Divergent host responses during primary simian immunodeficiency virus SIVsm infection of natural sooty mangabey and nonnatural rhesus macaque hosts. J Virol 2005;79(7):4043–54.
[276] Kaur A, Di Mascio M, Barabasz A, Rosenzweig M, McClure HM, Perelson AS, et al. Dynamics of T- and B-lymphocyte turnover in a natural host of simian immunodeficiency virus. J Virol 2008;82(3):1084–93.
[277] Laforge M, Campillo-Gimenez L, Monceaux V, Cumont MC, Hurtrel B, Corbeil J, et al. HIV/SIV infection primes monocytes and dendritic cells for apoptosis. PLoS Pathog 2011;7(6):e1002087.
[278] Goldstein S, Brown CR, Ourmanov I, Pandrea I, Buckler-White A, Erb C, et al. Comparison of simian immunodeficiency virus SIVagmVer replication and $CD4^+$ T-cell dynamics in vervet and sabaeus African green monkeys. J Virol 2006;80(10):4868–77.

[279] Kornfeld C, Ploquin MJ, Pandrea I, Faye A, Onanga R, Apetrei C, et al. Antiinflammatory profiles during primary SIV infection in African green monkeys are associated with protection against AIDS. J Clin Invest 2005;115(4):1082–91.
[280] Estes JD, Gordon SN, Zeng M, Chahroudi AM, Dunham RM, Staprans SI, et al. Early resolution of acute immune activation and induction of PD-1 in SIV-infected sooty mangabeys distinguishes nonpathogenic from pathogenic infection in rhesus macaques. J Immunol 2008;180(10):6798–807.
[281] Villinger F, Folks TM, Lauro S, Powell JD, Sundstrom JB, Mayne A, et al. Immunological and virological studies of natural SIV infection of disease-resistant nonhuman primates. Immunol Lett 1996;51(1–2):59–68.
[282] Pandrea I, Gaufin T, Brenchley JM, Gautam R, Monjure C, Gautam A, et al. Cutting edge: experimentally induced immune activation in natural hosts of simian immunodeficiency virus induces significant increases in viral replication and $CD4^+$ T cell depletion. J Immunol 2008;181(10):6687–91.
[283] Brandt L, Benfield T, Mens H, Clausen LN, Katzenstein TL, Fomsgaard A, et al. Low level of regulatory T cells and maintenance of balance between regulatory T cells and TH17 cells in HIV-1-infected elite controllers. J Acquir Immune Defic Syndr 2011;57(2):101–8.
[284] Chase AJ, Yang HC, Zhang H, Blankson JN, Siliciano RF. Preservation of FoxP3+ regulatory T cells in the peripheral blood of human immunodeficiency virus type 1-infected elite suppressors correlates with low $CD4^+$ T-cell activation. J Virol 2008;82(17):8307–15.
[285] Elahi S, Dinges WL, Lejarcegui N, Laing KJ, Collier AC, Koelle DM, et al. Protective HIV-specific $CD8^+$ T cells evade Treg cell suppression. Nat Med 2011;17(8):989–95.
[286] Ballesteros-Tato A, Leon B, Lund FE, Randall TD. $CD4^+$ T helper cells use CD154-CD40 interactions to counteract T reg cell-mediated suppression of $CD8^+$ T cell responses to influenza. J Exp Med 2013;210(8):1591–601.
[287] Favre D, Lederer S, Kanwar B, Ma ZM, Proll S, Kasakow Z, et al. Critical loss of the balance between Th17 and T regulatory cell populations in pathogenic SIV infection. PLoS Pathog 2009;5(2):e1000295.
[288] Pereira LE, Villinger F, Onlamoon N, Bryan P, Cardona A, Pattanapanysat K, et al. Simian immunodeficiency virus (SIV) infection influences the level and function of regulatory T cells in SIV-infected rhesus macaques but not SIV-infected sooty mangabeys. J Virol 2007;81(9):4445–56.
[289] Brenchley JM, Price DA, Schacker TW, Asher TE, Silvestri G, Rao S, et al. Microbial translocation is a cause of systemic immune activation in chronic HIV infection. Nat Med 2006;12(12):1365–71.
[290] Estes JD, Harris LD, Klatt NR, Tabb B, Pittaluga S, Paiardini M, et al. Damaged intestinal epithelial integrity linked to microbial translocation in pathogenic simian immunodeficiency virus infections. PLoS Pathog 2010;6(8):e1001052.
[291] Brenchley JM, Paiardini M, Knox KS, Asher AI, Cervasi B, Asher TE, et al. Differential Th17 CD4 T-cell depletion in pathogenic and nonpathogenic lentiviral infections. Blood 2008;112(7):2826–35.
[292] Raffatellu M, Santos RL, Verhoeven DE, George MD, Wilson RP, Winter SE, et al. Simian immunodeficiency virus-induced mucosal interleukin-17 deficiency promotes Salmonella dissemination from the gut. Nat Med 2008;14(4):421–8.
[293] Campillo-Gimenez L, Cumont MC, Fay M, Kared H, Monceaux V, Diop O, et al. AIDS progression is associated with the emergence of IL-17-producing cells early after simian immunodeficiency virus infection. J Immunol 2010;184(2):984–92.
[294] Mattapallil JJ, Douek DC, Hill B, Nishimura Y, Martin M, Roederer M. Massive infection and loss of memory $CD4^+$ T cells in multiple tissues during acute SIV infection. Nature 2005;434(7037):1093–7.
[295] Klatt NR, Estes JD, Sun X, Ortiz AM, Barber JS, Harris LD, et al. Loss of mucosal $CD103^+$ DCs and IL-17+ and IL-22+ lymphocytes is associated with mucosal damage in SIV infection. Mucosal Immunol 2012;5(6):646–57.

[296] Ciccone EJ, Greenwald JH, Lee PI, Biancotto A, Read SW, Yao MA, et al. CD4+ T cells, including Th17 and cycling subsets, are intact in the gut mucosa of HIV-1-infected long-term nonprogressors. J Virol 2011;85(12):5880–8.
[297] Sankaran S, Guadalupe M, Reay E, George MD, Flamm J, Prindiville T, et al. Gut mucosal T cell responses and gene expression correlate with protection against disease in long-term HIV-1-infected nonprogressors. Proc Natl Acad Sci USA 2005;102(28):9860–5.
[298] Hofer U, Schlaepfer E, Baenziger S, Nischang M, Regenass S, Schwendener R, et al. Inadequate clearance of translocated bacterial products in HIV-infected humanized mice. PLoS Pathog 2010;6(4):e1000867.
[299] Micci L, Cervasi B, Ende ZS, Iriele RI, Reyes-Aviles E, Vinton C, et al. Paucity of IL-21-producing CD4(+) T cells is associated with Th17 cell depletion in SIV infection of rhesus macaques. Blood 2012;120(19):3925–35.
[300] Pallikkuth S, Micci L, Ende ZS, Iriele RI, Cervasi B, Lawson B, et al. Maintenance of intestinal Th17 cells and reduced microbial translocation in siv-infected rhesus macaques treated with interleukin (IL)-21. PLoS Pathog 2013;9(7):e1003471.
[301] Adalid-Peralta L, Godot V, Colin C, Krzysiek R, Tran T, Poignard P, et al. Stimulation of the primary anti-HIV antibody response by IFN-alpha in patients with acute HIV-1 infection. J Leukoc Biol 2008;83(4):1060–7.
[302] Tavel JA, Huang CY, Shen J, Metcalf JA, Dewar R, Shah A, et al. Interferon-alpha produces significant decreases in HIV load. J Interferon Cytokine Res 2010;30(7):461–4.
[303] Vanderford TH, Slichter C, Rogers KA, Lawson BO, Obaede R, Else J, et al. Treatment of SIV-infected sooty mangabeys with a type-I IFN agonist results in decreased virus replication without inducing hyperimmune activation. Blood 2012;119(24):5750–7.
[304] Soumelis V, Scott I, Gheyas F, Bouhour D, Cozon G, Cotte L, et al. Depletion of circulating natural type 1 interferon-producing cells in HIV-infected AIDS patients. Blood 2001;98(4):906–12.
[305] Kamga I, Kahi S, Develioglu L, Lichtner M, Maranon C, Deveau C, et al. Type I interferon production is profoundly and transiently impaired in primary HIV-1 infection. J Infect Dis 2005;192(2):303–10.
[306] Malleret B, Maneglier B, Karlsson I, Lebon P, Nascimbeni M, Perie L, et al. Primary infection with simian immunodeficiency virus: plasmacytoid dendritic cell homing to lymph nodes, type I interferon, and immune suppression. Blood 2008;112(12):4598–608.
[307] Wilson EB, Yamada DH, Elsaesser H, Herskovitz J, Deng J, Cheng G, et al. Blockade of chronic type I interferon signaling to control persistent LCMV infection. Science 2013;340(6129):202–7.
[308] Teijaro JR, Ng C, Lee AM, Sullivan BM, Sheehan KC, Welch M, et al. Persistent LCMV infection is controlled by blockade of type I interferon signaling. Science 2013;340(6129):207–11.
[309] Hardy GA, Sieg S, Rodriguez B, Anthony D, Asaad R, Jiang W, et al. Interferon-alpha is the primary plasma type-I IFN in HIV-1 infection and correlates with immune activation and disease markers. PLoS One 2013;8(2):e56527.
[310] Stoddart CA, Keir ME, McCune JM. IFN-alpha-induced upregulation of CCR5 leads to expanded HIV tropism in vivo. PLoS Pathog 2010;6(2):e1000766.
[311] Rotger M, Dang KK, Fellay J, Heinzen EL, Feng S, Descombes P, et al. Genome-wide mRNA expression correlates of viral control in CD4+ T-cells from HIV-1-infected individuals. PLoS Pathog 2010;6(2):e1000781.
[312] Wu JQ, Dwyer DE, Dyer WB, Yang YH, Wang B, Saksena NK. Genome-wide analysis of primary CD4+ and CD8+ T cell transcriptomes shows evidence for a network of enriched pathways associated with HIV disease. Retrovirology 2011;8:18.
[313] Hyrcza MD, Kovacs C, Loutfy M, Halpenny R, Heisler L, Yang S, et al. Distinct transcriptional profiles in ex vivo CD4+ and CD8+ T cells are established early in human immunodeficiency virus type 1 infection and are characterized by a chronic interferon response as well as extensive transcriptional changes in CD8+ T cells. J Virol 2007;81(7):3477–86.

[314] Barblu L, Machmach K, Gras C, Delfraissy JF, Boufassa F, Leal M, et al. Plasmacytoid dendritic cells (pDCs) from HIV controllers produce interferon-alpha and differentiate into functional killer pDCs under HIV activation. J Infect Dis 2012;206(5):790–801.
[315] Machmach K, Leal M, Gras C, Viciana P, Genebat M, Franco E, et al. Plasmacytoid dendritic cells reduce HIV production in elite controllers. J Virol 2012;86(8): 4245–52.
[316] Cordeil S, Nguyen XN, Berger G, Durand S, Ainouze M, Cimarelli A. Evidence for a different susceptibility of primate lentiviruses to type I interferons. J Virol 2013;87(5):2587–96.
[317] Meier A, Chang JJ, Chan ES, Pollard RB, Sidhu HK, Kulkarni S, et al. Sex differences in the Toll-like receptor-mediated response of plasmacytoid dendritic cells to HIV-1. Nat Med 2009;15(8):955–9.
[318] Jacquelin B, Mayau V, Targat B, Liovat AS, Kunkel D, Petitjean G, et al. Nonpathogenic SIV infection of African green monkeys induces a strong but rapidly controlled type I IFN response. J Clin Invest 2009;119(12):3544–55.
[319] Bosinger SE, Li Q, Gordon SN, Klatt NR, Duan L, Xu L, et al. Global genomic analysis reveals rapid control of a robust innate response in SIV-infected sooty mangabeys. J Clin Invest 2009;119(12):3556–72.
[320] Spivak AM, Salgado M, Rabi SA, O'Connell KA, Blankson JN. Circulating monocytes are not a major reservoir of HIV-1 in elite suppressors. J Virol 2011;85(19):10399–403.
[321] Chomont N, El-Far M, Ancuta P, Trautmann L, Procopio FA, Yassine-Diab B, et al. HIV reservoir size and persistence are driven by T cell survival and homeostatic proliferation. Nat Med 2009;15(8):893–900.
[322] Descours B, Avettand-Fenoel V, Blanc C, Samri A, Melard A, Supervie V, et al. Immune responses driven by protective human leukocyte antigen alleles from long-term nonprogressors are associated with low HIV reservoir in central memory CD4 T cells. Clin Infect Dis 2012;54(10):1495–503.
[323] Ostrowski MA, Chun TW, Justement SJ, Motola I, Spinelli MA, Adelsberger J, et al. Both memory and $CD45RA^+/CD62L^+$ naive CD4(+) T cells are infected in human immunodeficiency virus type 1-infected individuals. J Virol 1999;73(8):6430–5.
[324] Deeks SG. HIV infection, inflammation, immunosenescence, and aging. Annu Rev Med 2011;62:141–55.
[325] Stacey AR, Norris PJ, Qin L, Haygreen EA, Taylor E, Heitman J, et al. Induction of a striking systemic cytokine cascade prior to peak viremia in acute human immunodeficiency virus type 1 infection, in contrast to more modest and delayed responses in acute hepatitis B and C virus infections. J Virol 2009;83(8):3719–33.
[326] Gattinoni L, Lugli E, Ji Y, Pos Z, Paulos CM, Quigley MF, et al. A human memory T cell subset with stem cell-like properties. Nat Med 2011;17(10):1290–7.
[327] Lugli E, Dominguez MH, Gattinoni L, Chattopadhyay PK, Bolton DL, Song K, et al. Superior T memory stem cell persistence supports long-lived T cell memory. J Clin Invest 2013;123(2):594–9.
[328] Klatt NR, Villinger F, Bostik P, Gordon SN, Pereira L, Engram JC, et al. Availability of activated $CD4^+$ T cells dictates the level of viremia in naturally SIV-infected sooty mangabeys. J Clin Invest 2008;118(6):2039–49.
[329] Paiardini M, Cervasi B, Reyes-Aviles E, Micci L, Ortiz AM, Chahroudi A, et al. Low levels of SIV infection in sooty mangabey central memory CD(4)(+) T cells are associated with limited CCR5 expression. Nat Med 2011;17(7):830–6.
[330] Saez-Cirion A, Hamimi C, Bergamaschi A, David A, Versmisse P, Melard A, et al. Restriction of HIV-1 replication in macrophages and $CD4^+$ T cells from HIV controllers. Blood 2011;118(4):955–64.
[331] Chen H, Li C, Huang J, Cung T, Seiss K, Beamon J, et al. $CD4^+$ T cells from elite controllers resist HIV-1 infection by selective upregulation of p21. J Clin Invest 2011;121(4):1549–60.

[332] Buzon MJ, Seiss K, Weiss R, Brass AL, Rosenberg ES, Pereyra F, et al. Inhibition of HIV-1 integration in ex vivo-infected CD4 T cells from elite controllers. J Virol 2011;85(18):9646–50.
[333] Rabi SA, O'Connell KA, Nikolaeva D, Bailey JR, Jilek BL, Shen L, et al. Unstimulated primary CD4+ T cells from HIV-1-positive elite suppressors are fully susceptible to HIV-1 entry and productive infection. J Virol 2011;85(2):979–86.
[334] Graf EH, Mexas AM, Yu JJ, Shaheen F, Liszewski MK, Di Mascio M, et al. Elite suppressors harbor low levels of integrated HIV DNA and high levels of 2-LTR circular HIV DNA compared to HIV+ patients on and off HAART. PLoS Pathog 2011;7(2):e1001300.
[335] Stewart GJ, Ashton LJ, Biti RA, Ffrench RA, Bennetts BH, Newcombe NR, et al. Increased frequency of CCR-5 delta 32 heterozygotes among long-term non-progressors with HIV-1 infection. The Australian Long-Term Non-Progressor Study Group. AIDS 1997;11(15):1833–8.
[336] Meyer L, Magierowska M, Hubert JB, Rouzioux C, Deveau C, Sanson F, et al. Early protective effect of CCR-5 delta 32 heterozygosity on HIV-1 disease progression: relationship with viral load. The SEROCO Study Group. AIDS 1997;11(11):F73–8.
[337] Casazza JP, Brenchley JM, Hill BJ, Ayana R, Ambrozak D, Roederer M, et al. Autocrine production of beta-chemokines protects CMV-Specific CD4 T cells from HIV infection. PLoS Pathog 2009;5(10):e1000646.
[338] Annunziato F, Galli G, Nappi F, Cosmi L, Manetti R, Maggi E, et al. Limited expression of R5-tropic HIV-1 in CCR5-positive type 1-polarized T cells explained by their ability to produce RANTES, MIP-1alpha, and MIP-1beta. Blood 2000;95(4):1167–74.
[339] Murayama Y, Mukai R, Inoue-Murayama M, Yoshikawa Y. An African green monkey lacking peripheral CD4 lymphocytes that retains helper T cell activity and coexists with SIVagm. Clin Exp Immunol 1999;117(3):504–12.
[340] Beaumier CM, Harris LD, Goldstein S, Klatt NR, Whitted S, McGinty J, et al. CD4 downregulation by memory CD4+ T cells in vivo renders African green monkeys resistant to progressive SIVagm infection. Nat Med 2009;15(8):879–85.
[341] Vinton C, Klatt NR, Harris LD, Briant JA, Sanders-Beer BE, Herbert R, et al. CD4-like immunological function by CD4− T cells in multiple natural hosts of simian immunodeficiency virus. J Virol 2011;85(17):8702–8.
[342] Chakrabarti LA, Lewin SR, Zhang L, Gettie A, Luckay A, Martin LN, et al. Age-dependent changes in T cell homeostasis and SIV load in sooty mangabeys. J Med Primatol 2000;29(3–4):158–65.
[343] Muthukumar A, Zhou D, Paiardini M, Barry AP, Cole KS, McClure HM, et al. Timely triggering of homeostatic mechanisms involved in the regulation of T-cell levels in SIVsm-infected sooty mangabeys. Blood 2005;106(12):3839–45.
[344] Taaffe J, Chahroudi A, Engram J, Sumpter B, Meeker T, Ratcliffe S, et al. A five-year longitudinal analysis of sooty mangabeys naturally infected with simian immunodeficiency virus reveals a slow but progressive decline in CD4+ T-cell count whose magnitude is not predicted by viral load or immune activation. J Virol 2010;84(11):5476–84.
[345] Milush JM, Mir KD, Sundaravaradan V, Gordon SN, Engram J, Cano CA, et al. Lack of clinical AIDS in SIV-infected sooty mangabeys with significant CD4+ T cell loss is associated with double-negative T cells. J Clin Invest 2011;121(3):1102–10.
[346] Veazey R, Ling B, Pandrea I, McClure H, Lackner A, Marx P. Decreased CCR5 expression on CD4+ T cells of SIV-infected sooty mangabeys. AIDS Res Hum Retroviruses 2003;19(3):227–33.
[347] Pandrea I, Apetrei C, Gordon S, Barbercheck J, Dufour J, Bohm R, et al. Paucity of CD4+CCR5+ T cells is a typical feature of natural SIV hosts. Blood 2007;109(3):1069–76.
[348] Pandrea I, Onanga R, Souquiere S, Mouinga-Ondeme A, Bourry O, Makuwa M, et al. Paucity of CD4+ CCR5+ T cells may prevent transmission of simian immunodeficiency virus in natural nonhuman primate hosts by breast-feeding. J Virol 2008;82(11):5501–9.

[349] Pandrea I, Parrish NF, Raehtz K, Gaufin T, Barbian HJ, Ma D, et al. Mucosal simian immunodeficiency virus transmission in African green monkeys: susceptibility to infection is proportional to target cell availability at mucosal sites. J Virol 2012;86(8):4158–68.

[350] Chahroudi A, Meeker T, Lawson B, Ratcliffe S, Else J, Silvestri G. Mother-to-infant transmission of simian immunodeficiency virus is rare in sooty mangabeys and is associated with low viremia. J Virol 2011;85(12):5757–63.

[351] Riddick NE, Hermann EA, Loftin LM, Elliott ST, Wey WC, Cervasi B, et al. A novel CCR5 mutation common in sooty mangabeys reveals SIVsmm infection of CCR5-null natural hosts and efficient alternative coreceptor use in vivo. PLoS Pathog 2010;6(8):e1001064.

[352] Henrich TJ, Kuritzkes DR. HIV-1 entry inhibitors: recent development and clinical use. Curr Opin Virol 2013;3(1):51–7.

[353] Holt N, Wang J, Kim K, Friedman G, Wang X, Taupin V, et al. Human hematopoietic stem/progenitor cells modified by zinc-finger nucleases targeted to CCR5 control HIV-1 in vivo. Nat Biotechnol 2010;28(8):839–47.

[354] Blanco-Melo D, Venkatesh S, Bieniasz PD. Intrinsic cellular defenses against human immunodeficiency viruses. Immunity 2012;37(3):399–411.

[355] van Manen D, Rits MA, Beugeling C, van Dort K, Schuitemaker H, Kootstra NA. The effect of Trim5 polymorphisms on the clinical course of HIV-1 infection. PLoS Pathog 2008;4(2):e18.

[356] An P, Duggal P, Wang LH, O'Brien SJ, Donfield S, Goedert JJ, et al. Polymorphisms of CUL5 are associated with $CD4^+$ T cell loss in HIV-1 infected individuals. PLoS Genet 2007;3(1):e19.

[357] Ross SR. Are viruses inhibited by APOBEC3 molecules from their host species? PLoS Pathog 2009;5(4):e1000347.

[358] An P, Johnson R, Phair J, Kirk GD, Yu XF, Donfield S, et al. APOBEC3B deletion and risk of HIV-1 acquisition. J Infect Dis 2009;200(7):1054–8.

[359] Simon V, Zennou V, Murray D, Huang Y, Ho DD, Bieniasz PD. Natural variation in Vif: differential impact on APOBEC3G/3F and a potential role in HIV-1 diversification. PLoS Pathog 2005;1(1):e6.

[360] Eyzaguirre LM, Charurat M, Redfield RR, Blattner WA, Carr JK, Sajadi MM. Elevated hypermutation levels in HIV-1 natural viral suppressors. Virology 2013;443(2):306–12.

[361] Vetter ML, Johnson ME, Antons AK, Unutmaz D, D'Aquila RT. Differences in APOBEC3G expression in $CD4^+$ T helper lymphocyte subtypes modulate HIV-1 infectivity. PLoS Pathog 2009;5(2):e1000292.

[362] Compton AA, Hirsch VM, Emerman M. The host restriction factor APOBEC3G and retroviral Vif protein coevolve due to ongoing genetic conflict. Cell Host Microbe 2012;11(1):91–8.

[363] Stremlau M, Owens CM, Perron MJ, Kiessling M, Autissier P, Sodroski J. The cytoplasmic body component TRIM5alpha restricts HIV-1 infection in old world monkeys. Nature 2004;427(6977):848–53.

[364] Luban J. TRIM5 and the regulation of HIV-1 infectivity. Mol Biol Int 2012;2012:426840.

[365] Lim SY, Rogers T, Chan T, Whitney JB, Kim J, Sodroski J, et al. TRIM5alpha modulates immunodeficiency virus control in rhesus monkeys. PLoS Pathog 2010;6(1):e1000738.

[366] Takeuchi JS, Perche B, Migraine J, Mercier-Delarue S, Ponscarme D, Simon F, et al. High level of susceptibility to human TRIM5alpha conferred by HIV-2 capsid sequences. Retrovirology 2013;10:50.

[367] Ylinen LM, Keckesova Z, Wilson SJ, Ranasinghe S, Towers GJ. Differential restriction of human immunodeficiency virus type 2 and simian immunodeficiency virus SIVmac by TRIM5alpha alleles. J Virol 2005;79(18):11580–7.

[368] Onyango CO, Leligdowicz A, Yokoyama M, Sato H, Song H, Nakayama EE, et al. HIV-2 capsids distinguish high and low virus load patients in a West African community cohort. Vaccine 2010;28(Suppl. 2):B60–7.

[369] Battivelli E, Migraine J, Lecossier D, Yeni P, Clavel F, Hance AJ. Gag cytotoxic T lymphocyte escape mutations can increase sensitivity of HIV-1 to human TRIM5alpha, linking intrinsic and acquired immunity. J Virol 2011;85(22):11846–54.

[370] Granier C, Battivelli E, Lecuroux C, Venet A, Lambotte O, Schmitt-Boulanger M, et al. Pressure from TRIM5alpha contributes to control of HIV-1 replication by individuals expressing protective HLA-B alleles. J Virol 2013;87(18):10368–80.

[371] Rihn SJ, Wilson SJ, Loman NJ, Alim M, Bakker SE, Bhella D, et al. Extreme genetic fragility of the HIV-1 capsid. PLoS Pathog 2013;9(6):e1003461.

CHAPTER

16

Humans Resembling Natural Hosts of SIV

Amalio Telenti

Institute of Microbiology, University Hospital, University of Lausanne, Lausanne, Switzerland

INTRODUCTION

HIV-1 and HIV-2 infections are the result of zoonosis from primate lentiviruses (SIVs); thus the relevance of comparing the pathogenesis of human infection with that of SIV natural hosts. This includes the understanding of differences in susceptibility to transmission and of differences in disease progression. Not all aspects that could be contrasted have been the subject of formal comparative research—significant attention has been devoted to differences in restriction of viruses, innate immunity responses, and immunopathogenesis. This chapter first discusses basic concepts on tolerance and restriction—two separate strategies in protection from disease. Then, I will review salient differences and similarities in lentiviral control and disease in primates, with a particular emphasis on the description of features of human infection that resemble that of the natural host.

RESISTANCE AND TOLERANCE

There are three main mechanisms of defense against pathogens: *avoidance*, *resistance*, and *tolerance* [1]. Discussion on *avoidance* is less germane to the discussion in this chapter. *Resistance* has been at the core of much research in infectious diseases. In contrast, there has been limited attention to the concept of *tolerance* as a powerful mechanism of defense against disease. Both concepts, resistance and tolerance, are important in the comparison of HIV-infected humans and the natural hosts of SIV.

Natural Hosts of SIV.
http://dx.doi.org/10.1016/B978-0-12-404734-1.00016-4

Resistance

Following mucosal exposure to HIV-1, only one or a few infectious particles successfully establish systemic infection. This finding indicates that at the earliest steps of infection there are several layers of barriers that limit viral transmission. A better understanding of the first infected target cells, the roles of antiviral restriction factors, and the possible contribution of innate effector cells to the control of viral replication during acute HIV-1 infection would lead to novel concepts in therapeutic interventions to reduce susceptibility to HIV-1.

Most comparative research in resistance to HIV-SIV infection has centered on the characteristics of the interferon response. Recognition of HIV-1 by Toll-like receptors expressed in dendritic cells [2,3] triggers the production of type I interferons and other proinflammatory and antiviral cytokines that not only initiate the innate immune response but also play a critical role in the cross-talk between innate and adaptive immunity [4,5]. However, the interferon response, while effective in the in vitro setting against HIV-1, is unable to achieve efficient control of HIV infection in vivo in infected individuals [6] except if provided exogenously [7]. The interferon response includes the expression of unique sets of proteins devoted to defense against exogenous retroviruses and control of endogenous retro-elements [8]. These effectors limit retroviral transmission across species. Some of these restriction factors are represented by complex families of genes (*APOBEC3, TRIM*), and others by single genes (*BST2*/ Tetherin and Sterile alpha motif and HD domain 1 (*SAMHD1)*). Together they act on discrete steps of the viral replication cycle (Figure 16.1). The recently described SAMHD1 [9,10] restricts HIV-1 infection in differentiated noncycling cells and particularly in cells of myeloid lineage (i.e., monocytes, macrophages, and dendritic cells) that play a key role in innate immune responses. Thus, SAMHD1 might impact the early steps of dissemination in the subepithelial mucosa and subsequent dissemination.

Proof of the long-standing conflict with retroviruses is the identification of positive selection in these host genes, and the countermeasures evolved by the pathogen: escape mutations and acquisition of retroviral accessory proteins. In rhesus macaques, TRIM5α selects SIV capsid escape mutants, a classic mechanism of resistance [11]. In humans, HIV-1 capsids appear to have reached an optimum of adaptation to huTRIM5α: the modest or negligible activity of various human alleles of this restriction factor no longer select for escape mutants [12]. Other restriction factors target conserved components of the virus, such as the viral RNA or envelope. Thus, viruses cannot become resistant by the acquisition of mutations in the targeted viral genes but have instead acquired a set of accessory proteins: Vif, Vpr, Nef, and (in some cases) Vpu or Vpx during coevolution with their hosts [13]. The lentiviral phylogeny in lagomorph, equine, small ruminant, bovine,

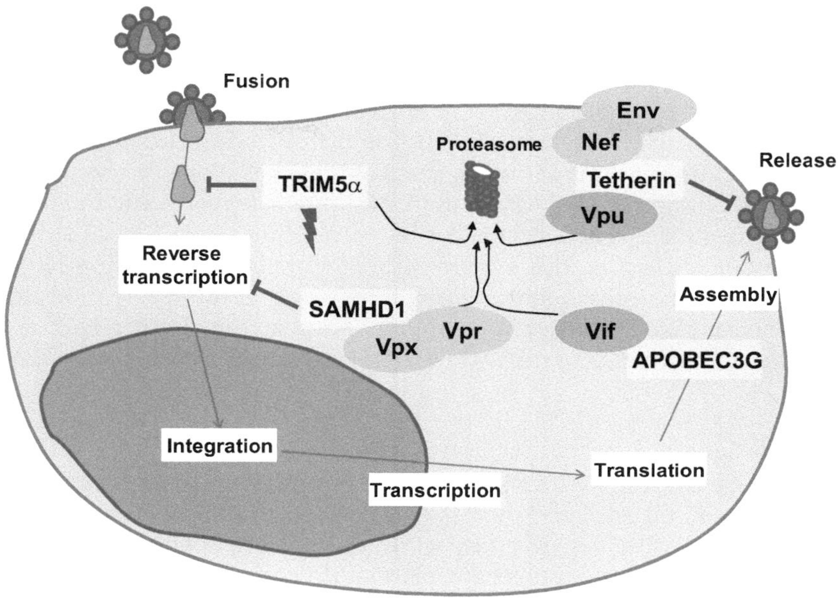

FIGURE 16.1 Host restriction factors and their viral countermeasures in HIV and SIVs. Host restriction factors are counteracted by retroviral proteins present in HIV-1, or in HIV-2 and simian immunodeficiency viruses, SIV. TRIM5α interacts with incoming retroviral capsids and may induce accelerated uncoating, proteasomal degradation, and triggering of the interferon response (dark gray thunderbolt). APOBEC3G is incorporated into virions and causes hypermutation of the viral genome in the next round of infection. Vif binds to APOBEC3G to induce its degradation. Tetherin inhibits the release of mature viral particles from the cell surface and is antagonized by Vpu of HIV-1, Nef of most SIVs, and Env of HIV-2. In myeloid cells, SAMHD1 blocks HIV-1 replication by hydrolyzing intracellular deoxynucleoside triphosphates, lowering their concentrations to below those required by the reverse transcriptase. HIV-1 has not evolved a viral antagonist to counter this restriction; however, Vpx of HIV-2 and Vpx or Vpr of certain SIV strains overcome the block by promoting SAMHD1 degradation.

feline, prosimian, and simian lineages reflects the capacity of lentiviruses to deploy accessory proteins that can assume homologous functions to counteract host factors [14]. Notably, recent evidence suggests that the capability of pandemic HIV-1 strains to switch from Nef to Vpu to antagonize human tetherin (which is naturally resistant to Nef) after zoonotic transmission may have been a prerequisite for the effective spread of HIV-1 [15,16].

In contrast to previously identified HIV-1 restriction factors, SAMHD1 does not meet a countermeasure developed by HIV-1 and remains active against this lentivirus. In contrast, HIV-2 and certain SIV strains encode the auxiliary protein Vpx that overcomes the block by SAMHD1 by promoting its degradation by the proteasome [9,10]. Correspondingly, SAMHD1 carries features of genetic conflict with Vpx [17,18], meaning that both host and pathogen-interacting proteins have evolved under positive selective pressure. Comparative genomics also reveals a conflict between SAMHD1 and Vpr for certain SIVs [18]. It is unclear whether there is any advantage for HIV-1 to counteract SAMHD1 and thus to infect dendritic cells and monocytes. The nonpermissiveness of these cells may be used by the virus as an escape strategy from immune system surveillance. In all, the host deploys a dedicated system for the intracellular detection of lentiviral infection. However, this system ultimately fails in controlling HIV-1. Discussion of the subsequent failure of the acquired immunity to protect from or clear HIV-1 infection is beyond the scope of this chapter. Failure to resist viral infection brings the strategy of "tolerance" into the discussion.

Tolerance

Not to be confused with the concept of immunological tolerance (nonrecognition of self-antigens), here, *tolerance* refers to a host defense mechanism that limits the negative impact of an infecting agent (Figure 16.2). This is achieved through mechanisms that do not modify pathogen burden but the degree of tissue damage cost by the host response or by the pathogen itself. The concept of tolerance as a defense strategy was only recently discussed in the field of animal infection [1,19,20]. It is an important concept, because tolerance has major implications for the epidemiology and the coevolution of host–parasite interactions [21]. In terms of epidemiology, the persistence of a high burden of pathogens will affect

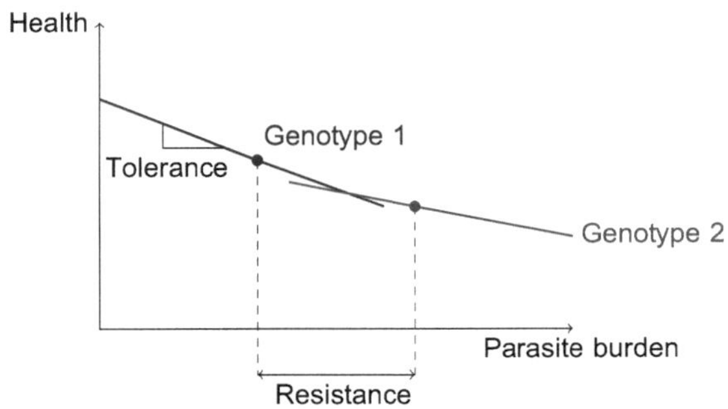

FIGURE 16.2 Contrasting effects of tolerance and resistance mechanisms of defense against pathogens. To distinguish between failed resistance and failed tolerance, it is important to measure pathogen burden and a relevant indicator of host fitness status. Resistance will be defined by differences in the average burden of the pathogens across genotypes (individuals of a species) or across species. Tolerance will be defined by differences in host fitness (e.g., disease progression, health) for a given pathogen burden. *With permission from R. S. Regoes based on Refs. [19,23].*

spread. In terms of evolutionary mechanisms of adaptation, tolerance will be a convenient mechanism for the host and the parasite and may lead to fixation of useful mutations. This is in sharp contrast with the features of restriction genes discussed in previous paragraphs. There, the evolutionary arm race results in characteristic features of positive selective pressure in host and pathogen genes at the interacting domains [22].

Studying tolerance can be challenging because it is expected to involve multiple mechanisms that help reduce the host vulnerability to damage [20]. In practical terms, for the HIV field, it will require the identification and measurement of events of tolerance, and thereafter the study of the basic processes. In particular, research will be required to identify mechanisms of reduction of host-induced damage and limitation of immunopathogenesis. These concepts will be of importance when discussing rare HIV-infected humans who, despite a high HIV-1 burden, maintain stable CD4+ T-cell counts for prolonged periods—a situation reminiscent of that of natural hosts such as infected sooty mangabeys.

PATTERNS OF HUMAN DISEASE

HIV-1 infection in humans results in profound differences in disease progression. Definitions of the different clinical and biological patterns have been previously summarized [24]. In the absence of antiviral treatment, disease progression to AIDS occurs on average in 8 years. However, some individuals achieve full control of viral replication and nonprogressive disease; and at the other end of the spectrum, other infected individuals progress rapidly to AIDS in less than 3 years—usually in the presence of high viral burden. More uncommon, there are patterns of progression where high levels of viral replication can be tolerated; these individuals are referred to as extreme viremic nonprogressors. A parallel can be observed between various disease patterns and infection profiles in non-human primates (NHPs) (Figure 16.3). The current understanding of these extremes of HIV-1 infection progression, and shared genetic, molecular, or immunopathogenic characteristics with the natural host of SIV infection, are discussed next.

Elite Controllers

A small proportion of HIV-1 infected individuals (<5–10% depending on the criteria) present a very slow progression toward AIDS (long-term nonprogressors), and exceptionally achieve full control of viral replication, as defined by viral load <50 copies/mL, or in research assays <1 copy of viral RNA/mL or undetectable viral load. Nonprogression of disease in these individuals has been explained in terms of virological,

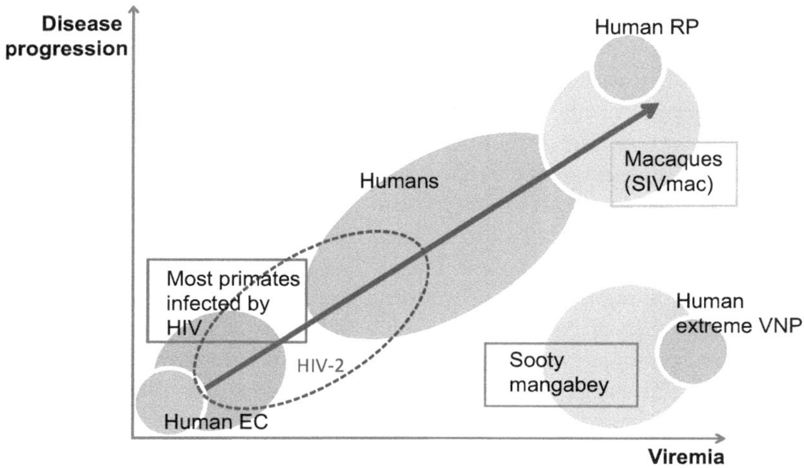

FIGURE 16.3 Schematic representation of the parallelism between human and non-human primate models of HIV/SIV pathogenesis. EC = elite controllers, RP = rapid progressors, VNP = extreme viremic non progressors. Indicated by the discontinuous line is also the pattern of HIV-2 infection. *Adapted with permission from G. Silvestri, Conference on Retroviruses and Opportunistic Infections 2010.*

immunological, and genetic determinants. Cited viral factors include a lesser fitness of circulating virions and higher rates of sequence defects in viruses and integrated provirus. In many cases, the reduced replication fitness reflects viral adaptation to host immune pressure. The best-characterized immunological factors relate to differences in cytotoxic CD8 T-cell responses. Differences in prevalence of the restricting HLA alleles in different human groups lead to HIV-1 adaptation to HLA at a population level [25].

The International HIV Controllers Study [26] compared the genetic data of 974 subjects defined as viral controllers and 2648 progressors. It specified unique amino acids in the HLA-B peptide binding groove as variants associated with viral control. These included Arg97, Cys67, Gly62, and Glu63, all in HLA-B. The study also implicated Ser77 in the HLA-A groove. HLA-B position 97 is also the most significant association with host control in African Americans [27], as well as independent effects at position 116 in the canonical F pocket, position 63 in the B pocket, and the nonpocket position 245, which is thought to influence CD8-binding kinetics.

The contemporary Major Histocompatibility Complex (*MHC*) class I repertoire of chimpanzees targets analogous conserved domains of HIV-1/SIV$_{cpz}$ to those targeted by human protective alleles *HLA-B*57* and *B*27*, in particular of the Gag protein [28]. The functional characteristics of the chimpanzee *MHC* repertoire might be the result of a selective sweep caused by lentiviruses [28]. Studies in rhesus macaques also showed significant correlations between specific MHC alleles and control of SIV_{mac} replication levels; most notably, MHC B-locus alleles *Mamu-B*08* and *B*17* are associated with elite control [29–31].

A recent primate model of elite control has been reported by Pandrea et al. [32]. Infection of rhesus macaques with SIV_{agm} infection results in robust acute infection, with high levels of viral replication and depletion of mucosal CD4+ T cells. However, a complete and durable control of SIV_{agm} replication follows the acute infection. This is defined as undetectable VLs in blood and tissues beginning 72–90 days post inoculation and continuing at least 4 years; seroreversion; progressive recovery of mucosal CD4+ T cells, with complete recovery by 4 years; normal levels of T-cell immune activation, proliferation, and apoptosis; and no disease progression [32]. Pandrea et al. [32] also showed that the apparent functional cure could be reverted by depleting CD8+ T cells, which resulted in transient rebounds of viremia, thus suggesting that control may be at least in part immune mediated. However, viral control was independent of MHC, APOBEC restriction, and Trim5 genotypes. This is proposed as a new animal model in which elite control can be established in all cases.

Rapid Progressors

The definition of HIV-1 rapid progression includes criteria such as accelerated immunosuppression soon after seroconversion (in the first 2–3 years), and in many cases, high levels of viremia [24,33]. Similar to the findings on elite controllers, limited data suggest that viral and host factors contribute to the severity of early disease [34]. There are, however, few such individuals in clinical cohorts [34,35]. Although up to 8% of individuals can be described as rapid progressors, the need to identify patients with a known date of infection (seroconverters), and the short window of clinical observation before antiretroviral treatment is initiated, limits the recruitment of such individuals into studies as well as the collection of relevant biological materials for analysis.

Dalmau et al. investigated the characteristics of primary isolates of a group of 46 rapid progressors [36]. This study identified a high prevalence (>25%) of viral variants that use CXCR4 as a coreceptor, or a mixture of variants that were dual tropic (CXCR4/CCR5). In contrast the frequency of alternative coreceptor use is very low in primary HIV-1 infection [37] and in individuals with common rates of progression. The same study identified higher in vitro replicative capacity of those primary isolates. The *MHC* locus also determines rapid progression: individuals homozygous for select HLA-A, -B, -C alleles are at particular risk, as are those who carry particular HLA-B alleles (e.g., HLA-B*35). There is a depletion of protective alleles among rapid progressors [24,35].

The kinetics of disease progression following experimental SIV infection of rhesus macaques in general correlates with rapid progressive infection in humans (Figure 16.3). As non-natural hosts, rhesus macaques present high levels of viremia and progressive depletion of peripheral and mucosal CD4+ T cells accompanied by immune dysfunction, microbial translocation, high persistent interferon stimulation, and ineffective immune control eventually leading to AIDS [38].

Viremic Nonprogressors

Extreme viremic nonprogressors represent an uncommon pattern of disease in humans. Estimates indicate that only 0.1% of HIV-infected individuals would fulfill a strict definition that includes these criteria: (1) >3 years follow-up with a median HIV viremia from >3 measurements >100,000 cp/mL, (2) HIV viremia consistently above 10,000 cp/mL, (3) CD4+ T-cell count above 350/μL, and (4) no HIV treatment during follow-up [35] (Figure 16.4). Thus, this pattern of infection and disease would match that of various natural hosts of SIV, such as sooty mangabeys and African green monkeys. Characteristics of infection in those NHPs that have been investigated in humans are the degree of compromise of the intestinal mucosal barrier and transcriptome features during chronic infection—in particular the pattern of deregulation of the interferon response.

Rotger el al. investigated the monocyte-expressed Lipopolysaccharide (LPS) receptor sCD14 [35] in rapid progressors and viremic nonprogressors. Their study identified higher plasma levels among rapid progressors

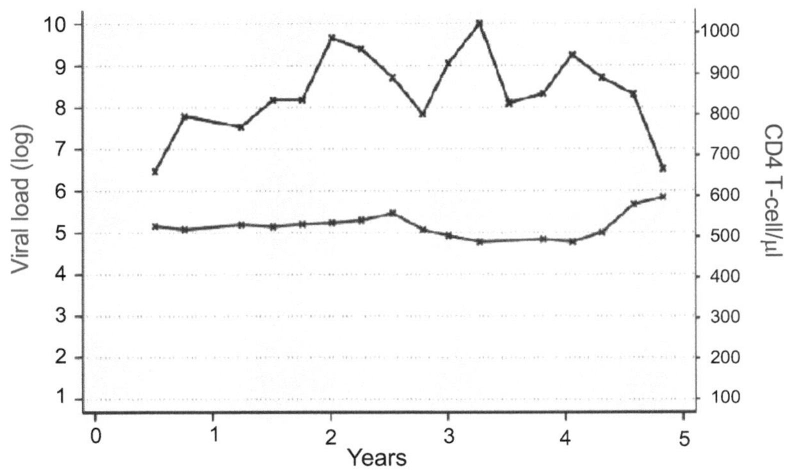

FIGURE 16.4 Individual viral load and CD4+ T-cell profiles of human viremic nonprogressors. HIV-1 viremia is shown in light gray, and CD4+ T-cell count in dark gray.

than for other clinical progression groups, while viremic nonprogressors presented low sCD14 plasma levels, a pattern consistent with less immunopathogenesis in these individuals. Similar observations were reported by Choudhary et al. [39] on viremic nonprogressors (with somehow lower viral loads). They identified a lower percentage of activated HLA-DR+ CD38+ CD4+ and CD8+ T cells and lower levels of proliferating Ki67-expressing CD4+ and CD8+ T cells in viremic nonprogressors compared to progressors. This study also evaluated the replication capacity of viral isolates from viremic nonprogressors. These isolates replicated to similar levels, shared the capacity to deplete CD4+ thymocytes or CD4+ T cells in secondary lymphoid tissue, and were equally cytopathic as isolates from standard progressors. Recently, Klatt et al. [40] described limited infection of central memory CD4+ T cells, but not of effector memory CD4+ T cells, to be associated with lack of progression in a group of six viremic individuals. In addition, and compared to "putative progressors," viremic nonprogressors appeared to express higher levels of Ki-67 in CD4+ T cells, a finding that the authors attributed to preserved CD4+ T-cell homeostasis. Furthermore, viremic nonprogressors exhibited decreased CCR5 expression on memory CD4+ T cells. These factors may contribute to maintenance of CD4+ T cells and lack of disease progression despite high levels of viremia.

Additional details are derived from transcriptome profiles. While rapid progressors shared a number of features with chronic SIV infection of rhesus macaques, in particular the high expression of an ISG and of immune activation markers, viremic nonprogressors presented transcriptome pattern consistent with lower immune activation—reminiscent of the observations in sooty mangabeys [41]. The study by Rotger et al. also proposed a limited set of genes that could be relevant to immunopathogenesis. *SOCS1* (Suppressor of cytokine signaling), involved in a negative feedback loop in the regulation of cytokine signal transduction signaled through the JAK-STAT5 was upregulated in the nonpathogenic SIV infection of sooty mangabeys and in human viremic nonprogressors. In contrast, *CASP1, CD38, LAG3,* and *TNFSF13B* were upregulated in the pathogenic SIV infection of rhesus macaque and in rapid progressors. These are known intermediates of inflammatory processes and apoptosis, and markers of immune exhaustion.

EXPOSED NONINFECTED INDIVIDUALS

Some individuals exposed to HIV-1 remain uninfected. Truncation of the coreceptor CCR5 is the only mechanism established to date conferring effective protection. Additional efforts during the past two decades led to the proposition of additional genetic factors that would account for resistance to infection among individuals with various mechanisms of exposure. However, only CCR5 variation has been consistently shown to confer protection and no other genetic variant has been convincingly replicated [42]. Specifically, a recent genome-wide association study in a highly exposed population of hemophiliacs failed to identify additional protective variants of large effect in individuals of Western European ancestry, beyond CCR5 [43]. In practical terms, this result means that similarly sized genome association studies are unlikely to be successful in cohorts in whom measurements of exposure is lower or difficult to ascertain (sexual transmission, intravenous drug use, mother-to-child transmission). Mirroring the results from genetic analysis, studies of mucosal immunity, or of HIV-specific humoral and T cell-mediated responses, have been described in exposed noninfected subjects (reviewed in [44]). It is not known whether such responses would be protective or represent a marker for exposure [44,45].

Highly exposed but noninfected NHPs have also observed in the natural host. Ma et al. recently reported that infection is highly prevalent in wild vervet populations from South Africa, where SIV_{agmVer} has long been present. However, they describe that while up to 80% of sexually active adult females are infected, the remaining 20% are uninfected in spite of the massive potential for exposure to SIV throughout their lives [46]. The study suggests that individuals in this cohort could be the subject of comprehensive genetic, immunologic, and virologic studies.

There are parallels in NHPs to the CCR5 gene deletion that protects humans from infection; however, NHP natural host are not frequently infected by viruses that use alternative entry coreceptors. In red-capped mangabeys, a 24 base pair in-frame deletion (CCR5Δ24) that abrogates surface expression of the receptor is present at an allelic frequency of 87% [47]. As a result, SIV_{rcm} uses CCR2b for entry and cannot use CCR5. CCR5Δ24 alleles are also present in sooty mangabeys, but with at very low allelic frequency [47,48]. Sooty mangabeys can also carry a two base pair deletion (CCR5Δ2) that results in a lack of surface expression of the coreceptor [49]. However, CCR5-null sooty mangabeys are susceptible to SIV_{smm} infection and exhibit robust viral replication. This is consistent with in vivo alternative coreceptor use by SIV_{smm}. In summary, humans and red-capped and sooty mangabeys are examples of convergent evolution resulting in nonfunctional mutant CCR5 alleles among primate species [49].

An animal model for highly exposed but noninfected individuals would be of great interest in view of the difficulty to approach this important question in the human population. However, such an animal model is not currently available [46], and there is an opinion that the SIV_{mac}/rhesus macaque is not relevant for such studies because the model was developed for pathogenesis and vaccine studies using strains selected for pathogenicity and not for transmission [46].

THE CASE OF HIV-2 INFECTION

Infection due to HIV-2 should be included in the discussion of human infections that resemble infection in animal models of disease [50]. HIV-2 is similar in sequence and structure to SIV from sooty mangabey. Infection in humans is thought to have entered the human population on at least eight separate occasions, yielding eight distinct HIV-2 groups. Only groups A and B are endemic, with the remainder being single-person infections (reviewed in [50]). No significant differences in disease progression, pathogenicity, or transmission have been observed across HIV-2 clades [50].

Although both HIV-2 and HIV-1 use the same repertoire of coreceptors and target CD4+ T cells, their pathogenesis is different. Humans infected with HIV-2 mostly evolve as long-term nonprogressors or viral controllers (Figure 16.3), while a minority develop disease indistinguishable from AIDS caused by HIV-1 [51]. Why HIV-2 and HIV-1 have distinct natural histories is not known, although differences in immune activation or viral proteins may play a role [50]. Recent emphasis is placed on the role of the accessory protein Vpx found in HIV-2. This protein counteracts host SAMHD1 [9], lifting its capacity to be restricted for replication in myeloid cells and possibly allowing an earlier triggering of the interferon response [52]. A number of authors propose that HIV-2 research could complement the study of HIV-1 controllers and of nonpathogenic SIV models. Study of HIV-2 is also an opportunity to explore the evolutionary adaptation of lentivirus to the human hosts and to identify correlates of protective responses.

CONCLUSIONS

There is a large diversity in the patterns of HIV-1 infection and disease progression in humans. Some patterns parallel SIV infection in the natural host and thus represent an opportunity to unravel basic aspects of HIV-1 pathogenesis in humans. The opposite is also true, in that progression to AIDS can occur in NHP natural hosts of SIV, both upon infection with their species-specific viruses as well as after cross-species transmission [53]. Comparative genomics, transcriptomics, and immunopathogenesis across human and NHPs constitute an opportunity to improve our understanding of human susceptibility to HIV-1 and support vaccine research.

Acknowledgments

Work in my laboratory is funded by the Swiss National Science Foundation.

References

[1] Medzhitov R, Schneider DS, Soares MP. Disease tolerance as a defense strategy. Science February 24, 2012;335(6071):936–41.

[2] Beignon AS, McKenna K, Skoberne M, Manches O, Dasilva I, Kavanagh DG, et al. Endocytosis of HIV-1 activates plasmacytoid dendritic cells via Toll-like receptor- viral RNA interactions. J Clin Invest November 1, 2005;115(11):3265–75.

[3] Meier A, Alter G, Frahm N, Sidhu H, Li B, Bagchi A, et al. MyD88-dependent immune activation mediated by human immunodeficiency virus type 1-encoded toll-like receptor ligands. J Virol August 2007;81(15):8180–91.

[4] Li Q, Estes JD, Schlievert PM, Duan L, Brosnahan AJ, Southern PJ, et al. Glycerol monolaurate prevents mucosal SIV transmission. Nature April 23, 2009;458(7241):1034–8.

[5] Altfeld M, Fadda L, Frleta D, Bhardwaj N. DCs and NK cells: critical effectors in the immune response to HIV-1. Nat Rev Immunol March 2011;11(3):176–86.

[6] Rotger M, Dang KK, Fellay J, Heinzen EL, Feng S, Descombes P, et al. Genome-wide mRNA expression correlates of viral control in CD4+ T-cells from HIV-1-infected individuals. PLoS Pathog 2010;6(2):e1000781.

[7] Pillai SK, Abdel-Mohsen M, Guatelli J, Skasko M, Monto A, Fujimoto K, et al. Role of retroviral restriction factors in the interferon-alpha-mediated suppression of HIV-1 in vivo. Proc Natl Acad Sci USA February 21, 2012;109(8):3035–40.

[8] Douville RN, Hiscott J. The interface between the innate interferon response and expression of host retroviral restriction factors. Cytokine October–November, 2010;52(1–2):108–15.

[9] Laguette N, Sobhian B, Casartelli N, Ringeard M, Chable-Bessia C, Segeral E, et al. SAMHD1 is the dendritic- and myeloid-cell-specific HIV-1 restriction factor counteracted by Vpx. Nature June 30, 2011;474(7353):654–7.

[10] Hrecka K, Hao C, Gierszewska M, Swanson SK, Kesik-Brodacka M, Srivastava S, et al. Vpx relieves inhibition of HIV-1 infection of macrophages mediated by the SAMHD1 protein. Nature June 30, 2011;474(7353):658–61.

[11] Kirmaier A, Wu F, Newman RM, Hall LR, Morgan JS, O'Connor S, et al. TRIM5 suppresses cross-species transmission of a primate immunodeficiency virus and selects for emergence of resistant variants in the new species. PLoS Biol 2010;8(8):e1000462.

[12] Rahm N, Gfeller D, Snoeck J, Martinez R, McLaren PJ, Ortiz M, et al. Susceptibility and adaptation to human TRIM5a alleles at positive selected sites in HIV-1 capsid. Virology July 2013;441(2):162–70.

[13] Gifford RJ, Katzourakis A, Tristem M, Pybus OG, Winters M, Shafer RW. A transitional endogenous lentivirus from the genome of a basal primate and implications for lentivirus evolution. Proc Natl Acad Sci USA December 23, 2008;105(51):20362–7.

[14] Gifford RJ. Viral evolution in deep time: lentiviruses and mammals. Trends Genet February 2012;28(2):89–100.

[15] Sauter D, Schindler M, Specht A, Landford WN, Munch J, Kim KA, et al. Tetherin-driven adaptation of Vpu and Nef function and the evolution of pandemic and nonpandemic HIV-1 strains. Cell Host Microbe November 19, 2009;6(5):409–21.

[16] Yang SJ, Lopez LA, Exline CM, Haworth KG, Cannon PM. Lack of adaptation to human tetherin in HIV-1 group O and P. Retrovirology 2011;8:78.

[17] Laguette N, Rahm N, Sobhian B, Chable-Bessia C, Münch J, Snoeck J, et al. Evolutionary and functional analyses of the interaction between SAMHD1 and the lentiviral Vpx protein. Cell Host Microbe February 2012;11(2):205–17.

[18] Lim ES, Fregoso OI, McCoy CO, Matsen FA, Malik HS, Emerman M. SAMHD1-degrading activity preceded the birth of Vpx in primate lentiviruses. Cell Host Microbe February 2012;11(2):194–204.

[19] Read AF, Graham AL, Raberg L. Animal defenses against infectious agents: is damage control more important than pathogen control. PLoS Biol December 23, 2008;6(12):e4.

[20] Ayres JS, Schneider DS. Tolerance of infections. Annu Rev Immunol 2012;30:271–94.
[21] Boots M. Fight or learn to live with the consequences? Trends Ecol Evol May 2008; 23(5):248–50.
[22] Daugherty MD, Malik HS. Rules of engagement: molecular insights from host-virus arms races. Annu Rev Genet December 15, 2012;46:677–700.
[23] Schneider DS, Ayres JS. Two ways to survive infection: what resistance and tolerance can teach us about treating infectious diseases. Nat Rev Immunol November, 2008;8(11):889–95.
[24] Casado C, Colombo S, Rauch A, Martinez R, Gunthard HF, Garcia S, et al. Host and viral genetic correlates of clinical definitions of HIV-1 disease progression. PLoS ONE 2010;5(6):e11079.
[25] Kawashima Y, Pfafferott K, Frater J, Matthews P, Payne R, Addo M, et al. Adaptation of HIV-1 to human leukocyte antigen class I. Nature April 2, 2009;458(7238):641–5.
[26] Study IHC. The major genetic determinants of HIV-1 control affect HLA class I peptide presentation. Science November 4, 2010;330:1551–7.
[27] McLaren PJ, Ripke S, Pelak K, Weintrob AC, Patsopoulos NA, Jia X, et al. Fine-mapping classical HLA variation associated with durable host control of HIV-1 infection in African Americans. Hum Mol Genet October 1, 2012;21(19):4334–47.
[28] de Groot NG, Heijmans CM, Zoet YM, de Ru AH, Verreck FA, van Veelen PA, et al. AIDS-protective HLA-B*27/B*57 and chimpanzee MHC class I molecules target analogous conserved areas of HIV-1/SIVcpz. Proc Natl Acad Sci USA August 24, 2010;107(34):15175–80.
[29] Loffredo JT, Friedrich TC, Leon EJ, Stephany JJ, Rodrigues DS, Spencer SP, et al. CD8+ T cells from SIV elite controller macaques recognize Mamu-B*08-bound epitopes and select for widespread viral variation. PLoS ONE 2007;2(11):e1152.
[30] Loffredo JT, Maxwell J, Qi Y, Glidden CE, Borchardt GJ, Soma T, et al. Mamu-B*08-positive macaques control simian immunodeficiency virus replication. J Virol August 2007;81(16):8827–32.
[31] Yant LJ, Friedrich TC, Johnson RC, May GE, Maness NJ, Enz AM, et al. The high-frequency major histocompatibility complex class I allele Mamu-B*17 is associated with control of simian immunodeficiency virus SIVmac239 replication. J Virol May 2006;80(10):5074–7.
[32] Pandrea I, Gaufin T, Gautam R, Kristoff J, Mandell D, Montefiori D, et al. Functional cure of SIVagm infection in rhesus macaques results in complete recovery of CD4+ T cells and is reverted by CD8+ cell depletion. PLoS Pathog August 2011;7(8):e1002170.
[33] Le Clerc S, Limou S, Coulonges C, Carpentier W, Dina C, Taing L, et al. Genomewide association study of a rapid progression cohort identifies new susceptibility alleles for AIDS (ANRS genomewide association study 03). J Infect Dis 2009;200(8):1194–201.
[34] Dalmau J, Puertas MC, Azuara M, Marino A, Frahm N, Mothe B, et al. Contribution of immunological and virological factors to extremely severe primary HIV type 1 infection. Clin Infect Dis January 15, 2009;48(2):229–38.
[35] Rotger M, Dalmau J, Rauch A, McLaren P, Bosinger SE, Marttinez R, et al. Comparative transcriptome analysis of extreme phenotypes of human HIV-1 infection and sooty mangabey and rhesus macaque models of SIV infection. J Clin Invest 2011;121:2391–400.
[36] Dalmau J, Rotger M, Erkizia I, Rauch A, Reche P, Pino M, et al. The CoRP Study Group. Highly pathogenic adapted HIV-1 strains limit host immunity and dictate rapid disease progression. AIDS April 12, 2014 [Epub ahead of print] PubMed PMID: 24732774.
[37] Chalmet K, Dauwe K, Foquet L, Baatz F, Seguin-Devaux C, Van Der Gucht B, et al. Presence of CXCR4-using HIV-1 in patients with recently diagnosed infection: correlates and evidence for transmission. J Infect Dis January 15, 2012;205(2):174–84.
[38] Chahroudi A, Bosinger SE, Vanderford TH, Paiardini M, Silvestri G. Natural SIV hosts: showing AIDS the door. Science March 9, 2012;335(6073):1188–93.

[39] Choudhary SK, Vrisekoop N, Jansen CA, Otto SA, Schuitemaker H, Miedema F, et al. Low immune activation despite high levels of pathogenic human immunodeficiency virus type 1 results in long-term asymptomatic disease. J Virol August 2007;81(16):8838–42.
[40] Klatt NR, Bosinger SE, Specht A, Lederman HM, Camerini D, Paiardini M, et al. Limited HIV infection of Central memory CD4+ T cells is associated with lack of progression in viremic individuals. In: Retroconference: Conference on Retrovirus and Opportunistic Diseases 2013. 2013: Abstract 176LB.
[41] Silvestri G, Sodora DL, Koup RA, Paiardini M, O'Neil SP, McClure HM, et al. Nonpathogenic SIV infection of sooty mangabeys is characterized by limited bystander immunopathology despite chronic high-level viremia. Immunity March, 2003;18(3):441–52.
[42] Telenti A, McLaren P. Genomic approaches to the study of HIV-1 acquisition. J Infect Dis November 1, 2010;202(Suppl. 3):S382–6.
[43] Lane J, McLaren PJ, Dorrell L, Shianna KV, Stemke A, Pelak K, et al. A genome-wide association study of resistance to HIV infection in highly exposed uninfected individuals with hemophilia A. Hum Mol Genet February 10, 2013;22(9):1903–10.
[44] Tomescu C, Abdulhaqq S, Montaner LJ. Evidence for the innate immune response as a correlate of protection in human immunodeficiency virus (HIV)-1 highly exposed seronegative subjects (HESN). Clin Exp Immunol May 2011;164(2):158–69.
[45] Ritchie AJ, Campion SL, Kopycinski J, Moodie Z, Wang ZM, Pandya K, et al. Differences in HIV-specific T cell responses between HIV-exposed and -unexposed HIV-seronegative individuals. J Virol April 2011;85(7):3507–16.
[46] Ma D, Jasinska A, Kristoff J, Grobler JP, Turner T, Jung Y, et al. SIVagm infection in wild African green monkeys from South Africa: epidemiology, natural history, and evolutionary considerations. PLoS Pathog January, 2013;9(1):e1003011.
[47] Chen Z, Kwon D, Jin Z, Monard S, Telfer P, Jones MS, et al. Natural infection of a homozygous delta24 CCR5 red-capped mangabey with an R2b-tropic simian immunodeficiency virus. J Exp Med 1998;188(11):2057–65.
[48] Palacios E, Digilio L, McClure HM, Chen Z, Marx PA, Goldsmith MA, et al. Parallel evolution of CCR5-null phenotypes in humans and in a natural host of simian immunodeficiency virus. Curr Biol July 30–August 13, 1998;8(16):943–6.
[49] Riddick NE, Hermann EA, Loftin LM, Elliott ST, Wey WC, Cervasi B, et al. A novel CCR5 mutation common in sooty mangabeys reveals SIVsmm infection of CCR5-null natural hosts and efficient alternative coreceptor use in vivo. PLoS Pathog 2010;6(8).
[50] de Silva TI, Cotten M, Rowland-Jones SL. HIV-2: the forgotten AIDS virus. Trends Microbiol December, 2008;16(12):588–95.
[51] Campbell-Yesufu OT, Gandhi RT. Update on human immunodeficiency virus (HIV)-2 infection. Clin Infect Dis March 15, 2011;52(6):780–7.
[52] Laguette N, Benkirane M. How SAMHD1 changes our view of viral restriction. Trends Immunol January, 2012;33(1):26–33.
[53] Pandrea I, Silvestri G, Apetrei C. AIDS in african nonhuman primate hosts of SIVs: a new paradigm of SIV infection. Curr Hiv Res January, 2009;7(1):57–72.

EPILOGUE

1

Final Remarks: What Have We Learned from Natural Simian Immunodeficiency Virus Hosts that Can Help HIV-Infected Humans?

Guido Silvestri, Aftab A. Ansari

Department of Pathology and Laboratory Medicine, Emory University, School of Medicine, and Yerkes National Primate Research Center, Atlanta, GA, USA

This book represents, we believe, the first major attempt to comprehensively review and discuss the large body of knowledge that has been gained, over the past two decades, on the topic of natural simian immunodeficiency virus (SIV) infections. In this final chapter, we will briefly summarize the main reasons why this knowledge is important to understand HIV/AIDS as a human disease, and what are the main theoretical and practical implications for the prevention and treatment of HIV infection.

The notion that SIV infection of natural host African monkey species is by-and-large nonpathogenic and results in minimal to no immunological abnormalities presents an obvious and striking contrast with the clinical course of HIV infection in humans, which almost invariably leads to AIDS if left untreated. The only exception to this paradigm is the SIVcpz infection of chimpanzees that is associated with a significant increase in mortality, although not to the levels seen in HIV-1 or SIVmac infections [1]. Importantly, the benign infection of natural hosts is clearly distinct from the rare cases of benign, nonprogressive HIV/SIV infections observed in a small subset of humans and macaques, defined as elite controllers, in which the absence of AIDS occurs in the context of very low levels of virus replication and is due, to a large extent, to the ability of the immune system to suppress virus replication [2]. Rather, some aspects of the basic biology of natural SIV infections, such as the low immune activation and the relative preservation of central memory and stem cell memory $CD4^+$ T cells from direct virus infection, appear to be present in an even rarer subset of HIV-infected individuals called "viremic nonprogressors", in which lack of disease progression and stable $CD4^+$ T-cell counts are associated with robust virus replication [3]. It must be noted that the discovery of some HIV-infected humans that respond to the infection similarly to natural SIV hosts emphasizes the direct relevance of these studies to human health.

When the striking difference in the clinical course of infection between HIV-infected humans (i.e., the overwhelming majority of them) and natural SIV hosts was first appreciated, a series of comparative phenomenological studies were designed and conducted to define the basis for this difference (reviewed in [4]). These earlier studies aimed at defining the mechanisms for the "AIDS-resistance" of natural hosts have provided a number of important insights on the pathogenesis of HIV infection that can be briefly summarized as follows. First, virus replication is not per se an automatic cause of immunodeficiency, as the viremia of natural SIV hosts is similar, if not higher, than in HIV-infected humans [5–7]. Second, the lack of pathogenicity of natural SIV infection does not involve a "superior" immunological control—mediated by either cytotoxic T-lymphocytes or neutralizing antibodies—which would result in strong suppression of virus replication (reviewed in [8]). Third, SIV infecting natural hosts is similar to HIV in terms of in vitro and in vivo cytopathicity and the ability to use $CD4^+$ T cells as main cellular targets [9–11]. It should be noted however that, while all natural SIVs tested so far appear to be clearly cytopathic, the presence of significant differences in the virus sequence between the various SIVs and HIV-1, which often involve accessory genes (i.e., *Nef, Vpr, Vpu, Vpx*, etc.) and typically result in specific functional differences in the corresponding gene products, may exert subtle but profound effects on both the level of cytopathicity and the pattern of infected cells in vivo [12]. Fourth, the discovery of key peculiar phenotypic aspects of natural SIV infection (i.e., low immune activation and limited infection of less differentiated subsets of memory $CD4^+$ T cells, i.e., central and stem cell memory cells, that appears to be in turn related to lower expression of the virus coreceptor CCR5 or downregulation of the CD4 molecule), suggests that chronic immune activation and infection/depletion of central memory $CD4^+$ T cells (T_{CM}) and stem cell memory $CD4^+$ T cells (T_{SCM}) are key contributors to AIDS pathogenesis in HIV-infected humans [7,13,14]. Indeed, in natural SIV hosts the resolution of immune activation, that occurs within weeks from the initial infection, and target cell restriction may favor each other in a virtuous cycle whose establishment is crucial to avoid the immune deficiency that follows HIV/SIV infections of humans and macaques [4].

The overall conceptual impact of these descriptive studies is hard to overestimate, and in fact we remain convinced that a full elucidation of how the mechanisms by which natural hosts have reached a pacific coexistence with their respective SIV strains is absolutely required to achieve a satisfactory understanding of the mechanisms of AIDS pathogenesis in HIV-infected humans. In this regard, it would be important that all stakeholders in contemporary HIV/AIDS research, including patients and activists, health care providers, basic and clinical scientists, major pharmaceutical industries, funding agencies, and, ultimately, the general public, are made aware that major progress in HIV prevention and therapy is unlikely to be achieved unless a substantial effort continues to be devoted to further our understanding of the basic mechanisms of AIDS pathogenesis. Indeed, few areas of basic HIV/AIDS pathogenesis research have been as productive in terms of providing paradigm shifting ideas and hypotheses as the field of natural SIV infection—which has often been referred to as "comparative AIDS research"—even despite the relatively low levels of dedicated funds and the small number of involved investigators. Simply put, it would be rather foolish for the community of scientists and research stakeholders to miss the tremendous opportunity of building on an experiment (or, rather, a series of experiments) that "mother nature" has conducted for us over thousands and thousands of years, and use the "results" of these natural experiments to develop novel ways to better treat and prevent HIV infection in humans.

In more practical terms, it is hoped, and in fact anticipated, that these conceptual advances will translate into novel approaches to prevent, treat, and potentially cure HIV infection in humans. For instance, identifying the mechanisms underlying the rapid resolution of immune activation—and thus the lack of chronic immune activation—in natural SIV hosts will likely provide targets for new "immunomodulatory" interventions to be used, in addition to standard antiretroviral therapy (ART), in HIV-infected individuals. It is no coincidence that, at the time in which low immune activation was identified as a key feature of natural SIV infections, the levels of chronic immune activation (and related inflammation) was found correlated with disease progression during the natural history of HIV infection and, perhaps even more significantly, with poor clinical and immunological response to ART [15]. It is possible, and even likely, that the mechanisms responsible for this residual immune activation and inflammation of ART-treated HIV-infected individuals with full suppression of virus replication largely overlap with the mechanisms that, when appropriately silenced, allow the establishment of a low immune activation status in natural SIV hosts. In this context, the identification of molecular pathways of immune regulation that are active in natural SIV hosts may have a significant impact in reducing or even eliminating the "residual" immune activation and inflammation that is associated with incomplete immune recovery as well as increased morbidity and end-organ disease during long-term antiretroviral treatment of HIV-infected individuals. Similarly, the discovery that the pattern of in vivo infected $CD4^+$ T-cell subsets may dramatically influence the progression to AIDS in HIV/SIV infections should prompt more investigations aimed at defining the determinants of $CD4^+$ T_{CM} and T_{SCM} infection in HIV-infected individuals and whether the level of CD4+ T_{CM} infection correlates with disease progression and/or residual morbidity in treated individuals. Intriguingly, the pattern of infected CD4+ T-cell subsets during the natural history of a primate lentiviral infection may be a crucial determinant of the size and overall stability of the reservoir of latently infected cells that persists for a very long time, if not indefinitely, when active virus replication is fully suppressed by ART. Understanding in full detail what cellular and molecular mechanisms are responsible for the specific in vivo infection of those CD4+ T-cell subsets that act as key virus reservoirs may greatly benefit from comparative studies conducted in species where these cells are either susceptible or resistant to both direct virus infection and the establishment of a truly latent infection. It is hoped that these types of studies will ultimately lead to the definition of novel therapeutic targets to specifically reduce, and eventually eliminate, the persistent virus reservoirs of ART-treated HIV-infected individuals that represent the most formidable obstacle toward a cure for the infection. Finally, as argued in the past, we believe that studies of natural SIV infections may provide useful data and concepts to inform the development of an AIDS vaccine [16]. While natural SIV hosts are clearly not resistant to virus transmission per se, some aspects of the biology of these infections have clear implications for the field of HIV prevention. In this regard, a striking feature of SIV infection in natural hosts is the rarity of mother-to-infant-transmission (MTIT). In HIV-1-infected humans, MTIT can occur in utero, during labor and delivery, and as a consequence of breastfeeding, with a total virus transmission rate of 35–40% in absence of ART. In stark contrast, MTIT of SIV in sooty mangabeys occurs at a rate of <7% and appears to be even less frequent in African green monkeys and mandrills [17,18]. A more complete understanding of the mechanisms responsible for this block to MTIT in natural hosts could significantly impact current treatment of pregnant HIV-infected women and their infants and possibly reduce the number of perinatal HIV infections worldwide [19].

While much more work needs to be done to complete our understanding of the basic virology and immunology of natural SIV infections, and to apply this knowledge to improve our ability to prevent and treat HIV infection in humans, we are more optimistic than ever about the tremendous conceptual and practical impact of this line of research. Natural SIV hosts are the unfortunate vehicle through which AIDS came to humanity, and we like to believe they will help us to find a way to make humanity fully resistant to AIDS, following those steps they were able to walk many, many years ago.

References

[1] Keele BF, Jones JH, Terio KA, Estes JD, Rudicell RS, Wilson ML, et al. Increased mortality and AIDS-like immunopathology in wild chimpanzees infected with SIVcpz. Nature July 23, 2009;460(7254):515–9.

[2] Walker BD, Yu XG. Unravelling the mechanisms of durable control of HIV-1. Nat Rev Immunol July 2013;13(7):487–98.
[3] Rotger M, Dalmau J, Rauch A, McLaren P, Bosinger SE, Martinez R, et al. Comparative transcriptomics of extreme phenotypes of human HIV-1 infection and SIV infection in sooty mangabey and rhesus macaque. J Clin Invest June 2011;121(6):2391–400.
[4] Chahroudi A, Bosinger SE, Vanderford TH, Paiardini M. Silvestri G.Natural SIV hosts: showing AIDS the door. Science March 9, 2012;335(6073):1188–93.
[5] Rey-Cuillé MA, Berthier JL, Bomsel-Demontoy MC, Chaduc Y, Montagnier L, Hovanessian AG, et al. Simian immunodeficiency virus replicates to high levels in sooty mangabeys without inducing disease. J Virol May 1998;72(5):3872–86.
[6] Broussard SR, Staprans SI, White R, Whitehead EM, Feinberg MB, Allan JS. Simian immunodeficiency virus replicates to high levels in naturally infected African green monkeys without inducing immunologic or neurologic disease. J Virol March 2001;75(5):2262–75.
[7] Silvestri G, Sodora DL, Koup RA, Paiardini M, O'Neil SP, McClure HM, et al. Nonpathogenic SIV infection of sooty mangabeys is characterized by limited bystander immunopathology despite chronic high-level viremia. Immunity March 2003;18(3):441–52.
[8] Silvestri G. Immunity in natural SIV infections. J Intern Med January 2009;265(1):97–109.
[9] Gordon SN, Dunham RM, Engram JC, Estes J, Wang Z, Klatt NR, et al. Short-lived infected cells support virus replication in sooty mangabeys naturally infected with simian immunodeficiency virus: implications for AIDS pathogenesis. J Virol April 2008;82(7):3725–35. http://dx.doi.org/10.1128/JVI.02408-07.
[10] Pandrea I, Ribeiro RM, Gautam R, Gaufin T, Pattison M, Barnes M, et al. Simian immunodeficiency virus SIVagm dynamics in African green monkeys. J Virol April 2008;82(7):3713–24.
[11] Klatt NR, Villinger F, Bostik P, Gordon SN, Pereira L, Engram JC, et al. Availability of activated $CD4^+$ T cells dictates the level of viremia in naturally SIV-infected sooty mangabeys. J Clin Invest June 2008;118(6):2039–49.
[12] Kirchhoff F. Immune evasion and counteraction of restriction factors by HIV-1 and other primate lentiviruses. Cell Host Microbe July 22, 2010;8(1):55–67.
[13] Beaumier CM, Harris LD, Goldstein S, Klatt NR, Whitted S, McGinty J, et al. CD4 downregulation by memory $CD4^+$ T cells in vivo renders African green monkeys resistant to progressive SIVagm infection. Nat Med August 2009;15(8):879–85.
[14] Paiardini M, Cervasi B, Reyes-Aviles E, Micci L, Ortiz AM, Chahroudi A, et al. Low levels of SIV infection in sooty mangabey central memory $CD4^+$ T cells are associated with limited CCR5 expression. Nat Med June 26, 2011;17(7):830–6.
[15] d'Ettorre G, Paiardini M, Ceccarelli G, Silvestri G, Vullo V. HIV-associated immune activation: from bench to bedside. AIDS Res Hum Retroviruses April 2011;27(4):355–64.
[16] Sodora DL, Allan JS, Apetrei C, Brenchley JM, Douek DC, Else JG, et al. Toward an AIDS vaccine: lessons from natural simian immunodeficiency virus infections of African nonhuman primate hosts. Nat Med August 2009;15(8):861–5.
[17] Pandrea I, Onanga R, Souquiere S, Mouinga-Ondéme A, Bourry O, Makuwa M, et al. Paucity of $CD4^+$ $CCR5^+$ T cells may prevent transmission of simian immunodeficiency virus in natural nonhuman primate hosts by breast-feeding. J Virol June 2008;82(11):5501–9.
[18] Chahroudi A, Meeker T, Lawson B, Ratcliffe S, Else J, Silvestri G. Mother-to-infant transmission of simian immunodeficiency virus is rare in sooty mangabeys and is associated with low viremia. J Virol June 2011;85(12):5757–63.
[19] Nduati R, John G, Mbori-Ngacha D, Richardson B, Overbaugh J, Mwatha A, et al. Effect of breastfeeding and formula feeding on transmission of HIV-1: a randomized clinical trial. JAMA March 1, 2000;283(9):1167–74.

Glossary

1. **Tregs** a subset of T cells phenotypically characterized by the cell surface expression of low densities of CD4 and CD127, high mean density of CD25, the expression of GARP and the intracellular expression of the transcription factor FoxP3. Functionally, the Tregs are defined as cells that control immune responses by a cell contact dependent nonmajor histocompatibility complex (MHC) restricted manner and known to play a role in immune tolerance.
2. **Th17 cells** a sublineage of $CD4^+$ T helper cells that produce the signature cytokine IL-17. The Th17 cells are associated with promoting a strong proinflammatory response by the secretion of chemokines and are known to be instrumental in defense against bacteria, fungi, and select viruses.
3. **Plasmacytoid dendritic cells (pDCs)** a subset of the dendritic cell lineage that serve as mobile sentinels that collect antigen from the peripheral tissues and carry them to secondary lymphoid organs via high endothelial venules to activate antigen specific T cells. They are characterized phenotypically as being $lineage^-$, $HLA\text{-}DR^+$, high mean densities of IL-3R α chain (CD123), BDCA-2, ILT-7 and are known to synthesize high levels of type 1 interferon.
4. **Vertical transmission** transmission of an infectious agent such as HIV or SIV from the mother to its offspring either during in utero, during birth or postpartum via breastfeeding is termed vertical transmission.
5. **Pegylated** the covalent conjugation of poly (ethylene glycol) (Peg) to proteins is termed pegylation. It is utilized to increase the in vivo half-life of the protein resulting in an increased biological activity of the protein.
6. **Elite controllers** HIV-1 infected individuals who maintain plasma viral loads of <50 copies/ml for prolonged periods of time accompanied by usually stable $CD4^+$ T-cell counts who have not been administered any antiretroviral therapy and who also exhibit extremely low rates of disease progression.
7. **Bacterial translocation** disruption of the gastrointestinal tract epithelial cell barrier that leads to leakage of gut microorganisms or their products into the systemic circulation is termed "bacterial translocation".
8. **Killer cell immunoglobulin-like receptors (KIRs)** these are molecules expressed on the cell surface of hematopoietic cells such as the natural killer (NK) cells that are highly polymorphic and are composed of either 1, 2, or 3 extracellular domains (D) and either a long (L) or short (S) intracellular chain and hence the nomenclature KIR1DL, 2DL, 3DL, and KIR1DS, etc. The natural ligands for the KIRs are the more conserved MHC-class I molecules. Whereas the ligation of the KIRs with a long intracellular chain leads to inhibition, the ligation of the short intracellular chain leads to NK cell activation.
9. **Villous atrophy** the lumen of the small intestinal wall consists of finger like projections termed villi, which are interspersed with crypts. The outermost layer of the villi and crypts are composed of epithelial cells which turnover every 2–3 days originating from progenitor cells that are located within the crypts. During HIV/SIV infection there is decreased proliferation of such epithelial cells due to direct/indirect effects of viral infection, which gradually leads to atrophy and thus the term "villous atrophy".

Index

Note: Page numbers followed by "*f*" indicate figures; "*t*," tables.

D

E

I

P

R

Color Plates

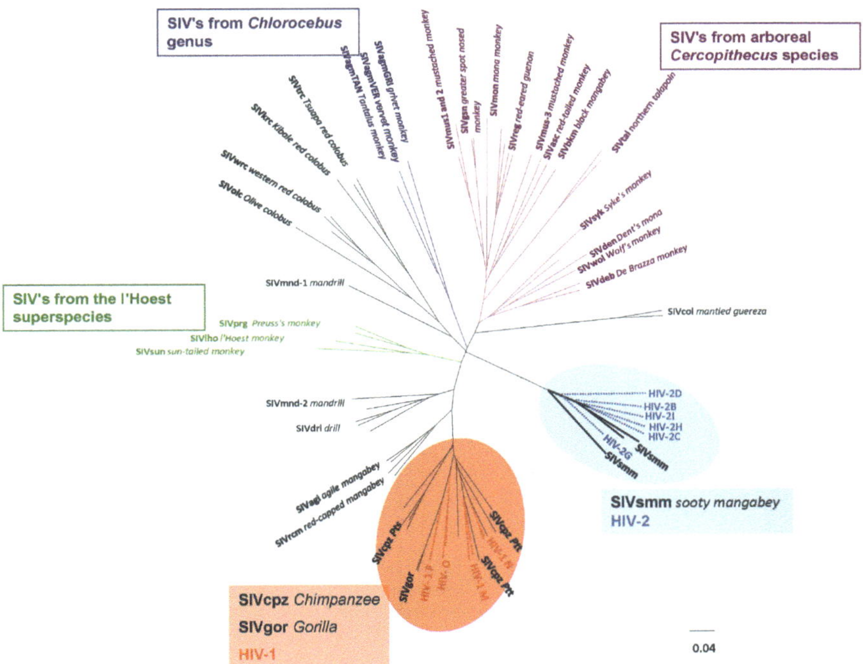

FIGURE 3.1 Genetic diversity and evolutionary history of the different lineages of HIV/SIV. Phylogenetic tree analysis using the neighbor joining method on a 294 bp fragment from the *pol* gene of different SIVs infecting non-human primates and HIVs infecting humans. Branch lengths are drawn to scale (the scale bar indicates 0.04 substitutions per site). Each NHP is infected with a species-specific SIV lineage. The different colors are used for clarity to discriminate the major SIV lineages. The different HIV-1 and HIV-2 lineages, which are interspersed with the SIVcpz/SIVgor and SIVsmm lineages, respectively, are indicated in red and blue circles, respectively.

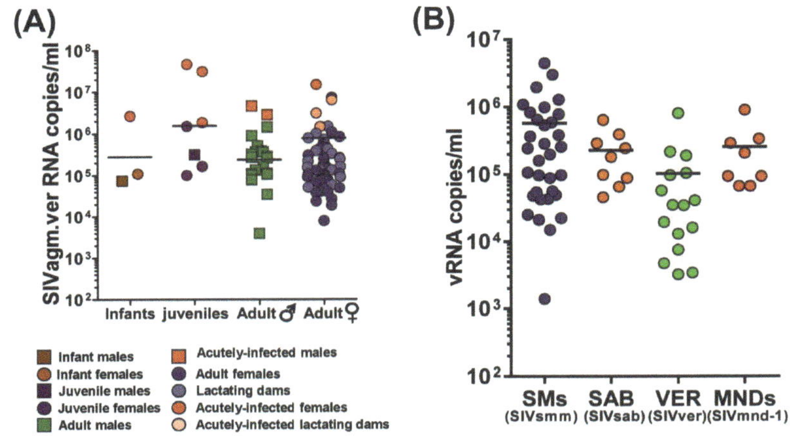

FIGURE 3.2 Comparison between SIV plasma viral loads in naturally infected vervets in the wild (A) and experimentally infected African NHPs of different species (B). In (A) VLs of lactating dams are depicted in lighter colors. Detection limit: 100 copies/ml.

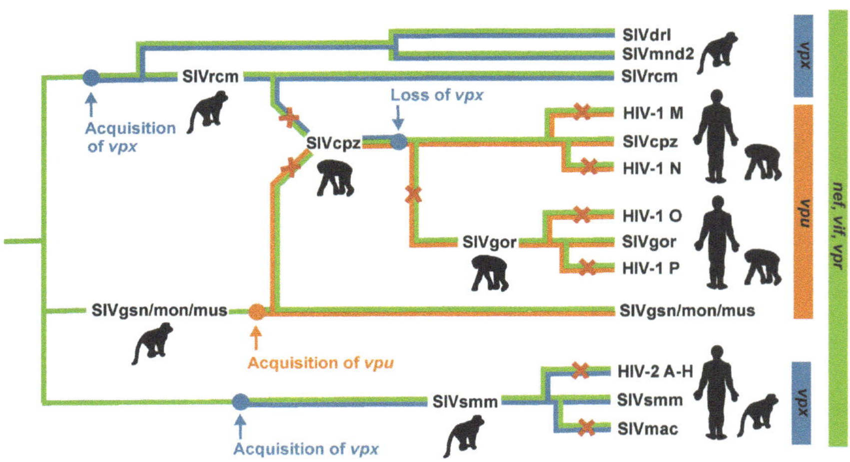

FIGURE 4.1 Gain and loss of accessory genes during primate lentiviral evolution. The cross-species transmission and recombination events that led to the emergence of *vpu* and *vpx* encoding primate lentiviruses are shown. Orange and blue indicate the presence of a *vpu* or *vpx* gene, respectively. All primate lentiviruses encode the accessory genes *vif*, *nef*, and *vpr* (green). Red crosses highlight cross-species transmissions.

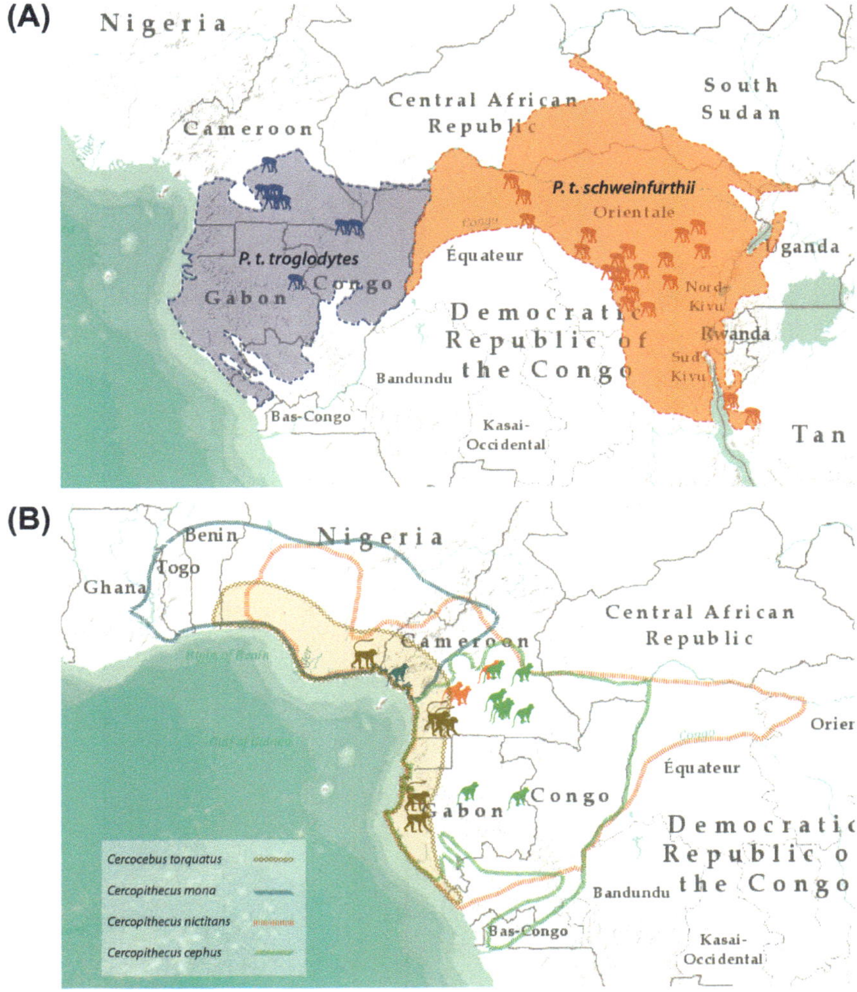

FIGURE 5.1 Ranges of the *Pan troglodytes troglodytes* (*P. t. troglodytes*) and *P. troglodytes schweinfurthii* (*P. t. schweinfurthii*) subspecies (A). Chimpanzee icons indicate communities found to have SIVcpz infection. (B) Ranges of the monkey species that may have contributed to the chimeric SIVcpz virus. Monkey icons indicate where SIV-infected animals have been identified.

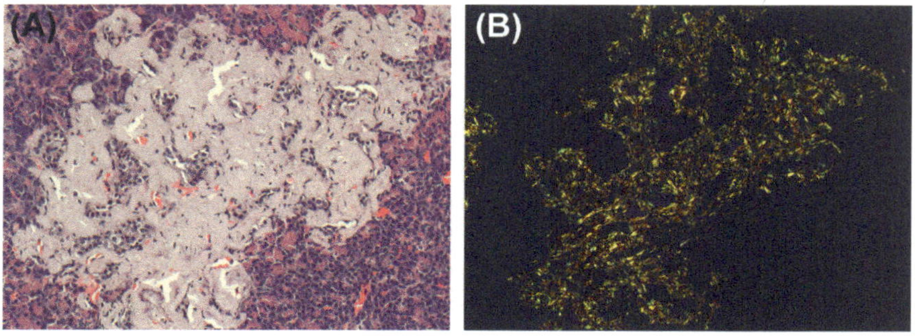

FIGURE 6.3 (A) Islets of Langerhans are replaced by abundant amorphous, homogenous, eosinophilic material (amyloid) during clinical diabetes mellitus. Hematoxylin and eosin (HE) stain. ×200. (B) Amyloid exhibits apple green birefringence with polarized light. Congo red stain. ×200.

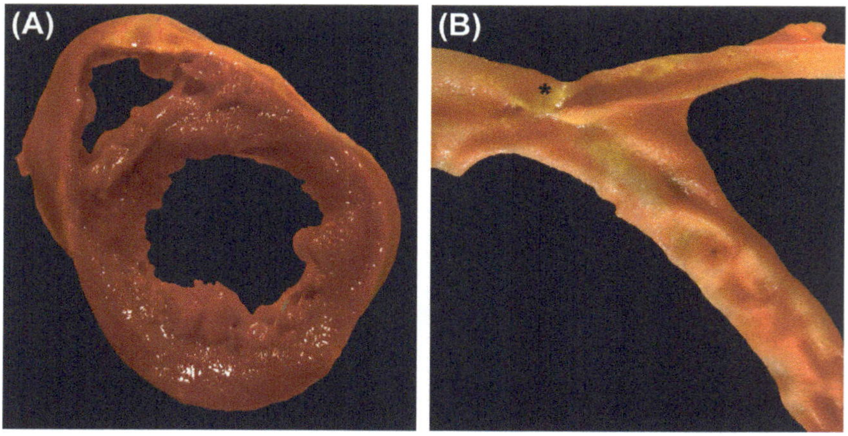

FIGURE 6.6 (A) Heart showing dilated cardiomyopathy with dilated left ventricle and thin left ventricular free wall. (B) There are variably sized pale yellow plaques on the intimal surface of the lower abdominal aorta and iliac arteries (aortoiliac bifurcation) due to atherosclerosis.

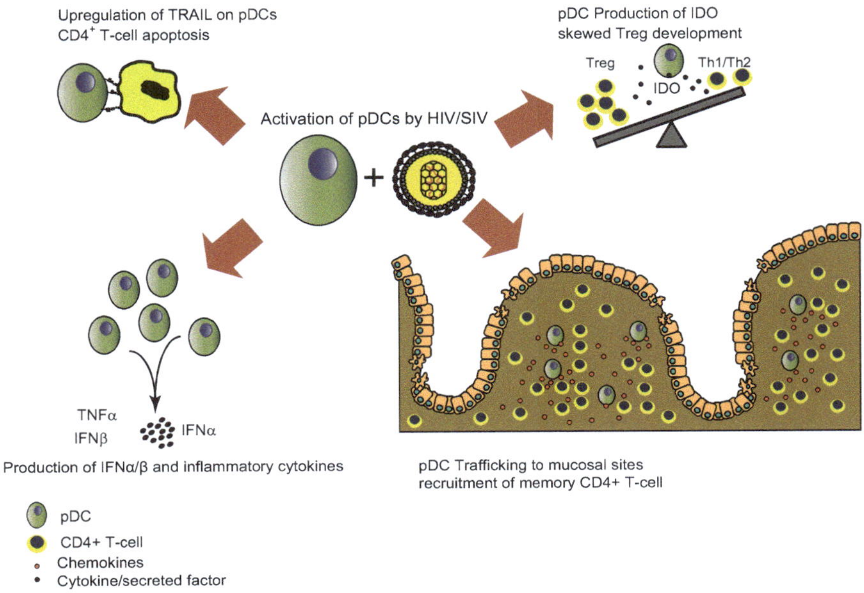

FIGURE 8.1 Current hypotheses of pDC-driven immune activation in HIV/SIV infection. Activation of pDCs by HIV/SIV leads to their activation and secretion of high levels of IFNα and -β and other inflammatory cytokines. Plasmacytoid dendritic cell activation by HIV or TLR7/8 ligands up-regulates surface expression of the TRAIL molecule, and TRAIL+pDCs are capable of driving apoptosis in $CD4^+$ T-cell lines by TRAIL-dependent pathways. Activated pDCs produce high levels of IDO that influence $CD4^+$ T cell maturation toward a Treg phenotype and inhibit classical $CD4^+$ T-cell responses. Finally, in early SIV infection, pDCs migrate to the vaginal and rectal mucosa, where they are associated with expression of ISGs and recruitment of $CD4^+$ T cells by chemokines such as MIP3α.

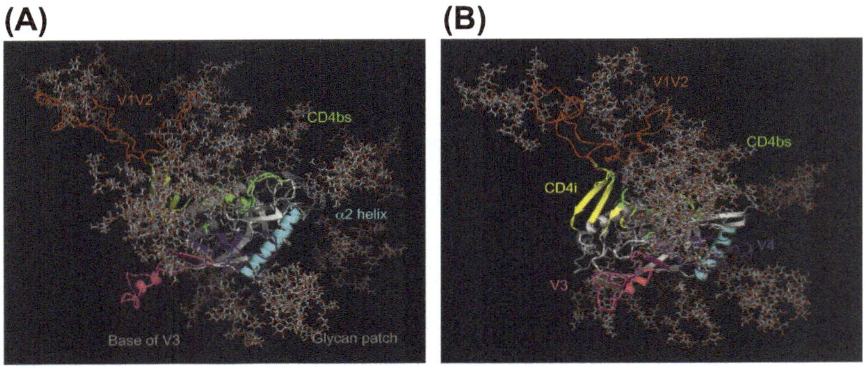

FIGURE 9.1 Neutralizing antibody targets on human immunodeficiency virus (HIV)-1 and HIV-2/simian immunodeficiency virus (SIV) in sooty mangabeys gp120. A homology-modeled, CD4-bound subtype A gp120 monomer, described previously was used to display targets of neutralizing antibodies that have been identified in HIV-1 (A), and HIV-2 or SIV (B) infections. A ribbon diagram of the gp120 crystal structure is shown in gray in two different orientations in A and B to visualize the corresponding targets. Carbohydrates are represented as sticks and are colored by element. Gp120 structural domains are color-coded as follows: V1V2 is red, V3 is magenta, V4 is blue, α2 helix is cyan, the CD4-induced (CD4i) bridging sheet is yellow, and residues comprising the conformational CD4 binding site (CD4bs) are green. For orientation, the α2 helix (cyan) is located on the outer domain of gp120; the CD4i region (yellow) is located in the inner domain. (A) Neutralizing antibody targets identified for HIV-1 infection are V1V2, CD4bs, α2 helix, a high mannose glycan patch on the outer domain, and conformational epitopes formed at the base of the V3. (B) In HIV-2/SIV infection, the targets identified thus far are within V1V2, CD4bs, CD4i, V4, and V3. Bridging sheet residues and CD4 contact residues were identified according to Decker et al. and Zhou et al. The web-based "glycoprot" tool (http://www.glycosciences.de/modeling/glyprot/php/main.php) was used to add carbohydrates to gp120. MacPyMol for Mac OS X was used to visualize the gp120 structure.

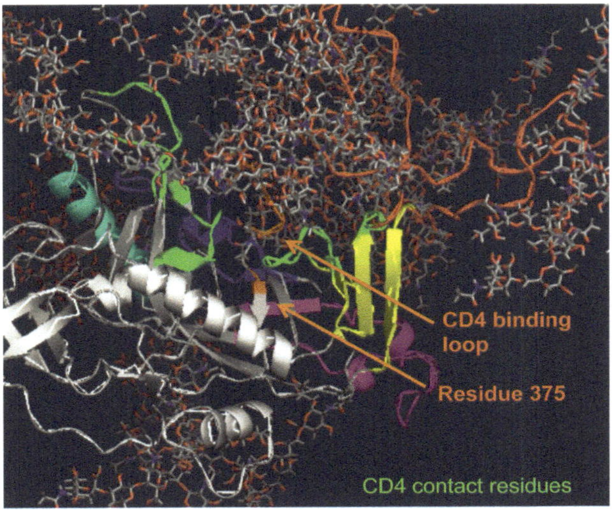

FIGURE 9.2 **The location of residue serine 375 and the CD4 binding loop on the gp120 monomer.** A homology-modeled, CD4-bound subtype A gp120 monomer, described previously, was used to display the location of the CD4 binding loop and residue Ser375 (based on the human immunodeficiency virus [HIV]-1 HXB2 numbering scheme), which in HIV-1 accommodates the Phe43 residue of the CD4 molecule. In simian immunodeficiency virus (SIV), this position is occupied by a bulky tryptophan residue. The diagram is zoomed into this region of the gp120 inner domain. A ribbon diagram of the gp120 crystal structure is shown in gray. Surrounding carbohydrates are represented as sticks and are colored by element. Gp120 structural domains are color-coded as follows: V1V2 is red, V3 is magenta, V4 is blue, $\alpha 2$ helix is cyan, the CD4-induced bridging sheet is yellow, and residues comprising the conformational CD4 binding site (CD4bs) are green. Orange color and arrows indicate the CD4 binding loop and residue Ser375. Bridging sheet residues and CD4 contact residues were identified according to Decker et al. and Zhou et al. The web-based "glycoprot" tool at (http://www/glycosciences.de/modeling/glyprot/php.main.php) was used to add carbohydrates to gp120. MacPyMol for Mac OS X was used to visualize the gp120 structure.

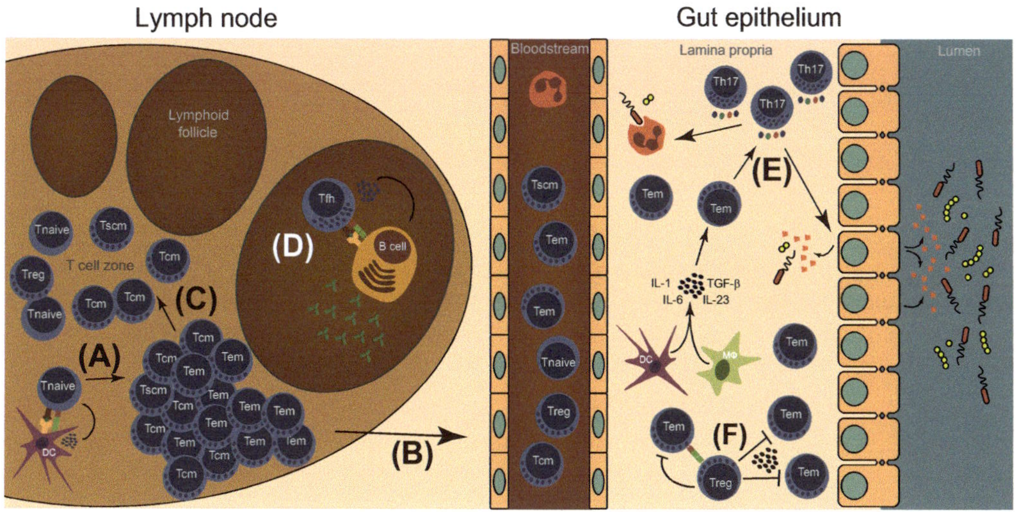

FIGURE 12.1 CD4+ T-cell development and differentiation in lymphoid and peripheral tissues. (A) Naïve CD4+ T cells become activated by encountering cognate antigen presented on class II MHC by professional antigen-presenting cells. (B) The antigen-experienced cell divides and differentiates mostly into short-lived effector and effector memory CD4+ T cells, which migrate to sites of inflammation. (C) A small subset of these cells differentiate into central memory and stem cell memory cells providing long-term immunological memory. (D) T follicular helper cells facilitate the maturation and survival of antigen-specific B cells. (E) Some CD4+ T cells in the gut mucosa differentiate into Th17 cells, which provide signals that help maintain gut mucosal integrity and recruit phagocytic cells to fight infection. (F) T regulatory cells help keep inflammation in check by providing anti-inflammatory signals.

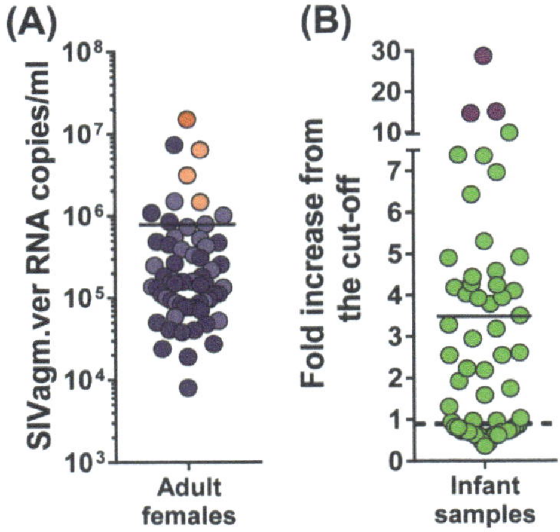

FIGURE 13.1 Massive offspring exposure to the SIV infecting the dams. (A) Massive exposure of the infants in the wild to the virus in milk was documented to occur by showing that the majority of the SIV-infected females were infected at the time of sample collection in a cross-sectional analysis. Plasma viral loads in both chronically and acutely infected females were high. Lactating dams are depicted as light circles. The samples identified as originating from acutely infected dams are illustrated as red (dark and light) circles. Detection limit, 100 copies/ml. (B) Massive in utero exposure of the offspring was indirectly documented by showing that a large proportion of the infants are seropositive for SIV, whereas the virus could only be amplified from a small fraction of infants (depicted as violet circles). This is highly suggestive for passively transmitted antibodies from their infected mothers, in agreement with the high levels of prevalence in the wild.

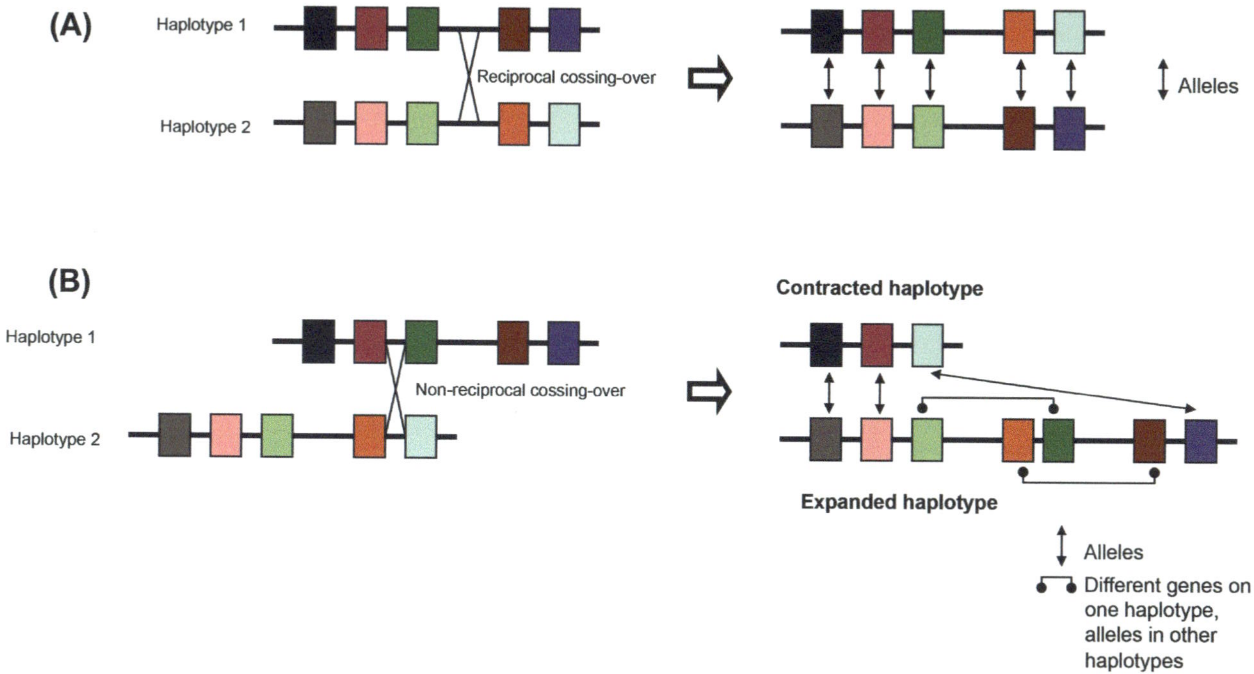

FIGURE 14.3 Scheme of reciprocal (A) and nonreciprocal (B) crossing-over events during meiosis, leading to stable and differential numbers of KIR genes in haplotypes.

FSC
www.fsc.org
MIX
Papier | Fördert
gute Waldnutzung
FSC® C083411

Kolibri 360 GmbH
Ferdinand-Jühlke-Straße 7
99095 Erfurt, Deutschland
produktsicherheit@kolibri360.de